Advances in Solid State Physics
Volume 42

Available Online

Advances in Solid State Physics is part of the Springer LINK service. For all customers with standing orders for Advances in Solid State Physics we offer the full text in electronic form via LINK free of charge. Please contact your librarian who can receive a password for free access to the full articles by registration at:

http://link.springer.de/orders/index.htm

If you do not have a standing order you can nevertheless browse through the table of contents of the volumes and the abstracts of each article at:

http://link.springer.de/series/assp/

There you will also find more information about the series.

Springer
Berlin
Heidelberg
New York
Barcelona
Hong Kong
London
Milan
Paris
Tokyo

Physics and Astronomy ONLINE LIBRARY

http://www.springer.de/phys/

Advances in Solid State Physics

Advances in Solid State Physics is a book series with a history of over 40 years. It contains the invited lectures presented at the Spring Meetings of the "Arbeitskreis Festkörperphysik" of the "Deutsche Physikalische Gesellschaft", held in March of each year. The invited talks are intended to reflect the most recent achievements of researchers working in the field both in Germany and worldwide. Thus the volume of the series represent a continuous documentation of most recent developments in what can be considered as one of the most important and active fields of modern physics. Since the majority of invited talks are usually given by young researchers at the start of their career, the articles can also be considered as indicating important future developments.

The speakers of the invited lectures and of the symposia are asked to contribute to the yearly volumes with the written version of their lecture at the forthcoming Spring Meeting of the Deutsche Physikalische Gesellschaft by the Series Editor early before the meeting.

Advances in Solid State Physics is addressed to all scientists at universities and in industry who wish to obtain an overview and to keep informed on the latest developments in solid state physics. The language of publication is English.

See also: http://www.springer.de/phys/books/assp/

Series Editor

Prof. Bernhard Kramer

I Institut für Theoretische Physik
Universität Hamburg
Jungiusstraße 9
20355 Hamburg
Germany
kramer@physnet.uni-hamburg.de

Bernhard Kramer (Ed.)

Advances in Solid State Physics 42

With 258 Figures

Springer

Prof. Bernhard Kramer (Ed.)
I Institut für Theoretische Physik
Universität Hamburg
Jungiusstraße 9
20355 Hamburg
Germany
kramer@physnet.uni-hamburg.de

Physics and Astronomy Classification Scheme (PACS): 60.00; 70.00; 80.00

ISSN print edition: 1438-4329
ISSN electronic edition: 1617-5034
ISBN 3-540-42907-7 Springer-Verlag Berlin Heidelberg New York

This work is subject to copyright. All rights are reserved, whether the whole or part of the material is concerned, specifically the rights of translation, reprinting, reuse of illustrations, recitation, broadcasting, reproduction on microfilm or in any other way, and storage in data banks. Duplication of this publication or parts thereof is permitted only under the provisions of the German Copyright Law of September 9, 1965, in its current version, and permission for use must always be obtained from Springer-Verlag. Violations are liable for prosecution under the German Copyright Law.

Springer-Verlag Berlin Heidelberg New York
a member of BertelsmannSpringer Science+Business Media GmbH

http://www.springer.de

© Springer-Verlag Berlin Heidelberg 2002
Printed in Germany

The use of general descriptive names, registered names, trademarks, etc. in this publication does not imply, even in the absence of a specific statement, that such names are exempt from the relevant protective laws and regulations and therefore free for general use.

Typesetting by the authors using a Springer TEX macro package
Final processing: DA-TEX · Gerd Blumenstein · www.da-tex.de

Cover concept using a background picture by Dr. Ralf Stannarius, Faculty of Physics and Earth Sciences, Institute of Experimental Physics I, University of Leipzig, Germany
Cover design: *design & production* GmbH, Heidelberg

Printed on acid-free paper   SPIN: 10858031   56/3141/mf - 5 4 3 2 1 0 -

Preface

The 2002 Spring Meeting of the "Deutsche Physikalische Gesellschaft" was held in Regensburg from March 25th to 29th, 2002. The number of conference attendees has remained remarkably stable at about 2800, despite the decreasing number of German PhD students. This can be taken as an indication that the program of the meeting was very attractive.

The present volume of the "Advances in Solid State Physics" contains the written versions of most of the invited talks, also those presented as part of the Symposia. Most of these Symposia were organized by several divisions in collaboration and they covered fascinating selection of topics of current interest.

I trust that the book reflects this year's status of the field in Germany. In particular, one notes a slight change in paradigms: from quantum dots and wires to spin transport and soft matter systems in the broadest sense. This seems to reflect the present general trend in physics. Nevertheless, a large portion of the invited papers – as well as the discussions at the meeting – concentrated on *nanostructured* matter.

Hamburg, April 2002   *Bernhard Kramer*

Contents

Part I. Quantum Dots

**High Magnetic Fields in Semiconductor Nanostructures:
Spin Effects in Single InAs Quantum Dots**
U. Zeitler, I. Hapke-Wurst, D. Sarkar, R. J. Haug, H. Frahm, K. Pierz,
and A. G. M. Jansen 3

1. Sample Preparation 4
2. Resonant Tunnelling 5
3. Zeeman Splitting 6
4. Fermi-Edge Singularities 9
5. Conclusion 11
References 12

On the Way to the II-VI Quantum Dot VCSEL
Thorsten Passow, Matthias Klude, Carsten Kruse, Karlheinz Leonardi,
Roland Kröger, Gabriela Alexe, Kathrin Sebald, Sven Ulrich,
Peter Michler, Jürgen Gutowski, Heidrun Heinke, and Detlef Hommel ... 13

1. Conventional ZnSe-Based Edge Emitting Laser Diodes 13
2. Optimization of CdSe Quantum Dot Structures
   for Application in a Laser Diode 14
   2.1. Quantum Dot Formation in the System CdSe/ZnSe 14
   2.2. Proof for Quantum Dots 17
   2.3. Optimization of CdSe Quantum Dot Stacks 17
   2.4. Optical Gain in CdSe Quantum Dot Stacks 18
   2.5. Electrically Pumped CdSe Quantum Dot Laser Diode 19
   2.6. Comparison of Quantum-Dot and Quantum-Well Laser Diodes .. 20

3. Distributed Bragg Reflectors and Microcavities 21
    3.1. High-Reflectivity Distributed Bragg Reflectors
        Using ZnSe/MgS Superlattices 21
    3.2. Monolithic Microcavities 22
4. Summary 24
References 24

ZnCdSe Quantum Structures Growth, Optical Properties and Applications
Martin Strassburg, O. Schulz, Matthias Strassburg, U. W. Pohl,
R. Heitz, A. Hoffmann, D. Bimberg, M. Klude, D. Hommel,
K. Lischka, and D. Schikora 27

1. Quantum Dots in the Active Region:
   0D Localisation and Mobility of Excitons 28
2. Optimisation of p-Contacts and p-Claddings 32
3. Optical Confinement 34
4. Conclusions 35
References 36

Part II. Optics

Photonic Crystals: Optical Materials for the 21st Century
K. Busch, A. Garcia-Martin, D. Hermann, L. Tkeshelashvili,
M. Frank, and P. Wölfle 41

1. Introduction 41
2. Photonic Band Structure Computation 43
3. Defect Structures in Photonic Crystals 46
4. Nonlinear Photonic Crystals 49
5. Conclusions 51
References 52

Quantum Optical Effects in Semiconductors
W. Hoyer, M. Kira, and S. W. Koch 55

1. Theory 56
2. Correlated Photons in Quantum-Well Emission 59
    2.1. Multiple Quantum-Well Systems 61
3. Quantum Effects in Microcavity Emission 63

4. Summary 65
References 66

Beryllium Chalcogenides: Interface Properties and Potential for Optoelectronic Applications
V. Wagner, J. Geurts, and A. Waag 67

1. Introduction 67
2. Optoelectronic Applications 70
3. ZnSe/BeTe Interfaces 72
   3.1. Structural Properties 72
   3.2. Optical Properties 73
   3.3. Type-II LEDs 74
4. CdSe/BeTe Interfaces 75
5. Summary and Outlook 77
References 78

ZnO MOVPE
A. Waag, Th. Gruber, Ch. Kirchner, D. Klarer, K. Thonke,
R. Sauer, F. Forster, F. Bertram, and J. Christen 81

1. Introduction 81
2. Experimental Details 82
3. Results and Discussion 83
4. Conclusions 89
References 90

Part III. Electron and Spin Transport

Electrical Spin Injection from Ferromagnetic Metals into GaAs
Manfred Ramsteiner, Haijun Zhu, Atsushi Kawaharazuka,
Hsin-Yi Hao, and Klaus H. Ploog 95

1. Introduction 95
2. Experimental Approach 96
   2.1. Sample Design 96
   2.2. Magneto-Electroluminescence 96
   2.3. Time-Resolved Photoluminescence 97

3. Injection from Fe into GaAs 97
4. Injection from MnAs into GaAs 101
5. Spin Relaxation Times 103
6. Spin Injection Model 105
7. Conclusions 105
References 106

Probing the Conduction Channels of Gold Atomic-Size Contacts: Proximity Effect and Multiple Andreev Reflections
E. Scheer, W. Belzig, Y. Naveh, and C. Urbina 107

References 118

Metal-Insulator Transitions at Surfaces
Michael Potthoff 121

1. Surface Phase Transitions 121
2. Metal-Insulator Transitions 122
3. Surface States 123
4. Mott Transition 124
5. Mean-Field Approach 125
6. Critical Regime 127
7. Conclusions 130
References 131

The Role of Contacts in Molecular Electronics
Gianaurelio Cuniberti, Frank Großmann, and Rafael Gutiérrez 133

1. Introduction 133
2. Charge Transport on the Molecular Scale 134
3. Method 136
4. Applications to Molecular Devices 138
   4.1. Focusing on the Bridge Molecule: Sodium Wires 138
   4.2. Focusing on the Leads: Carbon Nanotube Leads 140
   4.3. Focusing on the 'Molecule Plus Lead' Complex:
        A Pure Carbon Device 144
5. Discussion and Conclusions 146
References 147

Electronic Transport, Spectral Fine Structures, and Atom Clusters in Quasicrystals and Approximants
H. Solbrig and C. V. Landauro 151

1. Introduction 151
2. Transport Parameters from Spectral Information 152
3. Spectral Fine Structure with Icosahedral Clusters 154
   3.1. Iron Network Generates 100 meV Pseudogap 154
   3.2. Two Types of Narrow Resistivity Peaks 156
   3.3. The sp-Character of the Electron States 156
   3.4. Non-Ohmic Resistance Scaling 156
   3.5. Inverse Matthiessen Rule 157
4. Spectral Transport Model of the Quasicrystal 158
   4.1. Lorentzian Resistivity Model 158
   4.2. Transport Parameters of Bulk Quasicrystals 159
5. Conclusions 160
References 161

Full Counting Statistics of Mesoscopic Electron Transport
Wolfgang Belzig 163

1. Introduction 163
2. Method 164
3. Circuit Theory 166
4. Examples 168
   4.1. Quantum Contact 168
   4.2. Tunnel Junction 170
   4.3. Double Tunnel Junction 170
5. Conclusion 172
References 173

Nonequilibrium Transport through a Kondo-dot in a Magnetic Field
Peter Wölfle, Achim Rosch, Jens Paaske, and Johann Kroha 175

1. Magnetization and Conductance
   in Lowest Order Perturbation Theory 177
2. Leading Logarithmic Corrections to M and I 178
3. Resummation of Perturbation Theory in Nonequilibrium:
   A Poor Man's Scaling Approach 180
4. Conclusion 184
References 184

Part IV. Thin Films and Surfaces

Electrons, Phonons and Excitons at Semiconductor Surfaces
Johannes Pollmann, Peter Krüger, Albert Mazur, and Michael Rohlfing 189

1. Introduction 189
2. Electrons at Semiconductor Surfaces 190
   2.1. LDA Calculations of Surface Atomic and Electronic Structure .. 190
   2.2. Quasiparticle Surface Bandstructure Calculations 195
3. Phonons at Semiconductor Surfaces 197
4. Excitons at Semiconductor Surfaces 201
5. Conclusion 204
References 204

Diffuse Interface Model for Microstructure Evolution
Britta Nestler 207

1. Introduction 207
2. Multi-Phase-Field Model 210
3. Numerical Simulations 213
   3.1. Grain Structures 214
   3.2. Multiphase Alloy Systems 215
4. Conclusion 217
References 218

Rapidly Produced Thin Films: Laser-Plasma Induced Surface Reactions
Peter Schaaf, Ettore Carpene, Michael Kahle, and Meng Han 219

1. Introduction 219
2. Experimental 220
3. Results and Discussions 221
   3.1. Thin Films on Iron 221
   3.2. Thin Films on Aluminum 224
   3.3. Thin Films on Titanium 225
   3.4. Thin Films on Silicon 227
4. Summary and Outlook 229
References 229

Nanotechnology – Bottom-up Meets Top-down
O. G. Schmidt, Ch. Deneke, Y. Nakamura, R. Zapf-Gottwick,
C. Müller, and N. Y. Jin-Phillipp 231

1. Introduction 231
2. Two- and Three Dimensional Periodic Arrays
   of Self-Assembled Semiconductor Quantum Dots 232
3. Semiconductor Nanotubes 234
4. Conclusion 238
References 239

Local Ordering Processes in Ferroelectric, Glass-like and Modulated phases: An EPR Study
G. Völkel, N. Alsabbagh, J. Banys, H. Bauch, R. Böttcher,
M. Gutjahr, D. Michel, and A. Pöppl 241

1. Introduction 241
2. EPR Investigations on Ferroelectrics 242
3. The Proton Glass BP_xBPI_{1-x} Studied by ENDOR Spectroscopy 244
   3.1. The Quantum RBRF Ising Glass Model 245
   3.2. $BP_{0.15}BPI_{0.85}$ and $BP_{0.40}BPI_{0.60}$ 246
4. Modulated Phases in DMAGaS Evidenced by Q Band EPR 247
5. Conclusions 250
References 250

Part V. Superconducting Systems

Superconductivity and Non-Fermi Liquid Normal State of Itinerant Ferromagnets
Christian Pfleiderer 255

1. The Superconducting Ferromagnet $ZrZn_2$ 256
2. Non-Fermi Liquid Normal State of MnSi 259
3. Possible Role of Quantum Criticality 262
References 265

Infrared Conductivity and Superconducting Energy Gap in MgB_2
Andrei Pimenov 267

1. Introduction 267
2. Infrared Properties of MgB_2 268
3. Conclusions 277
References 278

MgB_2 Wires and Tapes: Properties and Potential
W. Goldacker, S. I. Schlachter, S. Zimmer, and H. Reiner 281

1. Introduction 281
   1.1. Crystal Structure, Thermal Expansion and Residual Strain of MgB_2 282
   1.2. Composition of Wires and Tapes 283
2. Experimental: Wire Preparation and Characterisation Methods 284
3. Transport Critical Currents 286
4. Effect of Mechanical Reinforcement 287
5. Critical Currents with Applied Axial Tensile Strain 289
6. Conclusions and Outlook 290
References 291

Electron-Phonon Coupling and Superconductivity in MgB_2 and Related Diborides
Rolf Heid, Klaus-Peter Bohnen, and Burkhard Renker 293

1. Introduction 293
2. Computational Details 294
3. Ground-State Properties 295
4. Lattice Dynamics 296
5. Electron-Phonon Coupling and Superconductivity 298
   5.1. Isotropic Limit 298
   5.2. Beyond the Isotropic Limit 301
6. Summary 303
References 303

Self-Organized Quasi-One Dimensional Structures in High-Temperature Superconductors: the Stripe Phase
Enrico Arrigoni, Marc G. Zacher, Rober Eder, Werner Hanke, and Steven A. Kivelson 307

1. Introduction 307
2. Role of Long-Range Coulomb Interaction in the Stripe Phase 308
3. Stripes from Angle-Resolved Photoemission Spectroscopy 311
   3.1. Technique 312
   3.2. Spectral Features 314
4. Summary and Conclusions 317
References 317

Theory of Superconductivity Due to the Exchange of Spin Fluctuations in Hole- and Electron-Doped Cuprate Superconductors: d-Wave Order Parameter
Dirk Manske and Karl H. Bennemann 319

1. Introduction 319
2. Theory 321
3. Results and Discussion 325
4. Summary 328
References 330

Part VI. Disordered Systems and Soft Matter

Anomalous Behavior of Insulating Glasses at Ultra-low Temperatures
Christian Enss 335

1. Introduction 335
2. Internal Friction 337
3. Dielectric Constant 339
4. Polarization Echoes 342
   4.1. Decay of Spontaneous Echoes 342
   4.2. Magnetic Field Dependence 344
5. Summary 345
References 345

Colloidal Suspensions – The Classical Model System of Soft Condensed Matter Physics
Hans-Hennig von Grünberg 347

1. Introduction – Why it is Worth Studying Colloids 347
2. Effective Forces – What Small Particles can do to Large Particles ... 349
3. Many-body Effects – The Whole is More than the Sum of its Parts .. 351
4. Phase-Behavior 355
    4.1. Experimental Findings and Current Controversies 355
    4.2. Pressure and how to Calculate it 356
References 358

Excitable Membranes: Channel Noise, Synchronization, and Stochastic Resonance
Peter Hänggi, Gerhard Schmid, and Igor Goychuk 359

1. Introduction 359
2. The Hodgkin-Huxley Model 361
3. Stochastic Version of the Hodgkin-Huxley Model 362
    3.1. Quantifying Channel Noise 362
    3.2. Numerical Integration 364
    3.3. Coherence Resonance and Synchronization 365
    3.4. Stochastic Resonance 367
4. Conclusions 368
References 369

Birth and Sudden Death of a Granular Cluster
Ko van der Weele, Devaraj van der Meer, and Detlef Lohse 371

1. Introduction 371
2. Flux Model 373
3. More than two Compartments: Hysteresis 375
4. Coarsening and Sudden Death 376
5. The Limit for $N \to \infty$ Compartments: Anti-Diffusion 380
6. Extensions and Applications 381
References 382

Interacting Neural Networks and Cryptography
Wolfgang Kinzel and Ido Kanter 383

1. Introduction 383
2. Dynamic Transition to Synchronization 384

3. Random Walk in Weight Space 386
4. Secret Key Generation 387
5. Conclusions 389
References 391

**Low Energy Dynamics in Glasses Investigated
by Neutron Inelastic Scattering**
Jens-Boie Suck 393

1. Introduction 393
2. Inelastic Neutron Scattering 394
3. Sample Preparation and Experiments 395
4. Results 396
   4.1. Rapidly and Slow Quenched NiPdP 398
5. Discussion 398
6. Conclusions 402
References 403

Part VII. Magnetism

Metallic Magnetism
Jürgen Kübler 407

1. Introduction 407
2. Ground-State Properties 407
3. Excited States and Thermal Properties 411
   3.1. Magnons 412
   3.2. Spin Fluctuations and the Magnetic Phase Transition 413
4. Conclusion 417
References 418

**Domain State Model for Exchange Bias:
Influence of Structural Defects on Exchange Bias in Co/CoO**
Bernd Beschoten, Andrea Tillmanns, Jan Keller, Gernot Güntherodt,
Ulrich Nowak, and Klaus D. Usadel 419

1. Introduction 419
2. Models for Exchange Bias 420
3. Domain State Model for Exchange Bias 422
4. Domain State Magnetization 425

5. Role of Twin Boundaries for Exchange Bias in Co/CoO 427
   5.1. Sample Preparation 427
   5.2. Structural Properties 427
   5.3. Magnetic Properties 428
6. Conclusions 430
References 430

Itinerant Ferromagnetism and Antiferromagnetism from a Chemical Bonding Perspective
Richard Dronskowski 433

1. Physics of Cooperative Magnetism in a Nutshell 433
2. Chemical Bonds from Band Structure Calculations 434
3. Three Myths of Chemical Bonding 436
4. Chemical Bonding and Energetics of α-Fe 437
5. Magnetic Recipe for Transition Metals and Alloys 439
6. Rational Syntheses of Magnetic Borides 441
7. Conclusion 443
References 443

Theory of Ferromagnetism in (III,Mn)V Semiconductors
Jürgen König 445

1. Diluted Magnetic III-V Semiconductors 446
2. Mean-Field Theory 446
3. Collective Spin Excitations 447
   3.1. Beyond Mean-Field Theory and RKKY Interaction 447
   3.2. Independent Spin-Wave Theory for Parabolic Bands 448
   3.3. Elementary Spin Excitations 449
   3.4. Comparison to RKKY and to Mean-Field Picture 450
   3.5. Spin-Wave Dispersion for Realistic Bands 451
   3.6. Spin Stiffness 452
4. Limits on the Curie Temperature 453
5. Monte-Carlo Approach 454
6. Magnetic Domains 454
References 456

Tetrahedral Quantum Magnets in One and Two Dimensions
Wolfram Brenig, Andreas Honecker, and Klaus W. Becker 457

1. Introduction 457
2. The Tetrahedral Chain 459
3. The Checkerboard Magnet 463
References 467

Part VIII. Applications

SiGe:C Heterojunction Bipolar Transistors:
From Materials Research to Chip Fabrication
H. Rücker, B. Heinemann, D. Knoll, and K.-E. Ehwald 471

1. Introduction 471
2. Effect of Carbon on Boron Diffusion 472
   2.1. Diffusion Experiment 472
   2.2. Coupled Diffusion of C and Si Point Defects 472
   2.3. Suppression of Transient Enhanced Diffusion 474
3. Heterojunction Bipolar Transistors 475
   3.1. Operation of HBTs 475
   3.2. Effect of B Outdiffusion from the SiGe Layer 476
4. SiGe:C BiCMOS Technology 477
   4.1. Modular Integration of High-Speed HBTs 477
   4.2. HBT Device Characteristics 479
   4.3. Emitter Scaling 479
   4.4. Yield 480
5. Conclusions 481
References 482

Transition Edge Sensors for Imaging X-ray Spectrometers
H. F. C. Hoevers 483

1. Introduction 483
2. Principles of a Voltage Biased Detector with a TES 484
3. Design and Performance of an X-ray TES Microcalorimeter 486
4. Development of Imaging Arrays for X-ray Spectroscopy 489
   4.1. Single Pixel Optimization 489
   4.2. Micromachining of the Pixel and Array Support Structure 490
   4.3. Electrical Read-Out of an Imaging Array 491
References 493

Charge Injection in Polymer Light-Emitting Diodes
T. van Woudenbergh, P. W. M. Blom, and J. N. Huiberts 495

1. Introduction 495
2. Hole Mobility of PPV 496
3. Mechanism of Charge Injection 497
4. PLED with an Injection Limited Hole Contact 499
5. Conclusions 503
References 503

Sensors and the Influence of Process Parameters and Thin Films
Hans-Reiner Krauss 505

1. Introduction 505
2. Design Method 505
3. Examples 506
   3.1. Pressure Sensors 506
   3.2. Air Quality Sensor 508
   3.3. Acceleration Sensor 510
4. Conclusion 512
References 513

Index 515

Part I

Quantum Dots

High Magnetic Fields in Semiconductor Nanostructures: Spin Effects in Single InAs Quantum Dots

U. Zeitler[1], I. Hapke-Wurst[1], D. Sarkar[1], R.J. Haug[1], H. Frahm[2], K. Pierz[3], and A.G.M. Jansen[4]

[1] Institut für Festkörperphysik, Universität Hannover,
   Appelstraße 2, 30167 Hannover, Germany
[2] Institut für Theoretische Physik, Universität Hannover,
   Appelstraße 2, 30167 Hannover, Germany
[3] Physikalisch-Technische Bundesanstalt Braunschweig,
   Bundesallee 100, 38116 Braunschweig, Germany
[4] Grenoble High Magnetic Field Laboratory, MPIF- CNRS,
   B.P. 166, 38042 Grenoble Cedex 09, France

Abstract. We present a prominent example how the influence of high magnetic fields can lead to spectacular field induced effects in a semiconductor nanostructure. We observe current steps in the I-V characteristics of a GaAs-AlAs tunnelling structure with self-assembled InAs quantum dots embedded in the AlAs barrier. The steps originate from resonant tunnelling through individual InAs quantum dots. In a magnetic field the Zeeman splitting of the quantized dot states leads to a splitting of each current step in two. The Landé factor deduced from these measurements is in the range $g = 0.6 \ldots 1.5$ depending on the size of the dot and the orientation of the magnetic field. In high magnetic fields ($B > 20$ T) the current steps evolve into extremely enhanced peaks. The effect observed is explained by a field induced Fermi-edge singularity caused by the Coulomb interaction between the tunnelling electron on the quantum dot and the partly spin-polarized Fermi sea in the Landau quantized three-dimensional emitter.

Over the last years several groups succeeded in performing single-electron tunnelling experiments through self-assembled InAs quantum dots (QDs) [1,2,3,4]. When a magnetic field is applied the spin degeneracy of the quantized energy states in an InAs QD is lifted and it is possible to resolve distinct spin states at low temperatures [5,6].

In this work we present our recent results on magneto-tunnelling experiments through self-assembled InAs QDs. We will show that we can deduce the Landé factor of a single InAs quantum dot and will analyse its dependence on the dot size and on the direction of the magnetic field applied [7,8]. In high magnetic fields ($B > 20$ T) we find strong singularities in the resonant tunnelling though an individual InAs QD [6]. They will be explained with a theoretical model considering the electrostatic potential experienced by the emitter electrons around the Fermi edge due to the charged QD. We will show that the partial spin polarization of the Landau quantized three-dimensional emitter causes extreme values of the edge exponent $\gamma > 0.5$ not observed until present and going far beyond the standard theory valid for $\gamma \ll 1$ [9].

1 Sample Preparation

Our samples are single barrier GaAs-AlAs-GaAs tunnelling structures with three-dimensional highly doped GaAs electrodes and self-assembled InAs QDs embedded in the middle of the AlAs barrier.

For the bottom GaAs electrode first a 1 μm highly doped GaAs (electron concentration $n^+ = 2 \times 10^{24}$ m^{-3}) is grown on a n^+-doped GaAs substrate at a substrate temperature of 600°C. This layer is followed by 10 nm n-doped GaAs ($n = 10^{23}$ m^{-3}), 10 nm n^--doped GaAs ($n^- = 10^{22}$ m^{-3}) and a 15 nm nominally undoped GaAs spacer layer. The doping sequence leads to the formation of a three-dimensional electron system up to the AlAs barrier with an electron concentration $n_e \approx 10^{23}$ m^{-3} at the GaAs-AlAs interface.

On top of the bottom electrode we deposit the first 5 nm of the AlAs barrier. Subsequently, the growth is interrupted and the substrate temperature is ramped down to 520°C. Then 1.8 monolayers of InAs are deposited directly onto the AlAs. Due to the strong lattice mismatch between AlAs and InAs self-assembled InAs QDs form. The dots are covered by another 5 nm AlAs and a top GaAs electrode is grown symmetrically to the bottom one.

Electric contacts are realized by annealing AuGeNi into the electrodes. At the same time, the metallic top contacts serve as an etch mask for the structuring of macroscopic tunnelling diodes with a pillar diameter of 40 - 100 μm containing about $10^6 - 10^7$ InAs QDs.

To characterize the geometric properties of the InAs QDs we have produced reference samples where the growth of the structures was interrupted directly after deposition of the InAs. Additionally, we have grown uncovered InAs QDs directly on GaAs. The geometric properties of these uncovered InAs QDs can be visualized with an atomic force microscope, the results are shown in Fig. 1 [4]. The QDs forming on AlAs are considerably smaller compared to dots grown on GaAs under identic growth condition. Moreover, whereas the dot size on GaAs does not depend on the InAs coverage, it increases with coverage when the dots are grown on AlAs (not shown). In other words, the dot density for InAs dots grown on GaAs increases with increasing InAs coverage whereas it remains approximately constant for InAs QDs on AlAs. We assign this behaviour, as well as the relatively small dot size of InAs QDs on AlAs, to a reduced In diffusion on the rough AlAs surface which leads to a nucleation of QDs at positions only depending on the surface morphology.

The structural properties of the dots do not change considerably when they are covered by AlAs. This fact is visualized in Fig. 1(a) where we show a transmission electron micrograph of a complete sample with InAs dots

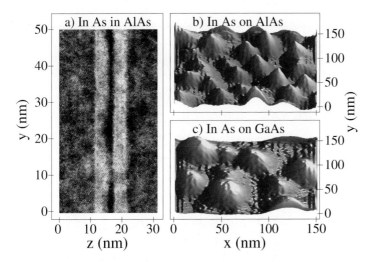

Fig. 1. Transmission electron micrograph of InAs dots embedded in a GaAs-AlAs-GaAs tunnelling device (a). For comparison, atomic force micrographs of uncovered reference samples are shown on the right panels with InAs quantum dots grown on GaAs (b) and on AlAs (c). (Figure taken from Ref. [4].)

embedded inside an AlAs barrier. For the samples used in our experiments the dots have a lateral diameter of 10-15 nm and a height of 3-4 nm.

2 Resonant Tunnelling

A current-voltage (I-V) characteristics of a typical tunnelling device containing self-assembled InAs QDs is shown in Fig. 2. For both bias directions we observed steps in the I-V curve which we attribute to resonant tunnelling through the InAs QDs [4,6]. For zero bias, all quantized states of the InAs QDs are situated above the Fermi energy of the emitter, E_F^{em}. When a finite bias voltage V is applied they start moving down energetically with respect to E_F^{em}. As shown in the schematic band diagram in Fig. 2 a step occurs whenever a dot state is aligned with E_F^{em}. From the magnetic field dependence of the onset voltages of the current steps we conclude that they can be identified with resonant tunnelling through the ground states of different InAs QDs.

Due to the finite height of the InAs QDs the bottom AlAs barrier is effectively thicker compared to the top barrier, see Fig. 1(c). As a consequence, the tunnelling current is largely determined by the transmission of the bottom AlAs barrier. For positive bias electrons tunnel through this barrier first and leave the dot nearly instantaneously through the top barrier, leaving the dot nearly always empty. In this single-electron tunnelling direction it is possible to access different quantized energy levels of an InAs QD and to probe the

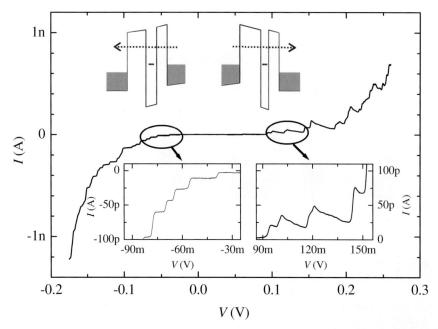

Fig. 2. Typical I-V characteristics of a GaAs-AlAs-GaAs tunnelling diode with embedded InAs QDs measured at $T = 0.35$ K. The top inset shows schematically the band structure for the two bias polarities. The bottom insets represent magnifications of the I-V curve for negative and positive bias

electrons in the emitter with the QD, see e.g. [10]. In contrast, for negative bias the tunnelling electrons are kept mostly in the dot and Coulomb charging effects become important. For this charging direction interactions with the emitter are negligible in the tunnelling current through the dot.

3 Zeeman Splitting

When applying a magnetic field the current steps originating from tunnelling through InAs QDs split up into two, see Fig. 3(a). This is due to the Zeeman splitting ΔE_Z of the quantized dot state as sketched in Fig. 3(c). A first current step occurs when the spin-down state of the dot is aligned with E_F^{em}, the second step is then due to the resonance of the spin-up level with E_F^{em}. The splitting between the two step is given by $\Delta V = g\mu_B B/e\alpha$ where g is the effective Landé factor of the InAs QD, μ_B is the Bohr magneton and α is a lever factor defined as the ratio of the voltage drop between the emitter and the dot and the total voltage applied. It is derived from the temperature dependence of the width of a current step at zero magnetic field caused by the thermal smearing of the Fermi edge in the emitter. Using $\alpha = 0.34$ for

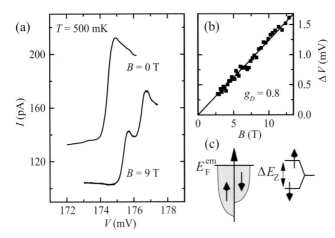

Fig. 3. (a) Zeeman splitting of the ground state in an InAs quantum dot visualized as the splitting of a current step observed at $B=0$ into two steps at $B=9$ T. (b) Voltage difference of the two spin-split current steps as a function of magnetic field. The line shows a linear fit corresponding to a Landé factor $g = 0.8$. (c) Schematic energy diagram of the emitter and an InAs QD in a magnetic field

the specific dot shown in Fig. 3 we extract $g = 0.8$ from a linear fit of ΔV as a function of B, see Fig. 3(b).

As sketched in Fig. 3(c) we propose in agreement with other experiments [5] that the sign of g is positive, i.e. the spin-down state is energetically situated below the spin-up state. In contrast, due to the negative g-factor in GaAs, the spin-down electrons in the emitter are energetically positioned above the spin up-electrons making the spin-up orientation the majority spin in high magnetic fields.

The fact that g for InAs QDs is positive can be deduced from the observation that the height of the first step associated with the energetically lower lying state in the dot increases with increasing temperature in high magnetic fields which is due a thermally induced higher occupation of the minority spin in the GaAs emitter with a negative g-factor [6].

We have measured the g-factor of numerous other InAs QDs [13], the results are compiled in Fig. 4. As can be clearly seen in the figure, g systematically increases with increasing onset voltage V_S of the corresponding current step. This hints to a systematically larger g-factor for smaller dots with a higher ground state energy. The absolute value of g as well as its dependence on the dot size can be explained qualitatively in the framework of a simple 3-band k-p model [11]. Here the Landé factor is given as

$$g = g_0 \left[1 - \frac{P^2}{3} \left(\frac{1}{E_g} - \frac{1}{E_g + \Delta_0} \right) \right] \quad (1)$$

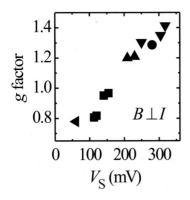

Fig. 4. Landé factor as a function of the onset voltage. For increasing onset voltages the corresponding dot size is decreasing

with $g_0 = 2$. $P^2 = 22$ eV is the interband matrix element for InAs and $\Delta_0 = 0.38$ eV is the valence-band spin-orbit splitting for InAs. In reality a small alloying of AlAs into the InAs QDs has to be considered which slightly reduced P^2 and Δ_0 by about 10%. This makes it reasonable to use $P^2 \approx 20$ eV and $\Delta_0 \approx 0.34$ eV for the further calculations.

The energy gap E_g between valence band electrons and conduction band holes in the InAs QDs can be determined from photoluminescence measurements to be in the range $E_g = 1.64$ eV - 1.76 eV [12]. Within this simple model this yields theoretically expected g-factors varying from 0.6 to 0.8 for the dots with gaps in this range, in reasonable agreement with the experimentally measured values.

The Landé factor of an InAs quantum does not only depend on the dot size but also on the orientation of the magnetic field. This can be expressed in a simple way phenomenologically as

$$g(\vartheta,\varphi) = \sqrt{(g_{[0\bar{1}1]}^2 \sin^2\varphi + g_{[0\bar{1}\bar{1}]}^2 \cos^2\varphi)\cos^2\vartheta + g_{[100]}^2 \sin^2\vartheta}. \quad (2)$$

The angles ϑ and φ are defined in the top panels of Fig. 5. Indeed, as shown in Fig. 5 the experimentally measured g-factor shows the expected behaviour, a finding also confirmed by systematic measurements on more dots [13].

The major effect of the g-factor anisotropy is observed when the magnetic field is tilted from the growth direction [100] into the [0$\bar{1}\bar{1}$] direction inside the growth plane, see Fig. 5(a) [7]. The measured g-factors $g_{[0\bar{1}\bar{1}]}$ and $g_{[100]}$ differ by about 30%. We assign this to a larger influence of size quantization effects when the magnetic field is perpendicular to the direction of the strongest confinement. Such a g-factor anisotropy was also predicted theoretically for non-spherical systems [14].

Astonishingly also a small but measurable g-factor anisotropy is observed when B is tilted inside the growth plane, with $g_{[0\bar{1}\bar{1}]}$ being about 10% larger than $g_{[0\bar{1}1]}$, see Fig. 5(b) [8]. This observation hints to a slightly elongated base of the InAs QDs along the [0$\bar{1}\bar{1}$] direction leading to a larger influence

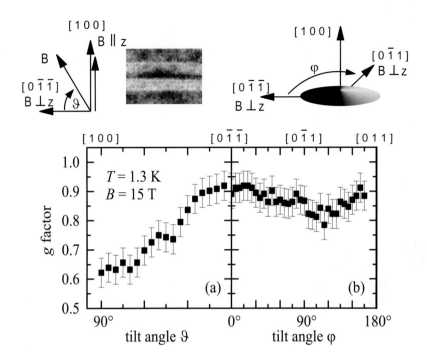

Fig. 5. Landé factor of one specific InAs QD as a function of the magnetic field direction. In (a) the field is tilted from the growth direction z into the growth plane x-y. In (b) B is turned by 180° inside the growth plane. The top panels show schematically the orientation of the magnetic field for the two configurations (a) and (b)

of size quantization effects on g when the magnetic field points along this direction.

4 Fermi-Edge Singularities

For moderate magnetic fields ($B < 10$ T, Fig. 3) the two current steps assigned to the tunnelling through a spin-split InAs QD state are comparable in height to half the step height at zero field. In large magnetic fields, however, they start to evolve into strongly enhanced peaks with a peak amplitude of more than one magnitude larger than the zero-field step height, see Fig. 6. The peaks are particularly pronounced for the spin orientation corresponding to the majority spin in the GaAs emitter.

The shape of these current peaks is characterized by a sharp ascent, with a width only limited by thermal broadening, and a moderate decrease towards higher voltages. The decrease of the current for $V > V_S$ can be described

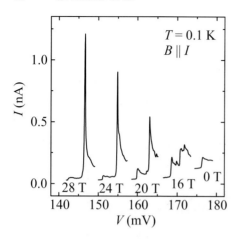

Fig. 6. I-V characteristics of a current step at $T \approx 0.1$ K for magnetic fields up to 28 T

within the framework of a Fermi-edge singularity (FES) [9], $I \propto (V - V_S)^{-\gamma}$, where V_S here is the voltage at the maximum peak current and γ is the edge exponent of the FES. The edge exponents γ extracted from the shape of the current peaks as a function of magnetic field are plotted in Fig. 7 for both spin directions.

An alternative method to determine γ uses temperature dependent measurements of the peak height of the FES. Here, the peak current I_0 scales as $I_0 \propto T^{-\gamma}$ [15]. The experimentally measured γ using this method for the majority spin are also shown in Fig. 7. It is not possible to extract the edge exponent for the minority spin directly from temperature dependent experiments. At high magnetic fields the observed increase of the current with increasing temperature is mainly caused by an additional thermal population of the minority spin in the emitter.

In order to understand the observed singularities quantitatively we have developed a theoretical model, details can be found in Refs. [6,13]. The key ingredients are the Landau quantization of the three-dimensional conduction electrons in the emitter and the Coulomb interaction between these electrons and the InAs quantum dot. In high magnetic fields, typically $B > 6$ T for our samples, all electrons are in the lowest Landau level. We observe the strongest singularities when the field is applied along the current direction. In this case the Landau quantized electrons can be described by quasi one-dimensional channels with momentum k along the tunnelling direction. The angular component of the single particle wave functions in the x-y-plane perpendicular to the tunnelling direction is quantized in channels $m \geq 0$.

Due to the small lateral size of the dot comparable to the magnetic length in the magnetic field range considered ($B = 10 \ldots 30$ T) the Coulomb interaction between the dot and the emitter electron rapidly decreases with m. The observed FES can then essentially be described by only considering electrons tunnelling from the $m = 0$ channel through the dot. Using this simplest as-

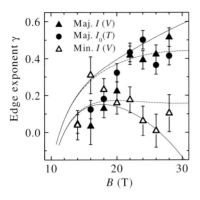

Fig. 7. Comparison of the experimentally measured edge exponents γ extracted from temperature dependence (circles) slope fitting (triangles) to the theoretical model (solid and dashed line). The majority spin in the emitter is shown with filled symbols, open symbols correspond to the minority spin. For the solid lines level broadening has been neglected, the dashed curves include an experimentally measured level broadening $\Gamma = 1.3$ meV [6]

sumption it is already possible to predict a high edge exponent of the order of unity for the FES observed experimentally for high magnetic fields [13].

Following the standard models on FESs [9,16,17,18] we have gone a step further and have developed a full theory which can directly calculate the edge exponent γ for the FES of an electron with spin σ tunnelling from a fully Landau quantized three-dimensional emitter through a small InAs quantum dot [6]. Including all channels m this model predicts even for a realistic sample with a finite Landau level broadening edge exponents as a high as $\gamma \approx 0.5$, going far beyond earlier theoretical considerations only valid for $\gamma \ll 1$ [9].

The resulting theoretically calculated edge exponents γ for both spin orientations are shown in Fig. 7. Already the simple model with no Landau level broadening ($\Gamma = 0$, solid lines) describes the experimentally measured edge exponents for both spin directions reasonably well. When a realistic Landau level broadening $\Gamma = 1.3$ meV is included (dashed lines) the field dependence of the edge exponent is smeared out in high fields and the experimentally measured edge exponents are reproduced even better. The basic features, however, remain unchanged. In particular, the edge exponent for the majority spin still shows a strong field dependence and reaches a very high value $\gamma \approx 0.5$ at the maximum field $B = 30$ T. The edge exponent related to the minority spin retains a moderate value $\gamma \approx 0.15$ for high magnetic fields.

5 Conclusion

In conclusion, we have studied the influence of a high magnetic field on resonant tunnelling through individual self-assembled InAs QDs. We have deduceD the g-factor of a single dot from these measurements. We have shown that g is positive and is systematically increasing with decreasing dot size. Additionally, g has been shown to feature a pronounced anisotropy with the smallest value when the magnetic field is applied into the direction of strongest confinement. Strongly enhanced current peaks appearing in high magnetic fields have been modelled as a field induced Fermi-edge singularity originating from the interaction between a localized charge and the electrons in the Landau quantized three-dimensional emitter. Edge exponents as high as $\gamma = 0.5$ have been both measured experimentally and described theoretically.

Acknowledgements

We would like to thank U. F. Keyser, A. Nauen J. Regul, and H. W. Schumacher for experimental assistance. The experiments in the Grenoble High Magnetic Field Laboratory were supported by the TMR Programme of the European Union under contract no. ERBFMGECT950077.

References

1. I. E. Itskevich, T. Ihn, A. Thornton, M. Henini, T. J. Foster, P. Moriarty, A. Nogaret, P. H. Beton, L. Eaves and P. C. Main, Phys. Rev. B **54**, 16401 (1996).
2. T. Suzuki, K. Nomoto, K. Taira and I. Hase, Jpn. J. Appl. Phys. **36**, 1917 (1997).
3. M. Narihiro, G. Yusa, Y. Nakamura, T. Noda and H. Sakaki, Appl. Phys. Lett. **70**, 105 (1997).
4. I. Hapke-Wurst, U. Zeitler, H. W. Schumacher, R. J. Haug, K. Pierz and F. J. Ahlers, Semicond. Sci. Technol. **14**, L41 (1999).
5. A. S. G. Thornton, T. Ihn, P. C. Main, L. Eaves and M. Henini, Appl. Phys. Lett. **73**, 354 (1998).
6. I. Hapke-Wurst, U. Zeitler, H. Frahm, A. G. M. Jansen, R. J. Haug, and K. Pierz, Phys. Rev. B **62**, 12621 (2001).
7. J.-M. Meyer, I. Hapke-Wurst, U. Zeitler, R. J. Haug, and K. Pierz, physica status solidi(b) **224**, 685 (2001).
8. I. Hapke-Wurst, U. Zeitler, R. J. Haug, and K. Pierz, Physica E **12**, 802 (2002).
9. K. A. Matveev and A. I. Larkin, Phys. Rev. B **46**, 15337 (1992).
10. P. C. Main, A. S. G. Thornton, R. J. A. Hill, S. T. Stoddart, T. Ihn, L. Eaves, K. A. Benedict and M. Henini , Phys. Rev. Lett. **84**, 729 (2000).
11. L. M. Roth und P. N. Argyres, in *Semiconductors and Semimetals (Volume 1: Physics of III-V compounds)* , p159 (Academic Press New York London,1966).
12. K. Pierz, Z. Ma, I. Hapke-Wurst, U. F. Keyser, U. Zeitler and R. J. Haug, Physica E, in press.
13. I. Hapke-Wurst, PhD Thesis, Hannover (2002).
14. A. A. Kiselev, E. L. Ivchenko, and U. Rössler, Phys. Rev. B **58**, 16353 (1998).
15. A. K. Geim, P. C. Main, N. La Scala, Jr., L. Eaves, T. J. Foster, P. H. Beton, J. W. Sakai, F. W. Sheard, M. Henini, G. Hill and M. A. Pate, Phys. Rev. Lett. **72**, 2061 (1994).
16. G. D. Mahan, Phys. Rev. **163**, 612 (1967); G. D. Mahan, *Many-Particle Physics* (Plenum, New York, 1981).
17. P. Nozières and C. T. De Dominicis, Phys. Rev. **178**, 1097 (1969).
18. K. D. Schotte and U. Schotte, Phys. Rev. **182**, 479 (1969).

On the Way to the II-VI Quantum Dot VCSEL

Thorsten Passow, Matthias Klude, Carsten Kruse, Karlheinz Leonardi, Roland Kröger, Gabriela Alexe, Kathrin Sebald, Sven Ulrich, Peter Michler, Jürgen Gutowski, Heidrun Heinke, and Detlef Hommel

Universität Bremen, Institut für Festkörperphysik,
P.O. Box 330 440, 28334 Bremen, Germany

Abstract. Formation mechanisms of quantum dots in the system CdSe/ZnSe are thoroughly analyzed in this paper. Defect free QDs are generated by segregation enhanced CdSe reorganisation and not by the Stranski-Krastanov growth mode. Stacking fault formation is enhanced in QD stacks and reduced by using strain compensating ZnSSe spacer layers. For a fivefold QD stack in a laser structure a T_0 value of about 1200 K up to 100 K was determined by threshold measurements. Electrically pumped lasing at room temperature was achieved above a threshold current density of $7.5\,\mathrm{kA/cm^2}$. Degradation measurements prove a higher stability of QDs against high current injection as compared to quantum wells. High reflectivities of above 99 % for undoped and p-type doped distributed Bragg reflectors based on ZnSe and MgS/ZnSe superlattices have been obtained. Monolithic vertical resonators possess a quality factor of about 100.

The material system CdSe/ZnSe is of great interest due to its possible application in optoelectronic devices emitting in the yellow to blue spectral region. Especially lasing operation in the green is hardly possible by other materials like InAs/GaAs, InP/GaP (infrared to red) and InN/GaN (blue to ultraviolett). Although continious wave (cw) ZnSe-based laser diodes working at room temperature have been achieved, the lifetime is limited to about 400 h due to the instability of the QW [1]. It is assumed that quantum dots (QDs) are less susceptible to defects. Hence the use of CdSe QDs as active region might be a solution for the stability problem. However, the formation of defect free QDs does not occur in the system CdSe/ZnSe by Stranski-Krastanov growth mode despite a similar lattice mismatch to InAs/GaAs of about 7 % [2]. Therefore, the fabrication of QDs requires special care. Vertical cavity surface emitting lasers (VCSELs) have besides other operational advantages a strongly reduced threshold current density. Thus the lifetime should be further strongly increased for VCSELs.

1 Conventional ZnSe-Based Edge Emitting Laser Diodes

We routinely fabricate egde emitting ZnSe-based laser diodes on GaAs substrates [3,4]. The standard laser structure is designed as double-heterostructure with separate confinement for the optical wave and the carriers with lattice matched MgZnSSe and ZnSSe for the cladding and wave guiding layers, respectively. A quaternary CdZnSSe QW in the center of the waveguide acts as light producing layer. Further details can be found in [3,4].

For the tests under current injection 10 μm Pd/Au contact stripes defined by the insulator Al_2O_3 are processed onto the laser structures. The laser bars with a length of about 1 mm are mounted onto Cu heat sinks with the epitaxial side facing upwards.

The tuning range for the emission wavelength of ZnSe based devices is mainly determined by the composition of the QW material. With a binary ZnSe QW emission a wavelength as short as 463 nm is achieved in the pulsed mode. Under DC current injection the shortest wavelength obtained so far is 490 nm (with a small addition of Cd) [5,6]. With optimized growth conditions of the CdZnSSe QW material – in terms of flux ratio – it is possible to grow QWs with a high Cd content and high optical as well as strucutural quality [4]. Thus we were able to realize laser emission around 560 nm in the cw mode as well as output powers of more than 1 W in the pulsed mode. In Fig. 1 the emission spectra of ZnSe-based laser diodes are shown. The only difference between those structures is the Cd composition in the QW. Figure 1 illustrates the potential of ZnSe-based devices.

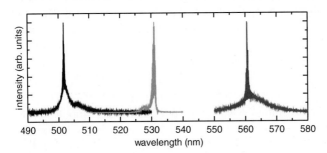

Fig. 1. Emission spectra of different ZnSe-based laser diodes. The only difference between the structures concerns the Cd composition of the QW material

2 Optimization of CdSe Quantum Dot Structures for Application in a Laser Diode

In this section we discuss the formation mechanism, properties and optimization of CdSe/Zn(S)Se QDs. Then we present the results of an electrically pumped CdSe QD laser diode in comparison to those of QW laser diodes.

2.1 Quantum Dot Formation in the System CdSe/ZnSe

During the growth of CdSe on ZnSe a transition from a two- to a three-dimensional surface is observed by means of reflection high-energy electron diffraction [8]. The growth mode change occurs between a thickness of about

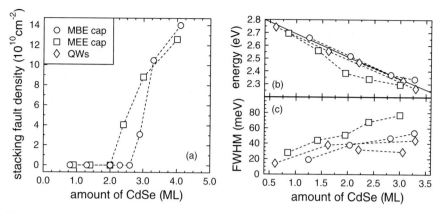

Fig. 2. (a) Stacking fault density determined by HRXRD, (b) PL (5 K) energy, and (c) PL (5 K) FWHM vs. amount of CdSe for samples where the CdSe is capped by conventional MBE or MEE. PL results (5 K) of CdSe QWs grown at low substrate temperature and calculated emission energies (*straight line*) for CdSe QWs are given for comparison

2 ML and 3 ML CdSe and is rather smooth. Figure 2a shows the stacking fault density determined by high resolution x-ray diffraction (HRXRD) [9] in dependence on the amount of CdSe in the samples. No stacking faults are found up to an amount of 2.6 ML for samples where the CdSe is simply capped by conventional molecular beam epitaxy (MBE) under stoichiometric conditions. However, high stacking fault densities in the order of 10^{10} cm^{-2} are found at higher CdSe deposits in agreement with other reports [10]. Thus, the growth mode transition is accompanied by a strong generation of stacking faults.

Spatially integrated photoluminescence (PL) spectra of the samples were measured using a HeCd-laser (325 nm), a $\frac{1}{4}$ m monochromator with 1200 mm^{-1} grating, and a LN$_2$-cooled CCD-camera. The results are presented in Fig. 2b and c. With higher CdSe content the emission shifts to lower energies. This shift is in good agreement to that calculated for CdSe QWs [11] in case of samples capped by use of conventional MBE. The full width at half maximum (FWHM) increases with increasing amount of CdSe. However, it remains below 50 meV up to 3.0 ML CdSe which is comparable to the case of CdSe QWs. Hence, the low temperature PL is more characteristic for QWs if the CdSe is overgrown by conventional MBE, even above the critical thickness of growth mode transition. This assumption is corroborated by the energy shift of the PL with increasing temperature which is "S-shaped" [2]. This is a typical finding for QWs with potential fluctuations [12]. The blue shift due to the thermalization of the localized states occurs at 50 K for 2.9 ML CdSe. It follows from these results that defect free QDs cannot be obtained in the Stranski-Krastanov growth mode in the system CdSe/ZnSe.

However, QDs are generated by segregation enhanced CdSe reorganization using migration enhanced epitaxy (MEE) for CdSe capping by ZnSe [2]. The strong effect of MEE is demonstrated in Fig. 2. The low temperature emission energy is red shifted compared to the samples with the cap layer grown by conventional MBE. This shift increases with an increasing amount of CdSe up to about 2.0 ML which is the critical amount for stacking fault formation in case of this particular sample design (see Fig. 2a). The maximum shift is as large as 130 meV. In addition, the FWHM is much larger for these samples, being as large as 67 meV already for 1.9 ML CdSe. The shape of the (see Fig. 3a) becomes Gaussian for more than 1.5 ML CdSe. This points to a critical amount of CdSe for quantum dot formation of about 1.5 ML. Thus there exists a window for the formation of defect-free QDs, however, this window is rather small.

MEE differs from conventional MBE by realizing a sequence of extremely group-II and group-VI rich conditions and a low growth rate of 0.025 ML/s in our case. The importance of the VI/II ratio is obvious from Fig. 3b. The FWHM is reduced from 50 meV for 1.7 ML CdSe to 36 meV for 2.5 ML CdSe by switching from group-II to group-VI rich conditions. In the latter case, the emission energy is in good agreement with theoretical values for CdSe QWs [11] while it is red shifted by about 60 meV for the former case. Cd surface segregation was observed in the system CdSe/ZnSe [13]. Low growth rates and cation rich conditions enhance surface segregation [14]. Thus we assume that the CdSe reorganisation occurs during the Zn deposition cycle for MEE because of a strongly enhanced surface diffusion of the segregating Cd atoms under Zn rich conditions and a higher Cd incorporation probability in Cd rich regions.

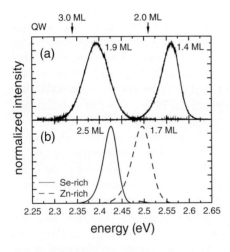

Fig. 3. Low-temperature (5 K) PL spectra for samples with the ZnSe overlayer grown by (**a**) MEE and (**b**) conventional MBE using Zn- (*solid*) or Se-rich (*dashed*) conditions. The amount of CdSe is given in the key. The arrows give theoretical values for CdSe QWs

2.2 Proof for Quantum Dots

In the following we will give clear evidence that we really obtained the formation of zero-dimensional QDs. To get access to single QDs, mesa structures with diameters down to 100 nm were defined by electron-beam lithography and chemical etching [15]. In Fig. 4 the μ-PL spectrum of a 120 nm large mesa structure taken at 5 K is displayed. Individual sharp lines with a typical PL linewidth of ∼ 0.5 meV are visible. In order to prove that the observed lines are due to electron-hole pair recombination in single QDs photon correlation measurements (details see [16]) on single lines have been performed. In the inset of Fig. 4, the normalized second order correlation function $g^{(2)}(\tau) = \langle I(t)I(t+\tau)\rangle/\langle I(t)\rangle^2$ with τ the delay time between photon pairs and $I(t)$ the measured intensity is exemplarily depicted for the PL line at 2.36 eV. The pump intensity corresponds to an excitation of the QD well below saturation. The second order correlation function shows a clear minimum at $\tau = 0$ and an exponential increase for negative and positive τ. This is a clear signature of photon antibunching. The photon correlation function can be fitted by an exponential function $g^{(2)}(\tau) = 1 - a \exp(-|\tau|/t_\mathrm{m})$, where a accounts for the background of uncorrelated photons present in the measurement and t_m corresponds to the exciton transition lifetime in the limit of low excitation power [17]. The fitted value of t_m is 3.2 ns. The fact that $g^{(2)}(0) = 0.38$, i.e., $g^{(2)}(0) < 0.5$ proves that the measured photon antibunching stems dominantly from a single quantum emitter.

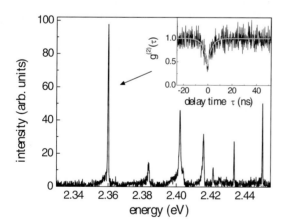

Fig. 4. μ-PL spectrum (5 K) of a small ensemble of CdSe/Zn(S,Se) quantum dots. Inset: Photon correlation function $g^{(2)}(\tau)$ measured for a single QD in a mesa under continuous optical excitation at 5 K. The fit is obtained by using a single exponential function with a decay time of 3.2 ns

2.3 Optimization of CdSe Quantum Dot Stacks

For laser diodes a high filling factor is necessary. In case of a QD matrix as active region this requires the use of quantum dot stacks. However, we found

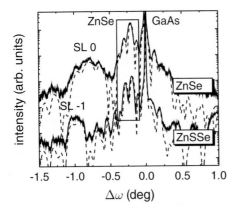

Fig. 5. Comparison of experimental (*solid*) and simulated (*dashed*) HRXRD (004) $\omega/2\theta$-scans. The spacer layer material is given in the key. The ZnSe signal is denoted by the box

an enhanced stacking fault formation in QD stacks with ZnSe spacer layers [18]. Thus we studied three-, five- and tenfold stacks for which the ZnSe spacer layers are replaced by tensily strained ZnSSe layers with a high S concentration. The ZnSSe layers compensate the highly compressive strain of the underlying CdSe QD sheet. Figure 5 shows the HRXRD profiles of a threefold stack with ZnSe spacer layers and of a fivefold stack with ZnSSe spacer layers containing nominal 26 % S. The stacks grown on GaAs(001) substrates with 2.0 ML CdSe per sheet and about 4.5 nm spacer layer thickness are embedded in a 50 nm thick ZnSe buffer and a 25 nm thick ZnSe cap layer. The influence of the stacking faults is strongest in the range of the ZnSe signal which has a double peak structure due to the phase shift which is introduced by the QD sheets [9]. The minimum between both peaks is less pronounced for high stacking fault densities compared to theoretical diffraction profiles obtained when assuming a perfect crystal. This effect is clearly visible for the threefold stack with ZnSe spacer layers indicating the high stacking fault density. However, no difference between the simulated and measured diffraction profile can be observed in the range of the ZnSe signal for the fivefold stack with ZnSSe spacer layers. Thus the stacking fault formation has been successfully reduced by introducing strain compensating spacer layers.

2.4 Optical Gain in CdSe Quantum Dot Stacks

We embedded a strain compensated fivefold QD stack into our conventional ZnSe-based laser structure (undoped and fully doped) as described before. Details of this particular structure can be found in [19,20].

The results discussed next are similar for the doped and undoped laser structure so that we restrict ourself to the exemplary presentation of the findings for the undoped structure. Gain spectra obtained by means of the variable stripe-length method (details see [21]) are shown in Fig. 6a at five different excitation densities excited via the waveguide. For densities beyond 70 kW/cm^2 the absorption turns over into amplification. The maximal modal

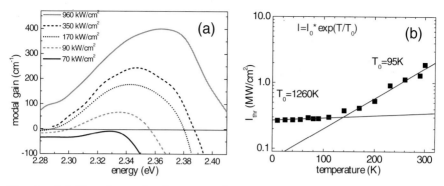

Fig. 6. (a) Experimental gain spectra for an undoped laser structure (5 CdSe-QD sheets) excited via the barrier at 10 K and different excitation intensities. (b) Temperature dependence of the threshold intensity for an undoped laser structure (5 CdSe-QD sheets) excited via the barrier. The two solid lines correspond to the fitted curves in the two temperature regions above and below 100 K

gain amounts to 400 cm^{-1} for an excitation density of 960 kW/cm^2. At room temperature (RT) we have achieved a modal gain of $\sim$ 150 cm^{-1} at an excitation density of 4.5 MW/cm^2. The internal loss is of the order of $\alpha \sim 20$ cm^{-1} which is comparable to the values given in [22]. A more detailed discussion can be found in [23].

Figure 6b shows the temperature dependence of the laser threshold intensity. The curve can be divided into two temperature regions. At low temperatures the threshold intensity stays nearly constant. Beyond 100 K, it shows an exponential increase with temperature. The two temperature regions can be fitted [24] by $I = I_0 \exp(T/T_0)$. The T_0 factor for temperatures below 100 K is rather high ($T_0 = 1260 \pm 180$ K) which is typical for strongly confined systems like QD structures [25]. For $T > 100$ K carriers are probably emitted from the deeply localized states into the barriers where they nonradiatively recombine ($T_0 = 95 \pm 18$ K).

2.5 Electrically Pumped CdSe Quantum Dot Laser Diode

At room-temperature clear electroluminescence from the devices is observed under current injection which so far has only been reported by us [26]. The emission occurs around 565 nm [19]. When operated below threshold a blue shift of the emission accompanied by an intensity increase is observed in the pulsed mode and at short pulse width below 100 ns only [27]. At higher driving currents a red-shift occurs together with a minimal intensity increase only which can be attributed to local heating in the device. Our conventional quantum well laser diodes emitting in the same spectral region do not exhibit these pulse-width dependent characteristics.

The unusual dynamics of the quantum dot laser prevents operation above threshold at pulse widths longer than 100 ns. However, below 100 ns electri-

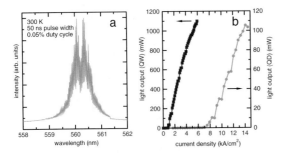

Fig. 7. (a) Emission spectra of the quantum dot laser when operated above threshold. Lasing occurs at 560 nm. (b) L/j characteristics of quantum well and quantum dot laser in comparison. Both devices are operated under a low duty cycle

cally pumped lasing is obtained for the first time for CdSe quantum dots [19]. In Fig. 7a the emission spectrum above threshold is shown. The lasing wavelength is 560 nm.

2.6 Comparison of Quantum-Dot and Quantum-Well Laser Diodes

Since the quantum dot stack was embedded in our standard laser structure it is possible to compare the device characteristics of quantum-dot and quantum-well lasers under identical conditions and at the same emission wavelength. In Fig. 7b such a comparison is exemplarily shown for the light output vs. driving current density. Whereas the quantum well laser possesses a threshold current density of 700 A/cm^2 and a maximum output power of 1.1 W, the characteristics of the quantum dot laser is not as good. Here a threshold current density of 7.5 kA/cm^2 and a maximum output of 100 mW is obtained. We expect a better performance of the quantum-dot laser for a stack with a higher dot density (more active material) and a sharper size distribution since the optical pumping experiments showed that the dots are in principle able to produce a high gain.

Although the quantum-dot laser exhibits a weaker performance when operated above threshold, it shows a higher stability in the LED mode as shown in Fig. 8. In this graph the results of a constant current degradation experiment are plotted. Based on the analysis of Chuang et al. [28] a $1/t$ long-term behaviour of the normalized light output is expected if the degradation mechanism is dominated by recombination enhanced defect reactions. In order to compensate for the lower efficiency of the quantum-dot laser it was operated at a higher driving current to obtain a comparable light output. As one can see the quantum dot laser has a significant longer lifetime as compared to the quantum-well laser – despite of the high driving current in the DC mode that gives rise to strong heating and is known to accelerate the device degradation [29]. However, the fast degradation of the quantum-well laser is also not fully understood at present but it is assumed that the influence of the highly strained quantum well might be responsible. Nevertheless, even compared to

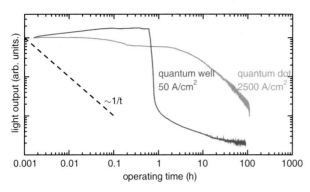

Fig. 8. Constant current degradation experiment performed on QW and QD laser structures under DC current injection. In order to achieve comparable light output powers the quantum dot laser was operated at a current density of $2.5\,\mathrm{kA/cm^2}$, whereas the quantum well laser was driven with $50\,\mathrm{A/cm^2}$

our quantum-well devices with emission around 520 nm a higher stability of the quantum dot laser is observed.

3 Distributed Bragg Reflectors and Microcavities

Up to now, optically pumped lasing operation of a II-VI VCSEL structure was achieved only when using a microcavity formed by dielectric distributed Bragg reflector (DBR) mirrors [30]. No electrically pumped VCSEL emitting in the yellow-blue spectral region has been realized yet. Monolithic microcavities suitable for the green-blue range with a reasonable cavity resonance have been achieved using quaterwave-thick layers of ZnSe and ZnSe/MgS superlattices (SLs) as the material for high and low refractive index, respectively [31]. Hence, we also use this approach.

3.1 High-Reflectivity Distributed Bragg Reflectors Using ZnSe/MgS Superlattices

Because of the short resonator length and therefore relatively low amplification compared to an edge emitter, reflectivities exceeding 99 % for the DBRs have to be achieved in order to realize a VCSEL. In addition, for device applications the epitaxial structure has to be fully strained to prevent the formation of dislocations that act as centers of nonradiative recombination and lead to fast degradation of the QW region during lasing operation. Furthermore, n- and p-type doped DBRs are necessary. The p-type doping might be a problem because the concentration of free holes decreases with decreasing refractive index of the material [33]. However, in order to reduce the number of Bragg mirror pairs, the difference in refractive index of low- and high-index materials should be as large as possible.

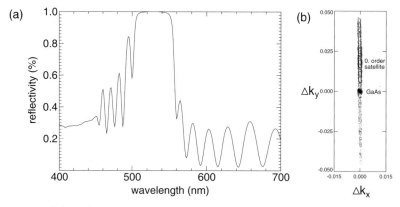

Fig. 9. (a) Reflectance spectrum of a DBR with 22 mirror pairs. A mirror pair consists of a ZnSSe layer as the high refractive index material and a short-period ZnSe/MgS-SL as the low index material. (b) Triple axis HRXRD reciprocal space map of the (224) reflection

The DBRs and microcavities presented here consist of Zn(S)Se for the high-index layers and of short-period ZnSe/MgS SLs for the low-index layers. Details concerning growth and characterization of the DBRs are described in [32]. In order to grow strained structures, the amount of sulphur in the high-index ZnSSe layers is about 6%. In Fig. 9a the reflectance spectrum of a DBR with 22 mirror pairs is depicted. The stop band centered around 528 nm has a flat shape with a reflectivity well above 99%. Furthermore, the structure is fully strained as confirmed by the triple axis HRXRD reciprocal space map of the (224) reflection shown in Fig. 9b. We obtained also p-type doping without any hint that the incorporation of N is disadvantageous for the structural quality and maximum reflectivity of the DBR [32]. These results show that fully doped monolithic microcavities with sufficient quality factors (Q-factors) for yellow-blue lasing operation at room temperature can be expected.

3.2 Monolithic Microcavities

After availability of reasonable DBR mirrors, the growth of complete VCSEL structures can be performed. The reflectance spectrum of a vertical resonator is depicted in Fig. 10a which is formed by a 17 period bottom mirror, a λ-thick ZnSe cavity and a 6 period top mirror. The cavity resonance at 513 nm has a FWHM of 9 nm, i.e., the Q-factor of the resonator is roughly 60. The results of temperature dependent PL measurements of a VCSEL structure with Q=100 containing a $Zn_{0.75}Cd_{0.25}S_{0.06}Se_{0.94}$ QW are shown in Fig. 10b. While the relative energy shift of PL emission of a typical ZnCdSSe QW (without resonator) is about 100 meV (4 K to 300 K), the cavity resonance peak (determined by reflectance measurements) is less temperature depen-

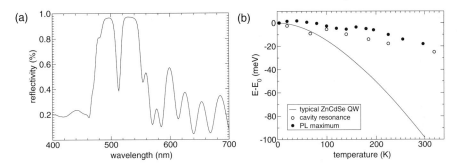

Fig. 10. (a) Reflectance spectrum of a resonator consisting of a 17 period bottom mirror, a λ-thick cavity and a 6 period top mirror. The cavity resonance at 513 nm is clearly pronounced. (b) Temperature dependence of the photoluminescence emission of a typical ZnCdSe quantum well (*straight line*), the cavity resonance (*open circles*) and the PL emission of the QW inserted into the microcavity (*filled circles*)

dent and shifts by roughly 25 meV only. The PL emission of a QW inserted into the resonator shows a temperature dependence similar to that of the cavity resonance. Hence, the resonance effectively guides the emission (495 nm at 4 K) of the QW.

In Fig. 11 cross section transmission electron microscopy (TEM) images of a resonator without QW (reflectance spectrum showed in Fig. 10a) are presented. The SLs in the structure are of good structural quality.

Fig. 11. Cross section TEM micrograph of a microcavity. The structure consists of a 17 periods bottom mirror, a λ-thick cavity and a 6 period top mirror. (**a**) overview, (**b**) magnification of a SL, (**c**) high resolution TEM picture of layers within SL consisting of a few monolayers ZnSe and MgS, respectively

4 Summary

The formation mechanism of QDs in the system CdSe/ZnSe was discussed. Evidence for 3D confinement up to at least 100 K in our structures was given. The reduction of stacking fault generation in QD stacks by use of strain compensating ZnSSe spacer layers was presented. A fivefold CdSe QD stack was applied as active region in a conventional seperate confinement laser structure. A high T_0 value of about 1200 K up to a temperature of 100 K was determined by means of threshold measurements. Electrically pumped lasing at room temperature was achieved above a threshold current density of $7.5\,\text{kA/cm}^2$ at a wavelength of about 560 nm. Degradation measurements yielded a higher stability against high current injection of the QD laser diode compared to conventional QW laser diodes. We presented DBRs consisting of Zn(S)Se and MgS/ZnSe SL pairs with high reflectivities of above 99 % obtained for undoped as well as for p-type doped structures. Monolithic vertical resonators were shown to possess a Q-factor of about 100. Thus we can fabricate all parts necessary for an electrically pumped II-VI QD VCSEL.

Acknowledgements

This work was supported by the Deutsche Forschungsgemeinschaft (Ho 1388/11 and Ho 1388/12) and VW-Stiftung (I/76 142). We thank G. Bacher and A. Forchel (University of Würzburg) for the fabrication of the mesa structures. We are grateful to K. Vennen-Damm for maintaining the MBE system.

References

1. E. Kato, H. Noguchi, M. Nagai, H. Okuyama, S. Kijima, A. Ishibashi: Electron. Lett. **34**, 282 (1998).
2. T. Passow, K. Leonardi, H. Heinke, D. Hommel, J. Seufert, G. Bacher, A. Forchel: phys. stat. sol. (b) **229**, 497 (2002).
3. M. Behringer, H. Wenisch, M. Fehrer, V. Großmann, A. Isemann, M. Klude, H. Heinke, K. Ohkawa, D. Hommel: Festkörperprobleme/Adv. Solid State Phys. **38**, 47 (1999).
4. M. Klude, G. Alexe, C. Kruse, T. Passow, H. Heinke, D. Hommel: phys. stat. sol. (b) **229**, 935 (2002).
5. D. C. Grillo, J. Han, M. Ringle, G. Hua, R. L. Gunshor, P. Kelkar, V. Kozlov, H. Jeon, A. V. Nurmikko: Electron. Lett. **30**, 2131 (1994).
6. N. Nakayama, S. Itoh, H. Okuyama, M. Ozawa, T. Ohata, K. Nakano, M. Ikeda, A. Ishisbashi, Y. Mori: Electron. Lett. **29**, 2194 (1993).
7. M. Behringer, K. Ohkawa, V. Großmann, H. Heinke, K. Leonardi, M. Fehrer, D. Hommel, M. Kuttler, M. Straßburg, D. Bimberg: J. Cryst. Growth **184/185**, 580 (1998).
8. K. Leonardi, D. Hommel, C. Meyne, J.-T. Zettler, W. Richter: J. Cryst. Growth **201/202**, 1222 (1999).
9. T. Passow, H. Heinke, J. Falta, K. Leonardi, D. Hommel: Appl. Phys. Lett. **77**, 3544 (2000).

10. D. Litvinov, A. Rosenauer, D. Gerthsen, H. Preis, S. Bauer, E. Kurtz: J. Appl. Phys. **89**, 4150 (2001).
11. M. Rabe, M. Lowisch, F. Henneberger: J. Cryst. Growth **184/185**, 248 (1998).
12. E. M. Daly, T. J. Glynn, J. D. Lambkin, L. Considine, S. Walsh: Phys. Rev. B **52**, 4696 (1995), and references therein.
13. T. Passow, H. Heinke, T. Schmidt, J. Falta, A. Stockmann, H. Selke, P. L. Ryder, K. Leonardi, D. Hommel: Phys. Rev. B **64**, 193311 (2001).
14. E. Chirlias, J. Massies, J. L. Guyaux, H. Moisan, J.Ch. Garcia: Appl. Phys. Lett. **74**, 3972 (1999), and references therein.
15. T. Kümmel, R. Weigand, G. Bacher, A. Forchel, K. Leonardi, D. Hommel, H. Selke: Appl. Phys. Lett. **73**, 3105 (1998).
16. C. Becher, A. Kiraz, P. Michler, A. Imamoglu, W. V. Schoenfeld, P. M. Petroff, Lidong Zhang, E. Hu: Phys. Rev. B **63**, 121312(R) (2001).
17. P. Michler, A. Imamoglu, M. D. Mason, P. J. Carson, G. F. Strouse, S. K. Buratto: Nature **406**, 968 (2000).
18. T. Passow, K. Leonardi, D. Hommel: phys. stat. sol. (b) **224**, 143 (2001).
19. M. Klude, T. Passow, R. Kröger, D. Hommel: Electron. Lett. **37**, 1119 (2001).
20. M. Klude, T. Passow, G. Alexe, H. Heinke, D. Hommel: Proc. SPIE **4594**, 260 (2001).
21. P. Michler, M. Vehse, J. Gutowski, M. Behringer, D. Hommel, M. F. Pereira Jr., K. Henneberger: Phys. Rev. B **58**, 2055 (1998).
22. I. L. Krestnikov, M. Straßburg, M. Caesar, A. Hoffmann, U. W. Pohl, D. Bimberg, N. N. Ledentsov, P. S. Kop'ev, Zh.I. Alferov, D. Litvinov, A. Rosenauer, D. Gerthsen : Phys. Rev. B **60**, 8695 (1999).
23. K. Sebald, P. Michler, J. Gutowski, R. Kröger, T. Passow, M. Klude, D. Hommel: phys. stat. sol. (b), in press (2002).
24. S. Bidnyk, T. J. Schmidt, Y. H. Cho, G. H. Gainer, J. J. Song, S. Keller, U. K. Mishra, S. P. DenBaars: Appl. Phys. Lett. **72**, 1623 (1998).
25. D. Bimberg, N. N. Ledentsov, M. Grundmann, N. Kirstaedter, O. G. Schmidt, M. H. Mao, V. M. Ustinov, A. Y. Egorov, A. E. Zhukov, P. S. Kop'ev, Zh.I. Alferov, S. S. Ruvimov, U. Gosele, J. Heydenreich: phys. stat. sol. (b) **194**, 159 (1996).
26. T. Passow, M. Klude, K. Leonardi, D. Hommel: in *Proceedings of the 25th International Conference on the Physics of Semiconductors*, M. Miura, T. Ando (Eds.) (Springer, Berlin 2001) pp. 1607–1608.
27. M. Klude, T. Passow, H. Heinke, D. Hommel: phys. stat. sol. (b) **229**, 1029 (2002).
28. S.-L. Chuang, A. Ishibashi, S. Kijima, N. Nakayama, M. Ukita, S. Taniguchi: IEEE J. Quantum Electr. **33**, 970 (1997).
29. M. Klude, M. Fehrer, V. Großmann, D. Hommel: J. Cryst. Growth **214/215**, 1040 (2000).
30. H. Jeon, V. Kozlov, P. Kelkar, A. V. Nurmikko, C.-C. Chu, D. C. Grillo, J. Han, G. C. Hua, R. L. Gunshor: Appl. Phys. Lett. **67**, 1668 (1995).
31. T. Tawara, H. Yoshida, T. Yogo, S. Tanaka, I. Suemune: J. Cryst. Growth **221**, 699 (2000).
32. C. Kruse, G. Alexe, M. Klude, H. Heinke, D. Hommel: phys. stat. sol. (b) **229**, 111 (2002).
33. H. Okuyama, Y. Kishita, T. Miyajima, A. Ishibashi, K. Akimoto: Appl. Phys. Lett. **67**, 904 (1994).

ZnCdSe Quantum Structures – Growth, Optical Properties and Applications

Martin Strassburg[1], O. Schulz[1], Matthias Strassburg[1], U.W. Pohl[1],
R. Heitz[1], A. Hoffmann[1], D. Bimberg[1], M. Klude[2], D. Hommel[2],
K. Lischka[3], and D. Schikora[3]

[1] Technische Universität Berlin, Institut für Festkörperphysik,
    10623 Berlin, Germany
[2] Universität Bremen, Institut für Festkörperphysik,
    28359 Bremen, Germany
[3] Universität Paderborn, Fachbereich 6 - Physik,
    33098 Paderborn, Germany

Abstract. ZnCdSe quantum structures are investigasted for the effect of exciton localisation on the potential for opto-electronic applications. The investigation on ZnCdSe quantum dots as the active material in a laser diode and their temperature dependece show a transition from 0D-like to 2D-like characteristics limiting their capabiblity for devices. Furthermore, optimisation of electrical contacts due to a post-growth increase of the p-type doping level and efficient index guiding allowing a substantial decrease of losses improving the lifetime of laser diodes more than 20 times.

Semiconductor quantum structures have been widely studied for the possibility to engineer electronic and optical properties being essential also for optimising opto-electronic devices. The fabrication of semiconductor (SC) light emitters and laser diodes (LDs) gives access to a large field of applications, e.g. laser TV and polymer optical fibre communication technology. Mixing of suitable discrete laser transitions will allow to cover the whole color spectrum. Thus, for the implementation of displays, projector devices and laser televison, blue, green and red emitting LDs are necessary. However, until now, the development of such devices is prevented by the lack of commercial green-emitting LDs with sufficient lifetime. Based on GaN and GaAs blue and red emitting LDs are realised, respectively. Especially, the introduction of zero-dimensional centres for light generation enabled the commercial breakthrough of UV and blue LDs [1] and lead to significant improvements in GaAs-based LDs [2]. The investigations of proper materials for the green-emitting LDs were focused on ZnCdSe and InGaN due to their optically properties (e.g., bandgap energy, oscillator strength,à). Although electrically pumped lasing was demonstrated in ZnSe-based quantum structures more than ten years ago [3], a breakthrough for green LDs was not achieved, yet.

The aim of this paper is to present and evaluate approaches and results leading to a significantly enhanced lifetime of ZnSe-based LDs. As it was shown by Sony Corp., which still holds the lifetime record of about 500 h for ZnSe LDs using a ZnCdSe QW as active region [4], the lifetime is limited by heat induced defect generation. In II-VI LDs, the heat production is much larger than in III-V LDs, because of the large series resistance, which originates from the non-ohmic contacts [5]. This heat accelerates the formation of so-called dark-line defects which cause a degradation of the active area and, hence of the whole device [6]. Thus, novel approaches to increase the lifetime of ZnSe-based LDs have to reduce the heat generation. Therefore, the whole device has to be improved to reduce the electrical and optical losses. Introducing quantum dot (QD) ensembles for the active region, novel contacts, and current- and wave-guiding, we will present improvements on several areas of ZnCdSe LDs. The impact on heat reduction is discussed and significantly lifetime extension is presented.

1 Quantum Dots in the Active Region: 0D Localisation and Mobility of Excitons

The active region of a ZnSe-based LD usually consists of single or multi-layered stacks of ZnCdSe QWs, where non-radiative processes lead to heat generation. The in-plane mobility allow charge carriers to migrate to defects, where non-radiative recombination contributes to the heating of the LD.

In recent years, the role of QDs as active region for ZnSe-based LDs has increased tremendously driven by enhanced linear and nonlinear optical properties compared to systems of higher dimensionality, and of course, by improving growth techniques. The introduction of QDs as active regions is advantageously, as it was demonstrated e.g., for red- and near-infrared-emitting LDs [2]. The extraordinary properties of QDs (e.g. discrete electronic levels, thermal stability of states, high (excitonic) gain, low density of states) enable the realisation of new principles in light emitting devices. It is known that the 3D confinement of carriers and excitons allows zero-phonon lasing [7] and exciton waveguiding [8], which provides for efficient optical confinement in a narrow spectral range. The enhanced wave-guiding could decrease optical losses. Therefore, one might expect a lowering of the lasing threshold and consequently a reduction of heat generation in such devices. However, ZnCdSe LDs with QDs in the active region, as shortly described above, have not been realised, yet. A possible explaination might be the insufficient localisation of excitons/carriers in such QD structures. Even, for structures showing optically pumped 0D lasing up to RT, insufficient p-conductivity of the cladding layers prevents succesful LD operation [9].

Why does 3D confinement of carriers/excitons not persist at RT in ZnCdSe QD structures? Therefore, the selforganised growth of II-VI QDs is evaluated. Due to the lattice mismatch between CdSe and ZnSe ($\Delta a/a > 7\ \%$), which is very similar to that of the InAs/GaAs system, and the respective band discontinuities a growth of type I QD heterostructures is enabled by selforganised island formation. Generally, high-quality QD structures have been demonstrated by numbers of research groups and commercial companies. Dense arrays of islands (up to several $10^{10}\,\mathrm{cm}^{-2}$) with island sizes ranging from about 10 nm to above 100 nm are grown by MBE in the Stranski-Krastanov mode [10,11,12].

Using appropriate growth conditions even higher island densities and smaller island sizes are realised by MBE and MOCVD techniques due to an inhomogeneous distribution of Cd in the deposited QW [13,14]. These inhomogeneities are generated by sub-monolayer growth and/or segregation of Cd during the capping procedure [15]. Such islands formed by Cd-rich regions in a ZnCdSe alloy QW have densities up to $10^{12}\,\mathrm{cm}^{-2}$. Obviously, the ZnCdSe system allowes for structures with dense arrays of nm-scaled islands providing 3D confinement for carriers and excitons. QD-like behaviour of such structures was confirmed e.g. by photo- and cathodoluminescence investigations performed at lower temperatures.

0D lasing mechanisms in II-VI QD structures have been extensively studied, especially at low temperatures. For an example, excitonic and biexcitonic gain and zero-phonon lasing are reported. In this way, and especially at low temperatures it was confirmed that ZnCdSe QDs meet theoretical predictions for novel laser structures. The large potential offered by ZnCdSe QDs for LDs is illustrated by an example, shown in Fig. 1. For a MOCVD-grown multilayered stack of ZnCdSe QDs in a ZnSSe matrix a low threshold density in optically pumped QD laser is realised. A lasing threshold density well below

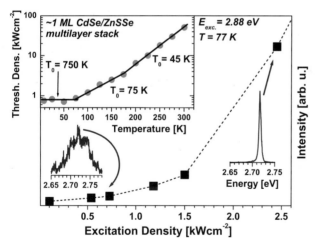

Fig. 1. Intensity of an optically pumped laser structure with CdSe/ZnSSe multilayered QDs in the active region. The spectra were recorded at the excitation densities marked by arrows. The inset shows the threshold density as a function of temperature. The decrease of characteristic temperatures T_0 with increasing temperature reflects the 0D- to 2D-like emission characteristics at low and elevated temperatures, respectively

1 kWcm^{-2} is observed, which is about five times less than in comparable QW structures [16]. The low threshold density is attributed to 0D excitons. This is confirmed by a high characteristic temperature T_0. In ideal QDs an infinite T_0 is expected, whereas in QW lasers T_0 is well below 100 K. The observed decrease in T_0 above T = 100 K is attributed to the delocalisation of excitons and carriers from the QDs. Thermal activation increases the exciton mobility and enables the transition in higher states, i.e. the QW, resulting in 2D-like excitons/carriers.

These experiments show, that localisation energy of excitons/carriers up to room temperature (RT) is problematically in ZnCdSe QD structures. Hence, it is essential to identify the origin and mechanisms of exciton/carrier mobility and, possible identifying means to preserve 0D behaviour up to RT.

Such investigations on the localisation and mobility of 0D and 2D excitons are presented, now. Localisation at interface roughness and fluctuations lead to 0D-like behaviour in QW structures. The influence of the respective confinement potentials on excitons and carriers has been investigated for CdSe- and ZnSe based QW structures for more than twenty years. However, the role of the exciton mobility, even in the case of quasi-3D confinement in QDs, i.e. for the localisation and delocalisation in 0D-like ZnSe- and CdSe-based structures, was met with increasing interest in recent years [17,18]. Though different theoretical approaches were introduced to explain such experiments, the most striking result was the persistent mobility of excitons and carriers in the presence of 0D localisation sites and, furthermore, even at low temperatures for which an ideal QD behaviour was expected. The mobility is caused by exciton or carrier transfer processes between QDs. Depending on the localisation energy of the exciton, such transfer might be more probable than radiative recombination. Such a behaviour can be described by a mobility edge representing the energy above which transfer and 2D-like behaviour occur even in QD structures. A trivial example for a mobility edge is given by the 2D wetting layer state, which is the first common electronic state for QDs grown in SK mode. In real QD structures, the mobility edge is observed well below the wetting layer energy and is attributed to the presence of a large number of energy states provided by a high-density of QDs.

Therefore, a redistribution of excitons is enabled among the QD ensemble from 0D sites with higher transition energies to 0D sites with smaller transition energies. At low temperatures the transfer is assigned to tunnel processes and, hence, is observed particularly in high-density QD ensembles. With increasing temperature, the difference between QD groundstate and barrier energies is decreased by thermal activation of excitons/carriers. Additionally, phonon-assisted escape becomes more probably and supports the lateral mobility of excitons/carriers.

Such transfer processes lead to a characteristic variation of PL decay time as a function of the detection energy. The decay time increases with the detection energy and shows saturation for energies given by QDs with large localisation energy for excitons. This behaviour originates from transfer processes being faster than the radiative decay of 0D excitons in the ZnCdSe system. Time-resolved investigations allow the determination of the radiative lifetime and the mobility edge in QD structures [18], as shown for an inhomogeneous broadened ZnCdSe-QD ensemble and a QW in Fig. 2. It is known that the radiative transition rate is proportional to the overlap of the carrier wavefunction. For strongly localised excitons the overlap remains constant. Hence, the radiative lifetime is expected to be independent of temperature for QD transitions. This is confirmed for the investigated QD structure. The QW reference sample shows an pronounced increase of radiative lifetime with temperature, when the thermal energy overcomes the localisation energy of excitons in QW inhomogeneities (e.g., islands formed by Cd fluctuations or thickness fluctuations of the QW). The temperature dependence of the mobility edge (see Fig. 2) reveals that only a significantly reduced amount of QDs provides 0D-like localisation up to elevated temperatures. Although the PL is still generated in 0D centres, what is confirmed by a parallel redshift of the PL maximum and the mobility edge, this pronounced redshift illustrates that with rising temperature more and more QDs allow the transfer to lower energy states. In comparison to that, in QW structures the mobility edge follows the temperature dependence of the bandgap [19].

Finally, the transfer of excitons/carriers between QDs results in a 2D-like behaviour. 0D-like lasing and the implementation of an excitonic waveguiding are not yet achieved at RT. Therefore, until now, the improvement of ZnCdSe-based LDs using QDs in the active region is by far not as high as expected. A significant reduction of losses and, thus, a lowering of parasitic heat generation is not demonstrated. Nevertheless, mechanisms determining exciton/carrier mobility are identified. The suppression of high-density energy states of QDs formed by Cd fluctuations is proposed and examined to reduce carrier mobility [20,21,22,23], and thus provide thermally more stable exciton/carrier localisation. The implementation of such QD structures

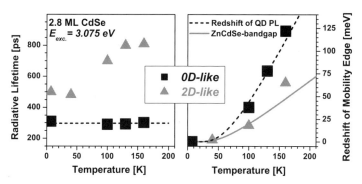

Fig. 2. Radiative lifetime and redshift of the mobility edge for (0D-like) quantum dot and (2D-like) quantum well structures as a function of temperature

might provide the expected properties of the active region, but they are still under investigation.

2 Optimisation of p-Contacts and p-Claddings

While the improvement of the active region still suffers from incompatibility to RT operation, the optimisation of the entire laser structure becomes indispensable. ZnSe-based lasers consist of a separate confinement heterostructure (SCH). The structure is generally grown on n-type GaAs-substrate. The n-side of the laser comprise ZnSe and ZnSSe buffer layers, a ZnMgSSe cladding layer and a ZnSSe waveguide. Free electron concentrations range from $2 \cdot 10^{18}$ cm^{-3} for binary to $4 \cdot 17^{17}$ cm^{-3} for quaternary compounds. The active region is a ZnCdSSe QW, which is followed by a p-ZnSSe waveguide, a p-ZnMgSSe cladding layer, a p-ZnSSe spacer and the p-contact composed of ZnSe, ZnSeTe and ZnTe. Free hole concentrations are between $1 \cdot 10^{18}$ cm^{-3} for binary and $6 \cdot 10^{16}$ cm^{-3} for quaternary layers. Details of the growth and design of the laser structures can be found elsewhere [24].

Present issues of LD improvement mainly focus on the poor electrical characteristics. While the n-type doping levels of the device are sufficiently high, the p-type doping, in particular of the ZnMgSSe cladding, is still a limiting factor in the performance of a laser diode. Furthermore, there exist no ohmic contacts to p-ZnSe due to its large bandgap energy and electron affinity that sum up to 6.7 eV thus exceeding the workfunction of metals [25]. Holes can hence only be injected by tunneling through the Schottky barrier. A high tunneling rate requires a narrow barrier width. Since the p-type doping of ZnSe is limited to about $2 \cdot 10^{18}$ cm^{-3} due to compensation effects [26], the achievement of higher acceptor concentration during MBE growth turned out to be not feasable. Thus, a strongly strained ZnTe/ZnSe-MQW structure was introduced [27,28]. To avoid strain induced defects and hence stronger compensation of the nitrogen acceptor in p-ZnSe, it is important to keep the thickness of the ZnTe-toplayer below 4 nm [29].

Finally, the solution can be either in-diffusion of additional acceptors or a reduction of the compensation level of the nitrogen dopant, which was used for p-type doping during MBE growth. In-diffused additional acceptors must have less compensation than the built-in nitrogen acceptor. Since both, Li and N form shallow acceptor states in II-VI compounds, Li$_3$N is considered as a promising candidate for co-doping. It was previously shown that ZnSe can be doped p-type in MOCVD using Li$_3$N [30]. Nevertheless, the applied growth temperature of 450 °C is too high for laser diodes due to the enhanced diffusion of group II elements at temperatures above 400 °C [31]. In this contribution, the characteristics of structures with post growth applied Li$_3$N are studied. Li$_3$N, Pd and Au are evaporated to form the p-electrode for ZnSe-based devices [32]. The diffusion of Li$_3$N is caused by the heat during metallisation and a subsequent rapid thermal annealing at 225 °C. In-diffusion

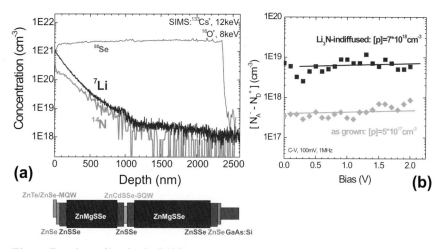

Fig. 3. Depth profile of a ZnCdSSe laser diode, demonstrating the increase of Li and N concentration due to indiffusion after deposition of a $Li_3N/Pd/Au$ electrode. The sample structure shown below corresponds to the same depth scaling and indicates the respective bandgaps (a). Enhancement of the free hole concentration of a p-type ZnSe:N layer with a p-ZnTe/ZnSe MQW (b)

of Li and N were proven by secondary ion mass spectroscopy (Fig. 3(a)). A simultaneous increase of the Li and N concentrations in the p-ZnMgSSe cladding of a LD is observed. The SIMS depth profiles reveal that Li and N diffuse in parallel into the p-side of the laser structure (except for a thin layer near the p-contact). This is a clear indication of complex formation. The ratio of Li and N lies between 2 and 3 in the entire p-side. Therefore we assume that Li and N are commonly diffusing into the lattice as a Li_xN-complex. The concentration of nitrogen in the p-ZnMgSSe cladding is enhanced to values between $7 \cdot 10^{18}$ cm^{-3} and $9 \cdot 10^{19}$ cm^{-3}. A further hint of Li_xN-complex formation is the pronounced improvement of the laser characteristics. Compared to lasers with standard Pd/Au electrodes, a significant reduction of the threshold current density from 250 A/cm^2 to below 50 A/cm^2 and a decrease of the operating voltage by 20 % down to 5.5 V accompanied by an increase of the differential quantum efficiency of 50 % lead to a lifetime extension by more than a factor of 20 with a total cw-lifetime of 9.5 h [33,34]. These improvements can be explained with an enhanced free hole concentration, leading to a current guiding below the stripe contact and hence an reduced threshold current density and a reduced Schottky barrier width lowering the operation voltage [35]. If Li and N would act as single acceptors, such large concentration of Li and N show strong compensation effects [36,37]. Thus, a dramatic deterioration of the laser performance would have been expected. Since the benefit of the Li_3N in-diffusion is so obvious, an enhancement of the free hole concentration is assumed due to Li_xN acting as an acceptor complex.

The increase of the p-type doping level is verified independently by C-V profiling. Therefore, the free hole concentration of an as grown test structure consisting of a p-ZnSe-layer and a standard ZnTe/ZnSe-MQW contact structure is compared to a similar structure with an additional in-diffused Li_3N layer. To contact the sample surface, the Li_3N/Pd/Au electrode must be removed by reactive ion etching after its evaporating and subsequent annealing. For C-V profiling, mercury is used to create Schottky contacts on the sample surface. The formation of the Schottky contact was confirmed by I-V profiling. While the free hole concentration of the as grown structure is $5 \cdot 10^{17}$ cm^{-3}, it is increased by more than one order of magnitude after Li_3N in-diffusion to $7 \cdot 10^{18}$ cm^{-3} (Fig. 3(b)). This demonstrates that the in-diffused Li_3N-complex acts as an acceptor.

3 Optical Confinement

Using the potential of lateral waveguiding in ZnSe-based LDs, losses and thus heat generation can be further reduced. In gain guided LD structures guiding of the lightwave is achieved in vertical direction by the waveguide layers which embed the active QW. Lateral guiding arises from the stripe contact geometry for providing a laterally structured carrier injection (Fig. 4(a)). To improve the performance of ZnSe-based LDs an additional lateral index guiding might be introduced to support the gain guiding (Fig. 4(b)). Such an optimised guiding of the lightwave can be obtained via laterally structured ion implantation. Ions with an appropriate energy to generate a maximal amount of vacancies near the active region are used, because the diffusion

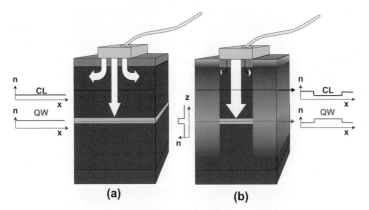

Fig. 4. Comparison of gain (a) and index guided laser structure created by ion implantation (b). The insets show the refractive index in the active region and the cladding. For details see discussion in the text

of group II components is enhanced by vacancies [31]. Hence the interfaces between the layers will be intermixed in irradiated regions. Details of the used implantation technique are described elsewere [38]. The diffusion of the Cd into the waveguide lowers the refractive index of the active region in the implanted and intermixed parts of the structure. Thereby, a small step of the refractive index is generated in the lateral direction. Simultaneously, the intermixing increases the refractive index of the waveguide and cladding in the implanted and intermixed regions of the structure. A small, negative step of the refractive index and thus an anti-waveguide is generated below and above the plane of the active region (Fig. 4(b)). The stripewidth of the laser is limited by the resist which forms the implantation mask and the intermixing of the stripe edges during ion implantation. Thus, a total stripewidth of $< 5\,\mu m$ can be achieved. Such small stripewidth is necessary for fundamental optical mode (TEM_{00}) emission. In addition to the refractive index variation, a blocking of the current spread occures due to defects in the intermixed regions (see white arrows in Fig. 4). These defects generate highly resistive material beneath the stripe. Such lasers are fully index guided. Since both, the light wave and the current are guided in such structures, losses will be reduced. Therefore, low threshold current densities and increased differential quantum efficiencies are expected for fully index guided lasers. The obtained results prove the applicability of ion implantation for ZnSe-based LDs. The current density is reduced by a factor of three down to 96 A/cm^2 and the differential quantum efficienciy is more than doubled [39,40]. Due to the reduced optical losses, higher modes are suppressed. Finally, the lifetime of LDs is extended by a factor of up to five.

4 Conclusions

We have presented that the characteristics of ZnCdSe QDs as material for the optically activ region are superior to that of corresponding QWs. Due to thermally activated transfer processes and week localisation these advantages vanish at RT. Since the mobility edge shows a strong redshift, even deep localising QDs behave 2D-like. Therefore, the increase of thermal stability is the main issue for the future development of ZnCdSe QDs. Using post growth enhancement of the p-type doping and lateral index guiding, electrical and optical losses in ZnCdSe QW lasers are reduced. Hence, remarkable low threshold current densities below 50 A/cm^2 are obtained. All these improvements lead to an increase of the cw-lifetime of a factor of more than 20 demonstrating the optimisation potential of ZnCdSe QW lasers.

Acknowledgement

The authors highly appreciate expert assistance of J. Krauser for ion implantation and J. Christen and Zh.I. Alferov for fruitful discussion.

References

1. S. Nakamura, Science **281**, 956 (1998).
2. for a review see D. Bimberg, M. Grundmann and N. N. Ledentsov, *Quantum Dot Heterostructures*, Wiley, (1999).
3. M. A. Haase, J. Qiu, J. M. DePuydt, and H. Cheng, Appl. Phys. Lett. **59**, 2992 (1991).
4. E. Kato, H. Noguchi, M. Nagai, H. Okuyama, S. Kijima, and A. Ishibashi, Electr. Lett. **34**, 282 (1998).
5. K. Kondo, H. Okuyama, and A. Ishibashi, Appl. Phys. Lett. **64**, 3434 (1994).
6. G. M. Haugen, S. Guha, H. Cheng, J. M. DePuydt, M. A. Haase, G. E. Höfler, J. Qiu, and B. J. Wu, Appl. Phys. Lett. **66**, 358 (1994).
7. N. N. Ledentsov, I. L. Krestnikov, M. V. Maksimov, S. V. Ivanov, S. V. Sorokin, P. S. Kop'ev, Zh.I. Alferov, D. Bimberg, and S. M. Sotomayor Torres, Appl. Phys. Lett. **69**, 1343 (1996).
8. Zh.I. Alferov, S. V. Ivanov, P. S. Kop'ev, A. V. Lebedev, N. N. Ledentsov, M. V. Maximov, I. V. Sedova, T. V. Shubina, and A. A. Toropov, Superl. and Microstr. **15**, 65 (1994).
9. A. V. Sakharov, S. V. Ivanov, S. V. Sorokin, I. L. Krestnikov, B. V. Volovik, N. N. Ledentsov, P. S. Kop'ev, Techn. Phys. Lett. **23**, 305 (1997).
10. M. Rabe, M. Lowisch, and F. Henneberger, J. Cryst.Growth **184/185**, 248 (1998).
11. D. Schikora, S. Schwedhelm, D. J. As, K. Lischka, D. Litvinov, A. Rosenauer, D. Gerthsen, M. Strassburg, A. Hoffmann, and D. Bimberg, Appl. Phys. Lett. **76**, 418 (2000).
12. K. Maehashi, N. Yasui, Y. Murase, T. Ota, T. Noma, and H. Nakashima, J. Electr. Mat. **29**, 542 (2000).
13. S. V. Ivanov, A. A. Toropov, S. V. Sorokin, T. V. Shubina, I. V. Sedova, A. A. Sitnikova, P. S. Kopev, Zh. I. Alferov H.-J. Lugauer, G. Reuscher, M. Keim, F. Fischer, A. Waag, and G. Landwehr, Appl. Phys. Lett. **74**, 498 (1999).
14. R. Engelhardt, U. W. Pohl, D. Bimberg, D. Litvinov, A. Rosenauer, and D. Gerthsen, J. Appl. Phys. **86**, 5578 (1999).
15. N. Peranio, A. Rosenauer, D. Gerthsen, S. V. Sorokin, I. V. Sedova, and S. V. Ivanov, Phys. Rev. B **61**, 16015 (2000).
16. F. Kreller, J. Puls, and F. Henneberger, Appl. Phys. Lett. **69**, 2406 (1996).
17. S. Yamaguchi, H. Kurusu, Y. Kawakami, Sz. Fujita, and Sg. Fujita, Phys. Rev. B **61**, 10303 (2000). see also: A. Klochikhin, A. Reznetsky, S. Permogorov, T. Breitkopf, M. Grün, M. Hetterich, C. Klingshirn, V. Lyssenko, W. Langbein, and J. M. Hvam, Phys. Rev. B **59**, 12947 (1999); L. E. Golub, S. V. Ivanov, E. L. Ivchenko, T. V. Shubina, A. A. Toropov, J. P. Bergman, G. R. Pozina, B. Monemar, and M. Wilander, Phys. Stat. Sol. (b) **205**, 203 (1998).
18. M. Strassburg, M. Dworzak, H. Born, R. Heitz, A. Hoffmann, M. Bartels, K. Lischka, D. Schikora, and J. Christen, Appl. Phys. Lett. **80**, 473 (2002).
19. K. P. O'Donnell, P. G. Middleton, in *Properties of Wide Bandgap II-VI-Semiconductors*, ed. R. Bhargava, INSPEC, emis Datareview series No. 17, 33 (1997).
20. M. Strassburg, M. Dworzak, A. Hoffmann, R. Heitz, U. W. Pohl, D. Bimberg, D. Litvinov, A. Rosenauer, D. Gerthsen, I. Kudryashov, K. Lischka, and D. Schikora, Phys. Stat. Sol. (a) **180**, 281 (2000).

21. E. Kurtz, M. Schmidt, M. Baldauf, S. Wachter, M. Grün, H. Kalt, C. Klingshirn, D. Litvinov, A. Rosenauer, and D. Gerthsen, Appl. Phys. Lett. **79**, 1118 (2001).
22. M. Strassburg, J. Christen, M. Dworzak, R. Heitz, A. Hoffmann, M. Bartels, K. Lischka, and D. Schikora, Phys. Stat. Sol. (b) **229**, 529 (2002).
23. T. Passow, K. Leonardi, H. Henke, D. Hommel, J.Seufert, G. Bacher, and A. Forchel, Phys. Stat. Sol. (b) **229**, 497 (2002).
24. M. Behringer, H. Wenisch, M. Fehrer, V. Grossmann, A. Isemann, M. Klude, H. Heinke, K. Ohkawa, and D. Hommel, Adv. Solid State Phys. **47**, 47 (1997).
25. Y. Koide, T. Kawakami, N. Teraguchi, Y. Tomomura, A. Suzuki, and M. Murakami, J. Appl. Phys. **82**, 2393 (1997).
26. P. M. Mensz, S. Herko, K. W. Haberern, J. Gaines, and C. Ponzoni, Appl. Phys. Lett. **63**, 2800 (1993).
27. Y. Fan, J. Han, L. He, J. Saraie, R. L. Gunshor, M. Hagerott, H. Jeon, A. V. Nurmikko, G. C. Hua, and N. Otsuka, Appl. Phys. Lett. **61**, 3160 (1992).
28. F. Hiei, M. Ikeda, M. Ozawa, T. Miyajima, and A. Ishibashi, and K. Akimoto, Electr. Lett. **29**, 878 (1993).
29. S. Tomiya, S. Kijima, H. Okuyama, H. Tsukamoto and T. Hino, S. Taniguchi, H. Noguchi, E. Kato, and A. Ishibashi, J. App. Phys. **86**, 3616 (1999).
30. T. Yasuda, I. Mitsuishi, and H. Kukimoto, Appl. Phys. Lett. **52**, 57 (1988).
31. M. Strassburg, M. Kuttler, U. W. Pohl, and D. Bimberg, Thin Solid Films **336**, 208 (1998).
32. M. Strassburg, O. Schulz, U. W. Pohl, and D. Bimberg, German Patent Nr. 19955280 (1999).
33. M. Strassburg, O. Schulz, U. W. Pohl, D. Bimberg, M. Klude, and D. Hommel, Electr. Lett. **36**, 878 (2000).
34. M. Strassburg, O. Schulz, U. W. Pohl, D. Bimberg, S. Itoh, K. Nakano, A. Ishibashi, M. Klude, and D. Hommel, IEEE J. Select. Top. Quant. Electr. **7**, 371 (2001).
35. B. Mroziewicz, M. Bugajski, and W. Nakwaski: *Physics of Semiconductor Lasers*, North Holland, (1991).
36. R. Heitz, B. Lummer, V. Kutzer, D. Wiesmann, A. Hoffmann, I. Broser, E. Kurtz, S. Einfeldt and J. Nürnberger, B. Jobst, D. Hommel, and G. Landwehr, Mat. Science Forum **182-184**, 259 (1995).
37. U. W. Pohl, G. H. Kudlek, A. Klimakov, and A. Hoffmann, J. Cryst. Growth **138**, 385 (1994).
38. O. Schulz, M. Strassburg, U. W. Pohl, D. Bimberg, S. Itoh, K. Nakano, A. Ishibashi, M. Klude, and D. Hommel, Phys. Stat. Sol. **180**, 213 (2000).
39. M. Strassburg, O. Schulz, U. W. Pohl, D. Bimberg, S. Itoh, K. Nakano, and A. Ishibashi, Electr. Lett. **36**, 44 (2000).
40. M. Strassburg, O. Schulz, U. W. Pohl, D. Bimberg, M. Klude, and D. Hommel, J. Cryst. Growth **214/215**, 1054 (2000).

Part II

Optics

Photonic Crystals:
Optical Materials for the 21st Century

K. Busch, A. Garcia-Martin, D. Hermann, L. Tkeshelashvili,
M. Frank, and P. Wölfle

Institut für Theorie der Kondensierten Materie, Universität Karlsruhe,
76128 Karlsruhe, Germany

Abstract. We outline a theoretical framework that allows qualitative as well as quantitative analysis of the optical properties of Photonic Crystals (PCs) derived from solid state theoretical concepts. Starting from advances in photonic band structure computations which we apply to structures containing dispersive components, we show how defect structures can be efficiently treated with the help of photonic Wannier functions. In addition, nonlinear PCs may be investigated by an appropriate multi-scale analysis utilizing Bloch waves as carrier waves together with an adaptation of $\boldsymbol{k}\cdot\boldsymbol{p}$-perturbation theory. This leads to a natural generalization of the slowly varying envelope approximation to the case of nonlinear wave propagation in PCs.

1 Introduction

Progress in Photonics is closely related to the development of optical materials with tailor made properties. Photonic Crystals (PCs) carry this principle to a new level of sophistication in the sense that the photonic dispersion relation and associated mode structure may be tailored to almost any need through a judicious design of these two-dimensional (2D) or three-dimensional (3D) periodic dielectric arrays. In particular, the choice of material composition, lattice periodicity and symmetry as well as the deliberate creation of defect structures embedded in PCs allows a degree of control over the properties of this novel class of optical materials that may eventually rival the flexibility in tailoring the properties of their electronic counterparts, the semiconducting materials.

The usefulness of PCs derives to a large extent from the fact that suitably engineered PCs may exhibit one or more photonic band gaps (PBGs) [1,2,3]. For instance, recent experiments have verified earlier theoretical predictions that three-dimensional (3D) PCs such as the inverse opals [4,5] exhibit frequency ranges over which ordinary linear propagation is forbidden irrespective of direction. The existence of these complete PBGs allows complete control over the radiative dynamics of active material embedded in PCs such as the complete suppression of spontaneous emission for atomic transition frequencies deep in the PBG [1] and leads to strongly non-Markovian effects such as fractional localization of the atomic population for atomic transition frequencies in close proximity to a complete PBG [6,7].

For a number of applications such as guided light in planar waveguide structures and most nonlinear wave mixing experiments it is sufficient to obtain control over the propagation of light within the corresponding plane of propagation. As a consequence, 2D-PCs and their 2D PBGs come into play. For such structures, advanced planar micro-structuring techniques borrowed from semiconductor technology can greatly simplify the fabrication process. Depending on the desired aspect ratio of sample depth (vertical direction) to lattice constant (transverse direction), high-quality 2D-PCs can be manufactured through plasma etching and lithography techniques [8,9,10] (aspect ratios up to 5:1) or through photo-electrochemically growing ordered macropores into silicon wafers [11,12] (aspect ratios up to 200:1). In the linear regime, PBGs in 2D-PCs offer novel passive optical guiding characteristics through the engineering of defects such as micro cavities and waveguides and their combination into functional elements such as wavelength add-drop filters [13,14]. Similarly, the incorporation of nonlinear materials into 2D-PBG structures creates the possibility for novel solitary wave propagation for frequencies inside the PBG, where ordinary linear propagation is forbidden. In the case of lattice-periodic Kerr-nonlinearities the threshold intensities and symmetries of these solitary waves depend on the direction of propagation [15,16], whereas in the case of nonlinear wave guiding structures embedded in a 2D-PBG material the propagation characteristics strongly depend on the nature of the waveguides [17].

As compared to the numerous applications of 2D- and 3D-PBGs, wave propagation in linear and nonlinear PCs for frequencies inside photonic bands has received far less attention. However, the recent discovery of super-refractive phenomena such as the super-prism effect [18] that are based on the highly anisotropic nature of iso-frequency surfaces in the photonic band structure suggest a number of potential applications specifically in optical telecommunication technology [19]. Therefore, they provide a valuable addition to the rich physics of wave propagation in PCs [20,21]. Similarly, in the context of nonlinear optical phenomena it is the tailoring of photonic dispersion relations and mode structures through judiciously designed PCs that allows to explore regimes for parameters such as group velocity, group velocity dispersion (GVD) and effective nonlinearities that hitherto have been virtually inaccessible. For instance, the existence of flat bands that are characteristic for 2D- and 3D- PCs and the associated low group velocities may greatly enhance frequency conversion effects [22] and may lead to improved designs for distributed-feedback (DFB) laser systems [22,23,24].

Any experimental exploration as well as technological exploitation of the huge parameter space provided by PCs has to be accompanied by a quantitative theoretical analysis in order to identify the most interesting cases and help to interpret the data as well as to find stable designs for successfully operating devices. In this manuscript, we provide an outline for a theoretical framework that allows to qualitatively as well as quantitatively determine the optical properties of PCs that is based on solid state theoretical concepts. Starting from photonic band structure computations (section 1), we show how defect structures can be efficiently treated with the help of photonic Wannier functions (section 2). Nonlinear PCs may be investigated by an appropriate multi-scale analysis that utilizes Bloch waves as carrier waves together with an adaptation of $\boldsymbol{k} \cdot \boldsymbol{p}$-perturbation theory. As an illustration, we discuss in section 3 the case of intensity dependent nonlinearities.

2 Photonic Band Structure Computation

Photonic band structure computations determine the periodic structures that exhibit PBGs and allow accurate interpretations of measurements. As a consequence, photonic band structure calculations represent an important predictive as well as interpretative basis for PC research and, therefore, lie at the heart of theoretical investigations of PCs. The goal of photonic band structure computation is the solution of the wave equation for the perfect PC, i.e., for a periodic array of dielectric material. For the simplicity of presentation we consider only 2D-PCs in the TM-polarized case for which the wave equation for the z-component of the electric field reads

$$\frac{1}{\epsilon_{\mathrm{p}}(\boldsymbol{r})} \left(\partial_x^2 + \partial_y^2 \right) E(\boldsymbol{r}) + \frac{\omega^2}{c^2} E(\boldsymbol{r}) = 0. \tag{1}$$

Here c denotes the vacuum speed of light and $\boldsymbol{r} = (x, y)$ denotes a two-dimensional position vector. The dielectric constant $\epsilon_{\mathrm{p}}(\boldsymbol{r}) \equiv \epsilon_{\mathrm{p}}(\boldsymbol{r} + \boldsymbol{R})$ is periodic with respect to the set $\mathcal{R} = \{n_1 \boldsymbol{a}_1 + n_2 \boldsymbol{a}_2; (n_1, n_2) \in \mathcal{Z}^2\}$ of lattice vectors $\boldsymbol{R}$ generated by the primitive translations $\boldsymbol{a}_i$, $i = 1, 2$ that describe the structure of the PC. Eq. (1) represents a differential equation with periodic coefficients and, therefore, its solutions obey the Bloch-Floquet theorem

$$E_{\boldsymbol{k}}(\boldsymbol{r} + \boldsymbol{a}_i) = e^{i \boldsymbol{k} \boldsymbol{a}_i} E_{\boldsymbol{k}}(\boldsymbol{r}), \tag{2}$$

where $i = 1, 2$. The wave vector $\boldsymbol{k} \in 1.\mathrm{BZ}$ that labels the solution is a vector of the first Brillouin zone (BZ) known as the crystal momentum. A straightforward way of solving Eq. (1) is to expand all the periodic functions into a Fourier series over the reciprocal lattice $\mathcal{G}$, thereby transforming the differential equation into an infinite matrix eigenvalue problem, which may be suitably truncated and solved numerically. Details of this plane wave method (PWM) for isotropic systems can be found, for instance, in [25] and for anisotropic systems in [26].

While the PWM provides a straightforward approach to computing the band structure of PCs, it also exhibits a number of shortcomings such as slow convergence associated with the truncation of Fourier series in the presence of discontinuous changes. Therefore, we have recently developed an efficient real space approach to computing photonic band structures [27]. Within this approach, the wave equation, Eq. (1), is discretized in real space leading to a sparse matrix problem. The solution of this algebraic problem is obtained by employing multi grid (MG) methods which guarantee an efficient solution by taking full advantage of the smoothness of the photonic Bloch functions. Even for the case of a naive finite difference discretization, the MG-approach easily outperforms the PWM and leads to a substantial reduction in CPU time. For instance, in the present case of 2D systems for which the Bloch functions are required we save one order of magnitude in CPU time as compared to PWM. Additional refinements such as a finite element discretization will further increase the efficiency of the MG-approach.

As an illustration of the improved characteristics of the MG-approach to photonic band structures, we consider the case of a square lattice (lattice constant a) of cylindrical posts in air. Furthermore, we assume that the cylinder material consists of an admixture of silicon ($\epsilon_m \approx 11.9$) and a certain volume fraction f of a material whose dielectric response in the frequency range of interest may be described by a two-level system $\epsilon_{tl}(\omega) = 1 + \omega_p^2/(\omega_a^2 - \omega^2 - i\gamma\omega)$. Here ω_p and ω_a describe the oscillator strength and the resonance frequency of the two-level system, respectively, while γ represents the line width of the resonance. As a consequence, the dielectric constant of the cylinders can be described via a Maxwell-Garnet effective dielectric constant

$$\epsilon_{cyl}(\omega) = \epsilon_m \left[1 + \frac{3f\alpha(\omega)}{1 - f\alpha(\omega)}\right], \tag{3}$$

where the polarizability $\alpha(\omega)$ of the two-level systems within the silicon matrix is given by $\alpha(\omega) = (\epsilon_{tl}(\omega) - \epsilon_m)/(\epsilon_{tl}(\omega) - 2\epsilon_m)$. Since the imaginary part of the cylinder dielectric constant $\epsilon_{cyl}(\omega)$ is much smaller than the real part, we perform the band structure computation using only the real part of $\epsilon_{cyl}(\omega)$ and treat the imaginary part perturbatively.

Any frequency dependence of the dielectric constant of the constituent materials presents a major problem to photonic band structure computation because the corresponding wave equation cannot be mapped onto a standard eigenvalue problem. Instead, the problem resembles a fix point problem

$$\frac{1}{\epsilon_p(\boldsymbol{r}, \omega)} \left(\partial_x^2 + \partial_y^2\right) E(\boldsymbol{r}) + \frac{\omega^2}{c^2} E(\boldsymbol{r}) = 0, \tag{4}$$

and, consequently, needs to be solved iteratively with an efficient photonic band structure solver providing the starting values and updates of the fix point frequencies. For the system under consideration, we approach the problem by determining the eigenfrequencies $\omega_{n\boldsymbol{k}}(\epsilon_c)$ of the PC for *fixed* dielectric constants ϵ_c of the cylinders. Repeating this calculation for several values of ϵ_c within the range of $\epsilon_{cyl}(\omega)$ as specified in Eq. (3) allows us to solve for the eigenfrequency-fix points. This principle is illustrated in Fig. 1, where we observe that there exist two distinct frequency regimes. For frequencies sufficiently far from the resonance of the two-level system, there

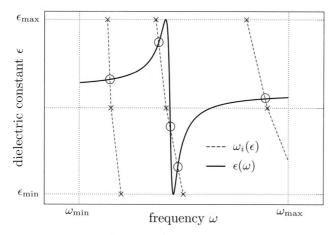

Fig. 1. Principle of the determination of the eigenfrequency-fix points for the frequency range of appreciable variation of the PC dielectric constant (solid line). The horizontal dashed lines indicate the values of the PC dielectric constant for which the eigenfrequencies have been evaluated via the MG-approach and crosses indicate the corresponding eigenfrequencies. The vertical dashed lines represent the variation of the eigenfrequencies with varying PC dielectric constant and the circles denotes the eigenfrequency-fix points

exists only one eigenfrequency-fix point, while near the resonance there exist three eigenfrequency-fix points. As an illustration, we choose the volume ratio $f = 0.05$ of two-level systems with resonance frequency $\omega_a a/2\pi c = 0.24$, oscillator strength $\omega_p = 0.7\omega_a$, and damping constant $\gamma = 0.025\omega_a$ such that the resonance frequency of the two-level system is inside the PBG close to the frequency of the first band for the undoped structure. The Bloch functions associated with this band are localized inside the dielectric cylinder (dielectric band) and are maximally affected by the variation of the cylinder dielectric constant with frequency. As a result we observe in Fig. 2 a maximal splitting of this band into one continuous band (corresponding to the solution that exists in both frequency regimes) and a localized structure inside the PBG, the shape of which resembles a bubble. However, despite their occurrence in the band gap, these modes are extended Bloch functions as they obey the Bloch-Floquet theorem for real k-vectors. This is a direct consequence of the "energy-dependent" potential $\epsilon_p(\mathbf{r}, \omega)$ and, to our knowledge, does not have an analogy in electronic crystals. This result suggests, that frequency-dependent cylinder dielectric constants that arise from, say, hot carrier generation in silicon through intense laser beams may allow the dynamical creation of propagating states inside the PBG that could be useful for a number of switching and routing applications. However, prior to considering any application of this effect, we have to investigate the consequences of the nonzero imaginary part of the cylinder dielectric constant. Owing to

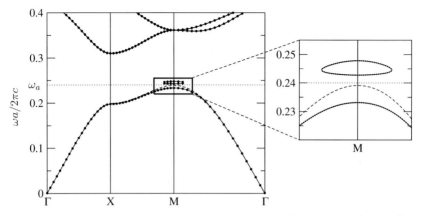

Fig. 2. Photonic band structure for a square lattice of cylinders ($r/a = 0.3$) in air, whose dielectric constant exhibits a frequency dependence characteristic of two-level systems embedded in silicon (for details on the parameters see the text). This leads to the formation of propagating modes inside the photonic band gap of the structure with undoped silicon cylinders. For reference, the band structure of the undoped PC is indicated by dashed lines

the fact, that for our model system the imaginary part of the cylinder dielectric function is much smaller than the corresponding real part, this may be done in a perturbative manner within the framework of a multi-scale analysis [24,28] (see also section 3). For instance, the calculated values of absorption lengths for the modes at the M-point are $L_{abs}(\omega_M = 0.233)/a = 36.93$, $L_{abs}(\omega_M = 0.243)/a = 3.41$, and $L_{abs}(\omega_M = 0.248)/a = 13.94$. These estimates suggest that the effect of dynamically generating propagating modes in a PBG via frequency-dependent dielectric functions should be observable.

3 Defect Structures in Photonic Crystals

In electronic micro-circuits, electrical currents are guided by thin metal wires where electrons are bound within the cross section of the wire by the so-called work function (confining potential) of the metal. As a result, electrical currents follow the path prescribed by the wire without escaping to the background. The situation is very different for optical waves. Although optical fibers guide light over long distances, micro-circuits for light based on fibers do not exist, because empty space is already an ideal conductor of light waves. The light in an optical fiber can easily escape into the background electromagnetic modes of empty space if the fiber is bent or distorted on a microscopic scale. PBGs in the band structure of PCs remove this problem by removing all the background electromagnetic modes over the relevant band of frequencies. Light paths can be created inside a PBG material in the form of engineered waveguide channels. The PBG localizes the light and prevents it from escaping the optical micro-circuit. To date, theoretical investigations of defect structures in PCs have almost exclusively been carried out using finite difference time domain (FDTD) methods [13,14]. However, applying general purpose methodologies such as FDTD or finite element (FEM) methods to defect structures in PCs largely disregards information about the underlying PC structure which is readily available from photonic band structure computation.

A more natural description of localized defect modes consists in an expansion of the electromagnetic field into a set of localized basis functions. As an illustration of how the theoretical description of localized defect states via a set of localized basis functions works, we want to discuss a number of defect structures embedded in a 2D-PC that consists of a square lattice of silicon dielectric posts ($r/a = 0.2$, $\epsilon_{cyl} = 11.9$) in air. The band structure for TM-polarized light for the defect-free PC exhibits a PBG between $\omega a/2\pi c = 0.281$ and $\omega a/2\pi c = 0.417$. The extended nature of Bloch functions of the defect-free PC suggests to obtain a generic set of localized basis functions through a lattice Fourier transform. The resulting Wannier functions are defined via

$$W_{n\boldsymbol{R}}(\boldsymbol{r}) = \frac{V_{\text{WSC}}}{(2\pi)^2} \int_{\text{BZ}} d^2k \, e^{-i\boldsymbol{k}\boldsymbol{R}} \, E_{n\boldsymbol{k}}(\boldsymbol{r}), \tag{5}$$

where V_{WSC} denotes the volume of the Wigner-Seitz cell. Furthermore, the Wannier function $W_{n\boldsymbol{R}}$ associated with band n and centered around lattice site $\boldsymbol{R}$ obey the orthonormality relation

$$\int_V d^2r \, W_{n\boldsymbol{R}}^*(\boldsymbol{r}) \, \epsilon_{\text{p}}(\boldsymbol{r}) \, W_{n'\boldsymbol{R}'}(\boldsymbol{r}) = \delta_{nn'} \delta_{\boldsymbol{R}\boldsymbol{R}'}, \tag{6}$$

where the integration is over all space.

As can be seen from an inspection of Fig. 3(a) it is a sobering exercise to compute the Wannier functions directly from the output of photonic band structure programs via Eq. (5). The poor localization properties and the erratic behavior of these Wannier functions originates in a phase indeterminacy of the Bloch functions

$$E_{n\boldsymbol{k}}(\boldsymbol{r}) \to e^{i\phi_n(\boldsymbol{k})} \, E_{n\boldsymbol{k}}(\boldsymbol{r}), \tag{7}$$

where $\phi_n(\boldsymbol{k})$ represents a global k-dependent phase function. Although fixing the random part of $\phi_n(\boldsymbol{k})$ by requiring the Bloch functions at the origin to be real-valued and non-negative, removes the erratic behavior of the Wannier functions to a large extent, this does not improve their localization properties. These difficulties have led a number of authors to adapt variations of the familiar empirical tight-binding parametrization to photonic band structures [29,30]. However, the success of empirical tight-binding parametrizations depends crucially on the existence of localized "orbitals" of the individual "atoms" that make up the crystal. As a consequence, its adaption to PCs presents major problems because bound states for a single dielectric scatterer simply do not exist and until now no tight-binding parametrization for TE-polarized radiation in 2D-PCs or electromagnetic waves in 3D-PCs has been obtained.

A solution to this unfortunate situation is provided by recent advances in electronic band structure theory. Marzari and Vanderbilt [31] have outlined an efficient scheme for the computation of optimally localized Wannier functions by determining numerically a phase transformation that minimizes an appropriate spread functional. In view of the translational properties of the Wannier functions $W_{n\mathbf{R}}(\mathbf{r}) = W_{n\mathbf{0}}(\mathbf{r} - \mathbf{R})$, this functional reads

$$\Omega = \sum_n \left[\langle n\mathbf{0}|r^2|n\mathbf{0}\rangle - (\langle n\mathbf{0}|\mathbf{r}|n\mathbf{0}\rangle)^2 \right] = \text{Min.} \tag{8}$$

Here we have introduced a shorthand notation for matrix elements according to $\langle n\mathbf{R}|r^2|n'\mathbf{R}'\rangle = \int_V d^2r W_{n\mathbf{R}}^*(\mathbf{r}) \, r^2 \, W_{n'\mathbf{R}'}(\mathbf{r})$. The result for the optimized Wannier function belonging to the first band of our model system is depicted in Fig. 3(b). The localization properties as well as the symmetries of the underlying PC are clearly visible. These optimally localized Wannier functions may be used as an expansion basis for the localized modes of defect structures embedded in PCs.

$$E(\mathbf{r}) = \sum_n \sum_{\mathbf{R}} E_{n\mathbf{R}} \, W_{n\mathbf{R}}(\mathbf{r}). \tag{9}$$

Inserting this expansion into the wave equation for the PC with added defect dielectric function $\Delta\epsilon(\mathbf{r})$

$$\left(\partial_x^2 + \partial_y^2\right) E(\mathbf{r}) + \frac{\omega^2}{c^2} (\epsilon_{\text{p}}(\mathbf{r}) + \Delta\epsilon(\mathbf{r})) E(\mathbf{r}) = 0, \tag{10}$$

leads to a generalized eigenvalue equation for the corresponding defect modes.

$$\sum_{n'} \sum_{\mathbf{R}'} \left(\delta_{nn'}\delta_{\mathbf{R}\mathbf{R}'} - \langle n\mathbf{R}|\Delta\epsilon|n'\mathbf{R}'\rangle\right) = \frac{c^2}{\omega^2} \sum_{n'} \sum_{\mathbf{R}'} \delta_{nn'} \beta_n(\mathbf{R} - \mathbf{R}'), \tag{11}$$

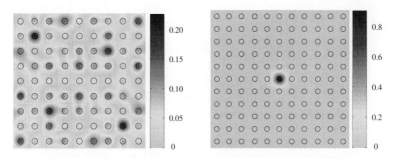

Fig. 3. Photonic Wannier functions for the first band obtained (a) by direct numerical integration of Eq. (5) (left panel). and (b) by minimizing the corresponding spread functional, Eq. (8) (right panel). The PC parameters are given in the text

where $\beta_n(\mathbf{R})$ is defined as

$$\beta_n(\mathbf{R}) = \frac{V_{\text{WSC}}}{(2\pi)^2} \int_{\text{BZ}} d^2k\, e^{i\mathbf{k}\mathbf{R}} \frac{\omega_{n\mathbf{k}}^2}{c^2}. \tag{12}$$

As an illustration of the efficiency of this approach we show in Fig. 4 the results for a cavity mode (removing a single cylinder, cavity frequency at $\omega a/2\pi c = 0.407$) and a waveguide structure (removing a row of cylinders, frequency of the waveguiding mode at $\omega a/2\pi c = 0.411$). This calculation used a 7x7 lattice and the 5 lowest bands resulting in a 245x245 matrix problem. Furthermore, the localization properties of the Wannier functions allow to neglect the overlap integrals once the respective Wannier centers are sufficiently far apart, which in turn leads to a sparse matrix problem. Therefore, optimally localized Wannier functions provide an efficient tool to study large defect structures embedded in PCs.

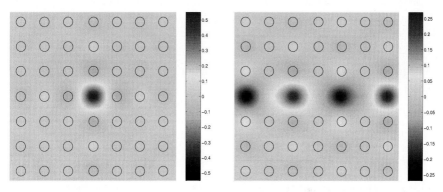

Fig. 4. Localized photonic defect modes for (a) a cavity, i.e., a single cylinder removed from the periodic structure (left panel) and (b) a waveguide, i.e., a row of cylinders removed from the periodic structure (right panel). The PC parameters are given in the text

4 Nonlinear Photonic Crystals

The existence of PBGs and tailoring of photonic dispersion relations and mode structures through judiciously designed PCs represent a novel paradigm for nonlinear wave interactions. To date, only a few works have been carried out for either Kerr-nonlinearities [15,16] or $\chi^{(2)}$-nonlinearities [22] in PCs. Moreover, the approximations involved seriously limit the applicability of these theories to real PCs. For instance, the study of Kerr-nonlinearities in 2D-PCs [15] has been limited to weak modulations in the linear index of refraction.

In this section, we outline a systematic approach to quantitative calculations of the optical properties of nonlinear PCs that is based on a multi-scale approach. We start by writing the corresponding wave equation in the time domain and introduce the nonlinear polarization $P_{\mathrm{NL}}(\mathbf{r},t)$, representing the nonlinear response of the system

$$\left(\partial_x^2 + \partial_y^2\right) E(\mathbf{r},t) - \frac{\epsilon_{\mathrm{p}}(\mathbf{r})}{c^2}\partial_t^2 E(\mathbf{r},t) = \frac{4\pi}{c^2}\partial_t^2 P_{\mathrm{NL}}(\mathbf{r},t). \tag{13}$$

Since optical nonlinearities are generally quite weak, Eq. (13) should be solved in a perturbative way taking into account that the effect of the nonlinearity accumulates only on time and spatial scales that are much slower than the natural scales of the underlying linear problem. Therefore, key simplifications to Eq. (13) arise from separating the fast from slow scales in space and time in the electromagnetic field $E(\mathbf{r},t)$. This separation is facilitated by formally replacing the space and time variables, $\mathbf{r}$ and t, with a set of independent variables $\mathbf{r}_n \equiv \mu^n \mathbf{r}$ and $t_n \equiv \mu^n t$. For instance, in this multi-scale approach [32] the time derivative is replaced according to

$$\partial_t = \partial_{t_0} + \mu \partial_{t_1} + \mu^2 \partial_{t_2} + \cdots, \tag{14}$$

and similar replacements of spatial and higher order derivatives. In addition, this is accompanied by a corresponding expansion of the electromagnetic field in powers of the formal expansion parameter μ, in order to construct a hierarchy of equations which effectively separate the different scales of the problem.

$$E(\mathbf{r},t) = \mu e_1(\mathbf{r}_0, \mathbf{r}_1, \cdots; t_0, t_1, \cdots) + \mu^2 e_2(\mathbf{r}_0, \mathbf{r}_1, \cdots; t_0, t_1, \cdots) + \cdots \tag{15}$$

In the present case, the fastest spatial scale, $\mathbf{r}_0$, corresponds to the wavelength of the electromagnetic waves propagating in the linear PC, and the fastest temporal scale, t_0, is associated with the optical period. As an illustration, we consider the case where at least one of the PC's constituent materials exhibits an intensity dependent index of refraction. For such Kerr-nonlinearities, the nonlinear polarization $P_{\mathrm{NL}}(\mathbf{r},t)$ is given by

$$P_{\mathrm{NL}}(\mathbf{r},t) = \chi^{(3)}(\mathbf{r})|E(\mathbf{r},t)|^2 E(\mathbf{r},t) \tag{16}$$

To first order in the expansion parameter μ, i.e., on the fast scale we obtain that the e_1-component of the electromagnetic field obeys Eq. (13) without the nonlinearity. Consequently, we make the ansatz

$$e_1(\mathbf{r}_0, \mathbf{r}_1, \cdots; t_0, t_1, \cdots) = a_{n\mathbf{k}}(\mathbf{r}_1, \cdots; t_1, \cdots) E_{n\mathbf{k}}(\mathbf{r}_0) e^{i\omega_{n\mathbf{k}} t_0}, \tag{17}$$

where the Bloch function $E_{n\mathbf{k}}(\mathbf{r}_0)$ represents a carrier wave and the envelope function $a_{n\mathbf{k}}(\mathbf{r}_1, \cdots; t_1, \cdots)$ has to be determined by explicitly considering the slower scales.

In second order in μ, we obtain that the envelope function $a_{n\bm{k}}$ is traveling with the group velocity $\bm{v}_{n\bm{k}}$ of the carrier wave

$$a_{n\bm{k}}(\bm{r}_1,\cdots;t_1,\cdots) \equiv a_{n\bm{k}}(\bm{z}_1;\bm{r}_2\cdots;t_2,\cdots), \tag{18}$$

where $\bm{z}_1 \equiv \bm{r}_1 - \bm{v}_{n\bm{k}}t_1$ and the group velocity $\bm{v}_{n\bm{k}}$ is given in terms of an expression familiar from $\bm{k}\cdot\bm{p}$-perturbation theory [27].

$$\bm{v}_{n\bm{k}} = -i\frac{c^2}{\omega_{n\bm{k}}}\int_{\mathrm{WSC}} d^2r_0\, E^*_{n\bm{k}}(\bm{r_0})\,\nabla_{\bm{r}_0}\,E_{n\bm{k}}(\bm{r_0}) = \frac{c^2}{\omega_{n\bm{k}}}\langle n\bm{k}| - i\nabla_{\bm{r}_0}|n\bm{k}\rangle. \tag{19}$$

Finally, in third order in μ, we obtain that the envelope function $a_{n\bm{k}}$ obeys a nonlinear Schrödinger equation

$$\begin{aligned}[i(\bm{v}_{n\bm{k}}\cdot\nabla_{\bm{r}_2} + \partial_{t_2}) + \nabla_{\bm{z}_1}\cdot\mathcal{M}_{n\bm{k}}\cdot\nabla_{\bm{z}_1}]\,a_{n\bm{k}}(\bm{z}_1;\bm{r}_2\cdots;t_2,\cdots) \\ + \alpha_{n\bm{k}}\,|a_{n\bm{k}}(\bm{z}_1;\bm{r}_2\cdots;t_2,\cdots)|^2\,a_{n\bm{k}}(\bm{z}_1;\bm{r}_2\cdots;t_2,\cdots) = 0,\end{aligned} \tag{20}$$

where the GVD tensor $\mathcal{M}_{n\bm{k}}$ and the effective nonlinearity $\alpha_{n\bm{k}}$ are

$$\begin{aligned}\bm{q}\cdot\mathcal{M}_{n\bm{k}}\cdot\bm{q} = |\bm{q}|^2\frac{c^2}{2\omega_{n\bm{k}}}\langle n\bm{k}|n\bm{k}\rangle - \frac{1}{2\omega_{n\bm{k}}}(\bm{q}\cdot\bm{v}_{n\bm{k}})^2 \\ +\frac{2c^4}{\omega_{n\bm{k}}}\sum_{m\neq n}\frac{\langle n\bm{k}| - i\bm{q}\cdot\nabla_{\bm{r}_0}|m\bm{k}\rangle\langle m\bm{k}| - i\bm{q}\cdot\nabla_{\bm{r}_0}|n\bm{k}\rangle}{\omega_{n\bm{k}}^2 - \omega_{m\bm{k}}^2},\end{aligned} \tag{21}$$

$$\alpha_{n\bm{k}} = 6\pi\omega_{n\bm{k}}\int_{\mathrm{WSC}} d^2r_0\,\chi^{(3)}(\bm{r_0})\,|E_{n\bm{k}}(\bm{r_0})|^4. \tag{22}$$

As a result of the foregoing analysis, we have obtained a generalization of the slowly varying envelope approximation. Within this approximation, the problem of pulse propagation in nonlinear PCs is mapped onto the problem of an envelope function propagating in an effective homogeneous medium with group velocity $\bm{v}_{n\bm{k}}$, GVD tensor $\mathcal{M}_{n\bm{k}}$, and effective nonlinearity $\alpha_{n\bm{k}}$ that are determined by the carrier wave, which, in turn, is given by a Bloch function of the linear PC. Therefore, the effective PC parameters can be obtained from band structure theory via Eqs. (19), (21), and (22) and quantitative investigations become possible. Furthermore, this framework of multi-scale analysis in conjunction with $\bm{k}\cdot\bm{p}$-perturbation theory can be applied to other nonlinearities such as non resonant $\chi^{(2)}$ media [33] or resonant distributed feedback lasing systems [24].

5 Conclusions

In summary, we have outlined a framework based on solid state theoretical methods that allows to quantitatively treat wave propagation in PCs. Photonic band structure computation of the infinitely extended PC provides the input necessary to efficiently obtain the properties of defect structures embedded in PCs via expansions into localized Wannier functions and allows the determination of effective parameters such as group velocity, GVD, and effective nonlinearities. Furthermore, we have shown that, in its own right, photonic band structure computation suggests novel tunability concepts for the band structure of PCs whose constituent materials exhibit a frequency dependence in their dielectric function.

Acknowledgments

We acknowledge the support by the DFG-Forschungszentrum Center for Functional Nanostructures (CFN) at the University of Karlsruhe. K.B., A.G-M., and L.T. acknowledge support from the Deutsche Forschungsgemeinschaft (DFG) under Bu 1107/2-1 (Emmy-Noether program). The work of M.F. has been supported by the DFG under Bu 1107/3-1 (Schwerpunktsprogramm SP 1113 *Photonische Kristalle*).

References

1. E. Yablonovitch, Phys. Rev. Lett. **58**, 2059 (1987).
2. S. John, Phys. Rev. Lett. **58**, 2486 (1987).
3. *Photonic Crystals and Light Localization in the 21st Century*, C.M. Soukoulis, Ed., NATO Science Series C **563**, Kluwer Academic (Dordrecht, Boston, London), 2001 .
4. A. Blanco et al., Nature (London) **405**, 437 (2000).
5. Yu.A. Vlasov et al., Nature (London) **414**, 289 (2001).
6. S. John and T. Quang, Phys. Rev. A **50**, 1764 (1994).
7. N. Vats, K. Busch, and S. John, Phys. Rev. A, in press (2002).
8. O. Painter et al., Science **284**, 1819 (1999).
9. P.L. Philips et al., J. Appl. Phys. **85**, 6337 (1999).
10. D. Labilloy et al., Phys. Rev. B **59**, 1649 (1999).
11. U. Grüning, V. Lehmann, and C.M. Engelhardt, Appl. Phys. Lett. **66**, 3254 (1995).
12. J. Schilling et al., Opt. Mat. **3**, S121 (2001).
13. S. Fan, P.R. Villeneuve, and H.A. Haus, Phys. Rev. Lett. **80**, 960 (1998).
14. S. Noda, A. Alongkarn, and M. Imada, Nature (London) **407**, 608 (2000).
15. N. Aközbek and S. John, Phys. Rev. E **57**, 2287 (1998).
16. N. Bhat and J. Sipe, Phys. Rev. E **64**, 056604 (2001).
17. S. Mingaleev, Y. Kivshar,and R.A. Sammut Phys. Rev. E **62**, 5777 (2000).
18. H. Kosaka, et al., Phys. Rev. B **58**, R10096 (1999).
19. H. Kosaka, et al., J. Lightwave Technology **17**, 2032 (1999).
20. B. Gralak, S. Enoch, and G. Tayeb, J. Opt. Soc. Am. A **17**, 1012 (2000).
21. M. Notomi, Opt. Quant. Elect. **34**, 133 (2002).
22. *Optical properties of Photonic Crystals*, K. Sakoda, Springer (Berlin, Heidelberg, New York), 2001.
23. C. Kallinger et al., Adv. Mat. **10**, 920 (1998).
24. L. Florescu, K. Busch, and S. John, J. Opt. Soc. Am. B, in press (2002).
25. K. Busch and S. John, Phys. Rev. E **58**, 3896 (1998).
26. K. Busch and S. John, Phys. Rev. Lett. **83**, 967 (1999).

27. D. Hermann et al., Optics Express **8**, 167 (2001).
28. D. Hermann, A. Garcia-Martin, K. Busch, and P. Wölfle, in preparation.
29. E. Lidorikis et al., Phys. Rev. Lett. **81**, 1405 (1998).
30. J.P. Albert et al., Phys. Rev. E **61**, 4381 (2000).
31. N. Marzari and D. Vanderbilt, Phys. Rev. E **56**, 12847 (1997).
32. *Perturbation Methods*, A.H. Nayfeh, Wiley (New York), 1973.
33. L. Tkeshelashvili, K. Busch, and P. Wölfle, in preparation.

Quantum Optical Effects in Semiconductors

W. Hoyer[1], M. Kira[2], and S. W. Koch[1]

[1] Department of Physics and Material Sciences Center, Philipps-University, Renthof 5, 35032 Marburg, Germany
[2] Laser Physics and Quantum Optics, Royal Institute of Technology (KTH), SCFAB, Roslagstullsbacken 21, 10691 Stockholm, Sweden

Abstract. Quantum optical effects in semiconductors are studied using a density-matrix approach which takes into account the many-body Coulomb interaction among the charge carriers, coupling to lattice vibrations, and the quantum nature of light. The theory provides a consistent set of equations which is used to compute photoluminescence spectra, predict the emission of squeezed light, investigate correlations between photons emitted by quantum-well structures, and to show examples where light-matter entanglement influences experiments done with classical optical fields.

During the last decades, both quantum optics and semiconductor physics have developed dramatically [1,2,3,4]. Advanced theories of quantum electrodynamics are now very successful in the description of the interaction of atomic systems with the quantized light field where the interaction between the atoms plays only a minor role. However, they can mostly be applied to describe dilute and only weakly interacting atomic gases since relatively simple models like few-level systems are used to describe the material. On the other hand, semiconductor physics is usually dominated by the Coulomb interaction between the carriers such that the majority of theories has focused on the correct description of the Fermionic many-body effects while the electromagnetic interaction has been limited mostly to classical fields. Only few recent approaches deal with the coupled semiconductor-photon system on a fully quantized level. Due to the overwhelming numerical complexity, most of these approaches have made use of rather strong approximations. For example, excitons are treated as perfect Bosons, which ignores the underlying Fermionic character of electrons and holes [5]. Also, a Green's functions approach has been used where the photon degree of freedom is eliminated by using a relation between absorption and luminescence spectra [6] which is strictly valid only under thermodynamic equilibrium conditions and for vanishing broadening [7].

Choosing a density matrix approach, we derive the equations of motion for the coupled carrier-photon-phonon system where the basic entities are Bloch electrons. In the first section, we present a short overview over this theoretical approach. We demonstrate where the quantum nature of light enters the well-known classical equations and how the fully quantum mechanical emission emerges. More explicitly, we show how our theory offers the possibility to compute photon correlations and mixed photon-carrier correlations and how it couples classical and quantum emission dynamics. As a consequence of this coupling, classical experiments can be influenced by quantum correlations and incoherently emitted light can show traces of the coherently exciting laser pulse.

1 Theory

For our investigations of quantum optical correlations, the starting point is the quantization of the electromagnetic field; a detailed description can be found for example in [8]. By solving the classical homogenous wave equation $\nabla^2 \boldsymbol{u_q}(\boldsymbol{r}) - n^2(\boldsymbol{r})|\boldsymbol{q}|^2 \boldsymbol{u_q}(\boldsymbol{r}) = 0$ in absence of the active material, we obtain a complete set of orthogonal eigenmodes for a system with an arbitrary background refractive index $n(\boldsymbol{r})$. Each of the modes is characterized by its wave vector $\boldsymbol{q}$ and the specific polarization direction. This basis can be used to expand the operator of the electric field

$$\hat{\boldsymbol{E}}(\boldsymbol{r},t) = \sum_{\boldsymbol{q}} i\mathcal{E}_{|\boldsymbol{q}|} \boldsymbol{u_q}(\boldsymbol{r}) b_{\boldsymbol{q}}(t) + \text{h.c.} \equiv \hat{\boldsymbol{E}}^{(+)}(\boldsymbol{r},t) + \hat{\boldsymbol{E}}^{(-)}(\boldsymbol{r},t) , \qquad (1)$$

where the operator $b_{\boldsymbol{q}}$ ($b_{\boldsymbol{q}}^\dagger$) destroys (creates) a photon in the mode $\boldsymbol{u_q}$. In the expansion, $\mathcal{E}_{|\boldsymbol{q}|}$ denotes the vacuum field amplitude $\sqrt{\hbar\omega_{|\boldsymbol{q}|}/(2\epsilon_0)}$ with $\omega_{|\boldsymbol{q}|} = c_0|\boldsymbol{q}|$ and the index $\boldsymbol{q}$ implicitly includes the polarization direction of the field. In the case of a constant refractive index, the eigenmodes are simply plane waves. Clearly, the analysis can easily be adjusted to different background geometries like Bragg reflectors or microcavities by simply computing the relevant field modes once at the very beginning of a calculation.

The classical properties of the field are described by the first order expectation values $\langle b_{\boldsymbol{q}} \rangle$. For example, $\langle \hat{\boldsymbol{E}}^{(\pm)}(\boldsymbol{r},t) \rangle$ at a given position is just a complex number such that it simultaneously has an exact phase and amplitude. The quantum mechanical features of light emerge as fluctuations in phase and amplitude. The simplest and often most dominant quantum features are given by photon-photon correlation functions

$$\Delta \langle b_{\boldsymbol{q}}^\dagger b_{\boldsymbol{q}'} \rangle = \langle b_{\boldsymbol{q}}^\dagger b_{\boldsymbol{q}'} \rangle - \langle b_{\boldsymbol{q}}^\dagger \rangle \langle b_{\boldsymbol{q}'} \rangle ,$$
$$\Delta \langle b_{\boldsymbol{q}} b_{\boldsymbol{q}'} \rangle = \langle b_{\boldsymbol{q}} b_{\boldsymbol{q}'} \rangle - \langle b_{\boldsymbol{q}} \rangle \langle b_{\boldsymbol{q}'} \rangle , \qquad (2)$$

where the classical contribution has been subtracted. The diagonal elements $\langle b_{\boldsymbol{q}}^\dagger b_{\boldsymbol{q}} \rangle$ determine the photon numbers of the system and $\langle \hat{\boldsymbol{E}}^{(+)}(\boldsymbol{r},t) \hat{\boldsymbol{E}}^{(-)}(\boldsymbol{r},t) \rangle$ gives the intensity of the field at position $\boldsymbol{r}$.

The equations of motion for these photonic correlation functions are coupled to the material excitations which are described in the basis of Bloch electrons and holes. The elementary creation and annihilation operators are $a_{\lambda,\boldsymbol{k}}^\dagger$ and $a_{\lambda,\boldsymbol{k}}$ for electrons in band λ with wavevector $\boldsymbol{k}$. In many applications, we investigate a two-band system and use the notation $v_{\boldsymbol{k}}$ and $c_{\boldsymbol{k}}$ for valence and conduction band, respectively. The simplest example of an electronic expectation value is the coherent microscopic polarization $P_{\bm{k}} = \langle v_{\bm{k}}^\dagger c_{\bm{k}} \rangle$ which is related to the total polarization density

$$\mathcal{P} = \sum_{\bm{k}} d_{\mathrm{cv}}^* P_{\bm{k}} + \mathrm{c.c.} \tag{3}$$

and is thus directly coupled to the light field via Maxwell's equations.

In the total Hamiltonian,

$$\hat{H} = \hat{H}_0 + \hat{H}_\mathrm{C} + \hat{H}_\mathrm{D} + \hat{H}_\mathrm{phon} , \tag{4}$$

we include the Coulomb interaction between the carriers $\hat{H}_\mathrm{C}$, the quantized light-matter interaction $\hat{H}_\mathrm{D}$ and the coupling between carriers and lattice vibrations $\hat{H}_\mathrm{phon}$ [9]. Additionally, H_0 gives the non interacting contributions of carriers, photons and phonons. Detailed discussions and the explicit form of the different parts of the total Hamiltonian can be found, e.g., in [10,11].

By computing the Heisenberg equations of motion for the relevant photon, phonon, carrier, and mixed correlation functions, a general hierarchy problem emerges where correlations are coupled to higher order correlations in an infinite chain. Once the different levels of correlations are established [12,11], we consistently truncate this hierarchy and obtain a closed set of equations.

Instead of presenting all the resulting equations explicitly, we focus on the generalized semiconductor Bloch and luminescence equations. Using the Heisenberg equation of motion $i\hbar\, \partial/\partial t\, \hat{O} = [\hat{O}, \hat{H}]$ with the Hamiltonian (4), we obtain

$$\begin{aligned}
i\hbar \frac{\partial}{\partial t} P_{\bm{k}} =\ & (\tilde{\epsilon}_{\bm{k}}^\mathrm{c} - \tilde{\epsilon}_{\bm{k}}^\mathrm{v}) P_{\bm{k}} - (1 - f_{\bm{k}}^\mathrm{e} - f_{\bm{k}}^\mathrm{h}) \Omega_\mathrm{R}(\bm{k}) \\
& + \sum_{\bm{k}',\bm{l}} V_{\bm{l}} \left(\Delta\langle v_{\bm{k}}^\dagger c_{\bm{k}'}^\dagger c_{\bm{k}'+\bm{l}} c_{\bm{k}-\bm{l}} \rangle + \Delta\langle v_{\bm{k}}^\dagger v_{\bm{k}'}^\dagger v_{\bm{k}'+\bm{l}} c_{\bm{k}-\bm{l}} \rangle \right. \\
& \left. - \Delta\langle c_{\bm{k}}^\dagger c_{\bm{k}'}^\dagger c_{\bm{k}'+\bm{l}} v_{\bm{k}-\bm{l}} \rangle^* - \Delta\langle c_{\bm{k}}^\dagger v_{\bm{k}'}^\dagger v_{\bm{k}'+\bm{l}} v_{\bm{k}-\bm{l}} \rangle^* \right) \\
& + \sum_{q_z, \bm{q}_\parallel} \Omega_\mathrm{SE}(\bm{q}) \left(\Delta\langle b_{q_z, \bm{q}_\parallel} c_{\bm{k}+\bm{q}_\parallel}^\dagger c_{\bm{k}} \rangle - \Delta\langle b_{q_z, \bm{q}_\parallel} v_{\bm{k}}^\dagger v_{\bm{k}-\bm{q}_\parallel} \rangle \right) ,
\end{aligned} \tag{5}$$

$$\begin{aligned}
i\hbar \frac{\partial}{\partial t} f_{\bm{k}}^\mathrm{e} =\ & \Omega_\mathrm{R}^*(\bm{k}) P_{\bm{k}} + \sum_{q_z, \bm{q}_\parallel} \Omega_\mathrm{SE}^*(\bm{q}) \Delta\langle b_{q_z, \bm{q}_\parallel}^\dagger v_{\bm{k}-\bm{q}_\parallel}^\dagger c_{\bm{k}} \rangle \\
& + \sum_{\bm{k}',\bm{l}} V_{\bm{l}} \left(\Delta\langle c_{\bm{k}}^\dagger c_{\bm{k}'}^\dagger c_{\bm{k}'+\bm{l}} c_{\bm{k}-\bm{l}} \rangle + \Delta\langle c_{\bm{k}}^\dagger v_{\bm{k}'}^\dagger v_{\bm{k}'+\bm{l}} c_{\bm{k}-\bm{l}} \rangle \right) - \mathrm{c.c.} ,
\end{aligned} \tag{6}$$

and a similar equation for the hole density. The first line of (5) contains the Hartree-Fock part and the second and third line the Coulomb-correlation part of the usual semiconductor Bloch equations. As usual, we have separated the true correlated contributions according to

$$\langle a_1^\dagger a_2^\dagger a_3 a_4 \rangle = \langle a_1^\dagger a_4 \rangle \langle a_2^\dagger a_3 \rangle - \langle a_1^\dagger a_3 \rangle \langle a_2^\dagger a_4 \rangle + \Delta\langle a_1^\dagger a_2^\dagger a_3 a_4 \rangle . \tag{7}$$

Furthermore, Ω_R denotes the renormalized Rabi frequency and $\tilde{\epsilon}_{\bm{k}}$ are the renormalized energies. The last line of (5) contains new correlations which are solely due to the quantum optical nature of the light. The correlations are probability amplitudes for scattering processes where a photon is absorbed while a charge carrier makes an intraband transition. In other words, these correlations result from the entanglement between photons and carrier densities. The interaction strength is determined by $\Omega_\text{SE}(\bm{q}) = i\mathcal{E}_{\bm{q}} |\bm{u}_{\bm{q}}| d_\text{cv}$. In the carrier equation, the quantum optical features result from the correlations between photons and polarization. More specifically, $\langle b^\dagger v^\dagger c \rangle$ gives the probability amplitude for a process where a photon is emitted under simultaneous recombination of an electron-hole pair. Therefore, these contributions directly lead to the depletion of densities via spontaneous emission. The Coulomb correlated parts in (5) and (6) lead to screening and energy renormalizations. For a consistent description, all four-point quantities must be included into the analysis.

The equation for the photon-number like correlations

$$i\hbar \frac{\partial}{\partial t} \Delta \langle b^\dagger_{q_z,\bm{q}_\parallel} b_{q'_z,\bm{q}_\parallel} \rangle = \hbar (\omega_{\bm{q}'} - \omega_{\bm{q}}) \Delta \langle b^\dagger_{q_z,\bm{q}_\parallel} b_{q'_z,\bm{q}_\parallel} \rangle$$
$$+ \Omega_\text{SE}(\bm{q}) \sum_{\bm{k}} \Delta \langle b_{q'_z,\bm{q}_\parallel} c^\dagger_{\bm{k}+\bm{q}_\parallel} v_{\bm{k}} \rangle - \Omega^*_\text{SE}(\bm{q}') \Delta \langle b^\dagger_{q_z,\bm{q}_\parallel} v^\dagger_{\bm{k}} c_{\bm{k}+\bm{q}_\parallel} \rangle \quad (8)$$

are coupled to photon assisted polarizations describing creation (destruction) of electron-hole transition amplitudes under simultaneous absorption (emission) of a photon. Both, photon correlations and photon assisted polarizations formally correspond to four-point quantities since with every emission or absorption of a photon the destruction or creation of an electron-hole *pair* is associated. In the same manner, the equations of motion for all the photon assisted terms as well as for the four-point carrier-carrier correlations must be derived and solved. We only mention explicitly the equation

$$i\hbar \frac{\partial}{\partial t} \Delta \langle b^\dagger_{q_z,\bm{q}_\parallel} v^\dagger_{\bm{k}} c_{\bm{k}+\bm{q}_\parallel} \rangle = \left[\tilde{\epsilon}^\text{c}_{\bm{k}+\bm{q}_\parallel} - \tilde{\epsilon}^\text{v}_{\bm{k}} - \hbar\omega_{\bm{q}} \right] \Delta \langle b^\dagger_{q_z,\bm{q}_\parallel} v^\dagger_{\bm{k}} c_{\bm{k}+\bm{q}_\parallel} \rangle$$
$$- \left(1 - f^\text{e}_{\bm{k}+\bm{q}_\parallel} - f^\text{h}_{\bm{k}} \right) \Omega_\text{ST}(\bm{k},\bm{q}) + \Omega_\text{SE}(\bm{q}) \sum_{\bm{k}'} \langle c^\dagger_{\bm{k}'} v_{\bm{k}'-\bm{q}_\parallel} v^\dagger_{\bm{k}} c_{\bm{k}+\bm{q}_\parallel} \rangle$$
$$+ i\hbar \frac{\partial}{\partial t} \Delta \langle b^\dagger_{q_z,\bm{q}_\parallel} v^\dagger_{\bm{k}} c_{\bm{k}+\bm{q}_\parallel} \rangle \bigg|_\text{IB} + i\hbar \frac{\partial}{\partial t} \Delta \langle b^\dagger_{q_z,\bm{q}_\parallel} v^\dagger_{\bm{k}} c_{\bm{k}+\bm{q}_\parallel} \rangle \bigg|_\text{CS} . \quad (9)$$

Equation (9) has an analogous structure as the semiconductor Bloch equation. The term $\Omega_\text{ST}(\bm{k},\bm{q})$ for example is the photon assisted version of the classical Rabi frequency and introduces the excitonic resonances in the luminescence spectrum. The source term

$$\sum_{\bm{k}'} \langle c^\dagger_{\bm{k}'} v_{\bm{k}'-\bm{q}_\parallel} v^\dagger_{\bm{k}} c_{\bm{k}+\bm{q}_\parallel} \rangle = f^\text{e}_{\bm{k}+\bm{q}_\parallel} f^\text{h}_{\bm{k}} + \Delta \langle c^\dagger_{\bm{k}'} v_{\bm{k}'-\bm{q}_\parallel} v^\dagger_{\bm{k}} c_{\bm{k}+\bm{q}_\parallel} \rangle \quad (10)$$

leads to emission dynamics even when classical fields and coherent polarization vanish.

Equations (8) and (9) constitute the semiconductor luminescence equations [11,13,14]. While additional higher order Coulomb scattering processes lead to screening and dephasing of the photon assisted polarizations, the quantum field provides additional coupling to the photon assisted densities which were also present in the last line of (5). It is through these intraband correlations that coherent and incoherent dynamics can become coupled [9,11,15].

2 Correlated Photons in Quantum-Well Emission

Conceptually, pure quantum effects can best be observed in situations where no classical light contibution exists, such that both classical fields and interband polarizations vanish. This situation can be achieved in semiconductors, e.g., shortly after an optical excitation high in the continuum. After such an excitation, coherent polarizations typically dephase on a timescale of the order of several tens of picoseconds. However, incoherent carrier densities are destroyed mostly via spontaneous emission in high quality structures and can thus exist up to the time of nanoseconds. Similar initial conditions can be provided by electrical injection of carriers. We can therefore assume the completely incoherent limit by setting all coherent quantities to zero to describe many realistic situations. In this case, only carrier densities and incoherent carrier-carrier correlations of the form $\Delta \langle c^\dagger c^\dagger cc \rangle$ and $\Delta \langle c^\dagger v^\dagger vc \rangle$ are coupled to the photon-assisted polarization via (9). Under steady-state conditions, the photoluminescence spectrum is proportional to the rate of incoming photons which according to (8) can directly be evaluated from the photon-assisted polarizations. In Fig. 1, a typical photoluminescence spectrum is shown. It is worthwhile to note that the strong excitonic resonance cannot be related directly to exciton populations. Actually, the magnitude of the exciton peak is almost independent of the exact form of the source term (10) and exists even in complete absence of excitonic population correlations [11,16].

In the right part of Fig. 1, we show another interesting quantity, namely the spectrum of the cross-correlations $I_{\mathrm{CC}} \equiv \frac{\partial}{\partial t} |\langle b^\dagger_{+q_z} b_{-q_z} \rangle|$ between photons emitted to the two opposite directions perpendicular to the quantum well. Although these cross-correlations are non-Hermitian quantities, there is a conceptually easy way to measure them. One possible experimental setup is shown schematically in Fig. 2. With the help of mirrors, photons emitted to both sides of the sample are redirected into a common detector. One of the paths can be varied in its optical length such that the corresponding detector operator is given by

$$d_{q_z} = \frac{1}{\sqrt{2}} \left(b_{q_z} + e^{i\phi} b_{-q_z} \right) , \qquad (11)$$

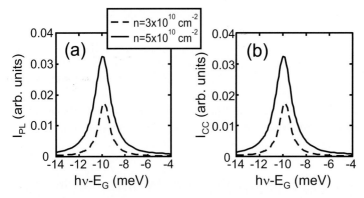

Fig. 1. Normalized photoluminescence spectrum of a single quantum well (**a**) compared to the spectrum of cross-correlations (**b**) for two different densities at a carrier temperature of 77 K

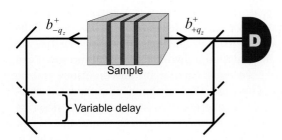

Fig. 2. Schematic setup of suggested interference measurement. Photons emitted at the two opposite sides of the sample are redirected into a common detector while the relative phase can be changed via a variable path length

where ϕ can be varied by varying the optical path. The detected signal is proportional to the photon flux, i.e., to the rate of

$$\langle d^\dagger_{q_z} d_{q_z} \rangle = \frac{1}{2} \left(\langle b^\dagger_{q_z} b_{q_z} \rangle + \langle b^\dagger_{-q_z} b_{-q_z} \rangle + 2\cos(\phi) |\langle b^\dagger_{q_z} b_{-q_z} \rangle| \right). \quad (12)$$

By varying ϕ, the cross-correlations can be determined.

In order to understand the fundamental reason for the existence of cross-correlations, we investigate the particular form of light-matter interaction. By assuming carrier confinement to the lowest quantum-well sub-band, the dipole Hamiltonian (see e.g. [11]) contains operator combinations of the form $b^\dagger_{q_z,\mathbf{q}_\parallel} v^\dagger_{\mathbf{k}} c_{\mathbf{k}+\mathbf{q}_\parallel}$. This interaction allows processes where photons are emitted by simultaneous recombination of an electron-hole pair. As a consequence of the missing translational symmetry, the z-component of the momentum is not conserved. In the classical regime, this symmetry breaking is known to lead to radiative decay of quantum-well polarization [17,18].

As a consequence of the non-conservation of momentum, the same state can symetrically emit light to both left and right propagating modes. This leads to entanglement between $\pm q_z$ modes. Therefore, the suggested experiment is in close analogy to the double-slit experiment [19] performed with a single electron or photon. Even though the light emission is completely incoherent, the photon entanglement allows interference, i.e., the observation of cross-correlations, at the detector.

2.1 Multiple Quantum-Well Systems

In multiple quantum-well structures, the analysis has to be generalized since different quantum wells can interact by the exchange of photons. This radiative coupling leads to modifications of absorption, reflection, transmission, four-wave mixing, etc., and has been observed for many different configurations. In the past, such radiative-coupling effects have mostly been studied for semi-classical fields [20,21,22,23].

In the previous section, we have shown how the non-conservation of momentum in thin quantum wells leads to the possibility of obtaining correlated photons from incoherent emission. In multiple quantum-well structures, the emission can be additionally altered via the local nature of emitters. In the following, we investigate in more detail the control of entanglement related cross-correlation properties in different multiple quantum-well systems. Therefore, we extend our theory to the treatment of multiple quantum-well structures in the presence of a quantum field under incoherent initial conditions.

In all cases of interest, the QWs are separated by a distance corresponding to either half or quarter of the optical resonance wavelength. Therefore the Coulomb interaction and tunneling between different quantum wells is negligible. In this case, the corresponding N-quantum well generalization of the Hamiltonian (4) is straight forward [11]. The approach leads to a new type of joint process where the recombination of an electron hole pair in one quantum well happens simultaneously with the excitation of an electron hole pair in another quantum well. These processes are mediated via the quantized light field and can lead to sub– or superradiant features in the light emission and depend on the distance between neighbouring quantum wells. In order to see true superradiant emission comparable to superradiance in atomic systems, one has to use relatively high carrier densities such that the exciton resonance is almost completely bleached. Even then, our theoretical estimates predict that one would need at least thousand radiatively coupled quantum wells [11].

More intriguing quantum optical effects can be anticipated in the cross-correlation studies based on the results of Sec. 2. We therefore compare the cross-correlations emitted by two structures consisting of four quantum-wells, spaced either $\lambda/2$ or $\lambda/4$ apart, where λ denotes the wavelength of the strong excitonic resonance. The result is shown in Fig. 3.

For quantum wells with a distance of $\lambda/2$, the cross-correlations are as large as the photoluminescence spectra very much like in the single quantum-well case. The interference measurement suggested above would therefore strongly depend on the optical path length. For quantum wells placed at a

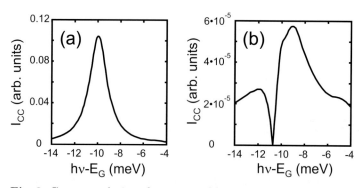

Fig. 3. Cross-correlations from a set of four quantum wells placed either (a) $\lambda/2$ or (b) $\lambda/4$ apart. The carrier density in all quantum wells is equal to $n = 5 \times 10^{10} \mathrm{cm}^{-2}$. The vertical units are arbitrary but in both cases identical

distance of $\lambda/4$, however, the cross-correlations vanish. Photons emitted to both sides of the MQW are therefore completely uncorrelated.

In order to explain the strong dependence of the cross-correlations on the quantum-well period, we investigate the basic properties of the quantized field in more detail. First, we recall that in the simplest case of a quantum well in free space the eigenmodes are just plane waves. Therefore, the electric field operator is given by

$$\hat{E} = \sum_{q_z, \boldsymbol{q}_\parallel} i \mathcal{E}_{|q|} e^{i \boldsymbol{q} \boldsymbol{r}} b_{q_z, \boldsymbol{q}_\parallel} + \mathrm{h.c.} \quad (13)$$

Since the momentum is not conserved in the direction perpendicular to the quantum well, the coupling to propagating and counter-propagating modes (i.e. with $\pm q_z$) is symmetrical. This becomes most obvious after introducing new field operators which combine propagating and counterpropagating modes. We define

$$B_{1,q_z} \equiv \frac{1}{\sqrt{2}} (b_{q_z} + b_{-q_z}) \;, \quad B_{2,q_z} \equiv \frac{1}{\sqrt{2}} (b_{q_z} - b_{-q_z}) \quad (14)$$

as new superpositions of two energetically degenerate photon operators. We have dropped the additional index $\boldsymbol{q}_\parallel$ for notational simplicity. These new operators allow us to express the field operator (13) as

$$\hat{E}(z) = i \sqrt{2} \sum_{q_z > 0} \mathcal{E}_{|q_z|} [\cos(q_z z) B_{1,q_z} + i \sin(q_z z) B_{2,q_z}] + \mathrm{h.c.} \; . \quad (15)$$

Figure 4 schematically shows how different quantum wells couple only to one of the field modes. If several quantum wells are positioned at $z = 0$ or at multiples of $\lambda/2$, their dynamics will only be coupled to the cosine mode. Due to the local character of the interaction, none of the quantum wells can couple to the sine mode. Therefore all correlations $\Delta \langle B^\dagger_{2,q_z} B_{2,q_z} \rangle$ and $\Delta \langle B^\dagger_{1,q_z} B_{2,q_z} \rangle$

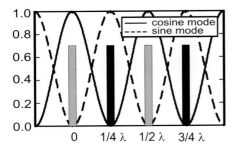

Fig. 4. The two linear combinations of propagating and counter-propagating modes shown together with a set of quantum wells

vanish. As a result, $\Delta\langle b^\dagger_{q_z} b_{-q_z}\rangle = \Delta\langle B^\dagger_{1,q_z} B_{1,q_z}\rangle/2 = \Delta\langle b^\dagger_{q_z} b_{q_z}\rangle$, and we find perfect cross-correlations for a periodicity of $\lambda/2$. For a periodicity of $\lambda/4$, we have two types of quantum wells positioned: i) at multiples of $\lambda/2$ and ii) at odd multiples of $\lambda/4$. Since the first group couples only to the cosine, the last only to the sine mode, both groups have completely independent emission dynamics. Mathematically, that means that also in this case $\Delta\langle B^\dagger_{1,q_z} B_{2,q_z}\rangle = 0$. This has direct consequences on photon numbers and cross-correlations; expressed with the new operators, they are

$$\Delta\langle b^\dagger_{q_z} b_{q_z}\rangle = \frac{1}{2}\left(\Delta\langle B^\dagger_{1,q_z} B_{1,q_z}\rangle + \Delta\langle B^\dagger_{2,q_z} B_{2,q_z}\rangle\right) = \Delta\langle b^\dagger_{-q_z} b_{-q_z}\rangle, \quad (16a)$$

$$\Delta\langle b^\dagger_{q_z} b_{-q_z}\rangle = \frac{1}{2}\left(\Delta\langle B^\dagger_{1,q_z} B_{1,q_z}\rangle - \Delta\langle B^\dagger_{2,q_z} B_{2,q_z}\rangle\right). \quad (16b)$$

Clearly, the number of photons emitted to the left and to the right is identical. And whenever there are quantum wells at even and odd multiples of $\lambda/4$, the cross-correlations depend on the number of quantum wells belonging to each subsystem. For equal numbers of identically excited quantum wells, the cross-correlations are zero as seen in Fig. 3.

3 Quantum Effects in Microcavity Emission

We saw already in Sec. 1 that in general coherent and incoherent dynamics become coupled via the quantum fluctuations of the light field. By detecting light in a direction different from the classical reflection and transmission directions, one ensures to detect only incoherently emitted light. Yet, signatures of the exciting field can be transferred to the incoherent light emission via the coherent photon carrier correlations of the form $\Delta\langle b^\dagger c^\dagger c\rangle$ [15]. Even transient amplitude squeezing has been predicted for this incoherently emitted light. To enhance the light-matter coupling we have investigated those quantum optical effects for quantum wells inside a microcavity, i.e., inside a resonator structure with mirror reflectivities close to unity and cavity lengths in the wavelength regime.

For the investigation of squeezing, the field quadratures $X_q = b_q^\dagger + b_q$ and $Y_q = (b_q^\dagger - b_q)/i$ are the most important quantities with their fluctuations

$$\Delta X_q^2 = \langle X_q X_q \rangle - \langle X_q \rangle \langle X_q \rangle = 1 + 2\Delta \langle b_q^\dagger b_q \rangle + 2\mathrm{Re}[\Delta \langle b_q b_q \rangle],$$
$$\Delta Y_q^2 = \langle Y_q Y_q \rangle - \langle Y_q \rangle \langle Y_q \rangle = 1 + 2\Delta \langle b_q^\dagger b_q \rangle - 2\mathrm{Re}[\Delta \langle b_q b_q \rangle] . \quad (17)$$

For incoherent excitation, the coherent terms and $\Delta \langle b_q b_q \rangle$ vanish such that $\Delta X_q^2 = \Delta Y_q^2 = 1 + 2\Delta \langle b_q^\dagger b_q \rangle$. The resulting fluctuations of both quadratures are equal and exceed the minimum uncertainty value of unity. When a coherent excitation is used, $\Delta \langle b_q b_q \rangle = |\Delta \langle b_q b_q \rangle| e^{i\phi(t)}$ is non-zero and the fluctuations are

$$\Delta X_q^2 = 1 + 2\Delta \langle b_q^\dagger b_q \rangle + 2|\Delta \langle b_q b_q \rangle| \cos(\phi(t)),$$
$$\Delta Y_q^2 = 1 + 2\Delta \langle b_q^\dagger b_q \rangle - 2|\Delta \langle b_q b_q \rangle| \cos(\phi(t)). \quad (18)$$

Thus, in the presence of a coherent excitation source the field quadratures become different, i.e., the fluctuations have a definite direction which is a signature of squeezing of the light. Since $\phi(t)$ is mainly determined by the optical frequency, i.e., $\phi(t) \approx 2\omega_q t$, we obtain squeezing which rotates with the optical frequency. Thus, the driving field forces the incoherent scattered light to carry phase information about the excitation process. If the coherent correlations $|\Delta \langle b_q b_q \rangle|$ are large enough, the squeezing can transiently go below the minimum uncertainty limit of 1.

Figure 5 shows how squeezing is enhanced up to 30% compared to the minimum uncertainty. On the left hand side, the maximum and minimum

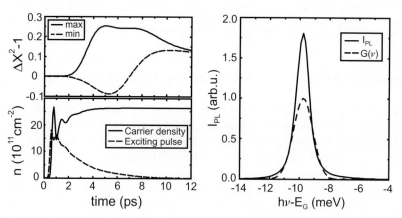

Fig. 5. Squeezed light in resonance fluorescence from a quantum well inside a microcavity. The upper left figure shows the time evolution of maximum and minimum of the quadrature fluctuations. The bottom figure gives density and exciting pulse. In the right figure, the spectral overlap between detector function $G(\nu)$ with a time resolution of 1.2 ps and the luminescence spectrum is shown

values of the quadrature fluctuations (17) computed for a realistic detector model with a time resolution of 1.2 ps are shown. Below, the exciting pulse and the resulting density evolution are displayed. The pulse must be sufficiently strong to cause Rabi oscillations in the carrier density before significant squeezing can be observed. The right hand side shows the energy resolution of our model detector for which we assume a time resolution of 1.2 ps. This resolution must be chosen carefully in order to guarantee maximal overlap with the photoluminescence spectrum in the frequency domain. For a similar calculation with better time resolution, Rabi oscillations could be observed in the field quadratures, but the squeezing was considerably weakened since photons over a much larger range of frequencies were collected.

The squeezing of light emission is one manifestation of quantum optical effects in semiconductors. According to (5), even the classical emission can show direct quantum contributions. Recent pump-probe experiments with microcavity systems have shown such new quantum features. There the pump generated correlations were detected by a weak probe which observed long living oscillations [3] as well as new spectral components [4] in probe reflection. The theory presented in this paper fully explains these experiments, cf. Fig. 6, and identifies the new features as entanglement of the carrier density and photon field.

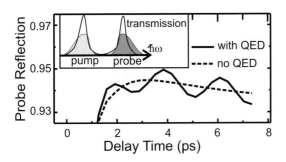

Fig. 6. Computation of probe signal. Comparison between full computation (solid) and computation without intraband correlations (dashed line). The inset shows the schematic pump sequence

4 Summary

In summary, this paper gives an overview of an approach to consistently treat photon and carrier correlations on the same level. The quantization of the light field allows us not only to investigate effects like spontaneous emission and modifications of the photon statistics but also the influence of quantum optical effects on the classical field. The first applications of this theory are reviewd in this paper, but many more interesting problems, such as photonic correlations, exciton-photon coupling effects, quantum emission in photonic crystal environments and many more will be studied in the future.

Acknowledgments

This work was supported by the Deutsche Forschungsgemeinschaft through the photonic crystal and the Leibniz programs and by the Humboldt Foundation and Max-Planck Society through the Max-Planck Research program. M.K. thanks the Swedish Research Council (TFR and NFR), the Göran Gustafssons foundation, and the Center for Parallel Computers.

References

1. E. Moreau, I. Robert, L. Manin, V. Thierry-Mieg, J.M. Gérard, I. Abram: Phys. Rev. Lett. **87**, 183601 (2001).
2. O. Benson, C. Santori, M. Pelton M, Y. Yamamoto: Phys. Rev. Lett. **84**, 2513 (2000).
3. Y.-S. Lee, T.B. Norris, M. Kira, F. Jahnke, S.W. Koch, G. Khitrova, H.M. Gibbs: Phys. Rev. Lett. **83**, 5338 (1999).
4. C. Ell, P. Brick, M. Hübner, E.S. Lee, O. Lyngnes, J.P. Prineas, G. Khitrova, H.M. Gibbs, M. Kira, F. Jahnke, S.W. Koch, D.G. Deppe, D.L. Huffaker: Phys. Rev. Lett. **85**, 5392 (2000).
5. S. Savasta, R. Girlanda: Journal of physics: Condensed matter **11**, 6045 (1999).
6. K. Hannewald, S. Glutsch, F. Bechstedt: Phys. Rev. Lett. **86**, 2451 (2001).
7. F. Jahnke, M. Kira, W. Hoyer, S.W. Koch: Phys. Status Solidi (b) **221**, 189 (2000).
8. C. Cohen-Tannoudji, J. Dupont-Roc, G. Grynberg: *Photons & Atoms* (Wiley, New York, 1989).
9. M. Kira, W. Hoyer, T. Stroucken, S.W. Koch: Phys. Rev. Lett. **87**, 176401 (2001).
10. H. Haug, S.W. Koch: *Quantum Theory of the Optical and Electronic Properties of Semiconductors*, 3rd ed. (World Scientific Publ., Singapore, 1994).
11. M. Kira, F. Jahnke, W. Hoyer, S.W. Koch: Prog. Quantum Electr. **23**, 189 (1999).
12. J. Fricke: Annals of Physics **252**, 479 (1996).
13. M. Kira, F. Jahnke, S.W. Koch, J.D. Berger, D.V. Wick, T.R. Nelson Jr., G. Khitrova, H.M. Gibbs: Phys. Rev. Lett. **79**, 5170 (1997).
14. G. Khitrova, H.M. Gibbs, F. Jahnke, M. Kira, S.W. Koch: Rev. Mod. Phys. **71**, 1591 (1999).
15. M. Kira, F. Jahnke, S.W. Koch: Phys. Rev. Lett. **82**, 3544 (1999).
16. M. Kira, F. Jahnke, S.W. Koch: Phys. Rev. Lett. **81**, 3263 (1998).
17. V.M. Agranovich, O.A. Dubowskii: JETP Lett. **3**, 223 (1966).
18. F. Tassone, F. Bassani, L.C. Andreani: Phys. Rev. B **45**, 6023 (1992).
19. D.F. Walls, G.J. Milburn: *Quantum Optics* (Springer, Berlin, 1994).
20. M. Hübner, J. Kuhl, T. Stroucken, A. Knorr, P. Thomas, S.W. Koch: Phys. Rev. Lett. **76**, 4199 (1996).
21. T. Stroucken, A. Knorr, C. Anthony, A. Schulze, P. Thomas, S.W. Koch, M. Koch, S.T. Cundiff, J. Feldmann, E.O. Göbel: Phys. Rev. Lett. **74**, 2391 (1995).
22. D.S. Citrin: Phys. Rev. B **50**, 5497 (1994).
23. L.C. Andreani: Phys. Lett. A **192**, 99 (1994).

Beryllium Chalcogenides: Interface Properties and Potential for Optoelectronic Applications

V. Wagner[1], J. Geurts[1], and A. Waag[2]

[1] Physikalisches Institut der Universität Würzburg,
Am Hubland, 97074 Würzburg, Germany
[2] Abteilung für Halbleiterphysik, Universität Ulm,
89081 Ulm, Germany

Abstract. Wide bandgap II-VI semiconductors are candidates for optoelectronic applications in the green-blue spectral range. Still existing lifetime problems are at least partly attributed to the high bond polarity and to doping limits of the conventional II-VI compounds, such as ZnSe. Beryllium chalcogenides have a considerable potential to overcome these problems. Therefore, binary, ternary or quaternary Be chalcogenide layers or superlattices may be applied in various regions of optoelectronic devices. We discuss their application as active layers, p-claddings and p-contacts, as well as BeTe-based type-II light emitting diodes. In these heterostructures and superlattices interfaces play a crucial role. Our studies of interface properties focus on the lattice-matched system ZnSe/BeTe, which forms a type-II band alignment with a very high band offset, and on the strongly mismatched system CdSe/BeTe, which forms quantum dots. The experimental methods are Raman spectroscopy of interfacial vibration modes and electromodulation spectroscopy of electronic transitions across the interfaces.

1 Introduction

The extensive study of wide bandgap II-VI semiconductors as candidates for optoelectronic applications in the green-blue spectral range was triggered about one decade ago by the development of a p-doping procedure for ZnSe [1]. However, in spite of tremendous research efforts by many groups, II-VI electronics still suffers from limited lifetimes, which up to now prevent commercial applications. These lifetime problems are strongly connected with heating effects in the devices due to limited efficiency. The efficiency limits originate especially from the difficulties in achieving ohmic p-contacts and from the limited p-doping of the cladding layers, together with a non-ideal stability of the active layers which leads to a breakdown of the devices. The degradation of the active layers is at least partly attributed to the bond polarity, which is for usual II-VI compounds such as ZnSe much higher than for III-V compounds such as e.g. GaAs.

These problems of common II-VI compounds for optoelectronics form the motivation for the application of Be-chalcogenides. Their application was initiated by Vèrié (1995), who predicted for this hitherto rather exotic class of chalcogenides an increased lattice stability, more specific a high shear modulus, due to its reduced ionicity with respect to the conventional II-VI compounds, such as ZnSe [2]. The reduced ionicity is due to the combination of the row-II cation Beryllium with the row-V anion Tellurium or the row-IV anion Selenium, which should result in a lower transfer of electronic charge towards the anion than it occurs e.g. for conventional II-VI-compounds such as e.g. ZnSe.

Experimentally, the bond polarity may be verified by Raman spectroscopy or Far-Infrared reflectance from the optical phonon vibration modes. The polarity induces the frequency splitting between the longitudinal optical (LO) and the transverse optical (TO) phonon, because the enhanced LO frequency originates from a dipole-induced macroscopic electric field. Along this scheme, we obtained from Raman spectroscopy and Far-Infrared reflectance for BeTe an effective charge transfer of 0.4e, which is far below the ZnSe value of 1.5e [4]. This result is in full agreement with our calculations of bond-charge distribution by density functional theory, which are shown in Fig. 1. Here, BeTe is compared with the conventional wide-gap II-VI compounds ZnTe and ZnSe on the one hand, and GaAs on the other. Obviously, for all compounds the bond charge density cumulates near the anion, but for BeTe this cumulation tendency is clearly lower than for ZnTe, and even much lower than for ZnSe. The BeTe density profile is rather similar to GaAs [5]. Thus, the Be-chalcogenides bridge the ionicity gap between the strongly ionic II-VI-compounds and the more covalent III-V ones.

BeTe has a direct bandgap of 4.2 eV [6] and an indirect transition at 2.8 eV. It is nearly lattice matched to GaAs and ZnSe, which should facilitate its incorporation into II-VI heterostructures on GaAs substrates. Furthermore, as all Tellurides, it allows a very heavy p-doping. In BeTe:N, doping levels up to $10^{20} cm^{-3}$ were detected [7]. Due to these favourable properties, Be-chalcogenides are considered as promising candidates for optoelectronics in several aspects, which will be discussed in section 2.

For the application of Be-chalcogenides in heterostructure or superlattice devices, the interfaces between different materials play a crucial role. Relevant material combinations are (i) (nearly) lattice matched systems in order to achieve a possibly high crystalline quality; (ii) strongly mismatched systems, which should lead to the formation of self-organized quantum dots. We discuss examples of both classes: the lattice-matched system ZnSe/BeTe, and the strongly mismatched system CdSe/BeTe, which forms self-organized quantum dots. The non-common-ion character of these systems offers as additional free parameter the choice of the interface bonds, i.e. Be-Se or Zn(Cd)-Te. The main experimental methods for interface analysis which are applied here are Raman spectroscopy of lattice vibrations and electromodulation spectroscopy of electronic transitions at interfaces. They allow a nondestructive analysis and enable the focussing on individual interfaces.

Beryllium Chalcogenides 69

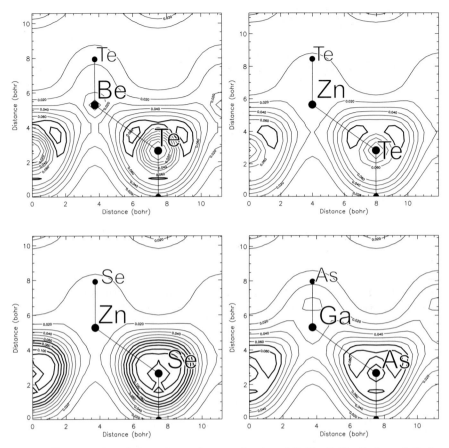

Fig. 1. Contour plot of the valence electron density of BeTe, ZnTe, ZnSe and GaAs

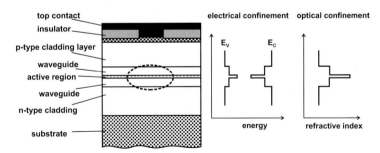

Fig. 2. Schematic set-up of a II-VI laser diode. Electrical and optical confinement is illustrated on the right hand side

2 Optoelectronic Applications

The layer sequence of a light emitting diode or a laser diode is shown in Fig. 2, together with the electronic bandgaps and the refractive index of the various layers, which govern the electrical and optical confinement, and thus the device efficiency. The motivation for the employment of Be chalcogenides in such an optoelectronic device implies various aspects: (i) cladding layers from quarternary Be-chalcogenides, such as BeMgZnSe; (ii) BeTe/ZnSe superlattices for p-contacts; (iii) ternary BeCdSe as active layers; (iv) BeTe-based type-II light emitting diodes.

The relation between the band gap and the lattice constant of the relevant II-VI-compounds is shown in Fig. 3. Obviously, the application of the quarternary chalcogenide BeMgZnSe as cladding layer allows a variation of the electronic band gap between 2.7 eV and 4.5 eV, while maintaining lattice matching to the GaAs substrate. For practical applications, only low admixtures of Be and Mg to the ZnSe are required in these layers (below 0.3). The composition control of this system by the variation of the group II-fluxes during MBE-growth is much easier than for e.g. MgZnSSe, in which also the group-VI fluxes have to be varied.

However, lattice matching to the substrate is not the only criterion to be fulfilled. In practice, the p-dopability decreases with increasing band gap. Thus, the required p-doping level of 10^{17}cm^{-3} limits the cladding layer gap to values below 2.9 eV. In turn, this implies for the active layer a band gap

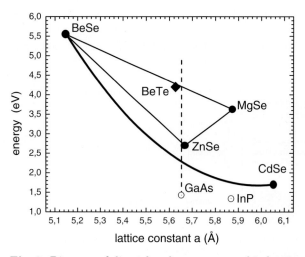

Fig. 3. Diagram of direct band gap versus cubic lattice constant at room temperature of selected II-VI compounds and III-V substrate materials. The quaternary area of BeMgZnSe and the ternary line of BeCdSe for bowing parameter 4.5eV is illustrated

below 2.7 eV in order to achieve the required confinement effect. Conventionally, a reduction of the band gap below the ZnSe-value is achieved by ternary CdZnSe. However, the implementation of the heavy cation Cd results in a compressive strain of the ternary CdZnSe active layer, which is a strong disadvantage from the viewpoint of stability. Recent calculations have shown that dislocations, which result from the metastable character of the substitutional N-ions in the p-cladding layer [8], preferentially migrate into compressively strained active layers. In order to avoid this effect, a tensile strain in the active layer would be very favorable. This tensile strain is achieved by BeCdSe active layers, which in addition offer the above mentioned increased lattice stability due to their reduced bond polarity. Due to the pronounced nonlinearity of the E_g-vs-composition behaviour, shown in Fig. 3, (bowing parameter = 4.5eV), the E_g value of BeCdSe amounts to about 2.5 eV (i.e. in the green-blue range) for layers which are moderately strained with respect to the GaAs-substrate. In this way, $Be_xZn_{1-x}Se$ with x-values about 0.46 forms an attractive alternative active-layer material. For active-layer stacks, consisting of five BeCdSe/ZnSe quantum wells, laser activity was achieved by optical pumping at a power density of 40 kW/cm^2 at 80K [9].

A rigorous approach for circumventing the metastability of the p-dopant is a complete avoiding of p-doped selenide layers, and employing instead telluride compounds, which are preferentially p-dopable. This concept may be realized by a superlattice of p-BeTe/i-ZnSe layers, as shown in Fig. 4. The position of the minibands may be tuned by the individual layer thicknesses. As an additional advantage this allows the realization of a depth-dependent miniband energy by a thickness gradient between adjacent layers. In this way, band-gap optimized p-contact layers can be realized, which bridge the valence band offset between ZnSe and the GaAs substrate [10].

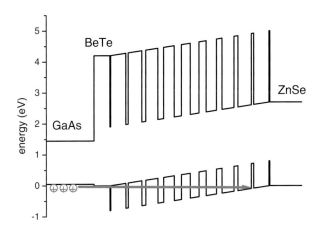

Fig. 4. Realization of a p-contact of ZnSe to GaAs by the use of a pseudo grading, i.e. short period ZnSe/BeTe-SL

3 ZnSe/BeTe Interfaces

While the fundamental electronic bandgap of ZnSe is direct and has a value of 2.7 eV, BeTe has its lowest direct bandgap at 4.2 eV [6]. The nearly lattice-matched system BeTe/ZnSe shows a type-II band alignment with a pronounced confinement of the electronic states in the ZnSe and of the hole states in the BeTe. The conduction band offset is 2.3 eV, the valence band offset is about 1 eV.

3.1 Structural Properties

Because BeTe and ZnSe have no elements in common, for the interface two possible chemical configurations exist: the bonds may be either Be-Se or Zn-Te. Because BeSe as well as ZnTe is strongly lattice-mismatched to BeTe and ZnSe ($|\Delta a/a| \approx 9\%$), both interfaces show a pronounced strain, which is compressive for Be-Se and tensile for Zn-Te. These different strain values and in addition the different interdiffusion tendency of the various elements are expected to result in characteristic differences in the structural and optical properties.

For studying the differences in the structural quality of both types of interfaces we used Raman spectroscopy. Fig. 5 shows confined longitudinal optical phonons in the ZnSe layers of ZnSe/BeTe superlattices with layer thicknesses 2.6nm/1.3nm. The observed numbers of confined modes is a measure for the interface quality. For Be-Se interfaces essentially only the first order gives a clear peak, while only weak indications of higher order modes appear. In contrast, for Zn-Te interfaces clear peaks up to the fifth order were observed, thus proving the superior quality of the latter interface type [4]. This result is in full agreement with transmission electron microscopy studies, which show a distinct non-planarity and a more pronounced intermixing for Be-Se interfaces [11].

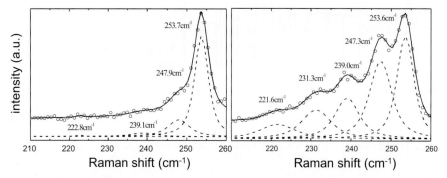

Fig. 5. Raman spectra of confined ZnSe LO-phonons of ZnSe/BeTe superlattices (2.6nm/1.3nm). Left side: Be-Se interfaces, right side: Zn-Te interfaces

3.2 Optical Properties

For analyzing the optical properties of the interfaces, which are governed by their electronic band alignment, we employed electroreflectance spectroscopy. This method offers several advantages: the structures in the spectra are centered at the electronic gaps and their lineshape is very sharp already at room temperature [12]. Moreover, the observation of higher bandgaps is in contrast to photoluminescence, which is essentially restricted to the fundamental gap. Due to the type-II band alignment of the BeTe/ZnSe system, spatially indirect transitions occur between the conduction band states of the ZnSe-layers and the valence band states of the BeTe-ones. Furthermore, due to the pronounced confinement the spatial overlap between the CB and VB wavefunctions is strongly confined to the interfaces. Therefore, very specific interface information is expected in the reflectance spectra.

"sdkfl sksdjfkjs lksr wejlkjerlt" "sdff"" The signals from the normal interfaces (i.e. ZnSe-on-BeTe) can be separated from the inverted ones (i.e. BeTe-on-ZnSe) in two different ways, which are shown in Fig. 6: (i) an in-depth electric field , which means a tilt of the band edge-vs.-depth diagram (Fig. 6a) leads to an energy split Δ between the transition energies from an electric quantum well state across the upper or the lower interface, respectively. The in-depth electric field may exist as an internal field in the heterostructure, or it may be applied externally. The latter case allows a controlled variation of the spectral distance of both transition signatures. Of course, the external field can also be applied for compensation of the internal field, thus enabling the determination of the original internal field value [13]. (ii) The different polarization of both transitions may be exploited. This polarization reflects the orientation of the bonds at the interfaces. Therefore the signal from the normal interfaces appears for [-110]-polarization, while for [110]-polarization the inverted interfaces are observed (Fig. 6b). The experimental results from a BeTe/ZnSe-stack with ZnTe interfaces is shown in Fig. 7. Obviously, the [-110]-spectrum shows especially at the fundamental indirect transition (i.e. e1-hh1) sharper features than the [110]-one. This indicates a superior quality of the normal interface with respect to the inverted one. The slight frequency

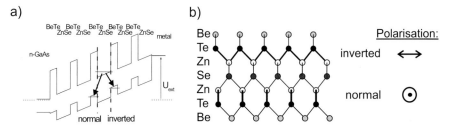

Fig. 6. Basic measurements modes to distinguish upper and lower interface signatures in ER: a) static E-field, b) light polarization

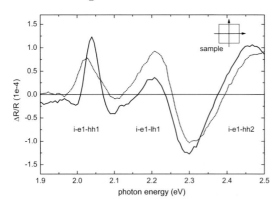

Fig. 7. ER spectra of a ZnSe/BeTe-SL with Zn-Te interfaces. Light polarization along [1̄10] (full line) and [110] (dotted line) corresponds to the normal and inverted interface, respectively

shift between the features of both spectra results from a small residual in-depth electric field.

3.3 Type-II LEDs

The polarized light which is emitted from the spatially indirect transition across the interface between a ZnSe quantum well and the adjacent BeTe-layer may be applied for optoelectronic purposes in LEDs. Fig. 8 shows such a type-II LED, in which the optical transition takes place. Note, that the p-doping here exclusively occurs in the BeTe, while for the n-doping only selenide layers are used. In this way, each II-VI compound has its favorable charge carrier type. An additional feature of this LED-type is the easy variation of the emitted wavelength, which strongly depends on the ZnSe quantum well thickness. E.g. a variation of the well thickness from 3 nm to 1 nm results in a wavelength shift from 640 nm to 515 nm, i.e. from the red to the green spectral region. With LEDs following this concept an external quantum efficiency up

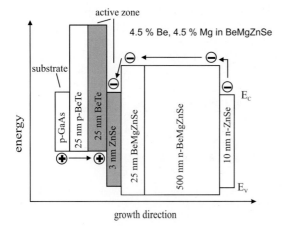

Fig. 8. Band diagram of a type-II LED utilizing the spatially indirect transition at the BeTe/ZnSe interface. The emission wavelength can be tuned by the design of the ZnSe quantum well

to 2 percent was achieved, while after an operation time of 1000 hours at 20 A/cm^2 no signs of degradation were observed [14].

4 CdSe/BeTe Interfaces

In contrast to the lattice-matched system ZnSe/BeTe, which was discussed in the previous section, CdSe/BeTe shows a pronounced lattice mismatch of 7 percent. Therefore it is an outstanding candidate for self-organized quantum dots, just like e.g. CdSe/ZnSe and InAs/GaAs. A unique property of CdSe/BeTe is the extremely strong electronic confinement due to the high direct BeTe bandgap of 4.2 eV. The formation of CdSe quantum dots is expected for layer thicknesses beyond about 1.5 monolayers. In order to study the interfaces and the onset of dot formation, we investigated stacks of 5x 1 ML CdSe embedded in BeTe by Raman spectroscopy. The conditions for the observation of localized vibration modes from the CdSe monolayers and the interfaces are very favorable in this material system, because the BeTe barriers have a very pronounced frequency gap between their optical and acoustic phonon modes due to the extreme mass asymmetry of Be and Te [5].

In Fig. 9 vibrational Raman spectra are shown for samples with the various possible interface configurations: only BeSe-interfaces (spectrum a), only CdTe-interfaces (spectrum c), and mixed interfaces, i.e. Be-Se bonds at the lower interface and Cd-Te bonds at the upper one (spectrum b), and opposite (spectrum d). Beside the modes from the GaAs-substrate, slightly below 300 cm^{-1}, the most prominent features are observed in the region of BeTe (near 500 cm^{-1}) and in the range of Cd-related vibrations (between 170 cm^{-1} and 220 cm^{-1}). In the upper spectra (a and b) both features are much less pronounced than in the lower ones (c and d), and a blueshift of the spectral position of the BeTe LO peak by 10 cm^{-1} occurs with respect its bulk position, which is observed in the spectra c and d. This indicates (i) an inferior quality of the BeSe interfaces, just as it was observed for ZnSe/BeTe; (ii) a dominant role of the lower interfaces. We attribute this behaviour to the tendency of Se to diffuse into the underlying BeTe [15].

For a detailed assignment of the Cd-related vibrations from the embedded monolayers and interfaces, we calculated the mode frequencies and displacement patterns, using a linear chain model, considering next neighbour interaction. The force constants were derived from the LO phonon frequencies of the constituent binaries BeTe, CdSe, and CdTe, taking into account their strain-induced variations. In order to consider possible thickness variations, also the eigenmodes for 2ML CdSe were calculated. For 1ML, two modes result: the most stronly confined one is centered at the Se-layer and involves further essentially only the adjacent Cd layers. It has a calculated frequency of 216 cm^{-1}, which is as expected close to the CdSe LO frequency. For the second 1ML-eigenmode the central Se-layer is at rest, and the main displacements are in the Cd-Te interface layers. The high Te-mass results

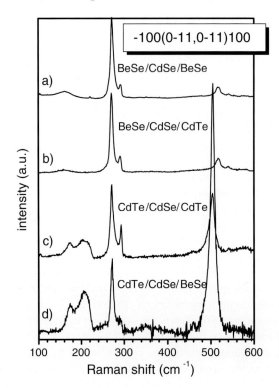

Fig. 9. Raman spectra of embedded CdSe-MLs in BeTe barriers with systematically varied interfaces (as indicated)

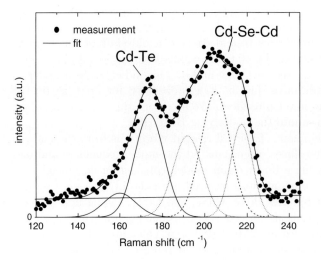

Fig. 10. Raman signature a stack of five single CdSe-ML embedded in BeTe barriers ($\bar{1}00(011,\text{unpol.})100$, T=77K, laser wavelength: 488nm). The solid line is the fit, consisting of Cd-Te vibrations (solid), 2-ML contributions (dotted) and a 1-ML contribution (dashed)

in a reduction of the mode frequency to 169 cm^{-1}. The calculation for the 2ML-case yields two Cd-Se-Cd-modes, a symmetric and an anti-symmetric one, whose frequencies are split symmetrically with respect to the former Se-centered mode: they amount to 204 cm^{-1} and 224 cm^{-1}. Furthermore, here also an additional mode exists which essentially involves the Cd-Te interface bonds. Its eigenfrequency is 164 cm^{-1}, which is very close to the value of the 1ML interface mode. Fig. 10 shows that the experimentally observed Raman structure is described very well as a superposition of vibrational contributions from 1ML- and 2ML-regions. This interpretation is underscored by Raman experiments at various laser wavelengths, which lead to different weighting of the scattering efficiencies of the 1ML- and 2ML-regions due to different resonance conditions. Tuning the laser photon energy between 2.5 eV and 2.71 eV clearly results in an a realtive enhancement of the peaks from the 1ML regions, because due to confinement effects their electronic levels are beyond those of the 2ML-regions [16].

Overall, our Raman results prove abrupt and well-defined interfaces for the case of Cd-Te-bonds. This is further underscored by high-resolution X-ray diffraction results, which show clear multi-quantum-well satellites and oscillation fringes, which directly correspond to the number of MQW-periods [17] and by transmission electron microscopy pictures, showing alternating 1ML- and 2ML-regions in each CdSe-layer, which exhibit a layer-to-layer correlation, resulting in a chessboard-like pattern.

5 Summary and Outlook

In summary,we have discussed various applications of Be-chalcogenide layers in II-VI optoelectronics, such as p-doped cladding layers, p-contacts, and active layers. Our studies of the structural and chemical interface properties of these layers by vibrational Raman spectroscopy showed that high-quality interfaces are achievable, even for embedded monolayers, but we also found that the interface quality strongly depends on the interfacial bond sequence. For the ZnSe/BeTe and CdSe/BeTe material system Zn-Te- and Cd-Te-bonded interfaces are far superior to Be-Se ones. In electroreflectance spectroscopy, spatially indirect electronic transitions at the interfaces were used to study band offsets, electric fields and interface quality.

Beside optoelectronics, presently a further intriguing field for the application of Be-chalcogenides opens up: by the incorporation of Manganese in the percent range they are well suited for magnetic purposes. In this way, they can continue the tradition of II-VI diluted magnetic semiconductors such as e.g. CdMnTe and ZnMnSe. First important achievements in this field are the realisation of spin injection into a semiconductor layer in a heterostructure which contained BeMgZnSe as a spin aligning layer [18], and the observation of ferromagnetic behaviour in p-doped BeMnTe [19]. In this way Beryllium chalcogenides can also play an important role in the very rapidly developping field of spintronics.

Acknowledgments

We acknowledge T. Muck, L. Hansen, G. Reuscher, S. Gundel, and S. Ivanov and coworkers for their support in experiments and calculations. Financial support by the Deutsche Forschungsgemeinschaft (SFB 410) and the Volkswagenstiftung, as well as by the INTAS Grant No. 97-31907 and RFBR is gratefully acknowledged.

References

1. S. Guha, J. M. DePuydt, M. A. Haase, et al., Appl. Phys. Lett. **63**, 3107 (1993).
2. C. Vèrié, in: Semiconductor Heteroepitaxy. Growth, Characterization and Device Applications, B. Gil, R.-L. Aulombard (Eds.) p.73 (World Scientific, Singapore, 1995).
3. W. Greiner, D.N. Poenaru: Cluster Preformation in
4. J. Geurts, V. Wagner, B. Weise, J.J. Liang, H. Lugauer, A. Waag, G. Landwehr in *Physics of Semiconductors*, D. Gershoni (Ed.), pdf-file 1264 on CD (Proc. 24th Int. Conf., Jerusalem 2-7 August, 1998).
5. V. Wagner, J. J. Liang, R. Kruse, S. Gundel, M. Keim, A. Waag, J. Geurts, phys. stat. sol. (b) **215**, 87 (1999).
6. K. Wilmers, T. Wethkamp, N. Esser, C. Cobet, W. Richter, V. Wagner, H. Lugauer, F.Fischer, T. Gerhard, M. Keim, M. Cardona, Journal of Electronic Materials **28**, 670 (1999).
7. H.-J. Lugauer, F. Fischer, T. Litz, A. Waag, U. Zehnder, W. Ossau, T. Gerhard, G. Landwehr, C. Becker, R. Kruse, J. Geurts, Mat. Sci. & Eng. **B43**, 88 (1997).
8. S. Gundel, W. Faschinger, Phys. Rev. B **65**, 035208 (2001).
9. S. V. Ivanov, O. V. Nekrutkina, S. V. Sorokin, V. A. Kaygorodov, T. V. Shubina, A. A. Toropov, P. S. Kop'ev, V. Wagner, J. Geurts, A. Waag, G. Landwehr, Appl. Phys. Lett. **78**, 404 (2001).
10. N. Yu. Gordeev, S. V. Ivanov, V. I. Kopchatov, I. I. Novikov, T. V. Shubina, N. D. Ilinskaya, P. S. Kopev, G. Reuscher, A. Waag, G. Landwehr, Semiconductors **35**, 1340 (2001).
11. T. Walter, A. Rosenauer, R. Wittmann, D. Gerthsen, F. Fischer, T. Gerhard, A. Waag, G. Landwehr, P. Schunk and T. Schimmel, Phys. Rev. B **59**, 8114 (1999).
12. D.E. Aspnes: in Handbook of Semiconductors **2**, T.S. Moss (Ed.), p.109 (North Holland, Amsterdam, 1980).
13. V. Wagner, M. Becker, M. Weber, M. Korn, M. Keim, A. Waag, J. Geurts, Appl. Surf. Sci. **166**, 30 (2000).
14. G. Reuscher, M. Keim, H.J. Lugauer, A. Waag, G. Landwehr, J. Cryst. Growth **214/215**, 1071 (2000).
15. V. Wagner, J. Wagner, T. Muck, L. Hansen, J. Geurts, S.V. Ivanov, phys. stat. sol. (b) **229**, 103 (2002).

16. V. Wagner, J. Wagner, T. Muck, G. Reuscher, A. Waag, J. Geurts, N. Sadchikov, S.V. Sorokin, S.V. Ivanov, P.S. Kop'ev, Appl. Surf. Sci. **175-176**, 169 (2001).
17. S.V. Ivanov, G. Reuscher, T. Gruber, T. Muck, V. Wagner, J. Geurts, A. Waag, G. Landwehr, T.V. Shubina, N.A. Sadchikov, A.A. Toropov, P.S. Kop'ev, in *Nanostructures: Physics and Technology*, p. 98 (Ioffe Physico-Technical Institute, St Petersburg, 2000).
18. R. Fiederling, M. Keim, G. Reuscher, W. Ossau, G. Schmidt, A. Waag, L. W. Molenkamp, Nature **402**, 787 (1999).
19. L. Hansen, D. Ferrand, G. Richter, M. Thierley, V. Hock, N. Schwarz, G. Reuscher, G. Schmidt, A. Waag, L.W. Molenkamp, Appl. Phys. Lett. **79**, 3125 (2001).

ZnO Metal Organic Vapor Phase Epitaxy: Present State and Prospective Application in Optoelectronics and Spin Electronics

A. Waag[1], Th. Gruber[1], Ch. Kirchner[1], D. Klarer, K. Thonke[1], R. Sauer[1], F. Forster, F. Bertram[2], and J. Christen[2]

[1] Abteilung Halbleiterphysik, Universität Ulm
Albert Einstein Allee 45, 89081 Ulm, Germany
[2] Abteilung Halbleiterepitaxie, Otto-von-Guericke Universität Magdeburg, Postfach 4120, 39016 Magdeburg

Abstract. Due to its band gap of 3.3 eV and exciton binding energy of 60 meV ZnO is a very interesting candidate for UV optoelectronics. Beyond that, the incorporation of magnetic impurities allows to fabricate semimagnetic semiconductors, which are potentially ferromagnetic at room temperature. Recent results on the MOVPE growth of ZnO will be reported and compared to the status of MBE-ZnO material. The optimization of the MOVPE process on various substrates will be described. Diethyl-zinc (DEZn) and an alcohole has been used as Zn- and oxygen precursor, respectively. The ZnO MOVPE layers have been analyzed by a variety of techniques, including HRXRD, PL, CL, CV, Hall and AFM.

1 Introduction

The wide band gap semiconductor ZnO is used widely for various applications such as transparent front contacts for solar cells, gas sensors, and surface acoustic wave devices [1]-[3]. The optoelectronic properties of ZnO are attracting increasing interest, which is primarily due to recent reports on p-type doping [4]-[7]. The band gap of ZnO of 3.3 eV at room temperature makes the material a promising candidate for optoelectronic devices in the blue and ultraviolet region. It appears especially favorable for such applications because of its large exciton binding energy of 60 meV, its low power threshold for optical pumping at room temperature [8] and the availability of bulk ZnO substrates, which offers the possibility of homoepitaxy for the fabrication of devices. Furthermore, the band gap of ZnO can be tuned over a large energy range by alloying it with CdO and MgO [9]. In order to accomplish optoelectronic devices, a reliable ZnO technology with the potential of fabricating p-type ZnO has to be developed. A major obstacle in achieving p-type material has been the high "intrinsic" n-type conductivity of nominally undoped ZnO, whose origin is usually thought to be due to point defects. Even high quality bulk material shows a carrier concentration of typically 10^{17} cm^{-3} [10]. Hydrogen [11] and structural defects, such as oxygen vacancies and zinc interstitials [12], have been suggested as shallow donors in ZnO.

In the latter case a decrease in background concentration with increasing layer quality can be expected.

Recent reports on the p-type doping of ZnO [4]-[7] in particular have lead to a revival of research on this material system. A lot of activities have so far been focusing on plasma-enhanced molecular beam epitaxy of ZnO. Layers of excellent structural and optical quality have been accomplished using e.g. MgO nucleation layers on sapphire substrates [13], or $ScAlMgO_4(0001)$ substrates [14]. ZnMgO-ZnO and ZnO-ZnCdO quantum well systems have been fabricated [9]-[26], and heavy n-type doping can readily be achieved [15]. However, despite various reports on the measurement of p-type conductivity in ZnO, a reliable p-doping technique is still to be developed.

MBE is a very versatile tool for the development of new material systems, since in-situ surface analysis like Reflection High Energy Electron Diffraction (RHEED) can be used. In comparison, however, the capability of Metalorganic Vapor Phase Epitaxy (MOVPE) is much higher in terms of scaling and high throughput. This is particularly important e.g. for LED production, where MBE can practically not meet the large scale production requirements.

Here, we report on the metal-organic chemical vapor deposition of ZnO. We focus on heteroepitaxial growth on c-plane Al_2O_3 and the influence of the VI/II ratio during growth on the properties of the layers. The layer quality has been investigated by various techniques to assess the structural, optical and eletronic properties of the material.

2 Experimental Details

The ZnO layers were grown in an AIXTRON 200 RF horizontal flow reactor. H_2 was used as the carrier gas in the process. For the MOVPE growth, diethylzinc, DEZn $((C_2H_5)_2Zn)$, and iso-propanol, i-PrOH $((CH_3)_2CHOH)$ have been used as the zinc and oxygen precursors, respectively. The relative supply of the oxygen and zinc precursors, i.e. the VI/II ratio, during growth has been varied from 18 to 60.

C-plane sapphire and $2\,\mu m$ thick GaN templates on c-plane sapphire were used for the heteroepitaxy, whereas commercially available ZnO substrates (Eagle Picher Inc.) were used for the homoepitaxy.
The growth rates have been determined by in-situ reflectometry and confirmed by systematic post-growth etching. Photoluminescence (PL) and high resolution x-ray diffraction (HRXRD) measurements were carried out to study the optical and structural properties. The electrical properties of the layers grown on sapphire were characterized by Hall measurements in Van der Pauw configuration. The properties of the layers grown on GaN templates and ZnO substrates were not accessible by this method, though, because of the high conductivity of these substrate materials themselves.

Considering the observed mosaicity of the layers, a Williamson-Hall analysis of the rocking curves of the (0002), (0004) and (0006) reflection was performed to determine the lateral coherence length, which has been interpreted as an averaged mosaic block size, and the averaged relative tilt angle between the mosaic blocks in the layers [16]. An analysis of the reciprocal space map of the ($11\bar{2}4$) reflection and reflectivity data gave insight into the strain situation of the ZnO layers. Furthermore annealing for 1 h at 800 °C has been applied for some of the samples to determine its effect on the layer properties.

The use of an alcohol as an oxygen source instead of e.g. O_2 or H_2O has been found to prevent severe pre-reactions at low reactor pressures, which leads to a strong deposition of particles upstream from the substrate [17]. It has been suggested that by using an alcohol and DEZn an intermediate product (an alkylzinc alkoxide) should form in the gas phase prior to the ZnO deposition on the substrate surface [18].

3 Results and Discussion

The reactor pressure has been found to be a crucial parameter for growing ZnO layers at reasonable growth rates, as can be seen in Fig. 1. No difference in growth rate for the different substrate materials used was observed.

The maximum growth rate could be obtained at a pressure of 400 mbar. The variation of the layer thickness has been determined to be below 10 % on a 2-inch wafer. The growth rate drops sharply for smaller pressures. We assume that the mean free path of the adducts DEZn and i-PrOH is too long for low pressures (equivalent to low concentrations of the adducts in the

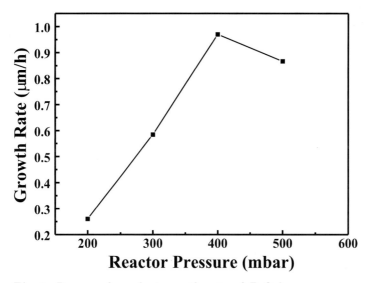

Fig. 1. Pressure-dependent growth rate of ZnO layers grown on sapphire at a growth temperature of 380 °C. The line is a guide to the eye only

reactor) resulting in a small formation of the intermediate alkylzinc alkoxide product necessary for the ZnO deposition. Increasing the pressure also leads to a decrease of the growth rate. A depletion of the gas phase can be inferred from the observation of a white powdery fall-out in the reactor upstream from the substrate. The layer thickness becomes very inhomogeneous and a growth rate can therefore not be determined for reactor pressures above 500 mbar.

Figure 2 shows $\Theta - 2\Theta$ scans of the (0002) reflection of ZnO layers grown on the three different substrate materials used: sapphire, GaN templates and ZnO substrates.

The structural quality can be inferred by the full widths at half maximum (FWHM) of 270 arcsec, 255 arcsec and 100 arcsec for layers grown on GaN templates, sapphire, and ZnO substrates, respectively. For the samples on GaN and ZnO these values are slightly higher than the FWHM of the GaN template (190 arcsec) and the ZnO substrate (64 arcsec) themselves, indicating that the growth process and/or growth start is not yet optimal. The results also show that the structural quality is mainly determined by the quality of the substrate material. The origin of the very weak third peak found in samples grown on GaN templates can be attributed to fully strained ZnO, with a lateral lattice constant equal to the GaN buffer layer. This has been verified by reciprocal space maps.

The PL spectra at 5 K of samples grown on the different substrates can be seen in Fig. 3. All spectra show a strong near band edge emission, which is

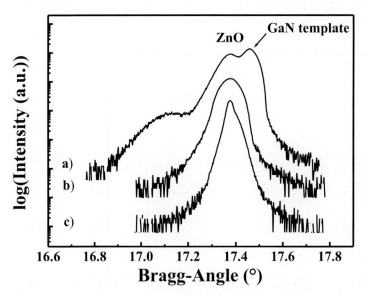

Fig. 2. $\Theta - 2\Theta$ scans of ZnO grown on a) GaN template, b) c-plane sapphire and c) ZnO substrate. For the film grown on GaN template an additional XRD peak from the GaN shows up

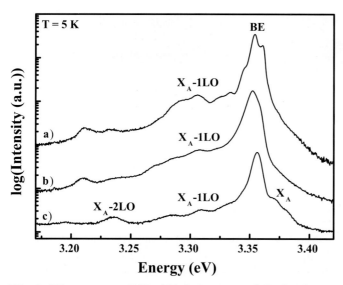

Fig. 3. PL spectra at 5 K of ZnO grown on a) ZnO substrate, b) GaN template and c) c-plane sapphire

dominated by a bound exciton (BE) signal at around 3.36 eV. The intensity is highest for homoepitaxially grown ZnO. The FWHM values are 8 meV, 7 meV and 5 meV for samples grown on sapphire, GaN and ZnO, respectively. It should be noted that the signal consists of several emission lines that were not clearly resolved in our measurements, with individual halfwidths below 4 meV. The reason for this broadened PL lines is assumed to be an inhomogenous strain distribution in the crystallites, which is corroborated by spatially resolved PL (not shown here). In the spectra the phonon replica of the free exciton ($X_A - (1 \text{ or } 2)\text{LO}$) with a phonon energy of 72 meV, as in bulk ZnO [19], can be identified. At higher temperatures up to four phonon replica can be observed. This and the absence of deep-level luminescence, usually ascribed to structural defects [20], is an indication of the optical and structural quality of the layers.

A typical surface morphology as measured by scanning electron microscopy is shown in Fig. 4 (left). In the SE image the typical hexagonal morphology is visible. Crystallites of different lateral sizes can be found. Here, the largest crystallites have a dimension of about 7 μm. On the facets of these bigger crystallites a number of small holes can be seen. The size of the pyramids depends on the growth conditions as well as on the substrate used, with a typical aspect ratio of 10 (ratio of pyramid diameter to film thickness). The facets of the pyramids are atomically flat, with a roughness on a side facet of only 1 nm (averaged over a $3 \times 3\,\mu\text{m}^2$ area).

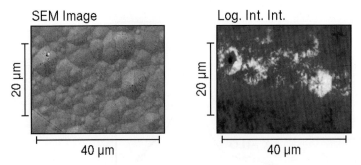

Fig. 4. (a) SE image and (b) cathodoluminescence intensity image of a ZnO layer

The cathodoluminescence of such layers is shown in Fig. 4 (right). The integral CL intensity (CLI) covers a range from 2.8 eV to 3.6 eV. In the CLI the crystallites show a pronounced 6-fold symmetry. The edges of the crystallites typically appear bright, and the top of each crystallite shows a dark contrast. This is a characteristic sign of screw dislocations growing in the center of each crystallite. The largest crystallites show highest CL intensity, indicating that the grain boundaries between crystallites play an important role as centers for non-radiative recombination.

The influence of the VI/II ratio during growth on the structural and electrical properties of ZnO layers grown on sapphire is shown in Fig 5. In the considered range of VI/II ratios between 18 and 60, the mobility (squares)

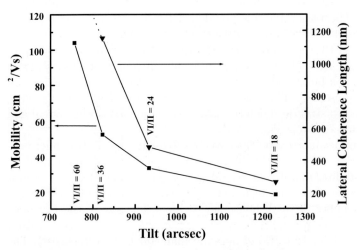

Fig. 5. Correlation of the mobility in ZnO layers grown on sapphire with their structural properties and its dependence on the VI/II ratio during growth. Squares refer to the mobility and triangles to the lateral coherence length

is inversely dependent on the tilt within the layers, which decreases linearly with the VI/II ratio, and a maximum mobility of more than $100\,\mathrm{cm^2/Vs}$ could be achieved. The tilt of the mosaic blocks in the ZnO layers has been obtained by employing a Williamson-Hall analysis [16]. On the other hand the lateral coherence length (triangles), which we interpret as the averaged grain size, is also inversely dependent on the tilt and thus increases linearly with the VI/II ratio. The determination of the lateral correlation length for ratios above 36 was not possible because the broadening of the rocking curves due to this effect became too small for a reliable interpretation. This result indicates that the grain boundaries are the limiting factor for the mobility in the layers, as they can act as potential barriers and scattering centers.

At room temperature the intrinsic carrier concentration of the layers grown with a VI/II ratio up to 36 varies around $(3-4) \times 10^{17}\,\mathrm{cm^{-3}}$. For higher ratios it increases by a factor of approximately 2 and remains almost constant beyond that. The increased carrier concentration is supposed to be due to either an increased incorporation of another donor species, or a variation in the point defect density. One possible origin for this behaiour is the incorporation of hydrogen, which can be driven out by annealing. Indeed, a reduction of the n-type carrier concentration has been seen after an annealing step at $500\,°\mathrm{C}$. This seems plausible taking into account that the oxygen precursor i-PrOH itself supplies hydrogen to the growth process and that hydrogen is used as the precursor carrier gas.

In thin films showing mosaicity, the carrier transport and in particular the Hall analysis can be governed by the properties of the grain boundaries, and is eventually no good measure of the material parameters themselves. In order to check this, we performed capacitance-voltage (CV) measurements on

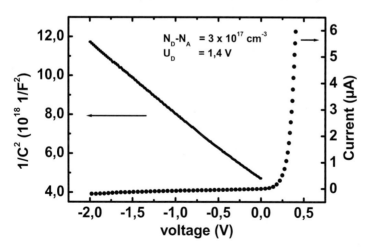

Fig. 6. Capacitance-Voltage characteristics of a ZnO epilayer grown on sapphire, with a thickness of $1.5\,\mathrm{\mu m}$. An effective donor concentration $N_D - N_A$ of $3.0 \times 10^{17}\,\mathrm{cm^{-3}}$ can be derived

ZnO epilayers. A Hg-prober has been used to implement Schottky contacts on ZnO. In Fig. 6, the CV characteristics of an undoped ZnO epilayer grown on sapphire substrate at a VI/II ratio of 48 is shown, together with an I-V-curve. According to CV results, the film has an effective donor concentration ($N_D - N_A$) of 3.3×10^{17} cm^{-3}, which is smaller than the carrier concentration of 1×10^{18} cm^{-3} derived by the Hall analysis. The discrepancies between CV and Hall analysis are systematic, with CV analysis giving a factor 2-5 lower carrier concentrations as compared to values derived from a Hall measurement. One reason for this could be the existence of a highly conductive interfacial layer between the ZnO and the substrate, comparable to the situation found in GaN [24]. This, however, can most likely not explain the large differences seen here. The mosaicity of the ZnO layer could also lead to a misleading Hall voltage, caused by grain boundary effects. Further work has to be done to clarify this point.

The optical quality of the layers grown on sapphire can also be correlated to the VI/II ratio. We find that the PL intensity increases several orders of magnitude with increasing VI/II ratio for as-grown samples as well as after an annealing step at 800 °C for the samples grown at low VI/II ratios. We suppose that again the grain boundaries can be made responsible for this dependence. As mentioned above, spatially resolved CL measurements show that the boundary regions are optically inactive. These regions seem to act as dark recombination centers and thus the integrated intensity in PL should increase as the averaged grain size increases, in agreement with our observations.

Furthermore the free exciton signals can be observed in reflectivity measurements for VI/II ratios above 24 in as-grown samples, indicating again the improved structural and optical quality of these layers (Fig. 7). Pronounced Fabry-Perot oscillations can also be identified in the low energy region, which proves the good interface and surface quality of the heterostructures grown. We determined the energy position of the free A- and B-exciton, X_A and X_B, using a simple two oscillator model [21].

The values of 3.371 eV and 3.381 meV obtained for the A- and B-exciton, respectively, are approximately 6 – 7 meV lower than for bulk material [22]. This hints at a tensile strain in the epilayers grown on sapphire. The conclusion is supported by a detailed analysis of a reciprocal space map of the ($11\bar{2}4$) reflex. It yields the lattice parameters of the ZnO layer to be $a = 3.258$ Å and $c = 5.195$ Å. The lattice constants a and c are considerably different from the respective bulk values of 3.250 Å and 5.204 Å, a being larger and c being smaller. These results confirm that the ZnO layers grown on c-Al$_2$O$_3$ exhibit tensile strain. It has been suggested that this strain might come from post-growth cooling due to the difference in the in-plane thermal expansion coefficients [23].

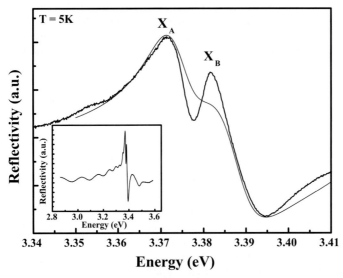

Fig. 7. Low temperature reflectivity spectrum, excitonic region and overall view (inset), of a ZnO layer grown on sapphire with a VI/II ratio of 48. A simple two oscillator model (thin line) has been applied to determine the energy positions of the free excitonic transitions X_A and X_B

4 Conclusions

In summary we have grown ZnO epilayers on c-plane sapphire, GaN templates and ZnO substrates by MOVPE using DEZn and i-PrOH. The VI/II ratio during growth has a large influence on the properties of the layers and is a crucial parameter for obtaining high quality ZnO. The best layers show a Hall mobility of more than $100\,\mathrm{cm^2/Vs}$, an n-type carrier concentration in the $10^{17}\,\mathrm{cm^{-3}}$ region, and good optical properties. If the VI/II ratio is chosen too low, the structural and optical properties are degrading. A post-growth annealing step can be applied, however, to improve the quality, especially the optical properties, of these layers considerably. The optical properties of layers grown under non-ideal conditions and undergoing a post-growth annealing step can be comparable to those exhibited by as-grown samples deposited under optimized conditions. This might be of importance if, for example, low-temperature growth is necessary for the fabrication of ternary compounds or p-doped layers. The strain in the ZnO layers grown on c-plane sapphire is found to be tensile. This posts a limitation on the maximum possible thickness of ZnO-based devices grown under these conditions on this substrate. The problem can be avoided by the use of high quality ZnO substrates or possibly by the development of high quality buffer layers, which is also expected to lead to a significant improvement of the structural properties of the layers [25,13].

In general, the potential of ZnO-based heterostructure lies in the field of optoelectronics in the UV. Compared to a GaN based technology, ZnO has some principal advantages: ZnO substrates are available, the exciton binding energy of ZnO is very high, and the material growth can be accomplished at much lower temperatures, with obvious technological advantages. ZnO can be processed by wet chemical etchants, in contrast to GaN. The material is piezoelectric, like GaN, and optically pumped lasing can be achieved even in ZnO surface-emitting nanowires [27]. As soon as a reliable p-type doping technique is available, the development of LEDs and also laser diodes should become possible.

In addition magnetic ions can be incorporated into ZnO isoelectronically – in contrast to GaN. This is of particular interest for the development of ferromagnetic semiconductors, with Curie temperatures above room temperature. Various theoretical predictions point to ZnMnO, ZnVO or ZnCoO [29], [28] being a transparent ferromagnet, having – under certain conditions – Curie temperatures above room temperature. The ferromagnetic phase is supposed to be stabilized by the exchange interaction between free carriers (electrons of holes) and the respective shells of the magnetic ions [29].

The magnetic properties could become interesting for a semiconductor based magnetoelectronics, where the combination of the band gap engineering and the engineering of magnetic properties – both based on the same semiconductor material – would lead to a substantial increase of freedom in the device design of magnetic RAMs, giant magnetoresistance (GMR) devices and others. It has recently been reported that n-type ZnCoO [30] and ZnVO [31] layers might be ferromagnetic between 280K and 350K.

It should be pointed out, however, that much more work has to be done in order to further improve the material quality and to exploit the interesting properties of ZnO in the future.

Acknowledgements

The authors would like to acknowledge the financial support of the BMBF as well as the Deutsche Forschungsgemeinschaft. We would also like to thank M. Haupt for the helpful discussions and technical assistance during the experiments.

References

1. B. Sang, K. Dairiki, A. Yamada, and M. Konagai, Jpn. J. Appl. Phys. **38**, 4983 (1999).
2. S. Pizzini, N. Butta, D. Narducci, and M. Palladino, J. Electrochem. Soc. **136**, 1945 (1989).
3. G. S. Kino and R. S. WAGER, J. Appl. Phys. **44**, 1480 (1973).
4. K. Minegishi, Y. Koiwai, Y. Kikuchi, K. Yano, M. Kasuga, and A. SHIMIZU, Jpn. J. Appl. Phys. **36**, L1453 (1997).

5. M. Joseph, H. Tabata, and T. KAWAI, Jpn. J. Appl. Phys. **38**, L1205 (1999).
6. Y. R. Ryu, S. Zhu, D. C. Look, J. M. Wrobel, H. M. Jeong, and H. W. WHITE, J. Cryst. Growth **216**, 330 (2000).
7. Eagle-Picher Technologies, LLC, Miami, Oklahoma, Press Release, January 7, 2002.
8. D. M. Bagnall, Y.F. Chen, Z. Zhu, T. Yao, S. Koyama, M. Y. Shen, and T. Goto, Appl. Phys. Lett. **70**, 2230 (1997).
9. A. Othomo, M. Kawasaki, T. Koida, H. Koinuma, Y. Sakurai, Y. Yoshida, M. Sumiya, S. Fuke, T. Yasuda, and Y. Segawa, Mater. Sci. Forum **264**, 1463 (1998).
10. D.C. Look, D.C. Reynolds, J.R. Sizelove, R.L. Jones, C.W. Litton, G. Cantwell, and W.C. Harsch, Solid State Commun. **105**, 399 (1998).
11. C. G. Van de Walle, Phys. Rev. Lett. **85**, 1012 (2000).
12. T. Yamamoto and H. Katayama-Yoshida, Jpn. J. Appl. Phys. **38**, L166 (1999).
13. Y. Chen, H. Ko, S. Hong, and T. Yao, Appl. Phys. Lett. **76**, 559 (2000).
14. A. Ohtomo, K. Tamura, K. Saikusa, K. Takahashi, T. Makino, Y. Segawa, H. Koinuma, M. Kawasaki, Appl. Phys. Lett. **75**, 2635 (1999).
15. B. M. Ataev, A. M. Bagamadova, A. M. Djabrailov, V. V. Mamedov, and R. A. Rabadanov, Thin Solid Films **260**, 19 (1995).
16. H. Wenisch, V. Kirchner, S. K. Hong, Y. C. Chen, H. J. Ko, and T. Yao, J. Cryst. Growth **227/228**, 944 (2001).
17. S. Oda, H. Tokunaga, N. Kitajima, J. Hanna, I. Shimizu, and H. Kokado, Jpn. J. Appl. Phys. **24**, 1607 (1985).
18. A. C. Jones, S. A. Rushworth, and J. Auld, J. Cryst. Growth **146**, 503 (1995).
19. A. W. Hewat, Solid State Commun. **8**, 187 (1970).
20. D. C. Reynolds, D. C. Look, B. Jogai, and H. Morkoc, Solid State Commun. **101**, 643 (1997).
21. C. F. Klingshirn, Semiconductor Optics, (Springer, Berlin 1995).
22. D. C. Reynolds, D. C. Look, B. Jogai, C. W. Litton, G. Cantwell, and W. C. Harsch, Phys. Rev. B **60**, 2340 (1999).
23. F. Vigue, P. Vennegues, S. Vezian, M. Laugt, and J. P. Faurie, Appl. Phys. Lett. **79**, 194 (2001).
24. D.C.Look, R.J.Molnar Appl. Phys. Lett. **70**, 3377 (1997).
25. Th. Gruber, C. Kirchner, and A. Waag, phys. stat. sol. (b) **229**, 841 (2002).
26. T. Makino, C.H. Chia, Nguyen T. Tuan, Y. Segawa, M. Kawasaki, A.Ohtomo, K.Tamura, H.Koinuma, Appl. Phys. Lett. **77**, 1632 (2000).
27. M.H. Hyang, S. Mao, H. Feick, H. Yan, Y. Wu, H. Kind, E. Weber, R. Russo, P. Yang Science **292** 1897 (2001).
28. K. Sato and H. Katayama-Yoshida phys. stat. sol. b **229**, 673(2002)
29. T. Dietl, H. Ohno, and F. Matsukura Phys. Rev. B **63**, 195205 (2001).
30. K. Ueda, H. Tabata, T. Kawai Appl. Phys. Lett. **79**, 989 (2001).
31. H. Saeki, H. Tabata, T. Kawai Sol. St. Commun. **120**, 439 (2001).

Part III

Electron and Spin Transport

Electrical Spin Injection from Ferromagnetic Metals into GaAs

Manfred Ramsteiner, Haijun Zhu, Atsushi Kawaharazuka,
Hsin-Yi Hao, and Klaus H. Ploog

Paul Drude Institute for Solid State Electronics
Hausvogteiplatz 5-7, 10117 Berlin, Germany

Abstract. Electrical injection of spin polarized electrons is observed at ferromagnetsemiconductor interfaces by analyzing the electroluminescence signal of GaAs/(In,Ga)As light emitting diodes (LED) with injection layers of the metals Fe or MnAs. The circular polarization degree of the emitted light reveals a spin polarization for recombining electrons of about 1.5% to 2%. Time-resolved data on spin relaxation times indicate that the actual spin injection efficiency from both Fe and MnAs into GaAs is about 5% to 6%. The underlying injection mechanism can be explained in terms of a tunneling process.

1 Introduction

Spintronic semiconductor devices utilize not only the charge, but also the spin of electrons to achieve novel functionalities [1]. The realization of such devices relies on the ability to inject a spin-polarized current into a semiconductor (SC), which remains still to be a challenge. The first successful experimental approaches are based on semiconducting injection layers, which have the severe disadvantage to be restricted to low temperatures [2,3,4,5]. As a major issue in the field of spintronics, the injection problem also stimulated a number of recent theoretical considerations [6,7,8,9,10,11]. The ferromagnetic metals (FM) Fe and MnAs are, in principle, excellent candidates for spin injection at room temperature. Furthermore, Fe-on-GaAs and MnAs-on-GaAs are extensively studied FM/SC hybrid systems. However, prior to our work [12], there have been no convincing experiments for spin injection from a FM into a SC. Concerning Fe on GaAs, the formation of a magnetically dead layer at the Fe/GaAs interface has been discussed as an obstacle for spin injection [13]. Moreover, theoretical work predicted that spin injection from any metal into a SC should be almost impossible considering a diffusive ohmic regime [6]. Nevertheless, electrical spin injection from Fe into semiconductors has been recently demonstrated [12,14]. The ferromagnetic metal α-MnAs is a further promising choice with a high Curie temperature ($T_C \approx 40°C$) and a relatively small coercitive field (50 Oe) [15]. Furthermore, the hybrid system MnAs/GaAs can be prepared by molecular beam epitaxy (MBE) with interfaces of extremely high quality [16,17,18]. However, spin injection from (Ga,Mn)As has so far been demonstrated only for diluted material with low Mn content, which is semiconducting (mostly p-type) and becomes ferromagnetic only at very low temperature [4,5].

We present here experimental evidence that spin injection from the ferromagnetic metals Fe and MnAs into the semiconductor GaAs is indeed possible. The spin polarization of injected electrons is detected by the circular polarization degree of the electroluminescence (EL) from n-i-p (In,Ga)As/GaAs light emitting diodes (LED) containing FM cap layers.

2 Experimental Approach

Our experimental evidence for spin injection from FM into SC relies on the appropriate sample design, the polarization analysis of EL measurements in external magnetic fields and the determination of spin relaxation times in (In,Ga)As quantum wells (QWs).

2.1 Sample Design

The LED device structures (cf. Fig. 1, left inset) were grown by molecular beam epitaxy (MBE) on p-GaAs(001) substrates with 500-nm-thick p-GaAs buffer layers [12]. The active region consists of two 4-nm-thick (In,Ga)As QWs separated by a 10-nm-thick GaAs barrier and sandwiched between two 50-nm-thick undoped GaAs spacer layers. On top of this intrinsic region, a 70-nm-thick n-GaAs layer was grown. The Fe or MnAs injection layers were deposited on the n-GaAs layer in separate MBE chambers. The growth conditions for the FM layers have been previously optimized with respect to the delicate FM/SC interfaces [17,19]. The 20-nm-thick Fe injection layer was deposited in an As-free metal MBE chamber and, finally, capped with a 10-nm-thick Al protection layer. The growth of the 50-nm-thick MnAs layer was carried out at a temperature of about 250°C and a growth rate of 20 nm/h [17]. By varying the As coverage of the GaAs surface, two epitaxial orientations of MnAs with respect to the substrate have been realized: Type A with MnAs[0001]∥GaAs[1$\bar{1}$0] and type B with MnAs[11$\bar{2}$0]∥GaAs[1$\bar{1}$0]. In both cases, the surface orientation is given by MnAs($\bar{1}$100)∥GaAs(001). For reference purposes, one part of each wafer was not capped with a FM layer, but subsequently covered with the nonmagnetic AuGe alloy layer. After metal electrode deposition, the epitaxial wafers were processed into 50-μm-wide mesa stripes defined by dry chemical etching and cleaved into pieces of 240 to 670 μm length. For time-resolved photoluminescence (TRPL) experiments, reference QW samples without p^+- and n^+-GaAs layers were grown under the same conditions.

2.2 Magneto-Electroluminescence

For the EL measurements, the LED was placed into a superconducting magnet system (OXFORD Spectromag 1000) with the temperature controlled in a continuous flow cryostat. The experiments were done in Faraday geometry, i.e., with the magnetic field direction parallel to the light propagation (cf. Fig. 1, left inset). The EL signal was collected from the wafer backside, dispersed in a single spectrograph (DILOR XY800) and detected by a charge-coupled device (CCD) array. The circular polarization was analyzed by passing the EL light through a photoelastic modulator (PEM) [20] and a linear polarizer with its analyzing direction rotated by 45° with respect to the optical axis of the PEM. The LED was operated with short current pulses (0.4 μsec pulse width at a frequency of 42 kHz) locked to the maximum or minimum phase shifts of the PEM. The degree of circular polarization is determined by $P = (I_+ - I_-)/(I_+ + I_-)$, where the right (left) circularly polarized component I_+ (I_-) is obtained for EL generation pulses locked to $+\lambda/4$ ($-\lambda/4$) phase shifts of the PEM [12].

2.3 Time-Resolved Photoluminescence

TRPL measurements were performed using a synchro-scan streak camera system in conjunction with a Ti:sapphire laser emitting 150 fs pulses at photon energies between 1.56 and 1.70 eV (repetition rate 76 MHz). The average excitation power density was about 10 Wcm^{-2}. The luminescence was dispersed by a single monochromator and focused onto the photocathode of the streak tube. The streak images were recorded by a cooled CCD array. The nominal temporal resolution of the synchro-scan system is 2 ps. The samples were mounted on the cold-finger of a He flow cryostat. An initial spin polarization of photo-excited carriers was created by pump pulses, which were circularly polarized (σ^+) by means of a quarter-wave plate [21,22]. The emitted PL light was analyzed into its right (I_+) and left (I_-) circularly polarized components using a second quarter-wave plate. The total carrier lifetime τ_R (which at low temperatures corresponds to the radiative recombination time) and the spin relaxation time τ_S have been obtained by fitting single exponential decay curves to the total PL intensity ($I_+ + I_-$) and $P = (I_+ - I_-)/(I_+ + I_-)$, respectively.

3 Injection from Fe into GaAs

The room-temperature EL spectrum of the LED with Fe injection layer, shown in Fig. 1, reveals one peak at 1.3 eV in accordance with the design of the active region. Since the EL peak width of 90 meV is larger than the expected heavy-hole/light-hole splitting, we assume contributions of both heavy-hole and light-hole transitions to the EL spectrum. For a given polarization degree of injected polarized electrons and unpolarized holes, recombination with the two different types of holes results in circular polarization degrees P of opposite signs (cf. right inset in Fig. 1). In both cases, the absolute value of P is the same and identical to the spin polarization of the

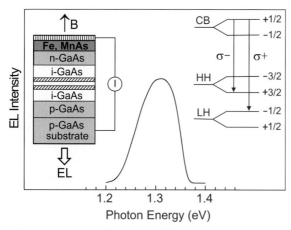

Fig. 1. EL spectrum of the LED with a Fe cap layer recorded at 300 K. Left inset: Device structure showing the direction of the magnetic field (B) and the emitted light (EL). Right inset: Circular polarization of the EL light from recombination of electrons with spin +1/2 for heavy-hole (σ^-) and light-hole transitions (σ^+)

radiatively recombining electrons [3]. The photon energy dependence of P, shown in Fig. 2 for large external magnetic fields ($|B| > 8$ T, where the saturation magnetization of the Fe layer is reached), reveals two sign reversals. The corresponding energies are defined by changeovers in the type of holes, which are predominantly involved in the recombination with elec-

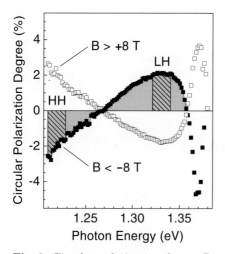

Fig. 2. Circular polarization degree P as a function of photon energy for the LED with a Fe cap layer at magnetic fields $B < -8$ T (full squares) and $B > +8$ T (open squares). The shaded areas indicate EL transitions of electrons with either predominantly heavy-holes (HH) or predominantly light-holes (LH)

trons. The low-energy range with negative P for $B < -8$ T (full squares in Fig. 2) is assigned to predominantly heavy-hole ground-state recombination. Consequently, the following energy range with positive P for $B < -8$ T is attributed to predominantly light-hole transitions. The second sign reversal is probably due to EL transitions involving an excited heavy-hole state. As expected, the sign of P changes when the direction of the magnetic field is reversed (full vs. open squares in Fig. 2).

As a proof for spin injection, we require that the polarization degree P as a function of an external magnetic field B follows the out-of-plane magnetization curve of the FM injection layers obtained independently by superconducting quantum interference device (SQUID) and spontaneous Hall effect measurements [23]. For the LED with Fe injection layer, the above assignments enable us to analyze both heavy-hole and light-hole transitions separately. The intensity component I_+ (I_-) of right (left) circularly polarized light has been determined by integrating over the corresponding energy ranges of the EL peak for the heavy-hole and light-hole contribution (cf. hatched areas in Fig. 2). The resulting polarization degrees P are shown in Fig. 3 as a function of the external magnetic field B together with the out-of-plane magnetization (solid lines) of the Fe layer. The circular polarization curves for heavy-hole (full squares) and light-hole (open squares) transitions follow over the whole magnetic field range the Fe magnetization and confirm the expected complementary behavior. This coincidence provides evidence that we observe injection of spin polarized electrons from Fe into GaAs with an efficiency of at least 2%.

It should be mentioned that the geometrical configuration shown in Fig. 1 (left inset) has been chosen to attain well-defined conditions for the demon-

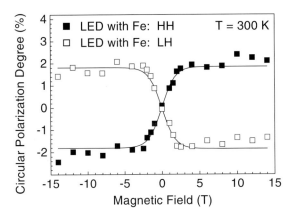

Fig. 3. Circular polarization degree P for heavy-hole (full squares) and light-hole (open squares) transitions as a function of external magnetic field measured at 300 K from the LED with a Fe cap layer. The magnetization curve of a thin Fe layer is shown for comparison in arbitrary units with two opposite signs (solid lines)

stration of spin injection. In this geometry, the selection rules relating the polarization degree of the EL signal to the one of the carriers are straightforward. However, for emission from a cleaved edge with the magnetic field direction perpendicular to the growth (quantization) direction, this relation becomes more complicated. Furthermore, ambiguities due to passing the EL light through the magnetic Fe layer are avoided. Due to the shape anisotropy, the magnetic field direction corresponds to the hard out-of-plane axis of magnetization with a negligible coercitive field (no hysteresis behavior). For actual spintronic device applications, in-plane magnetization should be chosen, where the magnetic hysteresis behavior can be utilized to sustain the spin injection properties without any external magnetic field.

Any artifact due to polarization-dependent reflection of the EL light at the Fe/GaAs interface is excluded, since the reflection properties at the Fe/GaAs interface cannot produce the complementary behavior found for the heavy-hole and light-hole transitions. Our results indicate furthermore that the low growth temperature chosen for the Fe deposition prevents the formation of a magnetically dead layer due to interdiffusion between Fe and GaAs, which destroys any spin information.

As shown in Fig. 4, the dependence of P on the external magnetic field has been found to be similar at low temperature (25 K), where the EL peak is rather narrow (33 meV), hence involving only heavy-hole transitions. For magnetic fields $|B| < 4$ T, the polarization degree P follows very closely the magnetization curve of thin Fe layers. To explain the deviation for magnetic fields $|B| > 4$ T, we have to consider, that the effect of spin injection is

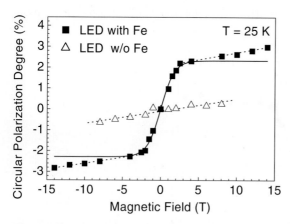

Fig. 4. Circular polarization degree P as a function of external magnetic field measured at 25 K from LEDs with (full squares) and without (open triangles) Fe cap layer. The out-of-plane magnetization curve of a thin Fe layer is shown for comparison in arbitrary units (solid line). The dotted lines are guides to the eye and indicate the contribution of the Zeeman splitting induced spin alignment in GaAs (see text)

expected to be superimposed on the impact of the spin alignment due to the Zeeman splitting in large external magnetic fields. The polarization of the EL light is influenced by the thermalization of both electrons and heavy holes into the energetically lowest Zeeman states in the semiconductor layers of the LED structure. No saturation is expected for this kind of spin alignment, if the corresponding Zeeman splittings are smaller than the thermal energy kT for the whole range of external magnetic fields [24]. This explains the observed deviation of the polarization degree P from the magnetization curve for magnetic fields $|B| > 4$ T. The signatures of spin injection are not observed for the reference LED without Fe cap layer. This LED reveals a clearly smaller circular polarization degree P with a linear dependence on the magnetic field as shown in Fig. 4 (open triangles). This magnetic field dependence can be explained simply by the Zeeman splitting induced spin alignment in the LED structure, without invoking any spin injection. The slope of the linear dependence is almost identical to that found for the LED with Fe cap layer in the magnetic field range $|B| > 4$ T, which strongly supports our conclusions. The influence of spin alignment due to the Zeeman splitting in the LED is expected to be less pronounced at elevated temperatures (cf. Fig. 3).

4 Injection from MnAs into GaAs

For the LED with MnAs injection layer, the dependence of P on the external magnetic field, shown in Fig. 5 (full squares), does not closely resemble the out-of-plane magnetization of the MnAs injection layer obtained by SQUID measurements. In this LED, the (In,Ga)As QWs in the active region were grown with an In content of 0.1, whereas an In content of 0.2 was chosen for

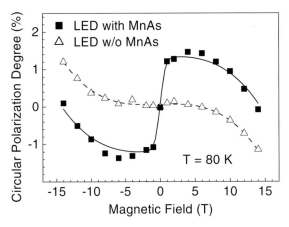

Fig. 5. Circular polarization degree P as a function of external magnetic field measured at 80 K from LEDs with (full squares) and without (open triangles) MnAs injection layer. The solid and dashed lines are guides to the eye

the QWs of the LED with Fe injection layer. It turns out that the impact of spin alignment due to the Zeeman splitting is more pronounced in the LED structure with the lower In content. This finding becomes evident from the magnetic field dependence of P for the reference LED without a MnAs an injection layer (open triangles in Fig. 5). This behavior is probably due to the specific spin relaxation times and g-factors in the QWs with lower In content, i.e. less strain in the (In,Ga)As layers. The non-linear dependence on the magnetic field can then be explained by assuming Fermi-Dirac distributions for electrons and heavy holes in quasi-equilibrium and relatively large g-factors in the (In,Ga)As QWs.

The possible contribution of spin injection from MnAs to the measured magnetic field dependence of P might be therefore masked by the superposition with the effect of the spin alignment due to the Zeeman splitting. Consequently, we subtract the P curve obtained for the reference LED from that of the LED with MnAs injection layer. Indeed, this net polarization, displayed in Fig. 6, follows the MnAs out-of-plane magnetization curve obtained by SQUID measurements. This finding provides evidence for successful spin injection from another FM. In this case, we achieved a somewhat lower degree of circular polarization degree of about 1.5% in the saturation range. If the symmetries of the conduction band wavefunctions in MnAs play an important role, the spin injection into GaAs may depend on the orientation of the MnAs injection layer. However, no significant difference has been found between MnAs layers with orientation of type A (full squares in Fig. 6) and type B (open squares in Fig. 6).

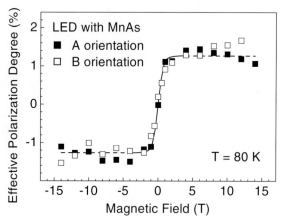

Fig. 6. Effective polarization degree P as a function of external magnetic field measured at 80 K from LEDs with MnAs injection layers of A orientation (full squares) and B orientation (open squares). The contribution due to the Zeeman splitting in the LED layers has been subtracted. The magnetization curve of a thin MnAs layer is shown for comparison in arbitrary units (solid line; dashed line is the continuation as a guide to the eye)

5 Spin Relaxation Times

As mentioned above, the polarization degree P for transitions involving only one type of holes is identical to the spin polarization of the recombining electrons in the (In,Ga)As QWs. However, if the spin relaxation time τ_S in the QWs [21,22] is considerably shorter than the carrier lifetime τ_R, the measured value of P represents a lower limit for the spin polarization of the injected electrons. From a simple rate equation model, it follows that the spin injection efficiency for our experiments would then be

$$\eta = (1 + \tau_R/\tau_S) \times P \qquad (1)$$

which defines a correction factor $\kappa = (1 + \tau_R/\tau_S)$. Here, we neglect the possible spin relaxation during the transport from the FM/GaAs interface to the (In,Ga)As QWs.

The crucial time constants τ_R and τ_S obtained by TRPL spectroscopy are displayed in Fig. 7(a) for a reference sample containing (In,Ga)As QWs with an In content of 0.2 (like in the LED with Fe injection layer). Whereas τ_S (open triangles) shows a monotonic decrease with increasing temperature, τ_R (full triangles) exhibits a pronounced maximum at about 100 K. The decrease of τ_R for temperatures above 100 K is attributed to an increasing contribution of non-radiative recombination. For the ratio τ_R/τ_S, displayed in Fig. 7(b), we find $1 < \tau_R/\tau_S < 2$ at temperatures below 50 K and above

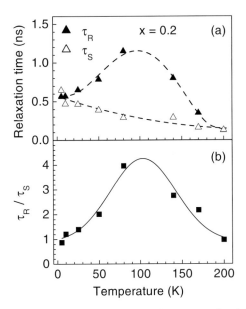

Fig. 7. Measured carrier lifetime τ_R (a: full triangles), spin relaxation time τ_S (a: open triangles) and the ratio τ_R/τ_S (b: full squares) as a function of sample temperature for a (In,Ga)As QW structure with an In content of 0.2

150 K. Consequently, with a correction factor $\kappa \approx 2.5$, we deduce an effective spin injection efficiency of $\eta \approx 5\%$ for our experiments with Fe injection layer.

For (In,Ga)As QWs with an In content of 0.1 (like in the LED with MnAs injection layer), the contribution of non-radiative recombination is found to be weaker, which manifests itself by a monotonic increase of τ_R over the whole temperature range as shown in Fig. 8(a). At the same time, τ_S reveals a relatively weak temperature dependence. Consequently, the resulting ratio τ_R/τ_S, displayed in Fig. 8(b), reaches values as large as 20 at temperatures around 250 K. For our experiments with a MnAs injection layer at 80 K, we obtain $\kappa \approx 4$, which leads to a spin injection efficiency of $\eta \approx 6\%$.

Our measurements reveal that the time constants τ_R and τ_S depend strongly on the In content in the (In,Ga)As QWs. Thus, the accuracy of the chosen correction factors κ depends crucially on the uncertainty in the In content of the QWs in the LED structures. It should be noted that the temperature dependencies of τ_S do not agree with the assumption that only one of the possible DP (D'Yakono-Perel), EY (Elliot-Yafet) or BAP (Bir-Aharonov-Pikus) mechanisms dominates [22].

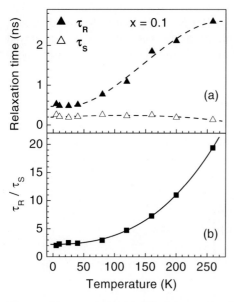

Fig. 8. Measured carrier lifetime τ_R (a: full triangles), spin relaxation time τ_S (a: open triangles) and the ratio τ_R/τ_S (b: full squares) as a function of sample temperature for a (In,Ga)As QW structure with an In content of 0.1

6 Spin Injection Model

Our findings seem to be in contradiction to the theoretical work predicting the spin injection efficiency from a metal into a semiconductor to be limited to less than 0.1% due to the resistance mismatch [6]. The reason for this apparent discrepancy might be given by the fact that our FM/SC layers form Schottky-type contacts, which give rise to tunneling under appropriate bias conditions [25]. Thus, electrons from the FM layer have to tunnel through the Schottky barrier before reaching the active region of the LED [26]. Such a tunneling process can lead to an enhanced spin injection efficiency, since it is not affected by the resistance mismatch [7]. Experimental evidence for spin-polarized electron tunneling has been reported by scanning tunneling microscopy [27,28]. However, recent theoretical work predicts on the basis of a ballistic picture that spin injection can be efficient also without tunnel barriers [8]. Thus, one important aim of further work is to experimentally verify, under which conditions tunneling is an important process for spin injection from FM into SC.

7 Conclusions

In conclusion, spin injection from the ferromagnetic metals Fe and MnAs into the semiconductor GaAs is demonstrated with an efficiency of about 5% to 6%, which has previously been considered to be impossible. The circular polarization degree of the electroluminescence from (In,Ga)As/GaAs LEDs with ferromagnetic injection layers is found to be a lower limit for the spin injection efficiency, since the spin relaxation times in the active regions of the LEDs are shorter than the carrier recombination times. The injection mechanism can be explained in terms of a tunneling process through a Schottky barrier. Further optimization of the ferromagnet-semiconductor interface as well as the device structure will make Fe and MnAs promising candidates for spin injection into semiconductors and pave the way for room-temperature operation of spintronics devices.

Acknowledgements

The authors would like to thank H.-P. Schönherr, M. Kästner, R. Hey, L. Däweritz, H. Kostial, and K.-J. Friedland for their efforts with the MBE growth, processing and spontaneous Hall effect measurements as well as H.T. Grahn for helpful discussions. Part of this work was supported by the Bundesministerium für Bildung und Forschung of the Federal Republic of Germany.

References

1. G.A. Prinz: Phys. Today **48**, 58 (1995).
2. R. Fiederling, M. Keim, G. Reuscher, W. Ossau, G. Schmidt, A. Waag, L.W. Molenkamp: Nature **402**, 787 (1999).
3. B.T. Jonker, Y.D. Park, B.R. Bennett, H.D. Cheong, G. Kioseoglou, A. Petrou: Phys. Rev. B **62**, 8180 (2000).
4. Y. Ohno, D.K. Young, B. Beschoten, F. Matsukura, H. Ohno, D.D. Awschalom: Nature **402**, 790 (1999).
5. S. Ghosh, P. Bhattacharya: Appl. Phys. Lett. **80**, 658 (2002).
6. G. Schmidt, D. Ferrand, L.W. Molenkamp, A. T. Filip, B.J. van Wees: Phys. Rev. B **62**, R4790 (2000).
7. E.I. Rashba: Phys. Rev. B **62**, R16267 (2000).
8. C.-M. Hu, T. Matsuyama: Phys. Rev. Lett. **87**, 066803 (2001).
9. D.L. Smith, R.N. Silver: Phys. Rev. B **64**, 045323 (2001).
10. H.B. Heersche, Th. Schaäpers, J. Nitta, H. Takayanagi: Phys. Rev. B **64**, 161307 (2001).
11. A. Fert, H. Jaffres: Phys. Rev. B **64**, 184420 (2001).
12. H.J. Zhu, M. Ramsteiner, H. Kostial, M. Wassermeier, H.-P. Schönherr, K.H. Ploog: Phys. Rev. Lett. **87**, 016601 (2001).
13. Y.B. Xu, E.T.M. Kernohan, D.J. Freeland, A. Ercole, M. Tselepi, J.A.C. Bland: Phys. Rev. B **58**, 890 (1998).
14. A.T. Hanbicki, B.T. Jonker, G. Itskos, G. Kioseoglou, A. Petrou: Appl. Phys. Lett. **80**, 1240 (2002).
15. M. Tanaka, J.P. Harbison, T. Sands, T.L. Cheeks, V.G. Keramidas, G.M. Rothberg: J. Vac. Sci. Technol. B **12**, 1091 (1994).
16. M. Tanaka, K. Saito, T. Nishinaga: Appl. Phys. Lett. **74**, 64 (1999).
17. F. Schippan, A. Trampert, L. Däweritz, K.H. Ploog: J. Vac. Sci. Technol. B **17**, 1716 (1999).
18. A. Trampert, F. Schippan, L. Däweritz, K.H. Ploog: Appl. Phys. Lett. **78**, 2461 (2001).
19. H.-P. Schönherr, R. Nötzel, W.Q. Ma, K.H. Ploog: J. Appl. Phys. **89**, 169 (2001).
20. M. Wassermeier, H. Weman, M.S. Miller, P.M. Petroff, J.L. Merz: J. Appl. Phys. **71**, 2397 (1992).
21. D. Hägele, M. Oestreich, W.W. Rühle, N. Nestle, K. Eberl: Appl. Phys. Lett. **73**, 1580 (1998).
22. A. Malinowski, R.S. Britton, T. Grevatt, R.T. Harley, D.A. Ritchie, Y. Simmons: Phys. Rev. B **62**, 13034 (2000).
23. K.-J. Friedland, R. Nötzel, H.-P. Schönherr, A. Riedel, H. Kostial, K.H. Ploog: Physica E **10**, 442 (2001); The magnetization curve shown in Fig. 3 has been obtained experimentally for the present Fe layers on GaAs(001).
24. Th. Wimbauer, K. Oettinger, A.L. Efros, B.K. Meyer, H. Brugger: Phys. Rev. B **50**, 8889 (1994).
25. B.T. Jonker, E.M. Kneedler, P. Thibado, O.J. Glembocki, L.J. Whitman, B.R. Bennett: J. Appl. Phys. **81**, 4362 (1997).
26. A. Hirohata, Y.B. Xu, C.M. Guertler, J.A.C. Bland, S.N. Holmes: Phys. Rev. B **63**, 104425 (2001).
27. S.F. Alvarado, P. Renaud: Phys. Rev. Lett. **68**, 1387 (1992).
28. V.P. LaBella, D.W. Bullock, Z. Ding, C. Emery, A. Venkatesan, W.F. Oliver, G.J. Salamo, P.M. Thibado, M. Mortazavi: Science **292**, 1518 (2001).

Probing the Conduction Channels of Gold Atomic-Size Contacts: Proximity Effect and Multiple Andreev Reflections

E. Scheer[1], W. Belzig[2], Y. Naveh[3], and C. Urbina[4]

[1] Fachbereich Physik, Universität Konstanz, 78457 Konstanz, Germany
[2] Departement Physik, Universität Basel, 4056 Basel, Switzerland
[3] Department of Physics, SUNY at Stony Brook, New York 11794, USA
[4] SPEC, CEA-Saclay, 91191 Gif-sur-Yvette Cedex, France

Abstract. We have investigated the electronic transport properties of Au quantum point contacts fabricated by the mechanically controllable breakjunction technique using non-linearities in the current-voltage characteristics induced by the proximity effect (PE) of superconducting aluminum electrodes. The PE shows up also in an hysteretic transition to the normal state when applying an external magnetic field. We find that the very smallest contact formed by a single atom between the electrodes contributes one channel to the transport as predicted by different theoretical models. The analysis of the transport channels of retraceable opening and closing procedures gives additional evidence that the transport channels are determined by the properties of the central atom.

Since the development of the scanning tunnelling microscope (STM) [1] it is not only possible to see, but also to manipulate and to measure the transport properties of individual atoms on surfaces [2]. By energy dependent measurements of the differential conductance a certain chemical information can be achieved [3]. The challenging aim of building up electronic circuits atom by atom with tailor-made properties, however, would require the detailed knowledge of the relation between the physical and chemical properties of the respective atoms and their conduction properties, a problem which has been addressed by different methods during the last years [4]. The most simple system for all investigations - including the present - is a one-atom contact between two metallic banks of the same element.

Electrical transport through such contacts is suitably described by the Landauer formalism, which treats it as a wave scattering problem. The transport properties of the contact connected to the leads are described by a set of transmission coefficients $\{T_n\}$ with $T_n \leq 1$. E.g. the conductance G of a contact is $G = G_0 \sum_n T_n$. Here, $G_0 = 2e^2/h$ is the conductance quantum [5]. Since in few-atom metallic point contacts the structure size is of the same order as the Fermi wavelength, such a contact has only a small number N of channels. If the set $\{T_n\}$ is known, many further transport properties, as e.g. shot noise, thermopower or superconducting properties of the contact can be deduced [6].

In order to test basic concepts of the quantum mechanical transport theory it is of particular interest to study metallic systems transmitting only one single channel with adjustable transmission coefficient. According to a quantum chemical model by Cuevas, Levy Yeyati and Martín-Rodero [7,8] the transmission coefficients of single-atom contacts are a function of the chemical properties of the metal and the atomic arrangement of the region around the central atom. Within their model only single atom contacts of monovalent metals as e.g. the alkali or noble metals should transmit one single channel. In particular for Au it has been predicted that the transmission coefficient of this single channel can achieve, in a perfectly ordered geometry of the central atom and its neighbors, a nearly saturated value of $T > 0.99$ [8]. We present here results of an experiment which allows to determine the transmission coefficients of this model substance with the help of proximity superconductivity induced by aluminum electrodes close to the contact region. We discuss the properties of the PE in a perpendicular magnetic field.

It has been shown [9] that the set $\{T_n\}$ of atomic-size constrictions is amenable to measurement using the strong non-linearities in the current-voltage (IV) characteristics of *superconducting* atomic contacts due to multiple Andreev reflections (MAR) [10]. Fig. 1 a) shows the numerical predictions for a single channel with transmission T [11,12,13,14]. The $i(V,T)$ curves present a series of sharp current steps at voltage values $eV = 2\Delta/m$, where m is a positive integer and Δ is the superconducting gap. Each one of these steps corresponds to an additional microscopic process of charge transfer as depicted in Fig. 1 b)-d). For example, the well-known non-linearity at $eV = 2\Delta$ arises when one electronic charge ($m = 1$) is transferred. The order $m = 2, 3, ...$ of a step corresponds to the number of electronic charges transferred in the underlying MAR process. Energy conservation imposes a threshold $eV \geq 2\Delta/m$ for the process of order m. For low transmission, the contribution to the current arising from the process of order m scales as T^m. The analysis of the transmission coefficients is performed by decomposing the experimental IVs of a particular contact into contributions of N independent terms with the $\{T_n\}$ (where $n = 1, ..., N$) as fitting parameters: $I(V) = \sum_{n=1}^{N} i(V, T_n)$. The important result of this analysis is that single-atom contacts of multivalent metals like Al, Pb or Nb always transmit more than one channel. The transmission coefficients of the channels depend sensitively on the exact atomic arrangement of the contact [9,15]. Since Au is not superconducting, the described method of determining conduction channels is not directly applicable. However, due to the so-called proximity effect (PE) [16,17], a finite piece of a non-superconducting metal in good contact with a superconducting metal adopts certain superconducting properties, e.g. it develops a "minigap" E_g in the quasiparticle excitation spectrum. As will be explained below, the appearance of the minigap is the basic property for determining the channel ensemble.

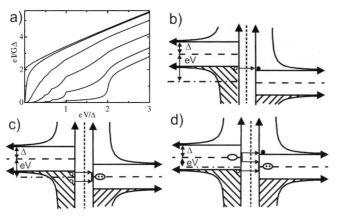

Fig. 1. a): Theoretical current voltage characteristics for a single-channel superconducting contact for different values of the transmission T (from bottom to top: 0.1,0.4,0.7,0.9,0.99,1 after [13]). b)-d): Representation of the transport processes of MAR at zero temperature: The quasiparticle (qp) densities of states for the left and right electrode forming the contact in a modified semiconductor model are shown. Qp's are represented as open circles when removed from an electrode and closed circles when added to an electrode. In addition the Cooper pairs (Cp's) are shown as ellipses on the Fermi levels (empty ellipses when a Cp is broken, ellipses filled with two circles when a Cp is formed) Energy conservation requires transport processes which are horizontal. The contact is represented by the vertical dotted line in the middle. Due to the contact with finite resistance it is possible to apply a voltage eV between the left and the right electrode, measured between the Fermi energies (dashed horizontal lines) of the electrodes. In the examples shown in b)-d) a positive voltage is applied to the right electrodes such that the charges are transported form the left to the right electrode. b) When the applied voltage exceeds $2\Delta/e$, transport of a single qp is possible, corresponding to the MAR process with $m = 1$. c) For voltages $V \geq 2\Delta/2e$ the simultaneous transport of two qp's, forming a Cp in the right electrode also contributes to the current ($m = 2$). d) For $V \geq 2\Delta/3e$ three qp's are transferred in a process where in the left electrode a Cp is broken, a new Cp is formed in the right electrode and a qp moved from the occupied states at the left electrode to the empty states in the right electrode

Experimentally, stable atomic-size contacts can be achieved with different techniques including STM [18,19] and mechanically controllable breakjunctions (MCB) [20]. Using such methods it has been shown that smallest stable contacts of gold have most often a conductance close to G_0. We have produced micro-fabricated breakjunctions [21] of Al, where the center part of the constriction consists of Au. Using shadow evaporation through a suspended mask we evaporate perpendicular to the substrate surface two Al electrodes separated by a gap of width $2L_N$. Without breaking the vacuum, two Au layers of thickness $d_{au}/2$ are evaporated at two different angles $\pm\ 10°$ in order to fill the gap and to form a continuous film (see Fig. 2). For sample parameters see Table 1. After lifting off the mask the underlying sacrificial

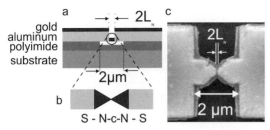

Fig. 2. a) Schematic side view of the samples. The metallic thin films are evaporated on top of a flexible substrate covered by a polyimide sacrificial layer. The latter is etched away to suspend the central bridge that connects the two large Al electrodes trough a small Au constriction. b) Schematic top view of the central part of the resulting system. Two nominally identical NS structures coupled through a constriction which can be controlled in-situ down to the atomic size. c) SEM micrograph of sample No. 1 (top view). The Al electrodes are the brightest regions. The narrow and dark skirt around them corresponds to the gold layer, through which the contact is established

Table 1. Sample number, thickness of the Al [Au] layer $d_{Al[Au]}$, elastic mean free path of the Au layer l_{Au}, mini-gap E_g [superconducting gap Δ for sample No. 4], critical temperature Θ_c, critical field B_c, spacing between Al electrodes as deduced by best fit to the theory of Ref. [17] $(L_N/\xi_S)_{fit}$ (assuming $\xi_S = 280$ nm) and as estimated from the micrographs $(L_N/\xi_S)_{exp}$, and best fit value Γ_{fit} and estimated value Γ_{exp} of the mismatch parameter. For sample No. 3 this detailed analysis has not been performed.

No.	d_{Al} (nm)	d_{Au} (nm)	l_{Au} (nm)	$E_g[\Delta]$ (μeV)	Θ_c K	B_c mT	(L_N/ξ_S) fit	(L_N/ξ_S) exp	Γ fit	Γ exp
1	300	20	45	160	1.21	6.35	0.16	0.1	0.6	0.54
2	400	30	25	140	1.21	5.05	0.8	0.4	0.2	0.4
3	400	40	35	125	1.05	5.02				
4	150	-	-	180	1.21	10.2	-	-	-	-

polyimide layer is partially dry etched to form a free-standing nanobridge over 2 μm length. For comparison we also produced several breakjunctions made of an continuous Al film of thickness 150 nm to 200 nm.

The bridge is broken at the constriction at very low temperatures $\Theta < 1$ K and under cryogenic vacuum conditions by controlled bending of the elastic substrate mounted on a three-point bending mechanism. Details of the sample preparation and measuring setup are given in [9,22,23]. As found in previous experiments [20] the conductance G decreases in steps of the order of G_0, with smaller steps within a plateau (see Ref. [9,15]). The last contact before the wire breaks (the breaking is indicated by the onset of exponential decrease of G with distance of the electrodes) is usually below 1 G_0. We attribute this observation to the disorder in the contact region induced during the ion etching process.

From experiments on pure Al samples [9] we estimate the superconducting gap $\Delta = 180\ \mu eV$, the diffusion constant $D = v_F l_{Al}/3 = 0.042\ m^2/s$ and the coherence length $\xi = (\hbar D/2\Delta)^{1/2} = 280$ nm. From the residual resistance ratio RRR $= R(300K)/R(4K)$ of the Au layers we deduce an elastic mean free path of $l_{Au} \approx 25 - 45$ nm.

For recording current-voltage characteristics (IVs) the motion of the bending mechanism can be stopped at any conductance value G. Fig. 3 shows the IVs of two Al-Au-Al samples and an Al sample obtained in the tunnel regime at a temperature of 50 mK.

A clear minigap $E_g < \Delta$ is observed for both samples and a maximum of the current slightly above the minigap. This maximum is a consequence of the deviation of the density of states (DOS) $\rho(E)$ from the BCS shape. Both, the spatial dependence of $\rho(E, x)$ (x is the transport direction) and the size of the minigap can be described using a one-dimensional diffusive model where the sample geometry is approximated by two normal metal-superconductor (NS) structures weakly coupled to each other via an opaque tunnel barrier ("hard wall") such that $\rho(E, x = L_N)$ at the tunnel barrier can be calculated neglecting the presence of the second NS structure. $\rho(E, x)$ is determined by solving the Usadel equation for the retarded Green's functions $G(E, x)$ and $F(E, x)$ numerically [17]. The DOS is given by $\rho(E, x) = \rho_0 \mathrm{Re} G(E, x)$ with the normal-state DOS ρ_0. The length of the superconductor is assumed to be infinite, the length of the normal metal is L_N, i.e. the bridge is broken symmetrically in the middle of the normal region. A second important parameter in the model is the "mismatch parameter" $\Gamma = \sigma_N/\sigma_S\ (D_S/D_N)^{1/2}$ with the conductivities $\sigma_{N(S)}$ and the electronic diffusion constants $D_{N(S)}$

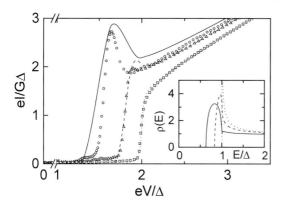

Fig. 3. Measured IVs (symbols) for the Al-Au-Al samples No. 1 (triangles), No. 2 (circles) and an Al sample (squares) when in tunnel regime. Also shown are the calculated IV of a BCS superconductor at $\Theta \leq 50$ mK (dotted line) and fittings to the IVs of samples No. 1 (dashed) and No. 2 (solid) according to Ref. [17]. Inset: DOS for a BCS superconductor (dotted) and samples No. 1 (dashed) and No. 2 (solid)

of the normal metal (superconductor). Both parameters can be estimated experimentally by the sample geometry and the RRR.

The tunnel IVs are calculated by autoconvoluting the DOS. The calculated $\rho(E, L_\mathrm{N})$ and IVs using those parameters are displayed in Fig. 3.

When the two electrodes are brought back into contact, the IVs show the subgap-structure due to MAR [12,13]. Fig. 4 displays several IVs (symbols) of single-atom contacts of samples No. 1 (left) and No. 2 (right). G of all examples are smaller than or comparable to the conductance quantum $G_0 = 2e^2/h$. For small G the maximum in the IV at $2E_\mathrm{g}$ is still observable and strong features appear at the sub-multiples $2E_\mathrm{g}/me$ ($m = 2, 3, ...$).

In order to describe the IVs quantitatively we generalize the theory of MAR by introducing the Andreev-reflection amplitude $A(E, L_\mathrm{N})$ calculated for our model geometry via

$$A(E, L_\mathrm{N}) = -i \frac{F(E, L_\mathrm{N})}{1 + G(E, L_\mathrm{N})}$$

into the scattering formalism of Ref. [12] and calculate the IVs for arbitrary transmission. This procedure implies that the PE inside one bank of the point contact is not altered when bringing the two banks into contact.

The lines in Fig. 4 are calculated within this model for a single channel with the transmission given in the figure caption. Note, that the calculation contains no free parameter since the transmission is given by the slope of the IV far above the gap.

Contrary to the observations for Al and other superconductors [9,15], we find for Au stable configurations in the contact regime whose transport properties can be explained by one single conduction channel. The transmission of this channel can be varied within a wide range of values ($0.15 < T < 0.98$) by rearranging the geometry of the central region (opening and closing again the contact).

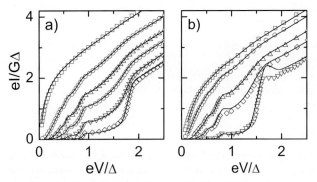

Fig. 4. a) [b)]: measured IVs (symbols) of six [five] contacts of sample No. 1 [No. 2] and best numerical fits for PE superconductors (lines) with the same parameters as in Fig. 3. The transmissions are from top to bottom: $T = 0.98, 0.85, 0.69, 0.55, 0.35,$ and 0.15 [$T = 0.93, 0.86, 0.68, 0.50,$ and 0.11]

Several authors have predicted within different approaches [8,20] that single-atom contacts of monovalent metals as, e.g. noble metals, should transmit electrons through one single channel only. The mechanism which gives rise to a single-channel transport in Ref. [8] is the formation of a resonance of the $6s$ band at the Fermi energy. The transmission of this channel is a function of the contact geometry. Transmission close to one is predicted only if the central atom is coupled equally to both electrodes, i.e. symmetric geometrical configuration. Disorder within several atomic layers from the central atom reduces the value of the transmission coefficient.

When further closing the contact to higher conductances a decomposition analysis as for the Al sample taking into account more channels has to be performed. When the closing is stopped at conductances around $2\ G_0$, stable reproducible configurations can be achieved, which allow for repeated breaking and closing of the contact with the same conduction properties. As an example we plot in Fig. 5 a representative subset of conductance vs. distance curves of an Al-Au-Al sample, measured below 100 mK during two days when changing the configurations. A magnetic field of $B = 10$ mT has been applied to produce linear IVs. The data shown as solid lines have been measured when continuously opening at a speed of 10-20 pm/s. For determining the channel ensemble (see lower panel of Fig. 5) two opening and closing cycles have been performed with stopping the motion at different positions and recording IV characteristics. The resulting T_n are plotted in the lower panel of Fig. 5 and their sum is additionally plotted as symbols in the upper panel.

Starting at $G \approx 2\ G_0$ from a three channel configuration with one well transmitted channel $T_1 \approx 0.8$ and two almost degenerate ones $T_2 \approx T_3 \approx 0.5$, the conductance jumps down to a short and tilted plateau at around $1.2\ G_0$, before a long last plateau with a reproducible substructure evolves. The sawtooth-like part of the plateau can be decomposed into the contributions of one widely open channel (which reveals the sawtooth behavior) and a smaller second one. The last part of the plateau has a conductance around $0.8\ G_0$ and can be described by a single conductance channel. After breaking, the contact is closed by an almost exponential increase of the conductance to about $G \approx 0.07\ G_0$ and a sudden jump to $G \approx 0.7\ G_0$, still with a single channel. After closing the contact further by about 0.1 nm, the contact jumps to one of two two-channel configurations (with different transmissions than the two-channel situation when opening). These configurations appear to be metastable since the contact remains in those states for different length intervals, which differ from repetition to repetition. The analysis of the channel ensemble reveals that one of the channels has the same transmission $T_1 \approx 0.6$ for both configurations while the second jumps between two very different values (0.3 and 0.8). A possible interpretation of this behavior would be one atom in a stable position and a second one alternating between two almost degenerate positions.

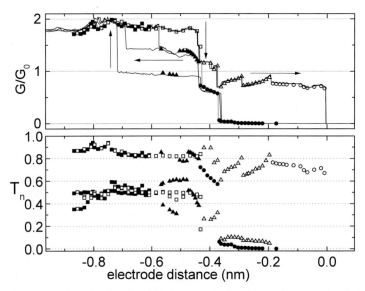

Fig. 5. Upper panel: Conductance as a function of electrode distance of Al-Au-Al sample No. 3 (with $E_g/e = 125\mu V$) when repeatedly opened and closed. The distance axis has been scaled by the exponential dependence of G in the tunnel regime. Due to hysteresis in the mechanical setup, an varying offset of about 0.1 nm has been subtracted such that the origin of the distance axis is for all repetitions at the position where the contact breaks. Lines: continuous motion at $v \approx 10 - 20$ pm/s and dc-conductance measurement with applied magnetic field of 10 mT for suppressing the superconductivity, symbols: sums of the transmission values when motion was stopped for recording IVs and determining the channel ensemble: Open symbols: opening, closed symbols: closing; circles (triangles, squares): one (two,three) channels. Not all data are shown. Lower panel: Transmission ensemble for opening (open symbols) and closing (closed symbols)

Finally the contact arrives at its initial stable plateau with three channels. The transmission values as well as the substructure of the plateau are reproducible within the accuracy of the conductance measurement in the continuous measurements and the determination of the channels ($\approx 3\%$). Although not all details of the substructure are observed in all of the ≈ 20 repetitions', the substructure of this stable plateau can be used to distinguish the extrinsic hysteresis of the mechanical setup and the intrinsic hysteresis of the atomic motion. From the distance values we have subtracted varying offsets of the order of 0.1 nm such that the contacts break at zero. The mechanical hysteresis has been determined such that the curvatures of the stable three-channel plateaus when opening and closing the contact overlay. From the agreement between subsequent conductance curves we estimate the relative accuracy of the distance axis to be about 2 %, the absolute scaling, however, is much less accurate: 20%. The remaining hysteresis between open-

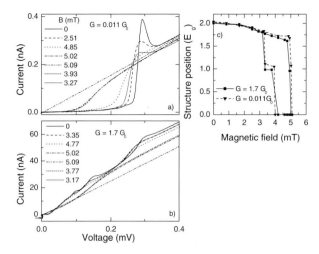

Fig. 6. a) and b) IV characteristics of two different configurations of sample No. 2 at 50 mK with applied perpendicular magnetic field. The field size has been varied according to the values given in the legend from top to bottom. The IVs of panel a) [b)] have been recorded for in tunnel [contact] regime with $G = 0.01\,G_0$ [$G = 1.7\,G_0$]. Panel c) shows the position of the maximum of dI/dV in units of the minigap $E_g = 160\,\mu eV$ at zero field as a function of the applied field. For a linear IV the voltage of the maximum was set to zero

ing and closing of ≈ 0.4 nm is of the order of the lattice spacing of gold and is thus an intrinsic effect of the atomic contact [24]. The reproducible situation was kept until the contact was closed thoroughly. These observations support the assumption that the conductance of the whole system is determined by the central atom and the coupling to its nearest neighbors.

Figs. 6 a) and b) show the evolution of the subgap structure with applied magnetic field for two different contacts of sample No. 2, one in the tunnelling regime, the other in the contact regime. In our experiment the field is applied perpendicular to the film plane. As the field size is increased, the excess current is suppressed, the current steps are slightly rounded and the peak positions are gradually shifted to lower voltages. For fields larger than 5.02 mT no submultiple current steps are observable and the apparent gap size measured by the position of the maximum of the derivative of the IV is reduced to almost half of its value at zero magnetic field.

This position is plotted in Fig. 6 c) as a function of the applied field. When a field of 5.09 mT or more is reached, the IV becomes linear with a slope corresponding to the sum of the transmissions determined in zero field. When lowering the field down to ≈ 4.0 mT, the IV remains linear. When further reducing the field, the IV suddenly switches to the intermediate behavior with half gap width as observed when raising the field. For fields smaller than ≈ 3.2 mT the observation of multiple Andreev reflections is recovered.

When reversing the field direction the same IV is observed for the same absolute value of the field, which proves that there is no residual field along the field axis. Effects of the earth magnetic field or spurious fields in different directions however cannot be excluded. A similar behavior is observed for contacts of sample No. 1 but with slightly higher absolute values of the critical fields. E.g. the IV becomes linear for fields above 6.35 mT, switches back to the intermediate state at 5.3 mT and to the fully superconducting state at 4.4 mT.

For explaining these observations we summarize briefly the behavior of a proximity superconductor in an external magnetic field (see e.g. [25] and references therein). A characteristic feature of proximity effect is the appearance of a so-called breakdown field instead of a critical field for superconductors at which the external field starts to penetrate the sample due to the reduced screening capability of the proximity system. Below a certain temperature $T^\star$, determined by the dimension and the effective penetration depth λ_N of the normal layer, the transition to the normal state becomes of 1st order, like for Type I superconductors. I.e. a "supercooling" and a "superheating" of the proximity superconducting state becomes possible between two limiting field scales $H_{sh} \approx \Phi_0/\lambda_N^2$ (superheating field) and $H_{sc} \approx \Phi_0/\xi(L_N)^2$, the supercooling field, where $\xi(L_N)$ describes an effective coherence length. The exact values of H_{sc} and H_{sh} depend on the degree of disorder of the electronic transport in the normal conductor, the NS interface and the sample dimensions.

Important for the discussion here is the fact that H_{sc} and H_{sh} depend differently on the strength of the PE in both electrodes. The important physical quantity is the magnetic penetrating the two electrodes. For simplicity, we model the difference in the PE at the point where the junction breaks by different dimensions of the normal metal. Since we do not control on the atomic level where the junction breaks it is likely that the bridge does not break exactly in the middle of the normal part, but that two parts with slightly different lengths L_{N1} and $L_{N2} = 2L_N - L_{N1}$ are formed. For $L_{N1} \neq L_{N2}$ the characteristic field scales will be different for both bridge arms. A quantitative description of the influence of the magnetic field is difficult because of the complicated shape of the samples. However, we can assume that the point contact spectra are mainly determined by the superconducting properties at the constriction where the voltage drops, i.e. depend on the length scales L_{N1} and L_{N2}.

With these approximations we interpret our observations as follows (see Fig. 7): With increasing magnetic field the structures in the IV are slightly rounded and shifted to lower voltages since the screening is reduced due to the finite elastic mean free path. However this effect is much smaller than observed for pure Al samples [23]. When the superheating field H_{sh2} of the longer bridge arm with length L_{N2} is reached, this part becomes normal, giving rise to an IV characteristic of a NS-junction, i.e. with a current onset

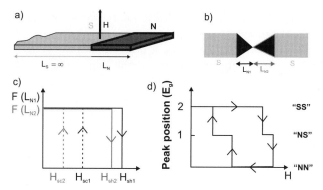

Fig. 7. Panel a) displays the model geometry of one bridge arm used for calculating the superconducting properties at the end of the normal metal where it makes contact with the other bridge arm. Panel b) represents schematically the sample geometry when the bridge is not broken exactly in the middle of the normal region. In panel c) we show schematically the development of the pair amplitudes at the constriction for both bridge arms as a function of the applied field and panel d) the resulting effective behavior of the samples in the different field ranges

at $eV \approx E_g$ and without MAR features. When further increasing the field, the superconductivity is completely suppressed when H_{sh1} of the shorter bridge arm is exceeded, i.e. the IV becomes linear. When reducing the field, both bridge arms remain normal conducting until the supercooling field H_{sc1} of the shorter bridge arm is reached. Below this field this bridge arm becomes superconducting again, giving rise to an NS-like IV. Only below H_{sc2} of the longer bridge arm the field is expelled from both parts and the IV with MAR structures is recovered. At zero external field the decomposition of the IV into the contribution of individual channels yields the same channel ensembles as before applying the field.

As already mentioned, a quantitative description of this behavior would require a more realistic modelling of the sample geometry. A rough estimate shows that a length difference of the bridge arms in the order of 10% to 20% of L_N would account for the observed difference in the field scales. For determining the density of states and Andreev reflection amplitudes in zero field (used for calculating the IVs in the contact regime) we assumed a symmetrically broken bridge. However, a length difference of the order of 20% would only slightly affect those results and has therefore been neglected.

To summarize, we have presented an investigation of the transport properties of gold tunnel and few-atom contacts having superconducting leads. The current-voltage characteristics in the tunnel regime strongly differ from what is observed for a pure BCS superconductor. From their analysis within the framework of the diffusive proximity effect, we extracted the quasiparticle density of states at the contact region and the corresponding Andreev reflection amplitude. This last ingredient was then used to extend the theory of

MAR to the case of a channel with arbitrary transmission between PE superconductors. Although this extended model does not account quantitatively for all the details of the IVs of few-atom contacts, it does explain the general trends. An important conclusion of this analysis is that the smallest gold contacts observable in the experiments accommodate a single conduction channel in accordance with theoretical predictions for single-atom contacts. The PE strongly influences the behavior in a perpendicular magnetic field where a hysteretic transition to the normal state is observed. We have demonstrated here, that it is possible to drive a particular contact reproducibly into the normal state and back to the superconducting state without changing $\{T_n\}$. We stress the high stability of the setup necessary for maintaining a particular contact stable during the measurement series.

We thank C. Bruder, M.H. Devoret, D. Esteve, J. M. van Ruitenbeek, G. Schön and C. Strunk for helpful discussions. We have enjoyed fruitful interaction with D. Averin, J. C. Cuevas and A. Levy Yeyati, and we thank them for providing us with their respective computer codes. This work was partially supported by the BNM (France) and the DFG (Germany).

References

1. G. Binnig et al., Appl. Phys. Lett. **40**, 178 (1982).
2. M.F. Crommie et al., Science **262**, 218 (1993).
3. G. Binnig and H. Rohrer, IBM J. Res. Dev. **30**, 355 (1986).
4. N.D. Lang, Phys. Rev. B **52**, 5335 (1995); C.C. Wan et al., Appl. Phys. Lett. **71**, 419 (1997); A. Yazdani et al., Science **272**, 1921 (1996).
5. R. Landauer, Philos. M. **21**, 863 (1970).
6. B. Ludoph and J.M. van Ruitenbeek, Phys. Rev. B **59**, 12290 (1999); H.E. van den Brom and J.M. van Ruitenbeek, Phys. Rev. Lett. **82**, 1526 (1999); B. Ludoph et al., ibid. **82**, 1530 (1999); M.F. Goffman et al., ibid. **85**, 170 (2000); R. Cron et al., ibid. **86**, 4104 (2001).
7. J.C. Cuevas et al., Phys. Rev. Lett. **80**, 1066 (1998).
8. A. Levy Yeyati et al., Phys. Rev. B **56**, 10369 (1997).
9. E. Scheer et al., Phys. Rev. Lett. **78**, 3535 (1997)
10. M. Octavio et al., Phys. Rev. B **27**, 6739 (1983).
11. G.B. Arnold, Journal of Low Temp. Phys. **68**, 1 (1987).
12. D. Averin and A. Bardas, Phys. Rev. Lett. **75**, 1831 (1995).
13. J.C. Cuevas et al., Phys. Rev. B **54**, 7366 (1996).
14. E.N. Bratus et al., Phys. Rev. B **55**, 12666 (1997).
15. E. Scheer et al., Nature **394**, 154 (1998).
16. D. Esteve in *Mesoscopic Electron Transport*, L.L. Sohn et al.(Eds.) (Kluwer, Dordrecht, 1997) pp. 375 and references therein.
17. W. Belzig et al., Superlattices and Microstructures **25**, 1251 (1999).
18. L. Olesen et al., Phys. Rev. Lett. **72**, 2251 (1994).
19. N. Agraït et al., Phys. Rev. B **47**, 12345 (1996).
20. J.M. van Ruitenbeek in *Mesoscopic Electron Transport*, L.L. Sohn et al. (Eds.) (Kluwer, Dordrecht, 1997) pp. 549 and references therein.
21. J.M. van Ruitenbeek et al., Rev. Sci. Inst. **67**, 108 (1996).

22. E. Scheer et al., Phys. Rev. Lett. **86**, 284 (2001).
23. E. Scheer et al., Physica B **280**, 425 (2000).
24. C. Untiedt et al., Phys. Rev. B **56**, 1251 (1997).
25. A.D. Zaikin, Solid State Commun. **41**, 533 (1982); W. Belzig et al., Phys. Rev. B **53**, 5727 (1996); A.L. Fauchère and G. Blatter, ibid. **56**, 14102 (1997); W. Belzig et al., ibid. **58**, 14531 (1998).

Metal-Insulator Transitions at Surfaces

Michael Potthoff

Lehrstuhl Festkörpertheorie, Institut für Physik,
Humboldt-Universität zu Berlin, Germany

Abstract. Various types of metal-insulator transitions are discussed to find conditions for which an ideal surface of a bulk insulator is metallic. It is argued that for the correlation-driven Mott metal-insulator transition the surface phase diagram should be expected to have the same topology as the phase diagram for magnetic order at surfaces: The corresponding linearized mean-field descriptions, a simplified dynamical mean-field theory of the Hubbard model and the Weiss mean-field theory for the Ising model, are found to be formally equivalent. A new kind of surface state appears in the low-energy part of the one-particle excitation spectrum as a precursor effect of the Mott transition.

The Mott metal-insulator transition at a crystal surface is a subject that touches different areas in solid-state theory which are usually treated as being disjoined: many-body theory of the correlation-driven metal-insulator transition, the general theory of surface phase transitions, and the theory of electronic surface states. It is the intention of the present paper to show that a corresponding combination of different concepts can be fruitful and allows for some new theoretical predictions.

1 Surface Phase Transitions

The large variety of novel and interesting phenomena in surface physics is closely related with the occurrence of surface phase transitions. As has been pointed out by Mills [1], the surface of a system may undergo a phase transition at a critical temperature $T_{c,s}$ being different from the bulk critical temperature T_c, i.e. the surface may undergo its own phase transition. Critical exponents, for example, can be defined and determined which are specific for the transition at the surface and which cannot be fully reduced to the bulk critical exponents [2,3]. Different kinds of surface phase transitions are conceivable and have been found, e.g. structural transitions, such as deviating geometrical order of the atoms near the surface of a single crystal (surface reconstruction), the loss of long-range crystalline order at the surface prior to a bulk melting transition (surface melting) or enrichment of one component at the surface of a solid binary alloy (surface segregation) [2]. Typical examples for surface phase transitions are also found among magnetically ordered systems: For example, the (0001) surface of ferromagnetic Gd is believed to have a Curie temperature which is higher than the bulk T_C [4].

Different types of surface phase transitions can be described in a qualitative but consistent way by means of classical Landau theory [2,3] or by lattice mean-field approaches which may be considered as coarse-grained realizations of the Landau theory. Especially, mean-field approaches to localized-spin models, such as the Ising or Heisenberg model, are frequently considered in this context [5]. For surface geometries there are a number of non-trivial results predicted by Landau or mean-field theory, such as temperature-dependent order-parameter profiles, which may give a surprisingly good description of experimental data (see Ref. [6], for example).

Typically, the surface undergoes the phase transition at the same temperature as in the bulk, $T_{c,s} = T_c$, if the local (structural, electronic, magnetic) environment remains unchanged, while $T_{c,s} > T_c$ if there is a perturbation Δ at the surface exceeding a certain critical value Δ_c. For $\Delta > \Delta_c$ and temperatures $T_c < T < T_{c,s}$ there is an ordered $D-1$-dimensional surface coexisting with a disordered D-dimensional bulk. More complicated phase diagrams are obtained in the case of multi-critical behavior, e.g. when the long-range order at the surface has a character different from the long-range order in the bulk (surface reconstruction, anti-ferromagnetic surface of a ferromagnetic bulk, etc). The Landau T–Δ phase diagram should be qualitatively correct whenever the $D-1$-dimensional system can support independent order [2,7].

2 Metal-Insulator Transitions

The concept of a surface phase transition and the corresponding Landau theory seems to extend straightforwardly to a certain kind of metal-insulator transitions, namely those which accompany an order-disorder thermodynamic phase transition (see Ref. [8] for an overview). The thermodynamic phase transition will be considered at the $T = 0$ quantum-critical point as due to ubiquitous thermal activation processes, a strict definition of a metal-insulator transition is possible at zero temperature only. It is well known that the formation of an ordered state may result in a gap for charge excitations as, for example, in the Peierls transition or in the transition to an anti-ferromagnetic state: Consider the typical example of a bipartite lattice and two-sublattice long-range order causing a doubling of the unit cell. For a non-degenerate band the splitting at the boundary of the new Brillouin zone will lead to a gap and, in the case of half-filling, to an insulating ground state.

Now, for sublattice order at the surface of a disordered bulk one would have the (naive) expectation that an insulating surface could coexist with a metallic bulk. This, however, is clearly impossible as a finite bulk density of states at the Fermi energy will always induce a non-zero, though possibly low density of states in the surface region. Likewise, the reverse scenario is impossible either: Namely, to realize a metallic surface phase of a bulk insulator caused by a thermodynamic phase transition, a disordered surface would have to coexist with an ordered bulk which, in general, is forbidden by strict arguments [2] (though a magnetic "dead-layer" scenario is found under somewhat exotic circumstances in a $D = 2$ q-state surface Potts model [9]).

Besides the thermodynamic transitions, there is a second important class of metal-insulator transitions, namely quantum-phase transitions [8]. Essentially these take place at $T = 0$ only and are not associated with any symmetry breaking. Important examples are the transition from a metal to a normal band insulator and the Mott-Hubbard transition from a metallic Fermi liquid to a Mott insulator. While the former can be understood within an independent-electron model, correlation effects are constitutive for the latter. Referring to a quantum-phase metal-insulator transition, it is very well feasible that the surface of a bulk insulator is metallic.

3 Surface States

Figure 1 shows a possible electron density of states for this situation. To have a metallic surface of an insulator, the density of states at the Fermi energy E_F must be finite at the surface while, with respect to the bulk states, E_F should lie within a band gap. Note that this necessarily implies the existence of a partially filled surface state: The appearance of a surface state at the Fermi energy is crucial to get a metallic surface of an insulator.

Two possible origins of electronic surface states are well known [10]: (i) Image-potential states may arise as Rydberg-like states in the long-ranged $-1/4z$ image potential which is due to the polarization charge that is induced by an electron approaching a *conductive* surface. The electron can be trapped between the image-potential surface barrier and the bulk barrier due to a bulk band gap. (ii) Crystal-induced surface states originate from the mere crystal termination. For an ideal unreconstructed crystal surface a further distinction between Tamm and Shockley states is meaningful: Shockley states appear within a hybridizational band gap which may open when the boundaries of two bulk bands have crossed as a function of decreasing lattice constant. Tamm-like surface states are due to the surface change of the one-electron potential and always lie near the bulk band from which they originate.

In fact, a crystal-induced surface state may give rise to a metallic surface phase as is demonstrated by the following two examples: Due the reconstruction of the Si(111)-(7 × 7) surface there is a partially occupied surface state which is consistent with the observed metallic behavior for this surface (cf. the discussion in Ref. [10]). A surface insulator-to-metal transition has been predicted for the ferromagnetic insulator EuO [11]: For temperatures below $T_C = 69$ K and decreasing, the majority $5d$ conduction band shifts towards the occupied $4f$-↑ bands thereby reducing the insulating gap. This so-called red shift is transmitted to an unoccupied Tamm state which is predicted to split off at the (100) surface from the lower edge of the conduction band. The energy difference between the surface state and the majority bulk band edge together with the T-dependent red shift is just sufficient to bridge the gap.

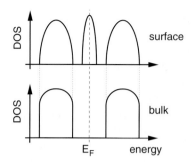

Fig. 1. Bulk and the surface density of states (schematic) in case of a metallic surface and an insulating bulk. $E_{\rm F}$: Fermi energy

Therefore, the surface state should become populated for low T which would imply a transition to a (half-)metallic state at the surface.

In any case the concept of a crystal-induced surface state is based on a model of effectively independent electrons. It is the detailed form of the one-electron potential that determines the energy position (lateral dispersion) of a surface state in the band structure. Hence, it is obvious that a surface state is not normally pinned to $E_{\rm F}$ (chemical potential); one may conclude that for a surface of a band insulator there is no *a priori* reason for a surface state to be partially filled.

4 Mott Transition

The question is whether or not a situation is conceivable where the appearance of the surface state is necessarily connected with the metal-insulator transition, i.e. where the surface state necessarily appears at $E_{\rm F}$. It is now clear that to this end one has to look for an electron-correlation effect.

Consider electrons in a narrow non-degenerate band interacting with each other via an on-site Coulomb repulsion U as described by the Hubbard model [12]. Any symmetry-broken phases will be excluded from the discussion. At half-filling $n = 1$ and for strong U, the system is then a paramagnetic insulator. The **k**-integrated one-electron excitation spectrum (DOS) of this so-called Mott-Hubbard insulator [8,13] has the same form as shown in the lower part of Fig. 1 where now the two peaks have to be interpreted as the lower and the upper Hubbard band separated by an energy of the order of U.

On the other hand, for $U = 0$ and for the weakly interacting case, the system is a normal Fermi liquid. At an intermediate interaction strength $U_{\rm c}$ of the order of the width W of the non-interacting DOS one thus expects a metal-insulator transition. This Mott transition is a prime example for a quantum phase transition at $T = 0$ which results from the competition between the electrons' kinetic energy $\sim W$ which tends to delocalize the electrons and their potential energy $\sim U$ which tends to localize them.

The Mott transition is a true many-body effect that cannot be explained by a simple perturbational approach. Even within the framework of simplified model Hamiltonians, as the Hubbard model, an ultimate theory of the Mott transition is still missing [8]. A decisive step forward, however, has been made in the last decade with the development of dynamical mean-field theory (DMFT) and its application to the Mott transition (see Sec. 5, for a review see Ref. [14], recent results can be found in Ref. [15]).

Within the DMFT one finds that the transition at $U = U_c$ is characterized by a diverging effective mass $m^* \to \infty$ or a vanishing quasi-particle weight $z \propto (m^*)^{-1} \to 0$, respectively. For $U < U_c$ but close to the critical point, the DOS has a three-peak structure, consisting of the two well-developed Hubbard bands as well as a narrow quasi-particle resonance at the Fermi energy with weight z – the DOS has the same form as the "surface DOS" shown by the upper part of Fig. 1.

It is conceivable that Fig. 1 describes a situation where the bulk of the system is a Mott-Hubbard insulator while the surface is in a metallic Fermi-liquid state. The quasi-particle resonance would then be a surface state (one-electron surface excitation) with a layer-dependent weight z_α decreasing exponentially with increasing distance from the surface. This surface state necessarily appears at the Fermi energy as it corresponds to low-energy excitations well known from quantum impurity systems (Kondo effect). The question is for which circumstances this coexistence of the Mott-Hubbard insulator and the metallic Fermi liquid can be realized.

5 Mean-Field Approach

As it is by no means obvious how to construct a (continuum) Landau theory for this problem, the method of choice is to formulate and evaluate a mean-field theory for an appropriate discrete lattice model. While for a magnetic phase transition one can resort to effective spin models such as the Ising model without any detailed knowledge of the electronic structure, the Hubbard model as the minimum model to describe the Mott transition includes spin as well as charge degrees of freedom and is thus much more complicated. Likewise it is much more complicated to find a proper mean-field theory. For example, Hartree-Fock theory, weak- and strong-coupling perturbational approaches or decoupling approximations for the Hubbard model are clearly inferior compared with the Weiss mean-field theory for the Ising model. The latter is non-perturbative, thermodynamically consistent and free of unphysical singularities in the entire parameter space. Since the Weiss theory becomes exact in the non-trivial limit of infinite spatial dimensions D [16], this may serve as a simple and precise characterization of what is a proper mean-field theory in general. One may therefore hope that the same limit will lead to a powerful mean-field approach in the case of the Hubbard model, too.

That the $D = \infty$ limit for a lattice fermion model is well-defined and non-trivial in fact, has been proven in the pioneering work of Metzner and Vollhardt [17]. Crucial is a proper scaling of the hopping $t \propto 1/\sqrt{D}$ to keep the dynamic balance between kinetic and potential energy. To convert the abstract definition of a proper mean-field theory into a useful concept for practical calculations, it has been important to realize that the $D = \infty$ Hubbard model can be mapped onto the single-impurity Anderson model (SIAM) as now one can profit from various methods available for impurity problems. This observation has been made by Georges, Kotliar and Jarrell [18,19]. The mapping is a self-consistent one which means that the parameters of SIAM depend on the one-particle Green function of the Hubbard model.

To study surface effects one has to consider a variant of the original Hubbard model. Using standard notations the Hamiltonian reads:

$$H = \sum_{i_\| j_\| \alpha \beta \sigma} t_{i_\|\alpha, j_\|\beta} c^\dagger_{i_\|\alpha\sigma} c_{j_\|\beta\sigma} + \frac{U}{2} \sum_{i_\|\alpha\sigma} n_{i_\|\alpha\sigma} n_{i_\|\alpha-\sigma} . \qquad (1)$$

Here $i_\|$ labels the sites within a layer α parallel to the surface. $\alpha = 1$ corresponds to the top surface layer. The model (1) differs from the original Hubbard model by the mere existence of the surface: The sole effect of the surface is to terminate the bulk. For any numerical calculation one has to assume a finite number of layers, $\alpha = 1, ..., d$, i.e. a film geometry, and to check the convergence of the results for large $d \to \infty$.

The generalization of DMFT for surface geometries has been developed by Potthoff and Nolting [20]. Again, the limit $D = \infty$ may serve as a guide to construct a powerful mean-field theory. Assuming uniform hopping parameters, $t_{i_\|\alpha, j_\|\beta} = t$ between nearest neighbors $(i_\|, \alpha)$ and $(j_\|, \beta)$, and using the same scaling $t \propto 1/\sqrt{D}$, the model itself as well as surface effects remain non-trivial for $D \to \infty$. As the different layers parallel to the surface must be treated as being inequivalent, the mapping procedure, however, becomes more complicated (see Fig. 2, left). The original many-body problem for a semi-infinite lattice with layers $\alpha = 1, 2, ..., d$ (with $d \to \infty$) is self-consistently mapped onto a *set* of impurity problems labeled by the same index $\alpha = 1, 2, ..., d$. Each SIAM can be treated independently to calculate the impurity self-energy $\Sigma_\alpha(\omega)$. There is, however, an indirect coupling which is mediated by the self-consistency cycle: Via the Dyson equation of the lattice model, the on-site Green function $G_\alpha(\omega)$ for a given layer α depends on *all* layer-dependent self-energies. The free Green function $\mathcal{G}_\alpha(\omega)$ of the αth SIAM which determines its one-particle parameters is then obtained from the DMFT self-consistency condition: $\mathcal{G}_\alpha(\omega)^{-1} = G_\alpha(\omega)^{-1} + \Sigma_\alpha(\omega)$.

Fig. 2 (right) shows the quasi-particle weight $z_\alpha = (1 - \Sigma'_\alpha(0))^{-1}$ as obtained from the DMFT using a standard ("exact diagonalization") method [21] to treat the different impurity problems. The critical interaction $U_c(d)$ of the $d = 17$ layer sc(100) film lies close to the bulk critical interaction $U_{c,\text{bulk}} \approx 16$ ($W = 12$ is the width of the free bulk DOS). In the metallic

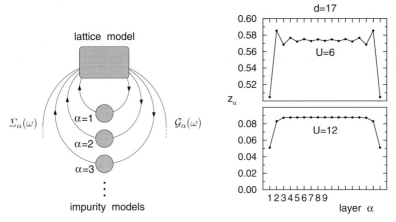

Fig. 2. *Left:* DMFT self-consistency cycle for the Hubbard model on a semi-infinite lattice with layer index $\alpha = 1, ..., d$ ($d \to \infty$). See text for discussion. *Right:* Layer-dependent quasi-particle weight z_α for a simple-cubic film of $d = 17$ layers with (100) surfaces. DMFT results for uniform nearest-neighbor hopping $t = 1$, half-filling $n = 1$, $T = 0$ and U as indicated

phase for $U < U_c(d)$ there is a quasi-particle resonance in the interacting local density of states for each layer with a finite weight z_α. As can be seen in the figure, the weight has a pronounced layer dependence. While for small U the profile has an oscillating character, it becomes monotonous for interaction strengths close to the transition. This is a typical result which is observed for films with different surface geometries and indicates a universal behavior of the critical profile for $U \to U_c(d)$. For both, $U = 6$ and $U = 12$, the surface quasi-particle weight z_1 is considerably lower than the bulk quasi-particle weight z at the film center. This result is plausible: Due to the reduced coordination number at the surface $q_1 < q$, the variance $\Delta_1 = q_1 t^2$ of the surface-layer DOS is reduced which implies the "effective" interaction $U/\sqrt{\Delta_1}$ to be stronger at the surface compared with the bulk. In this respect the surface is "closer" to the insulating phase. Yet, for $U \to U_c(d)$ all z_α vanish simultaneously and there is no surface transition.

6 Critical Regime

For systematic investigations of films with different (large) thicknesses, with different surface geometries and for different model parameters, a numerically exact evaluation of the DMFT requires an effort which is out of scale. Fortunately, a simplified treatment of the mean-field equations is possible at $T = 0$ for parameters close to the critical point as has been pointed out by Bulla and Potthoff [22]. This "linearized DMFT" (L-DMFT) is based on two plausible assumptions for the critical regime $U \to U_c$: (i) The effect of the two

Hubbard bands on the quasi-particle resonance can be disregarded and the resonance basically reproduces itself in the DMFT self-consistency cycle. (ii) The resonance has no internal structure and can be described by a one-pole approximation. Although these assumptions are approximate, the L-DMFT has successfully passed a number of tests which have been performed by comparing with the full theory and which show that the L-DMFT is well qualified to give quantitative estimates for critical interactions and critical profiles as well as the correct topology of phase diagrams [22,20,23].

Within the L-DMFT the mean-field equations reduce to algebraic equations for z_α which involve the electronic model parameters and the system geometry. For example, in the case of a surface geometry with $q_\parallel$ nearest neighbors within a layer, $q_\perp$ nearest neighbors within each of the adjacent layers and with uniform hopping t and interaction U except for the hopping $t_{11} \neq t$ within the surface layer $\alpha = 1$, the mean-field equations read:

$$z_1 = \frac{36}{U^2}(q_\parallel t_{11}^2 z_1 + q_\perp t^2 z_2),$$

$$z_\alpha = \frac{36 t^2}{U^2}(q_\parallel z_\alpha + q_\perp z_{\alpha+1} + q_\perp z_{\alpha-1}) \qquad (2)$$

for $\alpha = 2, 3, ..., \infty$. Equations of this type can be solved analytically or by simple numerical means.

Figure 3 shows the phase diagram for the Mott transition at two low-index surfaces of a semi-infinite $D = 3$ simple cubic lattice as obtained within the

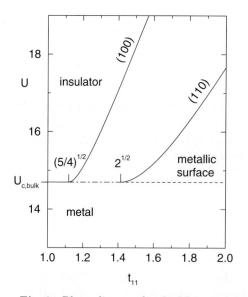

Fig. 3. Phase diagram for the Mott transition at the (100) and (110) surface of the sc lattice. t_{11} ($\geq t = 1$): modified hopping within the surface layer $\alpha = 1$. $U_{c,\text{bulk}}$: bulk critical interaction

L-DMFT. For the unperturbed surface ($t_{11} = t$) there is a single metal-insulator transition at the bulk critical interaction ("ordinary transition"). Within the L-DMFT $U_{c,\text{bulk}} = 6t\sqrt{q_\parallel + 2q_\perp} = 6\sqrt{6}$. An enhancement of the surface hopping $t_{11} > t$ which exceeds a certain critical value, leads to two critical interactions: At $U = U_{c,\text{bulk}}$ there is the transition of the bulk of the system irrespective of the state of the surface ("extraordinary transition"). The surface undergoes its own phase transition to the insulating state at a second critical interaction $U_{c,\text{surf}} > U_{c,\text{bulk}}$ ("surface transition"). The critical perturbation $t_{11,c}$ depends on the surface geometry. Multi-critical behavior is found for $t_{11} = t_{11,c}$ and $U = U_{c,\text{surf}} = U_{c,\text{bulk}}$ ("special transition").

For $U_{c,\text{bulk}} < U < U_{c,\text{surf}}$ a metallic surface is coexisting with an insulating bulk. As is demonstrated by Fig. 1 this implies the existence of a surface state. Fig. 4 shows that even for the unperturbed surface ($t_{11} = t$) in the metallic phase close to the ordinary transition there is a surface state and in the

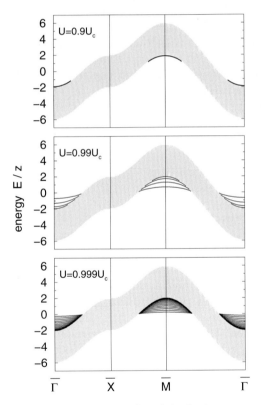

Fig. 4. Surface states (lines) in the low-energy part of the one-electron excitation spectrum and bulk continuum (grey region) in the $D = 2$ Brillouin zone for the unperturbed (100) surface ($t_{11} = t$) and different $U \to U_{c,\text{bulk}}$. Note that the energy scale is normalized with respect to the bulk quasi-particle weight z with $z \to 0$ for $U \to U_{c,\text{bulk}}$

limit $U \to U_{c,\text{bulk}}$ even an infinite number of surface states. These are split off from the bulk continuum of coherent low-energy excitations the width of which vanishes proportional to $z \propto U_{c,\text{bulk}} - U$ for $U \to U_{c,\text{bulk}}$. The critical profile of the quasi-particle weight in this limit, $z_\alpha \propto \alpha$, causes a strong surface perturbation of the low-energy electronic structure which drives the surface states. These surface states should be considered as a new kind of *correlation-induced* surface states as for the corresponding non-interacting model (with $t_{11} = t$) the occurrence of surface states is impossible.

Concluding, one finds a phase diagram with the same topology as predicted by the Landau theory of surface phase transitions. This analogy can even be made quantitative: Consider Weiss mean-field theory for ferromagnetic order in the Ising model on a semi-infinite lattice. For $T \to T_C$, the mean-field equation can be linearized, and one has $m_\alpha = (J/2T)(q_\parallel m_\alpha + q_\perp m_{\alpha+1} + q_\perp m_{\alpha-1})$ which by comparing with Eq. (2) immediately yields the following correspondences: $m_\alpha \Leftrightarrow z_\alpha$, $J/2 \Leftrightarrow 36t^2$ and $T \Leftrightarrow U^2$ where J is the coupling constant and m_α the layer-dependent magnetization. It is obvious that this analogy has a number of implications for the Mott transition.

For example, for a thermodynamic phase transition (analogously for the $T = 0$ Mott transition) there are two critical exponents that merely involve the critical temperature (interaction strength), the "shift exponent" λ_s and the "crossover exponent" ϕ [2,3]. They describe the trend of T_c (U_c) for films with thickness $d \mapsto \infty$ and the trend of $T_{c,\text{surf}}$ ($U_{c,\text{surf}}$) for the semi-infinite system near the special transition, respectively. Within the Laudau theory (linearized DMFT) one finds $\lambda_s = 2$ and $\phi = 1/2$.

7 Conclusions

Among different types of metal-insulator transitions at a surface of a single crystal, the correlation-driven Mott transition from a normal metal to a paramagnetic insulator is distinguished as it offers a comparatively simple route to a metallic surface phase of an insulating bulk. Formally, this is expressed by the equivalence between the respective linearized mean-field approaches to the Mott quantum-phase transition and to the thermodynamic (magnetic) phase transition. It should be stressed that the equivalence implies that all results of the Landau theory of surface phase transitions have a unique counterpart for the surface Mott transition. This includes phase diagrams, critical profiles of the quasi-particle weight, critical exponents and other critical behavior. In this way a comprehensive and consistent mean-field picture of the characteristics of the surface Mott transition is obtained.

Acknowledgements

The author would like to thank R. Bulla and W. Nolting for discussions and collaborations. The work has been supported by the Deutsche Forschungsgemeinschaft within the Sonderforschungsbereich 290.

References

1. D. L. Mills, Phys. Rev. B **3**, 3887 (1971).
2. K. Binder, In: *Phase Transitions and Critical Phenomena, Vol. 8*, ed. by C. Domb and J. L. Lebowitz (Academic, London, 1983).
3. H. W. Diehl, In: *Phase Transitions and Critical Phenomena, Vol. 10*, ed. by C. Domb and J. L. Lebowitz (Academic, London, 1986).
4. P. A. Dowben, M. Donath and W. Nolting, editors, *Magnetism and electronic correlations in local-moment systems: Rare-earth elements and compounds* (World Scientific, Singapore, 1998).
5. T. Kaneyoshi, I. Tamura, and E. F. Sarmento, Phys. Rev. B **28**, 6491 (1983); K. Binder and D. P. Landau, Phys. Rev. Lett. **52**, 318 (1984); C. Tsallis and E. F. Sarmento, J. Phys. C **18**, 2777 (1985); F. Aguilera-Granja and J. L. Morán-López, Phys. Rev. B **31**, 7146 (1985); P. J. Jensen, H. Dreyssé, and K. H. Bennemann, Surf. Sci. **269/270**, 627 (1992).
6. R. Pfandzelter and M. Potthoff, Phys. Rev. B **64**, 140405 (2001).
7. N. D. Mermin and H. Wagner, Phys. Rev. Lett. **17**, 1133 (1966).
8. F. Gebhard, *The Mott Metal-Insulator Transition* (Springer, Berlin, 1997).
9. R. Lipowsky, Z. Phys. B **45**, 229 (1982); J. Phys. A **15**, L195 (1982).
10. S. G. Davison and M. Streślicka, *Basic Theory of Surface States* (Clarendon, Oxford, 1992).
11. R. Schiller and W. Nolting, Phys. Rev. Lett. **86**, 3847 (2001).
12. J. Hubbard, Proc. R. Soc. London A **276**, 238 (1963); M. C. Gutzwiller, Phys. Rev. Lett. **10**, 159 (1963); J. Kanamori, Prog. Theor. Phys. (Kyoto) **30**, 275 (1963).
13. N. F. Mott, *Metal-Insulator Transitions* (Taylor and Francis, London, 1990).
14. A. Georges, G. Kotliar, W. Krauth, and M. J. Rozenberg, Rev. Mod. Phys. **68**, 13 (1996).
15. R. Bulla, T. A. Costi, and D. Vollhardt, Phys. Rev. B **64**, 045103 (2001); J. Joo and V. Oudovenko, Phys. Rev. B **64**, 193102 (2001).
16. F. Englert, Phys. Rev. **129**, 567 (1963).
17. W. Metzner and D. Vollhardt, Phys. Rev. Lett. **62**, 324 (1989).
18. A. Georges and G. Kotliar, Phys. Rev. B **45**, 6479 (1992).
19. M. Jarrell, Phys. Rev. Lett. **69**, 168 (1992).
20. M. Potthoff and W. Nolting, Phys. Rev. B **59**, 2549 (1999); Euro. Phys. J. B **8**, 555 (1999); Phys. Rev. B **60**, 7834 (1999).
21. M. Caffarel and W. Krauth, Phys. Rev. Lett. **72**, 1545 (1994).
22. R. Bulla and M. Potthoff, Euro. Phys. J. B **13**, 257 (2000).
23. Y. Ōno, R. Bulla, and A. C. Hewson, Euro. Phys. J. B **19**, 375 (2001); Y. Ōno, R. Bulla, A. C. Hewson, and M. Potthoff, Euro. Phys. J. B **22**, 283 (2001).

The Role of Contacts in Molecular Electronics[*]

Gianaurelio Cuniberti[1], Frank Großmann[2], and Rafael Gutiérrez[2]

[1] Max Planck Institute for the Physics of Complex Systems
01187 Dresden, Germany
cunibert@mpipks-dresden.mpg.de
[2] Institute for Theoretical Physics, Technical University of Dresden
01062 Dresden, Germany

Abstract. Molecular electronic devices are the upmost destiny of the miniaturization trend of electronic components. Although not yet reproducible on large scale, molecular devices are since recently subject of intense studies both experimentally and theoretically, which agree in pointing out the extreme sensitivity of such devices on the nature and quality of the contacts. This chapter intends to provide a general theoretical framework for modeling electronic transport at the molecular scale by describing the implementation of a hybrid method based on Green function theory and density functional algorithms. In order to show the presence of contact-dependent features in the molecular conductance, we discuss three archetypal molecular devices, which are intended to focus on the importance of the different sub-parts of a molecular two-terminal setup.

1 Introduction

The incessant development of single molecule techniques is forcing a paradigm shift in the many neighboring branches of nano-sciences. This process does not exclude the modeling and design of electronic devices. Novel fabrication methods that create metallic contacts to a small number of conjugated organic molecules allow the study of the basic transport mechanism of these systems and will provide direction for the potential development of molecular-scale electronic systems [1]. The concept is now realized for individual components, but the economic fabrication of complete circuits at the molecular level remains challenging because of the difficulty of connecting molecules to one another. A possible solution to this problem is 'mono-molecular' electronics, in which a single molecule will integrate the elementary functions and interconnections required for computation [2]. Indeed, the primary problems facing the molecular electronics designer are measuring and predicting electron transport. That is due to the fact that molecular electronics is strongly dependent on the quality and nature of the contacts [3]. Ideally, these contacts should be ohmic so that any non-linearity in the conductivity of the wire can be correctly attributed and studied. They must also be low in resistance

---

[*] This chapter is based on two invited talks (from GC and FG) given at the DPG Spring Meeting (March 2002; Regensburg, Germany)

to ensure that the properties measured are those of the molecule and not those of the molecule-contact interface. Moreover, the medium surrounding and supporting the molecule must be several orders of magnitude more insulating than the molecule itself because the contact area of the support with the electrical contacts is often much greater than that between the electrical contacts and the molecule [4].

Nevertheless, the contact problem can be turned into a challenge. Even with the intrinsic barrier that the contacts represent, barriers can be strategically used to favor the design of specific devices [5]. However, this requires a more detail account of the atomic structure of the interface. Green function and density functional theories [6] are the typical instruments to characterize transport through single molecules clamped between two metallic contacts. These very same instruments may even be adopted for calculating electromechanical switch behaviors [7] and current-induced forces [8] in molecular structures.

In this chapter, after a brief overview on charge transport on the molecular scale (Sect. 2), we provide, in Sect. 3, a general theoretical hybrid method based on Green function theory and density functional theory (DFT)-based algorithms. In order to show the presence of contact-dependent features in the molecular conductance, we introduce, in Sect. 4, three model molecular devices. The first is a sodium wire (Par. 4.1), where the role of contacts for a molecular bridge emerges clearly. However, the quality of contacts is not the only source of alteration of the molecular conductance. In the Par. 4.2, we show the peculiar effect that carbon nanotube leads might have on a contacted molecule. Finally, in the Par. 4.3, a pure carbon device, consisting of two carbon nanotube leads grasping a C_{60} molecule is studied in a parameter free DFT calculation.

2 Charge Transport on the Molecular Scale

In mesoscopic electron transport, many interesting interference related and quantization effects have been found in the past 20 years [9]. Much of the fundamental theory for mesoscopic systems can be taken over to the description of molecular scale conductance calculations. In both realms, a formulation that includes interference effects due to phase coherence as well as geometrical effects is needed. It was originally developed by Landauer [10] for a two-terminal geometry as displayed in Fig. 1. and further extended by Büttiker [11] to the multi-terminal case. The essential idea of the Landauer formulation is to relate the conductance to an elastic scattering problem and, ultimately, to transmission probabilities. The simplest way to derive this relationship is to consider a 'molecular' region connected to two ballistic leads, which are connected to electronic reservoirs at the chemical potentials μ_L, μ_R, see Fig. 1. It is assumed that electrons entering the reservoirs do completely lose their phase coherence. As stated in Ref. [12] assuming semi-

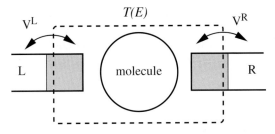

scattering region = 'extended molecule'

Fig. 1. Schematic representation of a two-terminal device. The scattering region (enclosed in the dashed-line frame) with transmission probability $T(E)$ is connected to semi-infinite left (L) and right (R) leads which end into electronic reservoirs (not shown) at chemical potentials μ_L, μ_R. By applying a small potential difference electronic transport will occur. The scattering region or molecule may include in general parts of the leads (shaded areas). This is necessary for the description of real systems, where the surface topology may be modified as a result of relaxation or reconstruction processes. This may introduce additional scattering due, e.g., to surface states

infinite leads is enough to warrant that no electron exiting the scattering region will reenter it with the same phase, so that an explicit modeling of the reservoirs is not necessary. In equilibrium $\mu_L = \mu_R$, but if an infinitesimal voltage $eV = \mu_L - \mu_R$ is applied a non-equilibrium situation is induced and a current will flow. The scattering region is characterized by the energy-dependent transmission coefficient $T(E)$. In the zero-temperature, linear response ($eV \to 0$) regime it is found [13] that the proportionality relation,

$$g = \frac{e^2}{\pi\hbar} T(E_F), \qquad T = \sum_{m,n=1}^{N_F} |t_{mn}|^2 \qquad (1)$$

holds. E_F is the Fermi energy of the whole system in equilibrium and the transmission amplitudes t_{mn} describe the scattering of one electron from channel n in the left lead to channel m in the right lead. They can be extracted, e.g., from the scattering matrix. The sums run over all N_F open channels at the Fermi level (whose number is assumed to be equal on both sides). Channels (transverse modes) appear due to the finite cross section of the leads which induces quantization of the electronic states perpendicular to the direction of current transport. For the special case of ideal transmission, i. e. $\sum_n |t_{mn}|^2 = 1 \ \forall m$, the conductance is simply proportional to N_F with the von Klitzing conductance quantum $g_K = e^2/(\pi\hbar)$ as the proportionality factor. Thus, g/g_K shows unit steps as a function of N_F. This is the well-known fact of conductance quantization [13], shown experimentally by van Wees et al. [14] and Wharam et al. [15].

In molecular conductance, in the case of strong coupling, it is the electronic structure of the molecule influenced by the leads that determines the

3 Method

For a general scattering region where inelastic effects are included, Meir and Wingreen [16] used non-equilibrium Green functions to derive an expression for the current which reduces to Eq. (1) in the elastic case. An advantage of their derivation is that an explicit connection to the Green function $\mathbf{G}$ of the *scattering region* dressed by the presence of the leads is established. The latter are introduced as self-energy corrections into the bare 'molecular' Green function $\mathbf{G}^{-1} = \mathbf{G}^{\mathrm{M}-1} + \mathbf{\Sigma}^{\mathrm{L}} + \mathbf{\Sigma}^{\mathrm{R}}$, The result for the transmission probability is then given by [16]

$$T(E) = 4\,\mathrm{Tr}\left\{\mathbf{\Delta}^{\mathrm{L}}(E)\mathbf{G}(E)\mathbf{\Delta}^{\mathrm{R}}(E)\mathbf{G}^{\dagger}(E)\right\}, \tag{2}$$

where $\mathbf{\Delta}^{\mathrm{L}}, \mathbf{\Delta}^{\mathrm{R}}$ are the spectral density describing the coupling of the scattering region to the $\alpha(=\mathrm{L,R})$-lead given by

$$\mathbf{\Delta}_{\alpha} = \frac{\mathrm{i}}{2}\left(\mathbf{\Sigma}_{\alpha}\left(E+\mathrm{i}0^{+}\right) - \mathbf{\Sigma}_{\alpha}^{\dagger}\left(E+\mathrm{i}0^{+}\right)\right),$$

and the trace is to be taken over states in the scattering region. The Green function $\mathbf{G}^{\mathrm{M}}$ is in general defined as the inverse operator $(E+\mathrm{i}0^{+} - \mathbf{H}^{\mathrm{M}})^{-1}$, for some suitable 'molecular' Hamiltonian $\mathbf{H}^{\mathrm{M}}$. Similar expressions have been derived by Fisher and Lee and Todorov, Briggs and Sutton [17]. In a seminal paper Szafer and Stone have derived the Landauer result, Eq. (1) from Kubo's linear response theory [18].

As mentioned above, only components of the Green function in the Hilbert subspace associated with the scattering region, which we will denote as 'the molecule' from now on to keep in mind that transport through molecular scale systems is the main issue to be addressed here, are needed. Notice, however, that the molecule may also include some atoms belonging to the leads, see Fig. 1. This will be the case when investigating *real* systems, where the surface atomic structure of the leads is explicitly taken into account (clean surfaces are usually energetically unstable, so that upon structural relaxation the surface topology may be modified and this will introduce additional scattering). From the full Green function of the open (infinite) system consisting of the leads plus the molecule it is possible to extract $\mathbf{G}$ using projector operator techniques [19]. In order to do so, one partitions the whole system into three components as shown in Fig. 1., where a left electrode, the molecule, and a right electrode are depicted. The full associated Hamiltonian matrix (in a suitable basis representation) can then be formally written as

$$\mathbf{H} = \begin{pmatrix} \mathbf{H}^{\mathrm{L}} & \mathbf{V}^{\mathrm{L,M}} & 0 \\ \mathbf{V}^{\mathrm{L,M}\dagger} & \mathbf{H}^{\mathrm{M}} & \mathbf{V}^{\mathrm{R,M}} \\ 0 & \mathbf{V}^{\mathrm{R,M}\dagger} & \mathbf{H}^{\mathrm{R}} \end{pmatrix}. \tag{3}$$

The matrices $\mathbf{V}^{\mathrm{L,M}}, \mathbf{V}^{\mathrm{R,M}}$ couple atoms belonging to the left(right) leads to the molecule, and it has been assumed that no direct lead-lead coupling exists. Notice that $\mathbf{H}^{\mathrm{L(R)}}$ are infinite dimensional sub-matrices. By means of a operator projecting onto the 'molecular' subspace, we can write the resulting M-dimensional matrix equation as:

$$\left(z\mathbf{S}^{\mathrm{M}} - \mathbf{H}^{\mathrm{M}} - \mathbf{\Sigma}^{\mathrm{L}}(z) - \mathbf{\Sigma}^{\mathrm{R}}(z)\right)\mathbf{G}(z) = \mathbf{1}, \quad z = E + \mathrm{i}0^{+}, \tag{4}$$

where $\mathbf{S}^{\mathrm{M}}$ is the overlap matrix for the general case of a non-orthogonal basis set. The energy-dependent self-energies $\mathbf{\Sigma}^{\mathrm{L}}, \mathbf{\Sigma}^{\mathrm{R}}$ include the coupling to the leads as well as information on the electronic structure of the leads. For the α-lead, they are given by

$$\mathbf{\Sigma}^{\alpha}(z) = \left(z\,\mathbf{S}^{\alpha,\mathrm{M}} - \mathbf{V}^{\alpha,\mathrm{M}}\right)^{\dagger} \mathcal{G}^{\alpha}(z)\left(z\,\mathbf{S}^{\alpha,\mathrm{M}} - \mathbf{V}^{\alpha,\mathrm{M}}\right). \tag{5}$$

The matrix $\mathbf{S}^{\alpha}$ is the overlap matrix element between molecule atoms and the α-lead atoms and $\mathcal{G}^{\alpha}(z)$ is the α-lead surface Green functions. Since the coupling matrices are in general short-ranged they will eliminate all contributions coming from atoms other than those closest to the molecule. Hence, only surface Green functions are usually needed.

We would like to stress that Eqs. (1) and (2) are only valid in the case that inelastic processes in the scattering region can be completely neglected. Otherwise no simple relationship between conductance and transmission can be obtained. A typical example where electron-electron interactions are decisive are quantum dots. There, the scattering region is weakly coupled to the leads so that the coupling-induced level broadening will be much smaller than the charging energy. Hence, electron interaction effects leading, e.g., to Coulomb blockade phenomena should be included in the description of quantum transport [20].

At this point we are led to the issue of characterizing the electronic structure of the molecule as well as of the leads. If we only focus on the essential physics, some kind of model Hamiltonians can be used [21]. However, if *real* situations are addressed where the knowledge of the detailed electronic structure is important, the use of more realistic computational schemes is unavoidable. From the point of view of electronic structure calculations three classes of approaches have been implemented for quantum transport calculations:

(i) Semiempirical or empirical tight-binding (TB) schemes, e.g. (extended) Hückel Hamiltonians, where the matrix elements are fitted to experiments or to first-principle calculations [22].

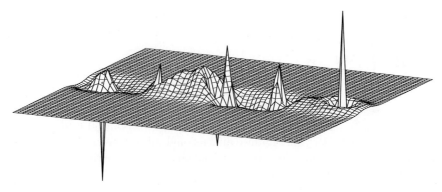

Fig. 2. A five sodium atom isolated wire. The HOMO density is plotted in arbitrary units

(ii) First-principles or *ab initio* approaches like Hartree-Fock and (DFT) [23].

(iii) Schemes which combine some elements of points (i) and (ii) in first-principles parametrized tight-binding Hamiltonians as it is the case for TB-DFT [7,24,25,26] methods.

Concerning the last class mentioned above, a computational scheme has been developed in Ref. [25] which combines a DFT-parametrized TB approach with the Landauer formalism to study the electronic transport properties of sodium atomic chains [24], small sodium clusters [25], carbon-based molecular junctions [7] as well as to simulate Scanning-Tunneling-Spectroscopy experiments on organic molecules [26]. The TB-DFT scheme relies on a representation of the electronic eigenstates of the system within a non-orthogonal localized basis set, usually taken as a valence basis. The many-body Hamiltonian is then approximately represented by a two-center tight-binding Hamiltonian. The matrix elements, however, are calculated numerically, avoiding the introduction of empirical parameters as in conventional TB approaches. We will now discuss some of the applications of this combined scheme.

4 Applications to Molecular Devices

4.1 Focusing on the Bridge Molecule: Sodium Wires

In this section, we review the numerical results for the resistance $\mathcal{R} = 1/g$ of sodium atomic wires as a function of the electrode-wire separation d and of the wire length [24]. The bond length in the wires was fixed at 6.00 a_B, which approximately corresponds to the equilibrium distance of a Na-dimer (d_eq=5.67 a_B). For wires with more than four atoms dimerization of the wire is expected due to a Peierls transition. Such effects will not be considered here.

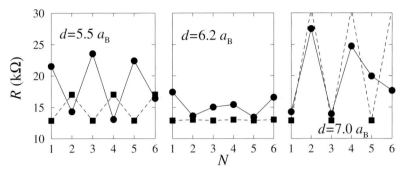

Fig. 3. Dependence of the resistance on the length of the atomic sodium wire for different electrode-wire separations. Dashed lines (connecting the squares) correspond to a resistance calculated with only the 3s-valence orbitals in the wire

In Fig. 3., the dependence of the resistance $\mathcal{R} = 1/g$ on the number of atoms in the chain is displayed for three different values of d. The result of Lang, who stated that $R_{N=1} > R_{N=2}$ [27], is only found, in our approach, in the case of strong coupling between the chain and the electrode. Concerning the coupling strength, there exists a critical value d_{crit} where both the single atom and the dimer have approximately the same resistance. This behavior can be understood by inspecting the transmission spectrum, as shown in Fig. 4. The value, the linear resistance of the wire acquires, depends sensitively on the position of the Fermi level E_F with respect to the modified eigenenergies of the wire. In order to distinguish between the bare eigenenergies we have displayed the free wire density of states (DOS) together with the corresponding $T(E)$ for two different values of the electrode-wire separation. Intuitively one would expect that E_F lies in the HOMO-LUMO gap for a Na-dimer (the HOMO is twice occupied) and would almost touch the singly occupied HOMO in the one atom case. This picture is, however, only exact in the case of a weak coupling to the electrodes, where the position of the eigenvalues of the wire remains approximately the same as for an isolated wire and the broadening induced by the coupling is smaller than the energy spacing between the eigenvalues.

For $d = 6.2 a_\mathrm{B}$, however, the eigenstates of Na_1 and Na_2 are strongly broadened and shifted by the coupling to the leads. The HOMO and LUMO (3-fold degenerate) of the single atom cannot be clearly resolved any more but evolve into a rather broad single peak. Especially at E_F the transmission for a single atom becomes smaller than for the dimer. With increasing distance the coupling to the electrodes is reduced and thus the renormalization and broadening of the eigenstates become weaker. At $d = 7.0 a_\mathrm{B}$, the HOMO and LUMO of the dimer are already 'resolved' and the transmission $T(E_\mathrm{F})$ within the gap is reduced.

In this section, we have introduced as possible bridge molecule a sodium wire. There the simplest assumption has been made for the lead self-energy

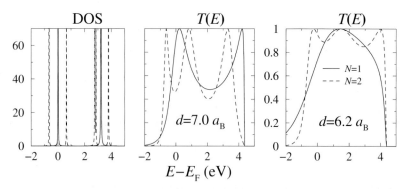

Fig. 4. The transmission coefficient as a function of energy for one and two atoms between the electrodes for two different electrode-wire separations. The left panel shows the DOS of the isolated wires. Only the low-energy part of the spectra is shown

entering in the calculation, a semi-infinite linear chain with a semi-elliptical spectral density Δ, obtained by Newns [28]. What are the effects which might arise from using nano-electrodes such as carbon nanotubes?

4.2 Focusing on the Leads: Carbon Nanotube Leads

Carbon nanotube (CNT) conductors have been in the focus of intense experimental and theoretical activity as another promising direction for building blocks of molecular–scale circuits [29,30]. Carbon nanotubes exhibit a wealth of properties depending on their diameter, on the orientation of graphene roll up, and on their topology, namely whether they consist of a single cylindrical surface (single–wall) or many surfaces (multi–wall) [31,32]. Carbon nanotubes have been recently used as wiring elements [30], as active devices [30,33], and, attached to scanning tunneling microscope (STM) tips, for enhancing their resolution [34]. With a similar arrangement the fine structure of a twinned DNA molecule has been observed [35]. However, CNT–STM images seem to strongly depend on the tip shape and nature of contact with the imaging substrate [36]. If carbon nanotubes are attached to other materials to build elements of molecular circuits, the characterization of contacts [37] becomes again a fundamental issue. This problem arises also when a carbon nanotube is attached to another molecular wire, a single molecule or a molecular cluster.

In this section we present analytic results for the transmission through a CNT-molecule-CNT system. In particular we analytically derive the spectral density of a single wall armchair carbon nanotubes, needed for calculating the transmission. A possible configuration is depicted in Fig. 5. Here a N atom molecule has been adopted as bridge molecule. However, the results we obtain are valid for any bridge *sub conditio* that the CNTs are contacting the molecular complex only via two *single* atomic contacts (labeled here as 1

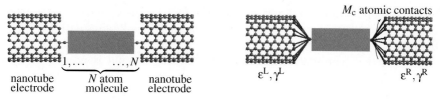

Fig. 5. Schematic representation of the N atom molecule–carbon nanotube hybrid with single (bottom) and multiple (top) contacts. on–site energies $\varepsilon^{\alpha=\mathrm{L,R}}$ are chosen to be zero

and N). For the system under investigation, where only the two atoms of the molecule are coupled to the leads, the formula for the transmission simplifies to [38,39]

$$T(E) = 4\,\Delta^{\mathrm{L}}(E)\Delta^{\mathrm{R}}(E)\left|G^{\mathrm{M}}_{1N}(E)\right|^{2}/\det(\mathbf{Q}), \tag{6}$$

where the spectral densities Δ^{L} and Δ^{R} are the only non-zero elements $(\boldsymbol{\Delta}^{\mathrm{L}})_{11}$ and $(\boldsymbol{\Delta}^{\mathrm{R}})_{NN}$, respectively, of the matrices $\boldsymbol{\Delta}$. The matrix element $\Delta^{\mathrm{L(R)}}$ is the spectral density of the left (right) lead. It is related to the semi-infinite lead Green function matrix $\mathcal{G}^{\mathrm{L(R)}}$. It is minus the imaginary part of the lead self-energies (per spin),

$$\Sigma^{\alpha} = \Lambda^{\alpha} - \mathrm{i}\,\Delta^{\alpha} = \sum_{m_{\alpha},m'_{\alpha}} \Gamma_{m_{\alpha}}\Gamma^{*}_{m'_{\alpha}}\mathcal{G}^{\alpha}_{m_{\alpha}m'_{\alpha}}, \tag{7}$$

with $\alpha = \mathrm{L, R}$. Owing to the causality of the self-energy, its real part Λ can be entirely derived from the knowledge of Δ via a Hilbert transform. Finally the determinant of $\mathbf{Q} = \mathbf{1} - \boldsymbol{\Sigma}\mathbf{G}^{\mathrm{M}}$ has to be calculated.

The rhs of Eq. (6) coincides with formulas used to describe electron transfer in molecular systems [21,40]. The above relationship between the Landauer scattering matrix formalism on the one side and transfer Hamiltonian approaches on the other side has been worked out in the recent past [41] showing, *de facto*, their equivalence. This enables us to make use of the formulas from a Bardeen-type picture in terms of spectral densities, which is often convenient for an understanding and analysis of the results obtained.

In calculating the spectral function, we make use of the assumption of identical left and right leads and drop the self-energy indices in Eq. (7). Since a π orbital representation was found to give good agreement with experiments (even quantitatively) [31], the Hamiltonian at hand can be assumed discrete. We can write the lattice Green function $G = (E + \mathrm{i}0^{+} - \mathbf{H})^{-1}$ in matrix form by rearranging the two dimensional n lattice coordinate with honeycomb underlying structure in the tight-binding Hamiltonian representation. The boundary conditions are imposed on two cuts parallel to a lattice bond so that the surface of a semi-infinite CNT contains 2ℓ atoms for a so-called (ℓ,ℓ) armchair CNT. We assume the x direction to be parallel to the tubes (and to the transport direction) and y to be the finite transverse coordinate (see Fig. 5.). The latter is curvilinear with n_y spanning the 2ℓ sites with periodic boundary conditions.

The lattice representation of the lead Green function is needed in the calculation of the self-energy contribution. It can generally be written by projecting the Green operator onto the localized state basis, $\psi_{k_x,k_y}(n_x = \text{border}, n_y) = \chi_{k_x}\phi_{k_y}(n_y)$, of the semi-infinite lead:

$$\mathcal{G}_{n_y n'_y}(E) = \langle n_y | (E + i0^+ - \mathbf{H})^{-1} | n'_y \rangle$$
$$= \sum_{k_x,k_y} \frac{\chi_{k_x}\phi_{k_y}(n_y)\chi^*_{k_x}\phi^*_{k_y}(n'_y)}{E + i0^+ - E_{k_x,k_y}}. \tag{8}$$

The eigenvalues of the tight-binding Hamiltonian

$$E_{\pm}\left(k_x^j, j\right) = \varepsilon \pm \gamma \sqrt{1 + 4\cos\left(\frac{j\pi}{\ell}\right)\cos\left(\frac{k_x^j a}{2}\right) + 4\cos^2\left(\frac{k_x^j a}{2}\right)}, \tag{9}$$

are obtained in a basis set given by symmetric ($+$) and antisymmetric ($-$) site configurations of the graphene bipartite lattice, corresponding to π and π^* orbitals, respectively [42]. The longitudinal momentum is restricted to the Brillouin zone, $-\pi < k_x^j a < \pi$, and the transverse wave number $1 \leq j \leq 2\ell$ labels 4ℓ bands, as many as the number of atoms in the unit cell of a (ℓ, ℓ) CNT. The two bands corresponding to $j = \ell$ are singly degenerate. They are responsible for the metallic character of armchair carbon nanotubes (these two bands cross at the Fermi level $E = \varepsilon$ for $k_x^\ell a = \pm 2\pi/3$). Also the two outermost bands corresponding to $j = 2\ell$ are singly degenerate, while the other remaining $(4\ell - 4)$ bands are collected in $(2\ell - 2)$ doubly-degenerate dispersion curves.

The single-particle Green function in a lattice representation for two sites belonging to the same sub-lattice can be written as

$$\mathcal{G}_{n_y n'_y}(E) = \frac{a}{2\pi\ell} \sum_{j,\beta} \int_{-\pi/a}^{\pi/a} dk_x^j \frac{\sin^2\left(k_x^j a\right) \varphi_j(n_y) \varphi_j^*(n'_y)}{E + i0^+ - E_\beta\left(k_x^j, j\right)},$$
$$= \frac{1}{2\ell} \sum_{j=1}^{2\ell} \varphi_j(n_y) \tilde{G}^j(E) \varphi_j^*(n'_y), \tag{10}$$

where $\varphi_j(n_y) = \exp(ik_y^j n_y a)$, with $k_y^j a = \pi j/\ell$, and $1 \leq j \leq 2\ell$. Note that in Eq. (10), n_y and n'_y should be either even or odd (that is they should belong to the same sub-lattice). The semi-infinite longitudinal Green function is given by

$$\tilde{G}^j(E) = \frac{a}{8\pi} \sum_{\beta=\pm} \int_{-\pi/a}^{\pi/a} dk_x^j \frac{\sin^2\left(k_x^j a/2\right)}{E + i0^+ - E_\beta\left(k_x^j, j\right)}.$$

The integral can be worked out analytically by extending k_x^j to the complex plane and adding cross-canceling paths (parallel to the imaginary axis) along the semi-infinite rectangle in the half plane $\mathrm{Im}\, k_x^j > 0$ and based on the interval between $-\pi/a$ and π/a. The closing path parallel to the real axis gives a real contribution linear in energy. This generalizes the approach by Ferreira et al. [43], recently adopted for obtaining an analytical expression for the diagonal Green function of infinite achiral tubes, to the case of semi-infinite CNTs. The determination of the poles inside the integration contour, given by

$$-2\cos\left(\frac{q_\beta^j a}{2}\right) = \cos\left(\frac{j\pi}{\ell}\right) + \beta\sqrt{\left(\frac{E-\varepsilon}{2\gamma}\right)^2 - \sin^2\left(\frac{j\pi}{\ell}\right)},$$

allows for the calculation of the residues and thus of the surface Green function. One finds

$$\tilde{G}^j(E) = \frac{1}{2\gamma}\frac{E-\varepsilon}{2\gamma}\left(1 + \mathrm{i}\frac{\sin\left(\frac{q_{\beta_*}^j a}{2}\right)}{\sqrt{\left(\frac{E-\varepsilon}{2\gamma}\right)^2 - \sin^2\left(\frac{j\pi}{\ell}\right)}}\right), \qquad (11)$$

where the choice of the contributing pole through the branch parameter $\beta_* = \mathrm{sign}(E-\varepsilon)$ has to be taken into account. The LDOS, obtained from the imaginary part of the surface Green function after Eq. (11) is plugged into Eq. (10), is shown in Fig. 6. It clearly differs from the LDOS of an infinite CNT as depicted for comparison in the right panel. As for the case of the SLT the pinning of the longitudinal wave function at the surface of the semi-infinite systems *cancels* all border zone anomalies when $q_\pm^j a$ matches multiples

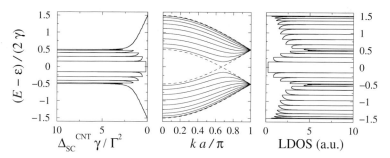

Fig. 6. Left panel: the normalized spectral density for a semi-infinite (ℓ,ℓ) CNT lead in the SC configuration; it corresponds to the LDOS at any atom site at the cut of the CNT lead. For comparison the dispersion relation and the LDOS of an infinite (ℓ,ℓ) CNT are shown in the middle and right panel respectively. Solid lines in the dispersion relation panel indicate doubly degenerate bands, dashed lines singly degenerate bands. Here $\ell = 10$, and on-site energies and hopping terms refer to $\alpha = \mathrm{L, R}$-leads

of 2π. In infinite SLTs these states are the *only* resonant states (van Hove singularities) so that the surface LDOS of a semi-infinite SLT never diverges. On the contrary, in CNTs there are states with zero group velocity outside the border zone which are responsible for the singularities of the spectral density of semi-infinite CNTs (left panel of Fig. 6.). The self-energy for a CNT lead is more complicate than the one for a SLT owing to the missing equivalence of the sites belonging to the two different sub-lattices. However, since the longitudinal part of the Green function, Eq. (11), is the same for all diagonal and off-diagonal terms of the surface Green function, the self-energy can still be cast into the form

$$\Sigma = \frac{1}{2\ell} \sum_{j=1}^{2\ell} \tilde{G}^j(E) \eta_{j/\ell}[\Gamma].$$

However, for the calculation of

$$\eta_{j/\ell}[\Gamma] = \left| \sum_{m=1}^{2\ell} \Gamma_m \varphi_j(m) \right|^2, \quad (12)$$

one has to specify the sub-lattice components of the transverse wave function and whether they belong to a bonding or anti-bonding molecular state. Again the distribution of the Γ_m contacts is needed in oder to calculate the weight η and thus the self-energy. Eq. (12) simplifies considerably in the SC case: $\eta = \Gamma^2$. Since η is uniform in j, the self-energy is simply proportional to the diagonal semi-infinite Green function and, as a consequence, the spectral density is proportional to the local density of states (Fig. 6.). The MC case ($\Gamma_m = \Gamma_{\text{eff}}/\sqrt{2\ell}$) is also easily tractable leading to a sum rule over the possible conducting channels. However, a direct proof is provided by the intuitive consideration that only the π-bonding state can contribute to the MC spectral density (all the other states have a non-constant spatial modulation provided, e.g., in Ref. [44]). Following our notation, the π-bonding state corresponds to $j = \ell$. The two different lead lattice structures carry the same physical information only in the MC limit case [38].

4.3 Focusing on the 'Molecule Plus Lead' Complex: A Pure Carbon Device

In this section we focus on the combination of CNT-leads with a realistic molecular cluster acting as the central molecule. Especially interesting is the case of a monovalent carbon cluster which makes the system an "all-carbon" device. Therefore, we studied a single C_{60} molecule bridging two single-wall metallic (5,5) carbon nanotubes (CNT). The CNT were taken symmetric with respect to the plane through the center of mass of C_{60} and perpendicular to the CNT cylinder axes (see left panel of Fig. 7.).

The central aim of [7] was to exploit the sensitivity of electron transport to the topology of the molecule/electrode interface in the proposed system.

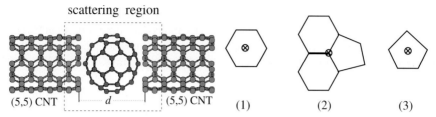

Fig. 7. Geometric configuration of the carbon molecular junction. A C_{60} molecule bridges two (5,5) CNTs. The right panel represents schematically the different orientations of C_{60} with respect to the surface cross-sections of the nanotubes (e.g. the left panel geometry corresponds to orientation (1)). The nanotube symmetry axis is depicted by a cross inside a circle

In this pure carbon system, charge transfer effects will be negligible. The Fermi level of the whole system will therefore lie within the HOMO-LUMO gap of the isolated C_{60}. Therefore, the electronic transport will be mainly mediated by the overlap of the tails of the molecular resonances within the HOMO-LUMO gap of C_{60}.

The key problem we addressed was how severely orientational effects do influence the electronic transport. To this end several possible orientations of the C_{60} (depicted by the polygon(s) facing the tube symmetry axis in the right panel of Fig. 7.) have been considered for a fixed distance between the molecule and the tubes. For the sake of comparison, structurally unrelaxed and relaxed molecular junctions were considered. The basic results are displayed in Fig. 8., for both relaxed and unrelaxed structures. Surprisingly, at fixed distance, just an atomic scale rotation of the highly symmetric C_{60} molecule induces a large variation of the transmission at the Fermi energy by several orders of magnitude. This can be seen in Fig. 8.(right panel) for three different orientations with maximum, minimum and one intermediate value of

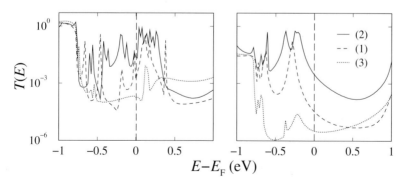

Fig. 8. Transmission results for both unrelaxed (left panel) and relaxed (right panel) configurations. The tube-tube distance d is fixed at 0.93 nm. Numbers indicate different molecular orientations as depicted in the right panel of Fig. 7

$T(E_\mathrm{F})$. As can be seen in Fig. 8.(left panel), neglecting relaxation decisively influences the transmission properties of the molecular junction. This shows up as a different and less smooth behavior of the transmission. The qualitative difference is related to the presence of dangling bond states on the CNT surfaces. Such states usually lie within a gap (a similar situation as that found, e.g., in semiconductor surfaces), in this case the HOMO-LUMO gap of the isolated molecule. They lead to the oscillatory behavior in the transmission for unrelaxed junctions. Upon relaxation these states are partly saturated or they rehybridize, moving away from the middle of the gap. However, some of them may still lie just above the HOMO or just below the LUMO of C_{60}, giving some contribution to the transmission within the gap.

The results for the relaxed structures reveal that, at the Fermi energy, the pentagon configuration (3) has a transmission lower by about *three orders of magnitude* than configuration (2). This fact could possibly be exploited in an electronic switching device on the nanoscale, as manipulation of fullerenes by using STM or atomic force microscope tips is becoming a standard technique in the field [45].

5 Discussion and Conclusions

Summing up our results, we can conclude that modeling transport at the molecular scale cannot avoid to go through an accurate structure calculation of the smallest element in the device, namely, the molecular bridge and part of the attached leads. *Ab initio* methods, although approximated on the DFT-LDA level description, provide thus a fundamental input to be integrated in standard quantum transport techniques. The hybrid quantum-transport–*ab-initio* method, reviewed in Sect. 3, provides the necessary playground for a precise description of linear transport in a two terminal device. The extention to transistor like configurations goes straightforwardly. On the other hand, the treatment of the non-equilibrium physics deserves special care, e.g. by using non-equilibrium Green functions and self-consistency arguments [46], which goes far beyond the limits of this short review.

Two main approximations have been tacitly assumed throughout this work: we have effectively employed (1) a single electron picture (2) in the limit of coherent transport. In doing so, we have been motivated by the fact that small molecules are typically well adsorbed to the leads providing a *strong coupling*. Here lies also the main difference with the 'artificial molecules' (obtained by confining the two dimensional electron gas at the interface of semiconducting heterostructures). In such mesoscopic devices, the electron states are strongly localized, thus the addition of any individual electron to the conducting island gives rise to the typical Coulomb staircase in the device *I-V* characteristics [20]. Finally, the C_{60} calculations (Par. 4.3) have shown that the molecule clamped between the two mesoscopic leads undergoes to strong structural modification. Neglecting this effect, *i.e.*, without properly relaxing the structure, leads to a substantial misestimation of the linear conductance. The modification of the flexible molecular structure under the effect of large bias voltages, and a direct computation of charge transfer problems are other important issues which were left out from the present work (for a recent account in this direction see [8]).

We can then conclude that different levels of investigation require different theoretical sophistication. A (semi)empirical description can be extremely useful in getting the flavor of the most qualitative effects, but can also be misleading. DFT codes may help in device parameter free calculations and have closer relevance to experiments. Besides that, it is evident that contacts can change dramatically the conductance profiles, and further progress in modeling is needed. It might indeed help to separate contact effects from 'molecular' effects. As far as leads are concerned, we have shown that low dimensional leads such as CNT do probe the conductance. Having a reliable model for the lead self-energy might be the key for cleaning (de-convoluting) spurious measurements.

Finally, we think that the addition of the richer physical environment that large molecules do experience might be easily scalable on top of the presented transport calculation scheme. This might be the case for molecular vibrations, and time dependent effects [47]. More difficult would be the extension to comprise electron-electron interactions [48,49] and non-equilibrium relaxation. Attempts to cope with this latter challenge have to include a self-consistent or combined treatment of electronic transport and structural optimization [8].

Acknowledgments

B. C. Möbius is gratefully acknowledged for fruitful discussions. We are indebted to M. Albrecht for providing us with Fig. 2. GC research at MPI is sponsored by the Schloeßmann Foundation. FG and RG gratefully acknowledge financial support by the DFG under FOR 335.

References

1. M. A. Reed, Proc. IEEE **87**, 652 (1999).
2. C. Joachim, J. K. Gimzewski, and A. Aviram, Nature **408**, 541 (2000).
3. G. Cuniberti, G. Fagas, and K. Richter, Acta Phys. Pol. **32**, 437 (2001).
4. K. W. Hipps, Science **294**, 536 (2001).
5. J. M. Seminario, C. E. De la Cruz, and P. A. Derosa, J. Am. Chem. Soc. **123**, 5616 (2001).
6. G. Cuniberti, E. De Micheli, and G. Viano, Commun. Math. Phys. **16**, 59 (2001); P. A. Derosa and J. M. Seminario, J. Phys. Chem. B **105**, 471 (2001).
7. G. Cuniberti et al., Physica E **12**, 749 (2002); R. Gutiérrez et al., Phys. Rev. B **65**, 113410 (2002).

8. M. Di Ventra, S. T. Pantelides, and N. D. Lang, Phys. Rev. Lett. **88**, 046801 (2002).
9. B. Kramer, Phys. Bl. **50**, 543 (1994).
10. R. Landauer, IBM J. Res. Develop. **1**, 223 (1957), reprinted in J. Math. Phys. **37**, 5259 (1996).
11. M. Büttiker, IBM J. Res. Develop. **32**, 317 (1988).
12. B. K. Nikolic, Phys. Rev. B **64**, 165303 (2001).
13. S. Datta, *Electronic Transport in Mesoscopic Systems* (Cambridge University Press, Cambridge, 1999).
14. B. J. van Wees *et al.*, Phys. Rev. Lett. **60**, 848 (1988).
15. D. A. Wharam *et al.*, J. Phys. C **21**, L209 (1988).
16. Y. Meir and N. S. Wingreen, Phys. Rev. Lett. **68**, 2512 (1992).
17. D. S. Fisher and P. A. Lee, Phys. Rev. B **23**, R6851 (1981); T. N. Todorov, G. A. D. Briggs, and A. P. Sutton, J. Phys.-Condens. Matter **5**, 2389 (1993).
18. A. D. Stone and A. Szafer, IBM J. Res. Develop. **32**, 384 (1988).
19. S. Priyadarshi, S. S. Skourtis, S. M. Risser, and D. N. Beratan, J. Chem. Phys. **104**, 9473 (1996).
20. *Single Charge Tunneling*, edited by H. Grabert and M. H. Devoret (Plenum Press, New York, 1992).
21. V. Mujica, M. Kemp, and M. A. Ratner, J. Chem. Phys. **101**, 6856 (1994).
22. C. M. Goringe, D. R. Bowler, and E. Hernández, Rep. Prog. Phys. **60**, 1447 (1997); S. Datta *et al.*, Phys. Rev. Lett. **79**, 2530 (1997); M. Magoga and C. Joachim, Phys. Rev. B **56**, 4722 (1997); E. G. Emberly and G. Kirczenow, Phys. Rev. B **58**, 10911 (1998); M. Paulsson and S. Stafström, J. Phys.-Condens. Matter **11**, 3555 (1999); L. Chico, L. X. Benedict, S. G. Louie, and M. L. Cohen, Phys. Rev. B **54**, 2600 (1996).
23. N. D. Lang and P. Avouris, Phys. Rev. Lett. **84**, 358 (2000); S. N. Yaliraki *et al.*, J. Chem. Phys. **111**, 6997 (1999); J. J. Palacios, A. J. Perez-Jimenez, E. Louis, and J. A. Verges, Phys. Rev. B **64**, 115411 (2001); H.-S. Sim, H.-W. Lee, and K. J. Chang, Phys. Rev. Lett. **87**, 096803 (2001); J. Taylor, H. Guo, and J. Wang, Phys. Rev. B **63**, 245407 (2001); P. S. Damle, A. W. Ghosh, and S. Datta, Phys. Rev. B **64**, 201403 (2001).
24. R. Gutiérrez, F. Großmann, and R. Schmidt, Acta Phys. Pol. **32**, 443 (2001).
25. R. Gutiérrez, F. Großmann, O. Knospe, and R. Schmidt, Phys. Rev. A **64**, 013202 (2001).
26. M. Toerker *et al.*, Submitted to Phys. Rev. B (2002).
27. N. D. Lang and P. Avouris, Phys. Rev. Lett. **81**, 3515 (1998).
28. D. M. Newns, Phys. Rev. **178**, 1123 (1969).
29. A. Karlsson *et al.*, Nature **409**, 150 (2001).
30. T. Rueckes *et al.*, Science **289**, 94 (2000); N. Yoneya, E. Watanabe, K. Tsukagoshi, and Y. Aoyagi, Appl. Phys. Lett. **79**, 1465 (2001).
31. R. Saito, G. Dresselhaus, and M. S. Dresselhaus, *Physical Properties of Carbon Nanotubes* (World Scientific Publishing Co. Pte. Ltd., London, 1998).
32. P. McEuen, Phys. World **13**, 31 (2000).
33. V. Derycke, R. Martel, J. Appenzeller, and P. Avouris, Nano Letters **1**, 453 (2001); H. W. C. Postma *et al.*, Science **293**, 76 (2001); R. Martel *et al.*, Appl. Phys. Lett. **73**, 2447 (1998).
34. H. Watanabe, C. Manabe, T. Shigematsu, and M. Shimizu, Appl. Phys. Lett. **78**, 2928 (2001). S. S. Wong *et al.*, Nature **394**, 52 (1998).

35. H. Nishijima *et al.*, Appl. Phys. Lett. **74**, 4061 (2000).
36. S. Akita, H. Nishijima, T. Kishida, and Y. Nakayama, Jpn. J. Appl. Phys. **39**, 7086 (2000); A. I. Onipko *et al.*, Phys. Rev. B **61**, 11118 (2000); A. L. Vázquez de Parga *et al.*, Phys. Rev. Lett. **80**, 357 (1998).
37. C. Thelander *et al.*, Appl. Phys. Lett. **79**, 2106 (2001); M. P. Anantram, S. Datta, and Y. Xue, Phys. Rev. B **61**, 14219 (2000); P. J. de Pablo *et al.*, Appl. Phys. Lett. **74**, 323 (1999).
38. G. Cuniberti, G. Fagas, and K. Richter, To appear in Chem. Phys. (2002).
39. J. Yi, G. Cuniberti, and M. Porto, Submitted to Phys. Rev. A (2002).
40. V. Mujica, M. Kemp, and M. A. Ratner, J. Chem. Phys. **101**, 6849 (1994).
41. A. Nitzan, Ann. Rev. Phys. Chem. **52**, 681 (2001); L. E. Hall, J. R. Reimers, N. S. Hush, and K. Silverbrook, J. Chem. Phys. **112**, 1510 (2000).
42. R. Saito, M. Fujita, G. Dresselhaus, and M. S. Dresselhaus, Phys. Rev. B **46**, 1804 (1992); P. R. Wallace, Phys. Rev. **71**, 622 (1947).
43. M. S. Ferreira, T. G. Dargam, R. B. Muniz, and A. Latgé, Phys. Rev. B **63**, 245111 (2001).
44. H. J. Choi and J. Ihm, Solid State Commun. **111**, 385 (1999).
45. C. Zeng, H. Wang, B. Wang, and J. G. Hou, Appl. Phys. Lett. **77**, 3595 (2000).
46. Y. Xue, S. Datta, and M. A. Ratner, To appear in Chem. Phys. (2002). E. H. Hauge and J. A. Støvneng, Rev. Mod. Phys. **61**, 917 (1989).
47. G. Cuniberti, A. Fechner, M. Sassetti, and B. Kramer, Europhys. Lett. **48**, 66 (1999).
48. H. J. Schulz, G. Cuniberti, and P. Pieri, in *Field theories for low-dimensional condensed matter systems*, edited by G. Morandi *et al.* (Springer, Berlin, 2000).
49. G. Cuniberti, M. Sassetti, and B. Kramer, Europhys. Lett. **37**, 421 (1997).

Electronic Transport, Spectral Fine Structures, and Atom Clusters in Quasicrystals and Approximants

H. Solbrig and C. V. Landauro

Technische Universität Chemnitz, 09107 Chemnitz, Germany

Abstract. *Ab-initio* results (ASA-LMTO) are presented for electronic transport properties of i-AlCuFe (1/1) (Cockayne model). We extract significant spectral information in terms of a model spectral resistivity. The final scaling of this model to the quasicrystal is guided by the experiment (thermopower versus temperature).

1 Introduction

In phase diagrams, quasicrystals (QCs) occupy small regions surrounded by other crystalline and amorphous phases with related high-temperature properties [1,2,3]. The most striking peculiarities of QCs appear at very low temperatures where the thermal energy window, $k_B T$, is narrow enough to resolve electronic spectral features that are due to extended quasiperiodicity. At high temperatures, however, the relationship of QCs and approximants results from equivalent conditions on the length scale of the atom clusters. There is experimental evidence for significant spectral structures around the Fermi energy, down to a few 10 meV [4]. Janot and de Boissieu [5] propose to treat QCs as arrays of clusters. Such arrays form virtual bound electron states [6] and give rise to 100-meV features in the spectral resistivities of the approximants α-AlMn(Si) and i-AlCuFe (1/1) [7,8]. Since experimental data are available from energy scales down to a few meV there are hopeful attempts to model spectral properties of QCs. Maciá [9] has proposed a state density model that considers spiky features in addition to a wide parabolic pseudogap with square-root wings. We have contributed [10,11] with a different concept that starts from the *ab-initio* calculated spectral conductivity of an approximant. Spectral features close to the Fermi energy that depend on the co-operation of typical atom clusters are supposed to persist in the QC. Such essential spectral information is extracted in terms of a Lorentzian resistivity model of the approximant. Rescaling to the quasicrystal is achieved upon fitting fine-structure sensitive model parameters to the experimental thermopower curve of the QC. Recently, Maciá [12] has proved that QCs with a related but two-dip spectral conductivity are suitable thermoelectric materials at the temperature of liquid nitrogen.

The present work treats two systems, i-AlCuFe and i-AlCuRu. The purposes of the following Sections: Section 2 comments on basic formulas. In Section 3 we characterize the electronic transport in i-AlCuFe (1/1). Section 4 presents Lorentzian resistivity models, and conclusions are drawn in Section 5.

2 Transport Parameters from Spectral Information

We calculate temperature-dependent transport parameters from spectral conductivities, $\hat{\sigma}_{xx}(\epsilon), \hat{\sigma}_{xy}(\epsilon) \ldots$, within the Chester-Thellung-Kubo-Greenwood (CTKG) formalism [13,14,15]. The $\hat{\sigma}_{\alpha\beta}(\epsilon)$ are either provided by *ab-initio* calculations or suitably set up as model conductivities. In the CTKG framework, the kinetic coefficients with $i, j = 1$ or 2,

$$\mathcal{L}^{ij}_{\alpha\beta}(T) = (-1)^{i+j} \int d\epsilon \, \hat{\sigma}_{\alpha\beta}(\epsilon) \, (\epsilon - \mu)^{i+j-2} \left(-\frac{\partial f^0(\epsilon, \mu, T)}{\partial \epsilon}\right), \quad (1)$$

play a central role. $f^0(\epsilon, \mu, T)$ is the Fermi-Dirac distribution function, and the chemical potential will be used in the low-temperature representation [16],

$$\mu(T) \approx \epsilon_F - T^2 k_B^2 \frac{\pi^2}{6} \left(\frac{d \, ln(\hat{n}(\epsilon))}{d\epsilon}\right)_{\epsilon=\epsilon_F} \equiv \epsilon_F - \xi T^2, \quad (2)$$

where ϵ_F, k_B, and $\hat{n}(\epsilon)$ denote the Fermi energy, the Boltzmann constant, and the electronic density of states (DOS). The kinetic coefficients are thus low-order moments of the spectral conductivities taken from the thermal energy window centered at $\mu(T)$. Certain applications below deal with only one transport direction. In such cases the subscripts α, β will be omitted. Within this scheme we obtain the electronic conductivity,

$$\sigma(T) = \mathcal{L}^{11}(T) \approx \hat{\sigma}(\epsilon_F) + \frac{\pi^2}{6} (\hat{n}(\epsilon_F) k_B T)^2 \left[\left(\frac{d}{\hat{n} \, d\epsilon}\right)^2 \hat{\sigma}(\epsilon)\right]_{\epsilon=\epsilon_F} \quad (3)$$

the thermopower with its Mott-like approximation,

$$S(T) = \frac{1}{|e|T} \frac{\mathcal{L}^{12}(T)}{\sigma(T)} \approx -\frac{\pi^2 k_B^2 T}{3|e|} \left(\frac{d \, ln(\hat{\sigma}(\epsilon))}{d\epsilon}\right)_{\epsilon=\mu}, \quad (4)$$

the electronic thermal conductivity,

$$K(T) = \frac{1}{|e|^2 T} \mathcal{L}^{22}(T) - T\sigma(T) \, S(T)^2, \quad (5)$$

and the Lorenz number,

$$L(T) = \frac{K(T)}{T \, \sigma(T)}, \quad (6)$$

where the Wiedemann-Franz limit, $L_0 = (3/2)(k_B/|e|)^2$, indicates a nearly-free electron transport regime. $S(T)$ and $\sigma(T)$ are completed by low-temperature Sommerfeld approximations. The Einstein relation on the energy surface, $\widehat{\sigma}(\epsilon) = (2\,e^2/V)\,\widehat{n}(\epsilon)\widehat{D}(\epsilon)$, guides to the electronic spectral diffusivity, $\widehat{D}(\epsilon)$. For a spectral representation of the normal Hall coefficient we start from the Bloch-Boltzmann result (cf. Ziman [17]) and eliminate all **k**-dependent quantities in favor of the spectral diffusivities, $\widehat{D}_{\alpha\beta}(\epsilon)$, and energy derivatives of $\widehat{\sigma}_{\alpha\beta}(\epsilon)$. Summarizing we arrive at

$$R_H(T) = \frac{1}{\sigma_{xx}(T)\sigma_{zz}(T)} \int d\epsilon\, \widehat{\sigma}_{x,yz}(\epsilon) \left(-\frac{\partial f^0(\epsilon,\mu,T)}{\partial \epsilon}\right), \qquad (7)$$

$$\widehat{\sigma}_{x,yz}(\epsilon) = |e|\left(\widehat{D}_{xz}(\epsilon)\frac{d}{d\epsilon}\widehat{\sigma}_{xz}(\epsilon) - \widehat{D}_{xx}(\epsilon)\frac{d}{d\epsilon}\widehat{\sigma}_{zz}(\epsilon)\right) \approx -Q\frac{d\widehat{\sigma}(\epsilon)}{d\epsilon}. \qquad (8)$$

Note that off-shell information is included due to the energy derivatives. In the following, we will use the approximate version in (8) that avoids off-diagonal transport quantities. This applies to systems such as i-AlCuFe where experiments [18,1] reveal correlated signs of $R_H(T)$ and $S(T)$ (cf. (4)). Q is a positive constant that accounts for a weak energy dependence of $\widehat{D}_{xx}(\epsilon)$ in (8). Several i-AlCuFe phases close to the quasicrystalline stoichiometry (12-13 at.% Fe) have almost the same electronic diffusivities [19].

In the *ab-initio* calculations we rely on the atomic-sphere approximation to the linear muffin-tin orbital method (ASA-LMTO, Andersen [20]), combined with 1/48 special **k**-sets [21] (i.e. up to 120 **k** points). This provides the energy eigenvalues, ϵ_i, the partial-wave decomposed LMTO band states, and the scattering phase-shifts of the effective atoms, all that to be used for the following purposes: (i) We calculate state densities on replacing the discrete levels by Gaussians (half width $\geq$10 meV). (ii) We calculate the velocity matrix, $\langle i|v_\alpha|j\rangle$, for application in the Kubo-Greenwood formula (KGF),

$$\widehat{\sigma}_{\alpha\beta}(\epsilon,\epsilon+\hbar\omega) = \frac{h|e|^2}{V}\sum_{ij}\langle i|v_\alpha|j\rangle\langle j|v_\beta|i\rangle\delta(\epsilon-\epsilon_i)\delta(\epsilon+\hbar\omega-\epsilon_j), \qquad (9)$$

where ω is the frequency of an external field (cf. Hobbs [22], Arnold [23]). The limit $\omega \to 0$ of (9) provides spectral conductivities, $\widehat{\sigma}_{\alpha\beta}(\epsilon)$. Optical conductivities, $\sigma_1(\omega)$, are obtained on taking an energy integral with the weight $(f^0(\epsilon,\mu,T) - f^0(\epsilon+\hbar\omega,\mu,T))/\hbar\omega$. Lorentzian level broadening (variable half widths Γ) simulates thermal transitions. (iii) We calculate the Landauer conductance of an infinite slab of approximant cells by means of a muffin-tin scattered-wave implementation of the Landauer method (MT-SW LM [24,25]). The resistance scaling with the thickness L reveals a trend towards a non-metallic transport regime. (iv) We calculate the atomic-sphere DOS at the central atom of a cluster where a part of the cluster atoms are omitted. This is accomplished by means of a muffin-tin scattered-wave cluster-recursion method (MT-SW RM [26]). Thus, the chemical origin of certain spectral features can be analyzed.

3 Spectral Fine Structure with Icosahedral Clusters

Our preference for the hypothetical Cockayne model (CM) of i-AlCuFe (1/1) [27] has two reasons. (i) It is strictly defined. Other models [28] are not fully characterized. (ii) It is intrinsically disposed to provide the well known high resistivities of the approximant/QC [29]. The CM has a cubical cell (a=12.3 Å, 80 Al + 32 Cu + 16 Fe) where four interpenetrating Bergman clusters bear each 3 Fe atoms (the b-Fe) in the inner icosahedral shell. The remaining 4 Fe atoms (the g-Fe) belong to the glue. They link the triangles of b-Fe to form an iron network without direct Fe-Fe contacts. It turns out that spectral fine structures on the 100 meV scale result from the co-operation of the icosahedral clusters, and we suppose that this co-operation survives in the QC.

3.1 Iron Network Generates 100 meV Pseudogap

Figure 1 shows the valence band of i-AlCuFe (1/1) (ASA-LMTO/KGF: 14 components, 120 special **k**-points, 20 meV half width). The most remarkable DOS feature in Fig. 1 is the 100 meV pseudogap (PSG), 260 meV above

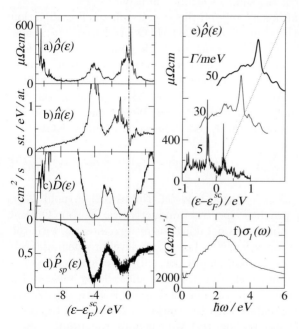

Fig. 1. i-AlCuFe (1/1), ASA-LMTO (20 meV half width, cf. text). Spectral curves of a) resistivity, b) state density, c) diffusivity, d) total sp-weight of the LMTO band states (total spd-weight=1), e) simulated thermal decay of resistivity fine structure, f) optical conductivity

the self-consistent Fermi level, ϵ_F^{sc}, just at the lower wall of a wide ($\sim$0.5 eV) Hume-Rothery pseudogap. Note that defects of structure/decoration can shift the actual Fermi energy away from ϵ_F^{sc} (experiments [1], LMTO calculations [8]). Apart from small shifts the PSG proves stable against several modifications (replacing the 1/48 **k** set, reducing to 3 components, adding empty spheres), but it appears less pronounced in former papers [31,32].

Narrow spectral features can arise from electronic resonances in cluster arrays [6]. For the CM and α-AlMn(Si), we have shown [8,7] that the fine structures close to ϵ_F^{sc} depend critically on the configurations of the networks of active atoms, Fe respectively Mn. In particular one-component decorations of the CM structure make no difference to common amorphous phases (cf. also Krajčí [33] for a model d-Al).

Figure 2a shows that part of the total DOS at the central b-Fe atom in a 16000-atom Cockayne cluster which is due to scattering paths strictly confined to the Fe network (2000 b/g-Fe cluster, MT-SW RM, 5 meV steps). In Fig. 2b the full approximant crystal is considered in the ASA-LMTO (10 meV half width). The omitted Al and Cu Atoms in Fig. 2a are accounted for by a small energy shift. Both curves suffer from artifacts. Fig. 2a exhibits spurious oscillations in the PSG due to the cluster surface whereas typical spikes are observed in Fig. 2b throughout the band. Spikes appear once an incomplete **k** sum is employed together with a dense sequence of energy points. They tend to disappear in large approximants [30]. Nevertheless, the PSG must be caused by extended multiple scattering in the Fe network.

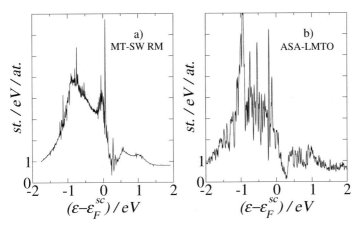

Fig. 2. DOS at b-Fe in i-AlCuFe (1/1), a) MT-SW RM (5 meV steps) with 2000 Fe atoms of the Fe network, b) ASA-LMTO with the full approximant (Γ = 10 meV)

3.2 Two Types of Narrow Resistivity Peaks

Both DOS pseudogaps are accompanied by resistivity peaks (Fig. 1a). The wide peak covers also that part of the d-Fe band where the diffusivity is low (Fig. 1c). The narrow peak belongs to the V-shaped PSG (Fig. 1b), and it proves stable against thermal perturbation (Fig. 1e). Note that increasing Lorentzian half width in the KGF simulates decreasing inelastic scattering time: $\Gamma = 1/(2\,\tau_{inel})$. We find [11] that the top resistivity at the narrow peak scales as $\rho_{max} \propto \Gamma^{-1/4}$ to be compared with the anomalous Drude law, $\sigma \sim \tau_{inel}^{2\beta-1}$ [34,35,36]. We conclude that thermal scattering with the states in the PSG simulates a sub-diffusive regime with $\beta = 3/8$. This statement concerns the character of states at fixed system size. Related statements [37,33] include size scaling. In the d-Fe band below ϵ_F^{sc} there are very narrow U-shaped conductivity gaps. One of them is caught in the resistivity peak near -200 meV with $\Gamma = 5$ meV in Fig. 1e. They disappear on increasing Γ. The ab-$initio$ optical conductivity $\sigma_1(\omega)$ of i-AlCuFe (1/1) is shown in Fig. 1f. We employ T $= 0$ K, and the actual Fermi energy is fixed at the energy of the narrow resistivity peak. At low frequencies, the curve reveals the signature of the corresponding conductivity pseudogap.

3.3 The sp-Character of the Electron States

In Fig. 1d each LMTO band state is characterized by its relative sp-weight in the cell, $\widehat{P}_{sp}(\epsilon) = (sp\text{-weight})/(spd\text{-weight})$ where $(spd\text{-weight}) = 1$. The low valence band is clearly sp-dominated, whereas the states towards the mid of the d-Cu band (-4 eV) have almost no sp-contributions, they are confined to the Cu subsystem. The Fermi energy, ϵ_F^{sc}, does not strictly follow the Hume-Rothery principle, it is found slightly below the DOS minimum PSG, just where both the sp-content and the diffusivity rise steeply (Fig. 1c). This bears some hints in view of phase stability. At ϵ_F^{sc}, forces that attempt to optimize the Fe network (more covalent, dominate below ϵ_F^{sc}) may be balanced by other forces that attempt to optimize the all-atom neighbor-shell order (more Hume-Rothery-like, dominate above ϵ_F^{sc}).

3.4 Non-Ohmic Resistance Scaling

We build a slab of nearly three approximant cells thickness ($L = 36.32$ Å, 383 atoms). Once a new atom is added we calculate the spectral conductance per cell cross section, $\widehat{G}(L, \epsilon, \mathbf{k}_\perp) = (2e^2/h)\,\mathrm{Tr}(\hat{t}^\dagger \hat{t})$, by means of the MT-SW LM (cf. Section 2). $\hat{t}(L, \epsilon, \mathbf{k}_\perp)$ is the amplitude transmission matrix. The electron energy, $\epsilon = \epsilon_F^{sc} - 76$ meV, is chosen slightly below ϵ_F^{sc} where a 27×27 scan of $\mathbf{k}_\perp$ in the (+,+) quadrant of the lateral Brillouin zone (BZ) reveals extreme transport regimes. The resistance, $\widehat{R} = \widehat{G}^{-1}$, of the complete slab is shown as a contour plot in Fig. 3a. $\widehat{R}_1$ corresponds to a resistivity of 200 $\mu\Omega$cm. Hence, around the gamma point, the slab resembles

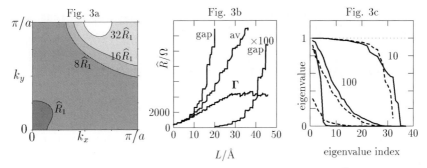

Fig. 3. A slab of i-AlCuFe (1/1) cells, thickness L, resistance $\widehat{R}$ at $\epsilon_F^{sc} - 76$ meV. a) $\widehat{R}(\mathbf{k}_\perp)$ at $L = 36.9$ Å (383 atoms), $\widehat{R}_1 = 5000\,\Omega \sim$ amorphous. b) $\widehat{R}(L)$ at $\mathbf{k}_\perp = (0,0)(\pi/a)$ (Γ), $\mathbf{k}_\perp = (0.6, 1.0)(\pi/a)$ (gap), BZ average (av), 480 atoms at the most. c) The 37 (at Γ, dashes), respectively 32 (at $(0.6,1.0)(\pi/a)$, solid) eigenvalues of the current matrix, $\widehat{t}^\dagger \widehat{t}$, after joining 10/100/(480 solid/200 dashes) atoms to the slab

amorphous metals. However, close to $\mathbf{k}_\perp^{gap} \equiv (0.6, 1.0)(\pi/a)$, a gap opens with up to 48 $\widehat{R}_1$. This interpretation is supported by L-resolved information up to $L = 45.8$ Å (480 atoms, Figs. 3b,c). In Fig. 3c we show the 32 (at $\mathbf{k}_\perp^{gap}$) respectively 37 (at Γ) eigenvalues of the Hermitian current matrix, $\widehat{t}^\dagger \widehat{t}$, after joining 10, 100 ... atoms to the slab. At $\mathbf{k}_\perp^{gap}$, no current-carrying channels are available beyond 100 atoms, and the resistance rises exponentially (Fig. 3b). At Γ, on the contrary, the transport regime beyond two approximant cells corresponds to a crystal slab in vacuum. Despite of such extremes, the BZ average stays Ohmic-like (Fig. 3b). After one cell relaxation, the average slope of the curve provides 690 $\mu\Omega$cm for the resistivity (cf. KGF with isotropic k-sampling, 610 $\mu\Omega$cm, Fig. 1a). Similar to α-AlMn(Si) [8] the resistance depends critically on the LSL stacking of iron planes.

3.5 Inverse Matthiessen Rule

Inverse Matthiessen rules (IMRs), $\sigma(T) = \widehat{\sigma}(\epsilon_F) + \Delta\sigma(T)$ with positive $\Delta\sigma(T)$, were observed among QCs [38] and among thin-film transient phases [39,3]. The available experimental data for QCs support $\Delta\sigma(T) \sim T^{1...1.5}$ [40] which fits well to $\sigma \sim T^{1/4} \sim \tau_{inel}^{-1/4}$ together with the Bloch-Grüneisen law, $\tau_{inel} \sim T^{-5}$ [32]. The latter scaling assumption deserves deeper foundation. Most important is the universality of $\Delta\sigma(T)$ in a group of phases. The lower part of the curve should be parabolic (cf. (3)). Universality requires that the second derivatives with respect to the number of states, $[(d/(\widehat{n}\,d\epsilon))^2 \widehat{\sigma}(\epsilon)]_{\epsilon=\epsilon_F}$, scale among the group members as $\widehat{n}(\epsilon_F)^{-2}$ (level spacing). This concerns the quantum mechanics inside the thermal energy window. If one adopts rigid-band arguments one can try the ansatz $\widehat{\sigma}(\epsilon) = A(\epsilon - \epsilon_1)^a$ together with $\widehat{n}(\epsilon) = B(\epsilon - \epsilon_1)^b$ in (3). A universal IMR at low temperatures is obtained with $a = 2$ (Lorentzian resistivity) and $b < 1$. We have shown that such conditions prevail in the PSG of i-AlCuFe (1/1) [11].

4 Spectral Transport Model of the Quasicrystal

The idea is not new: Find a spectral conductivity model that accounts for generic properties, and then obtain transport parameters by means of the CTKG formalism (cf. Enderby et al. [41] with the Anderson MIT). For QCs, this path has been followed by Maciá [9] who has considered self-similar fine structures of the state density as suggested by one-dimensional QCs. We suppose that the generic properties of icosahedral high-resistivity QCs arise from the special co-operation of the clusters [5], and that even low approximants should exhibit corresponding spectral signatures to be extracted and then rescaled to the appearance in the QCs.

4.1 Lorentzian Resistivity Model

We model the *ab-initio* calculated resistivities of the approximants by means of Lorentzians that are characterized by their heights, $1/(\pi\gamma)$, and positions δ with reference to ϵ_F^{sc}. The resistivity models of i-AlCuFe (1/1) and i-AlCuRu (1/1) [42] require two Lorentzians,

$$\widehat{\rho}_L(\epsilon) = A \left(\frac{\gamma_1/\pi}{(\epsilon - (\epsilon_F^{sc} + \delta_1))^2 + \gamma_1^2} + \alpha \frac{\gamma_2/\pi}{(\epsilon - (\epsilon_F^{sc} + \delta_2))^2 + \gamma_2^2} \right), \quad (10)$$

the first one for the wide peak in Fig. 1a, the second for the narrow peak (cf. Fig. 1a). A, α, δ_i, and γ_i are fitted to *ab-initio* results for the approximants. Our rescaling to the i-AlCuFe QC considers that the thermopower (4) is closely related to $\widehat{\sigma}(\epsilon)$. Hence, the following steps rely mainly on $S^{exp}(T)$ (Pierce et al. [29]): In step 1, δ_1 and δ_2 are fixed at the approximant values. In step 2, for the chemical potential, $\mu(T)$ (2), we suppose constant diffusivity, i. e. $(d(ln(\widehat{n})/d\epsilon)_{\epsilon=\epsilon_F} = (d(ln(\widehat{\sigma})/d\epsilon)_{\epsilon=\epsilon_F}$ that provides

$$\xi = -(|e|/2)(S(T)/T)_{T\to 0} \approx (S^{exp}(T)/T)_{4K} \approx 10^{-7} eV/K^2 .$$

In step 3, A, α, γ_1, and γ_2 are fitted as to fulfill two demands: $\widehat{\rho}(\epsilon_F) = \rho_{4K}^{exp}$, and minimum χ^2 for the approximation of the experimental thermopower up to 300 K. The large $\gamma_2 = 40$ meV accounts thus for the average performance of the i-$Al_{62.5}Cu_{25}Fe_{12.5}$ bulk QC up to 300 K. A corresponding low-temperature model should have γ_2 less than the approximant (44 meV). Common amorphous metals are properly described by one wide Lorentzian, cf. Table 1. However, close to quasicrystalline stoichiometries, on annealing at high temperatures, sequences of transient phases are formed that acquire successively transport properties of QCs. Such thin-film transient phases require an additional narrow Lorentzian [43]. Anomalous temperature curves

Table 1. Fitting parameters of Lorentzian spectral resistivities (10). x in the 3-rd column describes the relative preeminence of the narrow resistivity peak

Phase, ref. experiment	ρ_{4K} $\mu\Omega$cm	x	A $\mu\Omega$cmeV	δ_1 eV	γ_1 eV	α	δ_2 eV	γ_2 meV
am-Al$_{84}$Fe$_{16}$	162	-	1139	-1.0	1.6	0.0	-	-
i-AlCuFe (1/1)	610	1.6	580	-0.2	0.7	0.1	0.23	44
i-AlCuFe thin film [43]	2333	7.8	1047	-0.2	1.35	0.316	0.23	55
Al$_{62.5}$Cu$_{25}$Fe$_{12.5}$ b-QC [29]	8078	33.75	1047	-0.2	1.35	1.0	0.23	40
i-AlCuRu (1/1)	450	2.5	471.25	-1.1	1.0	0.15	0.275	60
i-Al$_{64}$Cu$_{20}$Ru$_{15}$Si$_1$ b-QC [1]	3310	27.14	471.25	-1.1	1.0	1.9	0.275	70
i-Al$_{68}$Cu$_{17}$Ru$_{15}$ b-QC [1]	5397	37.78	471.25	-1.1	1.0	1.7	0.275	45
i-Al$_{65}$Cu$_{20}$Ru$_{15}$ b-QC [1]	11585	78.125	471.25	-1.1	1.0	2.5	0.275	32

of transport parameters must occur once the real Fermi energy crosses a predominating narrow resistivity peak. We characterize this predominance by the ratio of the peak heights, $x \equiv \alpha(\gamma_1/\gamma_2)$. Table 1 demonstrates that x rises with increasing ρ_{4K}, i. e. if quasiperiodicity is improved (cf. Al-Cu-Fe).

4.2 Transport Parameters of Bulk Quasicrystals

The present parametrization scheme acts on the level of spectral quantities where arbitrary transport mechanisms can be included. Two-band models of quasicrystals [45] employ terms such as particles and holes which suppose sharp planes in **k**-space. Recently, both modeling procedures have been applied to thin-film transient phases [43].

Fig. 4 shows results for i-AlCuFe and i-AlCuRu bulk QCs compared with available experimental data (conductivity, thermopower [29,1]; optical conductivity [44]). We conclude:

(i) Simple spectral resistivity models with only two Lorentzians (10) reproduce the conductivities and the thermopowers of the considered phases at least up to 300 K. At high temperatures where the thermal energy window samples more extended spectral information additional Lorentzians may be required (e.g. i-AlCuRu (1/1)).

(ii) The optical conductivity of i-AlCuFe (Fig. 4c, Fermi energy $\epsilon_F = \epsilon_F^{sc} + \delta_2 + 10$ meV, temperature $T = 295$ K of the experiment [44]) is calculated within the Lorentzian model upon replacing $\widehat{\sigma}(\epsilon, \epsilon + \hbar\omega)$ (9) in the energy integal (cf. Section 2) by $1.8 \, [\widehat{\rho}_L(\epsilon) \times \widehat{\rho}_L(\epsilon + \hbar\omega)]^{-1/2}$. This means, no comparison of the compositions of states that are separated by $\hbar\omega$ takes place after this simplification. Obviously, the approximation breaks down above $\hbar\omega \approx 1.5$ eV. The correct result for the approximant (Fig. 1f) agrees much better with the experiment. We conclude that the compositions of states change drastically on the scale 2 eV.

(iii) We employ the simplified form of $\widehat{\sigma}_{x,yz}(\epsilon)$ in (8) with $Q = 2 \times 10^{-5}$ Am^2. This has the consequence that zeros of $S(T)$ and $R_H(T)$ will ever oc-

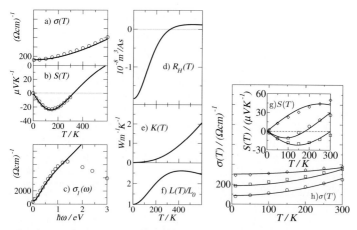

Fig. 4. Bulk quasicrystals, transport parameters derived from Lorentzian models (10) with Table 1, experimental results (circles), a) - f) i-AlCuFe (experiments [29] in a,b), [44] in c)) g-h) i-AlCuRu (experiments [1])

cur at closely related temperatures, such as 326 K and 322 K, respectively, in i-AlCuFe (Fig. 4b,d). Experimental results of the Hall coefficients (Fig. 4d) are not yet available. For the i-AlCuRu phases shown in Fig. 4g,h the experimental Hall coefficients [1] indicate that the simplified version of (8) may not be sufficient.

(iv) Measurements of thermal conductivities provide only the sum of ionic and the electronic contributions. Hence, the electronic thermal conductivity is commonly estimated from the electronic conductivity by means of (6) where the Wiedemann-Franz limit, L_0, of the Lorenz number is used. However, we have shown [11] that even the Lorenz number of i-AlCuFe (1/1) deviates markedly from L_0. The spectral model of the QC predicts enhanced temperature variation of the Lorenz number (Fig. 4f), and strongly increasing electronic thermal conductivity (Fig. 4e). We find $L(300K)/L_0 = 2.42$ and $K(300K) \sim 0.46$ W/m/K. The experimental result for the total thermal conductivity of i-AlCuFe is ~ 1.3 W/m/K [46].

5 Conclusions

We report *ab-initio* results for spectral electronic transport parameters of i-AlCuFe (1/1) which indicate that 100-meV spectral structures are generated in the iron sub-system on the length scale of the Bergman clusters. Such kind of spectral structure is supposed to survive in the quasicrystal in a rescaled form that accounts for the larger length scale of the cluster arrangement. A spectral transport model of the approximant extracts significant spectral features to be subsequently rescaled to the appearance in the quasicrystal. This rescaling is guided by the experimental thermopower of the quasicrystal. Our results support the concept that extended quasiperiodicity alone does not produce the observed transport anomalies of quasicrystals. They are rather due to both the positions of certain active atoms inside the typical clusters and the arrangement of these clusters.

Acknowledgments

We thank Prof. P. Häussler and his group for many exciting discussions, and to the Deutsche Forschungsgemeinschaft we owe thanks for financial support.

References

1. F. S. Pierce, P. A. Bancel, B. D. Biggs, Q. Guo, and S. J. Poon, Phys. Rev. B **47**, 5670 (1993).
2. U. Mizutani, J. Phys.: Cond. Matter **10**, 4609 (1998).
3. C. Roth, G. Schwalbe, R. Knöfler, F. Zavaliche, O. Madel, R. Haberkern, and P. Häussler, J. Non-Cryst. Solids **250-252**, 869(1999).
4. R. Escudero, J. C. Lasjaunias, Y. Calvayrac, and M. Boudard, J. Phys.: Cond. Matter **11**, 383 (1999).
5. C. Janot and M. de Boissieu, Phys. Rev. Lett. **72**, 1674 (1994).
6. G. Trambly de Laissardiére and D. Mayou, Phys. Rev. B **55**, 2890 (1997).
7. H. Solbrig and C. V. Landauro, Physica B **292**, 47 (2000).
8. H. Solbrig, C. V. Landauro, and A. Löser, Mater. Sci. Eng. A **294-296**, 596 (2000).
9. E. Maciá, Phys. Rev. B **61**, 8771 (2000).
10. C. V. Landauro and H. Solbrig, Mater. Sci. Eng. A **294-296**, 600 (2000).
11. C. V. Landauro and H. Solbrig, Physica B **301**, 267 (2001).
12. E. Maciá, Phys. Rev. B **64**, 94206 (2001).
13. G. V. Chester and A. Thellung, Proc. Phys. Soc. **77**, 1005 (1961).
14. R. Kubo, J. Phys. Soc. Jap. **12**, 570 (1957).
15. D. A. Greenwood, Proc. Phys. Soc. **71**, 585 (1958).
16. N.W. Ashcroft and N.D. Mermin, *Solid State Physics* (Saunders College Publishing, 1976).
17. J. M. Ziman, *Electrons and Phonons* (University Press, Oxford, 1960).
18. S.J. Poon, Advances in Physics **41**, 303 (1992).
19. A. Sahnoune, J. O. Ström-Olsen, and A. Zaluska, Phys. Rev. B **46**, 10629 (1992).
20. O. K. Andersen, Phys. Rev. B **12**, 3060 (1975).
21. D. J. Chadi and M. L. Cohen, Phys. Rev. B **8**, 5747 (1973).
22. D. Hobbs, E. Piparo, Girlanda R., and M. Monaca, J. Phys.: Cond. Matter **7**, 2541 (1995).
23. R. Arnold, Stabilisierendes Pseudogap und Streukonzept in nichtkristallinen Materialien. Ph. D. thesis, Technische Universität Chemnitz (1997).
24. R. Kahnt, J. Phys.: Cond. Matter **7**, 1543 (1995).
25. Andrè Löser, Master's thesis, Technische Universität Chemnitz-Zwickau (1996).

26. H. Solbrig, phys. stat. sol. (b) **139**, 223 (1987).
27. Eric Cockayne, Rob Phillips, X. B. Kan, S. C. Moss, J. L. Robertson, T. Ishimasa, and M. Mori, J. Non-Cryst. Solids **153-154**, 140 (1993).
28. T. Takeuchi, H. Yamada, M. Takata, T. Nakata, N. Tanaka, and U. Mizutani, Mater. Sci. Eng. A **294-296**, 340 (2000).
29. F. S. Pierce, S. J. Poon, and B. D. Biggs, Phys. Rev. Lett. **70**, 3919 (1993).
30. E. S. Zijlstra, T. Janssen, Europhys. Lett. **52**, 578 (2000).
31. G. Trambly de Laissardiére and T. Fujiwara, Phys. Rev. B **50**, 5999 (1994).
32. S. Roche and T. Fujiwara, Phys. Rev. B **58**, 11338 (1998).
33. M. Krajčí, J. Hafner, and M. Mihalkovič, Phys. Rev. B **65**, 24205 (2002).
34. D. Mayou, in *Lectures on quasicrystals*, F. Hippert and D. Gratias (Ed.), pp. 417 (Editions de Physique, Les Ulis-France, 1994).
35. C. Sire: in *Lectures on quasicrystals*, F. Hippert and D. Gratias (Ed.), pp. 505 (Editions de Physique, Les Ulis-France, 1994).
36. J. Bellissard and H. Schulz-Baldes, in *Proceedings of the 5th International Conference on Quasicrystals*, C. Janot and R. Mosseri (Ed.), pp. 439 (World Scientific, Avignon, 1995).
37. T. Fujiwara, in *Physical Properties of Quasicrystals*, pp. 169 (Springer-Verlag, Berlin, 1999).
38. D. Mayou, C. Berger, F. Cyrot-Lackmann, T. Klein, and P. Lanco, Phys. Rev. Lett. **70**, 3915 (1993).
39. R. Haberkern, K. Khedhri, C. Madel, and P. Häussler, Mater. Sci. Eng. A **294-296**, 475 (2000).
40. C. Gignoux, C. Berger, G. Fourcaudot, J.C. Grajeco, and H. Rakoto, Europhys. Lett. **39**, 171 (1997).
41. J. E. Enderby and A. C. Barnes, Phys. Rev. B **49**, 5062 (1994).
42. K. Sugiyama, T. Kato, K. Saito, and K. Hiraga, Philos. Mag. Lett. **77**, 165 (1998).
43. C. Madel, Elektronische Transporteigenschaften von amorphem und quasikristallinem Al-Cu-Fe. Ph.D. Thesis, Technische Universität Chemnitz (2000).
44. C. C. Homes, T. Timusk, and X. Wu, Phys. Rev. Lett. **67**, 2694 (1991).
45. R. Haberkern and G. Fritsch, in *Proceedings of the 5th International Conference on Quasicrystals*, C. Janot and R. Mosseri (Ed.), pp. 460 (World Scientific, Avignon, 1995).
46. A. Perrot, J. M. Dubois, M Cassart, and J. P. Issi, in *Proceedings of the 5th International Conference on Quasicrystals*, C. Janot and R. Mosseri (Ed.), pp.588 (World Scientific, Avignon, 1995).

Full Counting Statistics of Mesoscopic Electron Transport

Wolfgang Belzig

Departement für Physik und Astronomie, Universität Basel,
Klingelbergstr. 82, 4056 Basel, Switzerland

Abstract. Full Counting Statistics is the theory of quantum statistical properties of transport phenomena. It can be accessed by a modification of the standard Keldysh-Green's function technique, and is concisely formulated in terms of a circuit theory of mesoscopic electron transport. We summarize the basics of the theoretical methods and discuss two examples: a single-channel quantum contact and a double tunnel junction. For both examples we compare the transport statistics for the cases of normal electrons and Andreev reflection.

1 Introduction

Quantum transport is an inherently probabilistic process. Thus, the current is in general fluctuating in time. These fluctuations, provided they are nonthermal, contain information about the nature of the underlying transport mechanism. Most theoretical calculations and experiments study the time-averaged current or, more recently, the current noise (see e.g. [1,2]). However, the complete information on the statistics of transport can only be obtained from *all* correlators of the current. This was first noted by Levitov and coworkers [3,4,5], who adopted the terminology *full counting statistics* from quantum optics. The knowledge of the full counting statistics is therefore of fundamental interest, which provides the motivation to obtain it for various mesoscopic transport problems.

The full counting statistics (FCS) for a two-terminal device is defined as the probability $P_{t_0}(N)$ that N charges are transferred in a time interval t_0. Obviously, this quantity contains informations about interactions between many-particle processes. Examples are superconductors, in which particles are transported as Cooper pairs at subgap energies, or quantum Hall edge states with fractionally charged quasiparticles. Further interest arises from statistical particle properties, like the Fermi-Dirac statistics or Bose-Einstein statistics. In multi-terminal structures interference effects, like Aharonov-Bohm oscillations, obviously can play a role, too.

In the next two sections we will present first a relatively simple, nevertheless very powerful, approach developed by Nazarov [6,7,8] to access the full counting statistics in typical mesoscopic electron systems. This approach is based on an extension of the Keldysh-Green's function method, and can be embedded in an elegant circuit theory of mesoscopic electron transport. As simple applications of this method in Section 4 we calculate the transport statistics for two concrete two-terminal devices, where one terminal is a normal metal (N) and the other terminal is either a normal(N) or a superconducting (S) metal. In Section 4.1 we show that the full counting statistics of a single-channel conductor is binomial, both in the NN-case and in the NS-case. In Section 4.3 we focus on the FCS of two serially connected tunnel junctions. We show that Fermi-correlations and proximity induced pairing correlations, lead to a non-standard statistics and to distinct differences between the NN and the NS case. The results are summarized in Section 5.

2 Method

The most obvious implementation of a full counting statistics would be a definition of an operator of the accumulated current $\hat{N}(t) = \int_0^t \hat{I}(t')dt'$ and the FCS by the expectation value $P_{t_0}(N) = \langle \delta(N - \hat{N}) \rangle$. This procedure has, however, turned out to be not useful, the reason being that it is difficult to take care of the non-commutativity of current operators at different times. One way out is to define the FCS via the *cumulant generating function*(CGF) $S(\chi)$ by [6,7]

$$e^{-S(\chi)} \equiv \sum_N P_{t_0}(N)e^{iN\chi} = \langle \mathcal{T} e^{i\frac{\chi}{2} \int_0^{t_0} dt \hat{I}(t)} \tilde{\mathcal{T}} e^{i\frac{\chi}{2} \int_0^{t_0} dt \hat{I}(t)} \rangle . \tag{1}$$

Here $\mathcal{T}(\tilde{\mathcal{T}})$ denotes (anti-)time ordering and $\bar{I}(t)$ the operator of the current through a certain cross section. Such an expectation value can be implemented on the Keldysh contour, which makes it possible to use standard diagrammatic methods [9].

To connect the CGF with a perturbation theory we consider the nonlinear response of our electronic circuit to the time-dependent perturbation

$$H_i(t) = \frac{1}{2}\chi(t) \int d^3x \Psi^\dagger(x) \hat{j}(x) \Psi(x) . \tag{2}$$

The field $\chi(t)$ has different signs on the upper and lower part of the Keldysh contour, which makes it different from a classical vector potential. The operator j is the current density through a cross section C, depicted in Fig. 1. Since we are aiming at the total charge counting statistics, we will assume that χ is nonzero only in a finite time interval $[0, t_0]$. The unperturbed system evolves according to a Hamiltonian H_0. The equation of motion for the Green's function subject to $H = H_0 + H_i(t)$ reads

$$\left[i\frac{\partial}{\partial t} - \hat{h}_0 - \frac{1}{2}\chi(t)\hat{j}(x) \right] G(\chi, t, t') = \delta(t - t') , \tag{3}$$

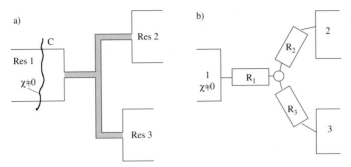

Fig. 1. A mesoscopic system. a) The actual layout. Three reservoirs are coupled by diffusive wires. The number of charges traversing the cross section C in terminal 1 are counted. b) The circuit theory approximation to a). The diffusive metal wires are modeled as three resistances, connected to a common central node

The relation of the Green's function (3) to the CGF (1) is obtained from a term-by-term comparison of the diagrammatic expansion. One finds the simple relation

$$-ie\frac{\partial S(\chi)}{\partial \chi} = t_0 I(\chi),\tag{4}$$

where the current $I(\chi)$ is obtained from the χ-dependent Green's function just like the usual current. Note that in this way we can find the statistics of *any* observable [10].

For a mesoscopic transport problem there is a particular simple way to access the full counting statistics of the current. Usually the device is separated into reservoirs (or terminals) and an active part, the first providing boundary conditions, and the second being responsible for the resistance. An example is shown in Fig. 1a. Let us consider the equation of motion for the Green's function *inside* a terminal for the following parameterization of the current operator in 2

$$\hat{j}(x) = (\boldsymbol{\nabla} F(x))\hat{\boldsymbol{p}}/m.\tag{5}$$

$F(x)$ is chosen such that it changes from 0 to 1 across a cross section C, which intersects the terminal, but is of arbitrary shape. The change from 0 to 1 should occur on a length scale Λ, which should obey $\lambda_F \ll \Lambda \ll l_{imp}, \xi_0$ (Fermi wave length λ_F, impurity mean free path l_{imp}, and coherence length ξ_0. Now we use standard manipulations to reduce (3) to its quasiclassical version [11]. This is usually a very good approximation, since all currents in a real experiment are measured in normal Fermi-liquid leads. The Eilenberger equation in the vicinity of the cross section reads

$$\left[-i\boldsymbol{v}_F\boldsymbol{\nabla} + \frac{\chi}{2}\boldsymbol{v}_F(\boldsymbol{\nabla} F(x))\check{\tau}_K + \check{\sigma}_{\text{rest}}, \check{g}\right] = 0.\tag{6}$$

Here $\check{\tau}_K$ is the matrix of the current operator (specified below), and $\check{\sigma}_{\mathrm{rest}}$ contains time derivatives and the usual self energy (pair potential, impurities, phonon selfenergy, etc.). Close to the cross section the counting field dominates, and we can neglect the selfenergy $\check{\sigma}_{\mathrm{rest}}$. The counting field can then be eliminated by the gauge-like transformation

$$\check{g}(x,\chi) = e^{i\chi F(x)\check{\tau}_K/2} \check{g}(x,0) e^{-i\chi F(x)\check{\tau}_K/2}. \tag{7}$$

Applying the diffusive approximation in the rest of the terminal, this leads to the following transformation of the terminal Green's function in contrast to the case without counting field

$$\check{G}(\chi) = e^{i\chi\check{\tau}_K/2} \check{G}(0) e^{-i\chi\check{\tau}_K/2}. \tag{8}$$

In this way the counting field can entirely be incorporated into a *modified boundary condition* imposed by the terminal onto the mesoscopic system. Note, that it follows from (6) and (5), that the counting field for a particular terminal vanishes from the equations of motion in the rest of the system.

The generalization of this method to the counting statistics for multiterminal structures was performed in [14]. The surprisingly simple result is, that one has to add a separate counting field for each terminal, in which charges are counted.

What are the achievements of this method? We should emphasize that it does not simplify the solution of a specific transport problem, i.e. we still have to know the solution corresponding to the Hamiltonian H_0. If this solution is not known, the counting field makes this situation not easier. Rather, the method opens a very general way to obtain the FCS, if a method to find the average currents, i.e. for $\chi = 0$, is known. In the next section we will introduce such a method, the circuit theory of mesoscopic transport. Initially it was invented to calculate average currents only, however the method to obtain the FCS introduced in this section is straightforwardly included.

What is the price to pay? Loosely speaking, the method to obtain the average currents has to be sufficiently general. Usually the absence of a field, which has different signs on the upper and lower part of the Keldysh contour, allows some simplification. For example, in the Keldysh-matrix representation all Green's functions can be brought into a tri-diagonal form, which is obviously simpler to handle than the full matrix. The method above does not allow this simplification anymore. Or, in other words the counting rotation (8) destroys the triangular form. Thus, the price we have to pay for an easy determination of the FCS is that we need a method, which respects the full Keldysh-matrix structure in all steps. The circuit theory, which we describe in the next section fulfills this requirement.

3 Circuit Theory

A concise formulation of mesoscopic transport is the so-called circuit theory [12,13]. Its main idea, borrowed from Kirchhoff's classical circuit theory, is to represent a mesoscopic device by discrete elements. These approximate the layout of an experimental device with arbitrary accuracy, provided one chooses enough elements. The accuracy can be made arbitrarily high, of course, but in practice one has to find the optimum between a small grid size and the computational effort.

Below we will treat only very basic circuits, for which we summarize the necessary ingredients below. As terminals we will consider normal metals or superconductors. A normal metal at chemical potential μ and temperature T is described by a Green's function

$$\check{G}_N = \begin{pmatrix} \hat{\sigma}_3 & \hat{K}_N \\ 0 & -\hat{\sigma}_3 \end{pmatrix} \quad , \quad \hat{K}_N = 2 \begin{pmatrix} 1 - 2f(E) & 0 \\ 0 & 1 - 2f(-E) \end{pmatrix}, \tag{9}$$

with Fermi distribution $f(E) = (\exp((E-\mu)/T) + 1)^{-1}$. A superconducting terminal at chemical potential $\mu_S = 0$ is described by

$$\check{G}_S = \begin{pmatrix} \hat{R} & (\hat{R} - \hat{A})\tanh\frac{E}{2T} \\ 0 & \hat{A} \end{pmatrix}, \tag{10}$$

where

$$\hat{R} = \begin{pmatrix} g_R & f_R \\ f_R & -g_R \end{pmatrix} \quad \text{and} \quad \hat{R}^2 = \hat{1}. \tag{11}$$

The advanced Green's function $\hat{A}$ can be written in a similar form. The spectral functions of the superconducting contact have to be determined for the concrete problem. For example in a BCS-superconductor they are given by $f_{R(A)} = (1 - g_{R(A)}^2)^{1/2} = \Delta/((E \pm i0) - \Delta^2)^{1/2}$.

If the charges in a terminal are counted, we have to apply a counting rotation (8) to the terminal Green's function. The couting-rotation matrix has the form $\check{\tau}_K = \hat{\sigma}_3 \bar{\tau}_1$, where $\hat{\sigma}_i(\bar{\tau}_i)$ denote Pauli-matrices in Nambu(Keldysh)-space. Note, that the presence of $\bar{\tau}_1$ in $\check{\tau}_K$ destroys the tri-diagonal form of the terminal Green's functions (9) or (10). The counting rotation also contains a χ-dependent phase shift due to the presence of $\hat{\sigma}_3$, which cannot be gauged away. This makes the straight application of the method of Section 2 to an equilibrium supercurrent a complicated problem of quantum measurement theory [8]. In the following we will circumvent this problem by considering only transport in the absence of a supercurrent.

The central element of the circuit theory is the arbitrary connector, characterized by a set of transmission coefficients $\{T_n\}$. Its transport properties are described by a *matrix current* found in [13]

$$\check{I} = -\frac{e^2}{\pi} \sum_n \frac{2T_n [\check{G}_1, \check{G}_2]}{4 + T_n (\{\check{G}_1, \check{G}_2\} - 2)}. \tag{12}$$

Here $\check{G}_{1(2)}$ denote the Green's functions on the left and the right of the contact. We should emphasize that the matrix form of (12) is crucial to obtain the FCS, as mentioned previously.

If the circuit consists of more than one connector, the transport properties can be found from the circuit theory by means of the following circuit rules. We take the Green's functions of the terminals as given and introduce for each internal node an (unknown) Green's function. The two rules determining the transport properties of the circuit completely are

1. $\check{G}^2 = \check{1}$ for the Green's functions of all internal nodes.
2. The total matrix current in a node is conserved: $\sum_i \check{I}_{ij} = 0$, where the sum goes over all nodes or terminals connected to node j and each matrix current is given by (12).

Having determined the Green's functions on all nodes, we find the total CGF by integrating all currents into the terminals over their respective counting fields (see [14] and [15] for more details). Below we demonstrate this procedure for a double tunnel junction.

4 Examples

4.1 Quantum Contact

We first consider the most basic resistor: a contact characterized by a set of transmission eigenvalues $\{T_n\}$. The CGF can then be found from the relation $\partial S(\chi)/\partial \chi = (-it_0/4e^2) \int dE \, \text{Tr}(\check{\tau}_K \check{I}(\chi))$. Using the fact that $[\check{A}, \{\check{A}, \check{G}_2\}] = 0$ for all matrices with $\check{A}^2 = \check{1}$, it is easy to verify that under the trace in (12) $(\partial/\partial\chi)\{\check{G}_1(\chi), \check{G}_2\} = i\check{\tau}_K [\check{G}_1(\chi), \check{G}_2]$. We can therefore integrate (12) with respect to χ and obtain [8]

$$S(\chi) = -\frac{t_0}{2\pi} \sum_n \int dE \, \text{Tr} \ln \left[4 + T_n \left(\{\check{G}_1(\chi), \check{G}_2\} - 2 \right) \right] . \tag{13}$$

Equation (13) contains the statistical properties of all types of superconducting constrictions.

Using (12) for two normal contacts (with occupations $f_{1(2)} = 1/(\exp((E - \mu_{1(2)})/\Theta) + 1)$, where Θ is the temperature, and a counting field χ in terminal 1) we obtain the seminal result of [4,5]

$$S(\chi) = -\frac{t_0}{\pi} \sum_n \int dE \ln \left[1 + T_n B_1 \left(e^{i\chi} - 1 \right) + T_n B_{-1} \left(e^{-i\chi} - 1 \right) \right] . \tag{14}$$

Here we introduced the occupation factors $B_1 = f_1(1 - f_2)$ for a tunneling event from 1 to 2 and B_{-1} for the reverse process. The terms with *counting factors* $e^{\pm i\chi} - 1$ obviously correspond to charge transfers from 1 to 2 (2 to 1). At zero temperature and $\mu_1 - \mu_2 = eV$ the integration can easily be evaluated and we obtain

$$S(\chi) = -\frac{et_0|V|}{\pi} \sum_n \ln \left[1 + T_n \left(e^{i\chi \text{sgn} V} - 1 \right) \right] . \tag{15}$$

The corresponding statistics for a single channel with transparency T is binomial

$$P_{t_0}(N) = \binom{M}{N} T^{|N|} (1-T)^{M-|N|} \quad \text{for} \quad NV > 0, \tag{16}$$

and zero otherwise. Here we have introduced the number of attempts $M = et_0 V/\pi$, which is the maximal number of electrons that can be sent through one (spin-degenerate) channel in a time interval t_0.

The FCS of an SN-contact characterized by a set of transmission eigenvalues can easily be obtained from (13). Evaluating the trace in (13) the CGF can be expressed as [16]

$$S(\chi) = -\frac{t_0}{2\pi} \sum_n \int dE \ln\left[1 + \sum_{q=-2}^{2} A_{nq}\left(e^{iq\chi} - 1\right)\right]. \tag{17}$$

The coefficients A_{nq} are related to a charge transfer of $q \times e$. For example a term $(\exp 2i\chi - 1)$ corresponds just to an Andreev reflection process, in which two charges are transfered simultaneously, leading eventually to doubled shot noise [17]. Explicit expressions for the various coefficients are given in the Appendix. For the BCS case, they reduce to the results of [16]. In the fully gapped single-channel case at energies $E \ll \Delta$ only terms corresponding to Andreev reflection remain and the CGF becomes

$$S(\chi) = -\frac{et_0|V|}{\pi} \ln\left[1 + R_A\left(e^{i2\chi} - 1\right)\right], \tag{18}$$

where $R_A = T^2/(2-T)^2$ is the probability of Andreev reflection. The corresponding statistics is binomial

$$P_{t_0}(2N) = \binom{M}{N} R_A^N (1-R_A)^{M-N}, \quad P_{t_0}(2N+1) = 0. \tag{19}$$

It is interesting to see how the CGF for normal transport, i. e., (14), emerges. Putting $f_{R,A} = 0$ and $g_{R,A} = \pm 1$ the coefficients (34) and (35) can be written as

$$A_{\pm 1} = B_{\pm 1}^+ + B_{\pm 1}^- - 2B_{\pm 1}^+ B_{\pm 1}^- - B_{\pm 1}^+ B_{\mp 1}^- - B_{\pm 1}^- B_{\mp 1}^+,$$
$$A_{\pm 2} = B_{\pm 1}^+ B_{\mp 1}^- \; ; \; B_1^{\pm} = TB_1(\pm E), B_{-1}^{\pm} = TB_{-1}(\pm E).$$

Then the argument of the ln factorizes in positive and negative energy contributions

$$\ln\left[1 + \sum_{q=-2}^{2} A_q\left(e^{iq\chi} - 1\right)\right] = \sum_{i=\pm} \ln\left[1 + \sum_{q=-1}^{1} B_q^i\left(e^{iq\chi} - 1\right)\right]. \tag{20}$$

Integrating both terms over energy, they give the same contribution, yielding (14).

4.2 Tunnel Junction

The counting statistics of a tunnel junction between normal contacts can be obtained from an expansion of the matrix current (12), if all transparencies are small. The matrix current takes the form

$$\check{I}_{\text{tun}} = \frac{g}{2}\left[\check{G}_1, \check{G}_2\right]. \tag{21}$$

Here the matrix current depends only on the total conductance $g = (e^2/\pi)\sum_n T_n$ of the contact. The corresponding CGF can be obtained by a direct expansion of (14) for the normal case:

$$S(\chi) = -\frac{gt_0 V}{e}(e^{i\chi} - 1) \equiv -\bar{N}(e^{i\chi} - 1). \tag{22}$$

The corresponding statistics is Poissonian

$$P_{t_0}(N) = e^{-\bar{N}}\frac{\bar{N}^N}{N!}. \tag{23}$$

Thus, the individual charge transfers are completely uncorrelated and the statistics is determined by a single parameter $\bar{N}$, related to the average current by $\bar{I} = e\bar{N}/t_0$. In a superconducting junction at subgap-energies the current vanishes and it thus makes no sense to compare.

4.3 Double Tunnel Junction

We now consider a diffusive island (or chaotic cavity) connected to two terminals by tunnel junctions [15,18]. This provides not only an example of a non-standard statistics, but is also an interesting application of the circuit theory. The layout is shown in Fig. 2.

The central node is described by an unknown Green's function $\check{G}_c$. We have two matrix currents entering the node, which obey a conservation law:

$$0 = \check{I}_1 + \check{I}_2 = \frac{1}{2}\left[g_1\check{G}_1 + g_2\check{G}_2, \check{G}_c\right]. \tag{24}$$

Employing the normalization condition $\check{G}_c^2 = 1$ we find the solution

$$\check{G}_c = \frac{g_1\check{G}_1 + g_2\check{G}_2}{\sqrt{g_1^2 + g_2^2 + g_1 g_2\{\check{G}_1, \check{G}_2\}}}. \tag{25}$$

Similar to the general case, we can integrate the current $I(\chi) \sim \text{Tr}\check{\tau}_K\check{I}_1$ and obtain the CGF

$$S(\chi) = -\frac{t_0}{4e^2}\int dE \text{Tr}\sqrt{g_1^2 + g_2^2 + g_1 g_2\{\check{G}_1, \check{G}_2\}}. \tag{26}$$

This result is valid for all types of contacts between normal metals and superconductors.

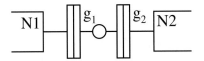

Fig. 2. Circuit model of a double tunnel junction

We evaluate the trace for two normal leads and find

$$S(\chi) = -\int \frac{t_0 dE}{2e} \sqrt{(g_1+g_2)^2 + 4g_1g_2\left(B_1(e^{i\chi}-1) + B_{-1}(e^{-i\chi}-1)\right)}. \tag{27}$$

At zero temperature and $\mu_1 - \mu_2 = eV > 0$ we find the result of [19],

$$S(\chi) = -\frac{t_0 V}{2e}\sqrt{(g_1+g_2)^2 + 4g_1g_2(e^{i\chi}-1)}. \tag{28}$$

This form of the CGF has two relatively simple limits. If the two conductances are very different (e.g. $g_1 \ll g_2$), we return to Poissonian statistics:

$$S(\chi) = -\frac{t_0 V g_1}{e}(e^{i\chi}-1). \tag{29}$$

On the other hand, in the symmetric case $g_1 = g_2 = g$, we find [19]

$$S(\chi) = -\frac{t_0 V g}{e}(e^{i\chi/2}-1). \tag{30}$$

The probabilities are non-Poissonian

$$P_{t_0}(N) = e^{-2\bar{N}}\frac{(2\bar{N})^N}{(2N)!}, \tag{31}$$

with the average number of transfered particles $\bar{N} = t_0 V g/2e$.

The CGF (26) for the transport between a normal metal and a superconductor yields (at zero temperature, for simplicity) [15,18]

$$S(\chi) = -\frac{t_0 V}{e\sqrt{2}}\sqrt{g_1^2 + g_2^2 + \sqrt{(g_1^2+g_2^2)^2 + 4g_1^2 g_2^2(e^{i2\chi}-1)}}. \tag{32}$$

Thus, the effect of the superconducting proximity effect is two-fold: charges are transferred in units of $2e$ (indicated by the π-periodicity) and another square root is involved in the CGF, resulting from the higher order correlations. In the limit that both conductances are very different (e.g. $g_1 \ll g_2$) we obtain again Poissonian statistics

$$S(\chi) = -\frac{t_0 V}{e}\frac{g_1^2}{g_2}\left(e^{i2\chi}-1\right). \tag{33}$$

This corresponds to uncorrelated transfers of pairs of charges. This can eventually lead to positive crosscorrelations in a beam-splitter geometry [15].

5 Conclusion

In this article we have reviewed some recent progress in the theory of *full counting statistics*. A relatively simply, nevertheless very powerful framework was presented, which allows to access the FCS of a large variety of mesoscopic structures. This so-called *circuit theory* can be used to obtain the statistical properties of all kinds of multi-terminal superconductor–normal-metal heterostructures.

Here, we have treated a few simple, nevertheless instructive examples. In particular, we have compared normal electron- and Andreev reflection-transport in a single-mode conductor and a double tunnel junction. In all cases Andreev transport occurs in pairs of electrons, which can be deduced already from simple particle conservation arguments: Cooper-pairs coming from the superconductor at subgap energies can leave only through the normal metal, at which the electrons therefore arrive in pairs. However, correlations between these pairs are in general different from the correlations in the normal case. This follows from the different forms of the cumulant generating functions.

Finally we would like to point to some future directions of research. Some works have started to explore the FCS of multi-terminal structures with particular focus on correlations between different terminals [14,15,18,20]. Time dependent transport (electron turnstile, quantum pumps) has been studied in a number of works already (see e.g. [21]). Also connections to photon counting or photon transport have been made [22,23]. Thus, *full counting statistics* is a rapidly emerging field, in which we expect to see many surprises.

I acknowledge discussions with J. Börlin, C. Bruder, M. Büttiker, G. Burkart, D. Loss, Yu. V. Nazarov, and P. Samuelsson. This work was supported by the Swiss NSF and the NCCR "Nanoscience".

Appendix: Cumulant Generating Function for the Single Channel Contact

We present the coefficients in the CGF for a single channel of transparency T. We assume the superconductor to be in equilibrium and the normal metal at a chemical potential μ_N. The occupation factors are then $f_\pm^N = f_0(\pm E - \mu_N)$ and $f_\pm^S = f_0(\pm E)$, where f_0 is the Fermi-Dirac distribution. The coefficients take the form

$$A_1 = T(1-T/2)\frac{2(g_R - g_A)}{(2-T(g_R-1))(2-T(g_A+1))} \times \left[f_+^N(1-f_+^S) + f_-^N(1-f_-^S)\right] \quad (34)$$

$$+2T^2\frac{1-f_R f_A - g_R g_A}{(2-T(g_R-1))(2-T(g_A+1))} \times \left[(f_+^N - f_-^N)(f_+^S - f_-^S)\right.$$

$$\left.(1-(f_+^N - f_-^N)(f_+^S - f_-^S)) + 2(f_+^S - f_-^S)^2(1-f_+^N)(1-f_-^N)\right],$$

$$A_2 = \frac{T^2}{2}f_+^N f_-^N \frac{1 + f_R f_A - g_R g_A - (f_+^S - f_-^S)^2(1 - f_R f_A - g_R g_A)}{(2-T(g_R-1))(2-T(g_A+1))}. \quad (35)$$

The coefficients A_{-n} are obtained from (34) and (35) by substituting $f_+^{S(N)} \leftrightarrow (1 - f_-^{S(N)})$, i.e. interchanging eletron- and hole-like quasiparticles.

References

1. Ya. M. Blanter and M. Büttiker, Phys. Rep. **336**, 1 (2000).
2. X. Jehl et al., Phys. Rev. Lett. **83**, 1660 (1999); X. Jehl et al., Nature **405**, 50 (2000); A. A. Kozhevnikov, R. J. Schoelkopf, and D. E. Prober, Phys. Rev. Lett **83**, 1660 (2000); P. Dieleman et al.; Phys. Rev. Lett. **79**, 3486 (1997); T. Hoss et al., Phys. Rev. B **62**, 4079 (2000).
3. L. S. Levitov and G. B. Lesovik, JETP Lett. **58**, 230 (1993).
4. L. S. Levitov, H. W. Lee, and G. B. Lesovik, J. Math. Phys. **37**, 4845 (1996).
5. H. Lee, L. S. Levitov, and A. Yu. Yakovets, Phys. Rev. B **51**, 4079 (1996).
6. Yu. V. Nazarov, Ann. Phys. (Leipzig) **8**, SI-193 (1999).
7. W. Belzig and Yu. V. Nazarov, Phys. Rev. Lett. **87**, 067006 (2001).
8. W. Belzig and Yu. V. Nazarov, Phys. Rev. Lett. **87**, 197006 (2001).
9. J. Rammer and H. Smith, Rev. Mod. Phys. **58**, 323 (1986).
10. Yu. V. Nazarov and M. Kindermann, cond-mat/0107133.
11. G. Eilenberger, Z. Phys. **214**, 195 (1968); A. I. Larkin and Yu. N. Ovchinnikov, Sov. Phys. JETP **26**, 1200 (1968); K. D. Usadel, Phys. Rev. Lett. **25**, 507 (1970).
12. Yu. V. Nazarov, Phys. Rev. Lett. **73**, 134 (1994).
13. Yu. V. Nazarov, Superlattices Microst. **25**, 1221 (1999).
14. Yu. V. Nazarov and D. Bagrets, cond-mat/0112223.
15. J. Börlin, W. Belzig, and C. Bruder, cond-mat/0201579.
16. B. A. Muzykantskii and D. E. Khmelnitskii, Phys. Rev. B **50**, 3982 (1994).
17. V. A. Khlus, Sov. Phys. JETP **66**, 1243 (1987); M. J. M. de Jong and C. W. J. Beenakker, Phys. Rev. B **49**, 16070 (1994); K. E. Nagaev and M. Büttiker, *ibid.* **63**, 081301(R) (2001).
18. J. Börlin, Diploma thesis (University of Basel, 2002).
19. M. J. M. de Jong, Phys. Rev. B **54**, 8144 (1996).
20. F. Taddei and R. Fazio, Phys. Rev. B **65**, 075317 (2002).
21. A. Andreev and A. Kamenev, Phys. Rev. Lett. **85**, 1294 (2000); Yu. Makhlin and A. Mirlin, *ibid.* **87**, 276803 (2001); L. S. Levitov, cond-mat/0103617.
22. C. W. J. Beenakker and H. Schomerus, Phys. Rev. Lett. **86**, 700 (2001).
23. M. Kindermann, Yu. V. Nazarov, and C. W. J. Beenakker, Phys. Rev. Lett. **88**, 063601 (2002).

Nonequilibrium Transport through a Kondo-dot in a Magnetic Field

Peter Wölfle, Achim Rosch, Jens Paaske, and Johann Kroha

Institut für Theorie der Kondensierten Materie, Universität Karlsruhe,
76128 Karlsruhe, Germany

Abstract. Electron transport through a quantum-dot in the Coulomb blockade regime is modeled by a Kondo-type hamiltonian describing spin-dependent tunneling and exchange interaction with the local spin. We consider the regime of large transport voltage V and magnetic field B with $\max(V, B) \gg T_K$, the Kondo temperature, and show that a renormalized perturbation theory can be formulated describing the local magnetization M and the differential conductance G quantitatively. Based on the structure of leading logarithmic corrections in bare perturbation theory we argue that the perturbative renormalization group has to be generalized to allow for frequency dependent coupling functions. We simplify the full RG equations in the spirit of poor man's scaling and calculate M and G in leading order of $1/\ell n[(V,B)/T_K]$.

The transport of electrons through a quantum dot in the limit of weak coupling to the leads is dominated by Coulomb interaction effects, forcing integral electron charge on the dot [1]. Adding or removing electrons requires a large energy, the charging energy E_c, and thus transport is blocked (Coulomb blockade). Only if energy levels of the addition spectrum of the dot are in the range between the chemical potentials of the two leads transport is possible. The differential conductance as a function of a gate voltage shifting the energy levels of the dot is then characterized by maxima separated by Coulomb blockade valleys. In the case that the total spin of the dot is nonzero, however, the antiferromagnetic exchange interaction of this local spin with the conduction electron spins in the leads gives rise to a Kondo resonance in the local density of states at the Fermi level. Electron transport may then take place via resonance tunneling, thus removing the Coulomb blockade [2,3]. The dramatic effect of the formation of the Kondo resonance at low temperatures in filling up the Coulomb blockade valley in the differential conductance, eventually restoring maximum conductance, has been seen in a number of experiments [4,5,6,7,8].

The Kondo resonance appears as a reaction of the conduction electron system to a continuous sequence of spin flip processes involving the local spin. From the point of view of the conduction electron system a local spin flip amounts to a sudden change of the impurity potential formed by the local spin. Since the scattering states of the conduction electron system before and after the switching are orthogonal, the system takes infinite time to adjust to the new situation (orthogonality catastrophe, [9]). In the course of this adjustment the system forms particle-hole excitations in numbers increasing with an inverse power of excitation energy. Since this process is interrupted when the next spin flip occurs, the infrared singularity in the spectrum associated with the orthogonality catastrophe is cut-off and a smooth resonance peak at the Fermi level is formed instead.

Powerful methods have been developed to describe this strong coupling problem, among them methods based on the renormalization group idea, in a simplified form [10] ("poor man's scaling") and in a full numerical implementation by Wilson [11], exact methods for calculating the thermodynamic quantities using the Bethe ansatz [12] and conformal field theory [13], and resummations of perturbation theory using auxiliary particle representations [14] to access dynamical quantities.

Necessary prerequisites for the existence of the Kondo effect are (1) the spin degeneracy of the ground state of the quantum dot, and (2) quantum coherence over sufficiently long time periods. The characteristic energy scale against which both of these perturbations have to be measured is the Kondo temperature $T_K \simeq D\sqrt{JN_0}\exp(-\frac{1}{2N_0 J})$, where J is the exchange coupling constant, N_0 is the conduction electron density of states per spin at the Fermi level and D is the conduction band width.

A minimal spin exchange interaction model of a quantum dot features three exchange coupling constants, $J_{LL}(J_{RR})$ coupling the local spin to the left (right) lead and $J_{LR} = J_{RL}$ describing the spin-dependent tunneling through the dot. Within a standard Anderson impurity model, characterized by hopping amplitudes t_L, t_R from the dot to the left and right lead, respectively, one obtains by way of a Schrieffer-Wolff transformation $J_{\gamma\gamma'} = t_\gamma t_{\gamma'}/|\epsilon_d|$, $\gamma, \gamma' = L, R$ where ϵ_d is the energy of the dot level closest to the Fermi level. Within such a model one has the relation $J_{LR} = (J_{LL}J_{RR})^{1/2}$. More complex models of a quantum dot allowing for several orbitals and additional interactions may lead to less restrictive relations between the J's [16]. Of particular interest are quantum dots with $J_{LR} \ll J_{LL} = J_{RR}$, as in that case a new fixed point of two-channel character governs the behavior of the system over a wide range of energies [15], but probably not down to the lowest energies [17,18], depending on the applied bias voltage V.

In this paper we consider the effect of a large difference in the chemical potentials μ_L, μ_R in the two leads, generated by a bias voltage $V = \mu_L - \mu_R$. We model the leads as non-interacting electron systems. The voltage is driving a steady state-current I. We assume the temperature T to be uniform over the sample. At sufficiently low temperatures, $T \leq T_K$, the effect of the spin flip processes will be to generate resonance structures in the local density of states at both Fermi edges, μ_L and μ_R. At the same time, at finite bias voltage V, tunneling of electrons through the dot creates a momentary nonequilibrium state in the leads which is equilibrated by dissipative processes. In this way the levels in the dot are broadened by inelastic processes even at $T = 0$, leading to a suppression of the Kondo effect. We will show that at least for sufficiently large J_{LR}, say $J_{LR} = J_{LL} = J_{RR}$, the system remains in the regime where renormalized perturbation theory is applicable.

Transport through a Kondo system has been considered by several authors. Early on, Appelbaum [19] calculated the differential conductance $G(V, B)$ for large bias voltage V in a finite magnetic field B, assuming erroneously, however, that the local spin remains in thermal equilibrium. König et al. [20] calculated $G(V, B)$ for an Anderson model, employing a certain resummation of perturbation theory. Brief accounts of the work to be presented below have been published elsewhere [15,21].

1 Magnetization and Conductance in Lowest Order Perturbation Theory

We study the following Kondo-type model Hamiltonian of a quantum dot:

$$H = \sum_{\mathbf{k},\sigma,\alpha=L,R} (\varepsilon_{\mathbf{k}} - \mu_\alpha) c^\dagger_{\alpha\mathbf{k}\sigma} c_{\alpha\mathbf{k}\sigma} - B S_z$$

$$+ \frac{1}{2} \sum_{\mathbf{k},\mathbf{k}',\sigma,\sigma',\alpha,\alpha'=L,R} J_{\alpha'\alpha} \mathbf{S} \cdot (c^\dagger_{\alpha'\mathbf{k}'\sigma'} \tau_{\sigma'\sigma} c_{\alpha\mathbf{k}\sigma}) , \qquad (1)$$

where $\mathbf{S}$ is the local spin operator on the dot (assumed to be $S = \frac{1}{2}$) and $\boldsymbol{\tau}$ is the vector of Pauli matrices. The $c^+_{\alpha\mathbf{k}\sigma}$ create conduction electrons with momentum $\mathbf{k}$ and spin σ in lead α. The chemical potential shifts induced by the bias voltage are given by $\mu_{L,R} = \pm V/2$ for the left (L) and right (R) lead. The exchange energies $J_{\gamma\gamma'}$, $\gamma = L, R$, will be assumed to have the symmetry $J_{RR} = J_{LL}$ and $J_{LR} = J_{RL}$. We shall use the dimensionless coupling constants $g_d = N_0 J_{RR}$ and $g_{LR} = N_0 J_{LR}$, with N_0 the local density of states at the Fermi energy, assumed to be flat in the accessible regime $|\omega| \leq V, B$. A magnetic field is included, inducing a Zeeman splitting of the magnetic levels of the local spin $\omega_\gamma = -\gamma B/2$, $\gamma = \pm 1$. The effect of the magnetic field on the conduction electrons gives only rise to particle-hole asymmetry terms, which are usually small and may be neglected.

In lowest order perturbation theory the current through the dot is given by the Golden Rule expression

$$I = \frac{\pi e}{4\hbar} \int d\omega \sum_{\gamma=\pm 1} n_\gamma \Big[g_1^2 f_{\omega-\mu_L}(1 - f_{\omega-\mu_R})$$

$$+ 2 g_2^2 f_{\omega-\mu_L}(1 - f_{\omega-\mu_R-\gamma B}) \Big] - (L \leftrightarrow R) , \qquad (2)$$

where g_1 and g_2 are the dimensionless coupling constants for spin-nonflip and spin-flip interaction, $g_1 = g_2 = N_0 J_{LR}$, $f_{\omega-\mu_\alpha}$ is the Fermi function in the lead $\alpha = L, R$, and γB is the Zeeman energy transfer taking place in a spin flip process. At low temperatures, $T \ll V, B$, the product of Fermi functions limits the energy integration to the window $\mu_R < \omega < \mu_L$ and $\mu_R + \gamma B < \omega < \mu_L$, respectively. The current is seen to depend on the occupation $n_\gamma = \frac{1}{2}(1+\gamma M)$ of the local spin states, where M is the magnetization.

For sufficiently small voltage, $V \ll T$, the local spin system is in thermal equilibrium. In the opposite limit, $V \gg T$, the stationary current through the dot drives the system out of equilibrium. The occupation numbers n_γ are then determined by the rate equation $\frac{\partial}{\partial t} n_\gamma = C_\gamma\{n_{\gamma'}\} = 0$, where C is the collision integral. This leads to the condition (assuming $J_{LR} = J_{LL}$ here)

$$n_\uparrow \sum_{\alpha,\alpha'=L,R} \int d\omega g_2^2 f_{\omega - \mu_\alpha} \left(1 - f_{\omega - \mu_{\alpha'} - B}\right) =$$
$$= n_\downarrow \sum_{\alpha,\alpha'=L,R} \int d\omega g_2^2 f_{\omega - \mu_\alpha} \left(1 - f_{\omega - \nu_{\alpha'} + B}\right) . \qquad (3)$$

The spin-flip coupling constant g_2 cancels out of this equation, yielding the nonequilibrium magnetization in the limit $J \to 0$,

$$M = 2 / \left[\coth \frac{B}{2T} + \frac{\sinh \frac{B}{T} - \frac{V}{B} \sinh \frac{V}{T}}{\cosh \frac{B}{T} - \cosh \frac{V}{T}}\right] . \qquad (4)$$

In the limit $V \to 0$, the equilibrium result $M = \tanh \frac{B}{2T}$ is recovered, whereas in the limit $V \gg B, T$, $M = \frac{2B}{V}$, independent of temperature. The result (4) has been obtained independently in [22,24].

2 Leading Logarithmic Corrections to M and I

In order to calculate higher order contributions in perturbation theory in J we switch now to a more formal description in terms of nonequilibrium Green's functions. We find it convenient to represent the local spin operator in pseudo fermion (PF) language [23],

$$\mathbf{S} = \frac{1}{2} \sum_{\gamma\gamma'} f_\gamma^+ \boldsymbol{\tau}_{\gamma\gamma'} f_{\gamma'} , \qquad (5)$$

where f_γ^+ creates a pseudofermion of spin $\gamma = \uparrow, \downarrow$ at the dot. The projection onto the physical sector of Hilbert space, with pseudofermion occupation number $Q = \sum_\gamma f_\gamma^+ f_\gamma = 1$, is done by adding a term λQ to the Hamiltonian and taking the limit $\lambda \to \infty$. This means that the PF system is taken in the low density limit. The Feynman diagrams of perturbation theory in the exchange interaction will therefore have one PF loop at most.

The local magnetization can be calculated in terms of the PF Green's function $G_\gamma^<(\omega)$ as $M = \sum_{\gamma=\pm 1} \gamma \int \frac{d\omega}{2\pi i} G_\gamma^<(\omega)$, where $G^<$ is found by solving a quantum kinetic equation. In steady state one finds

$$G_\gamma^<(\omega) \Gamma_\gamma(\omega) = \Sigma_\gamma^<(\omega) A_\gamma(\omega) \qquad (6)$$

where $A_\gamma(\omega)$ is the PF spectral function, $\Sigma_\gamma^<(\omega)$ is the lesser component of the self-energy and $\Gamma_\gamma(\omega)$ its imaginary part. In general Eq. (6) is an integral equation for $G_\gamma^<(\omega) = in_\gamma(\omega)A_\gamma(\omega)$, which, however, can be approximated in a controlled way by making use of the sharply peaked form of $A_\gamma(\omega)$ to give an algebraic equation for $n_\gamma(0)$. In this form Eq. (6) corresponds to the rate equation discussed above, with higher order processes included in the transition amplitudes. The diagrams of the self-energy up to third order in J are depicted in Fig. 1a. Note that the lines represent Keldysh matrix Green's functions dressed with arrows in all topologically different ways. The result including leading logarithmic corrections is found as

$$M = \frac{Z}{\coth \frac{B}{2T} \left[\frac{Z}{2} + g_-^2 B(1 + \mathcal{L}(B)) \right] + g_{LR}^2 (c(B) + c(-B))} \quad (7)$$

where

$$Z = 2g_+^2 B(1 + \mathcal{L}(B)) \\ + 2g_{LR}^2 \left[2B + (V+B)\mathcal{L}(V+B) - (V-B)\mathcal{L}(V-B) \right] \quad (8)$$

and we use the abbreviations $g_\pm^2 = g_d^2 \pm g_{LR}^2$, $L(B) = 2g_d \ln \frac{D}{|B|}$, and $c(B) = \coth \frac{V+B}{2T}[(V+B)(1+\mathcal{L}(V+B)) + V\mathcal{L}(V) + B\mathcal{L}(B)]$. We note that logarithmic corrections of order $g^3 \ell n D$ to the self-energies lead to corrections of order $g \ell n D$ to M. These corrections are larger than the leading logarithmic corrections in equilibrium ($V \to 0$), which are of order $g^2 \ell n D$. We have neglected these sub-leading corrections in (7) and note in passing that they can be obtained by including the energy shifts of the PF induced by ReΣ. On the other hand, the effects of ImΣ are important as will be discussed below.

The current through the dot may be expressed in terms of a spin-spin correlation function. Defining the conduction electron charge density operator

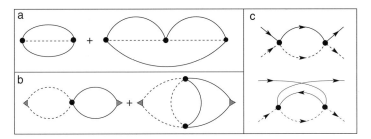

Fig. 1. Feynman diagrams for (a) PF self-energies, (b) current and (c) vertices entering the 1-loop RG equation. PF (electron-) propagators are displayed as dashed (full) lines

in lead α at the dot, $\hat{n}_\alpha = e\sum_{\mathbf{k},\sigma} c^+_{\alpha\mathbf{k}\sigma} c_{\alpha\mathbf{k}\sigma}$, the current operator $\hat{I}$ can be expressed as

$$\hat{I} = -\frac{\partial}{\partial t}\hat{n} = \frac{ie}{\hbar}J_{LR}\mathbf{S}\cdot\sum_{\mathbf{k},\mathbf{k}',\sigma,\sigma'}\left(c^+_{L\mathbf{k}\sigma}\tau_{\sigma\sigma'}c_{R\mathbf{k}\sigma'} - c^+_{R\mathbf{k}\sigma}\tau_{\sigma\sigma'}c_{L\mathbf{k}'\sigma'}\right). \quad (9)$$

The average current is given in terms of the Keldysh component of the contour-ordered correlation function

$$D_{LR}(t,t') = -i\langle\hat{T}_c\Big\{\sum_{\sigma,\sigma'} c^+_{L\mathbf{k}'\sigma'}(t)\tau_{\sigma'\sigma}c_{R\mathbf{k}\sigma}(t)\cdot\mathbf{S}(t')\Big\}\rangle \quad (10)$$

as

$$I = -\frac{e}{\hbar}J_{LR}\mathrm{Re}\Big\{D^K_{LR}(t,t)\Big\}. \quad (11)$$

In Fig. 1b the Feynman diagrams for D^K_{LR} to order J^3 are shown. A long but straightforward calculation yields the current including leading logarithmic corrections

$$I = \frac{e\pi}{2\hbar}g^2_{LR}\Big[\frac{3}{2}V + V\mathcal{L}(V) + (V+B)\mathcal{L}(V+B)$$
$$+ (V-B)\mathcal{L}(V-B) - M(c(B) - c(-B))\Big] \quad (12)$$

In the limit of zero magnetic field, Eq. (12) reduces to $I = \frac{e\pi}{4\hbar}g^2_{LR}3V(1+2L(V))$, in agreement with [25].

3 Resummation of Perturbation Theory in Nonequilibrium: A Poor Man's Scaling Approach

Even for small coupling constants g and for sufficiently large V and B, such that $V, B \gg T_K$, but still in the scaling regime $V, B \ll D$, such that $g\ell n D/|V| \ll 1$, bare perturbation theory converges slowly. It is necessary to sum the leaving logarithmic contributions in all orders of PT. In the equilibrium state a powerful method is available to perform this resummation in a controlled way: the perturbative renormalization group method [10]. It makes use of the fundamental idea that a change of the cut-off D can be fully absorbed into a redefinition of the coupling constants g. As long as the running coupling constant $g(D)$ is small, the change of g under an infinitesimal change of D, $\partial g/\partial \ell n D$, may be calculated in perturbation theory. It is well known that in the equilibrium Kondo problem the coupling constant is found to grow to infinity, thus leaving the perturbative regime beyond $g \sim 1$. A nonperturbative treatment is then necessary, as shown by Wilson [11] in his pioneering work on the numerical RG. In the nonequilibrium situation, the RG flow is cut-off by inelastic processes already within the weak coupling regime, such that perturbation theory is valid [25,15]. This is true at least for the case

$J_{LR} = J_{LL}, J_{RR}$, considered here. While the discussion of the RG flow in [15] was a qualitative one, motivated by results obtained in the so-called "Non-crossing approximation" for the Anderson model, and applied only to the limit $B = 0$, here we consider the RG formulation on a more fundamental level. A different and considerably more involved real-time RG scheme has been developed by Schoeller and König [26].

First we observe that a straightforward renormalization of g like in the equilibrium situation is not possible for $V \gg T_K$. In Eq. (7) logarithmic corrections appear (even in the limit $B \to 0$) as $2ln(D/V)$ in the denominator, but as $ln(D/V) + ln(D/T)$ in the numerator. This is related to the fact that the energy integrals in the numerator are confined to the vicinity of the Fermi energy, whereas in the denominator the energy integral covers a finite range of width V. This suggests that the frequency dependence of the couplings becomes important. In order to understand how these frequency dependencies are generated on the lowest level, i.e. in one-loop order, it is sufficient to analyze the behavior of the vertex corrections shown in Fig. 1c under a change of cut-off. Logarithmic corrections in perturbation theory are generated by PF - conduction electron bubbles containing the product of a real part of a PF Green's function, $1/(\omega \pm B/2)$, and the Keldysh component of the local conduction electron line, $-2\pi i N_0 \tanh[(\omega - \mu_\alpha)/2T]$. If the energy of the PF - conduction electron intermediate state is within the interval $[-D, D]$ the process will contribute to the renormalization of the coupling function $g(\omega)$, otherwise it will not. In general the coupling functions depend on three frequencies (taking energy conservation into account). Using the fact that the spectral function of the PFs is sharply peaked at $\omega = \mp B/2$, we set two of the three frequencies to $\omega = \mp B/2$, keeping only one frequency variable. In addition, the frequency dependence may be neglected within the running band-width, $|\omega| < D$. The spin structure of the general coupling functions is given by two invariant amplitudes, $\tilde{g}_2$ for spin flip and $\tilde{g}_1$ for spin non-flip processes

$$g^{\alpha\sigma,\omega;\alpha'\bar{\sigma},\omega-\gamma B}_{\gamma,-\gamma B/2;\tilde{\gamma},B/2} = (\tau^x_{\gamma\tilde{\gamma}}\tau^x_{\sigma\bar{\sigma}} + \tau^y_{\gamma\tilde{\gamma}}\tau^y_{\sigma\bar{\sigma}})\tilde{g}_2(\omega - \gamma B/2)$$

$$g^{\alpha\sigma,\omega;\alpha\sigma,\omega}_{\gamma,-\gamma B/2;\gamma,-\gamma B/2} = \tau^z_{\alpha\alpha}\tau^z_{\gamma\gamma}\tilde{g}_1(\omega) , \qquad (13)$$

where $g^{a;a'}_{b;b'}$ denotes the coupling function for conduction electrons of spin σ and energy ω in lead $\alpha (a = (\alpha, \sigma, \omega))$ interacting with a pseudo fermion in state $b = (\gamma, \omega_f)$ and going into states a', b'. The two frequency-dependent running coupling functions $\tilde{g}_1(\omega)$ and $\tilde{g}_2(\omega)$ obey the following flow equations,

$$\begin{aligned}\frac{\partial \tilde{g}_1(\omega)}{\partial \ln D} = &-\frac{\tilde{g}_2(\frac{B+V}{2})^2}{2}\left(\Theta_{\omega+B+\frac{V}{2}} + \Theta_{\omega-B-\frac{V}{2}}\right)\\ &-\frac{\tilde{g}_2(\frac{B-V}{2})^2}{2}\left(\Theta_{\omega-B+\frac{V}{2}} + \Theta_{\omega+B-\frac{V}{2}}\right)\end{aligned} \qquad (14)$$

$$\frac{\partial \tilde{g}_2(\omega)}{\partial \ln D} = -\frac{\tilde{g}_1(\frac{V}{2})\tilde{g}_2(\frac{B+V}{2})}{2}\left(\Theta_{\omega+\frac{B+V}{2}} + \Theta_{\omega-\frac{B+V}{2}}\right)$$
$$-\frac{\tilde{g}_1(\frac{V}{2})\tilde{g}_2(\frac{B-V}{2})}{2}\left(\Theta_{\omega-\frac{B-V}{2}} + \Theta_{\omega+\frac{B-V}{2}}\right),$$

with initial condition $\tilde{g}_1(\omega) = \tilde{g}_2(\omega) = JN_0$. In the limit $V, B \to 0$, Eq. (14) reduces to the well-known scaling equations (8).

So far the flow equations (14) reflect the nonequilibrium physics in two ways: they describe the renormalization of the coupling constants at the (up to four) different resonant energies, and they account for the variation with frequency in the regions between and somewhat outside the Fermi edges. There is, however, still one additional effect missing: the finite relaxation of the spins even in the limit $T \to 0$. Within the PF representation this effect shows up as a finite imaginary part of the PF self-energy. The relaxation is modified by vertex corrections, but is certainly not canceled altogether [15]. The relaxation rate may be interpreted as due to spin relaxation processes, leading to inelastic broadening of the local spin levels, thus cutting off the RG flow. While different rates exist for the different coupling functions, in leading approximation we may neglect the differences between them and assume a single rate Γ, which we identify with the transverse spin relaxation rate $\Gamma = 1/T_2$.

Assuming the RG flow to be stopped at the scale Γ, we replace the step functions Θ_ω in Eq. (14) by $\Theta(D - \sqrt{\omega^2 + \Gamma^2})$. The decay rate Γ has to be calculated self-consistently with the solution of the flow equations (14). Starting from the golden rule expression for the transverse relaxation rate,

$$\Gamma = \frac{\pi}{4\hbar} \sum_{\alpha,\alpha'=L,R,\gamma=\uparrow,\downarrow} \int d\omega \Big[\tilde{g}_1(\omega)^2 f_{\omega-\mu_\alpha}(1 - f_{\omega-\mu_{\alpha'}})$$
$$+ \tilde{g}_2(\omega - \gamma B/2)^2 f_{\omega-\mu_\alpha}(1 - f_{\omega-\mu_{\alpha'}-\gamma B})\Big] \quad (15)$$

the renormalized Γ is obtained by replacing $g_{1,2}$ by $\tilde{g}_{1,2}(\omega)$ as determined from the solution of (2).

In Fig. 2 the running coupling functions $\tilde{g}_1(\omega)$ and $\tilde{g}_2(\omega)$ are shown for different values of D. One can see how a Kondo resonance structure begins to form at the two Fermi energies, each one split by twice the Zeeman energy B. The growth of $\tilde{g}_{1,2}$ stops, however, due to the relaxation effects embodied in Γ, such that $\tilde{g}_{1,2}(\omega) \ll 1$ for all ω and D.

We are now ready to calculate further physical quantities. The renormalized value of the magnetization is obtained by substituting $\tilde{g}_{1,2}(\omega)$ in place of $g_{1,2}$ in the Golden Rule expression (3). In Fig. 3b we show the fully renormalized result for M as a function of V/B for different values of B.

The charge current I is calculated from Eq. (2) inserting the renormalized coupling functions. Fig. 3a shows a comparison of the differential conductance $G(V) = dI/dV$ obtained in this way, with the bare result (2) and the PT result (12).

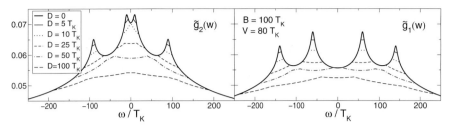

Fig. 2. Renormalized coupling constants $\tilde{g}_1(\omega)$ (right panel) and $\tilde{g}_2(\omega)$ (left panel) for $B = 100 T_K$ and $V = 80 T_K$ and for different values of the cut-off D

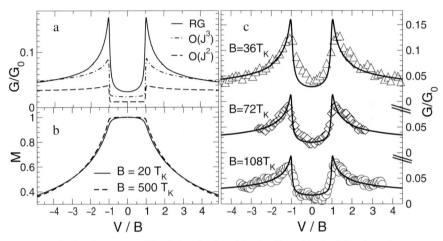

Fig. 3. a) Conductance $G(V/B)$ calculated in leading (dashed line) and next-to-leading (dot-dashed line) order PT compared to the result of perturbative RG (solid line) for $B = 100 T_K$, $D = 10^4 T_K$, $T_K = D\sqrt{g}e^{-1/2g}$). b) Local magnetization $M(V/B, B/T_K)$ of a symmetric dot for fixed magnetic field $B = 20, 500 \, T_K$. c) Experiments by Ralph and Buhrman [27] (symbols) on transport through metallic point contacts in magnetic fields $0.85, 1.7, 2.55$ T ($B = 36, 72, 104 \, T_K$ where $T_K \approx$ 30mK [27]). Assuming that the corresponding point contact is described by a single-channel ($J_{LR} = \sqrt{J_{LL} J_{RR}}$) Kondo model, $\frac{J_{RR}}{J_{LL}} \approx 4.2$ is determined from $G(V=0, B=0, T=50\text{mK}) = \frac{4 J_{LL} J_{RR}}{(J_{LL}+J_{RR})^2} G_\text{sym}(T/T_K)$, where G_sym is known exactly from NRG calculations. This fixes *all* parameters for our RG calculation (solid lines) which uses a straightforward generalization of [21] for $J_{LL} \neq J_{RR}$. As the (B-dependent) background is not known experimentally, we subtract $\Delta G = G_B - 5.2 \cdot 10^{-5} G_0 \frac{V}{T_K}$, where G_B is fitted to our results at large V. Our calculations are at $T = 0$; the experimental $T = 50$mK leads to an extra small broadening at $V = B$

The peak structures in $G(V, B)$ appearing at $V = \pm B$ have been detected in experiment. In order to reach large values of B/T_K, it is necessary to have relatively low T_K. This happened to be the case in transport through metallic point contacts, containing a magnetic impurity [27]. In Fig. 3c the result of our theory using the value of $T_K \simeq 30mK$ given in [27] is compared to the experimental data, after subtracting a background contribution [21]. The agreement is seen to be excellent.

4 Conclusion

We considered the transport of electrons through a Kondo dot in the regime of large transport voltage V and in the presence of a magnetic field B, such that $\max(V, B) \gg T_K$ is satisfied. A finite difference of the chemical potentials in the two leads, $\mu_L - \mu_R = V$, opens an energy window for inelastic processes even at $T = 0$, leading to a broadening Γ of the spin sublevels of the dot and providing a cut-off for the scaling towards the Kondo fixed point. A finite magnetic field induces a local magnetization M at the dot, which for $V \gg T_K, T$ is independent of temperature, being solely controlled by the voltage V. This can be seen on an elementary level by considering the rate equations for the spin occupation number. The fact that M is controlled by V has been overlooked in early work from the sixties on this problem [19]. The differential conductance is likewise affected. We calculated the leading logarithmic corrections to M and G and found that the structure of these log-terms changes in nonequilibrium. An analysis of the reasons for these changes lead us to conjecture a renormalization group formulation allowing for frequency dependent coupling functions [21]. We derived a set of RG equations within a poor man's scaling approximation, which for once reproduce the bare PT result, but then may be integrated to give the fully renormalized result. The integration is done including the self-consistently determined cut-off Γ. The results obtained in this way are valid up to corrections of order $1/\ell n[(V, B)/T_K]$. It is clear that the fully renormalized result features much more pronounced peaks than either Eq. (2) or (12), demonstrating the necessity of the RG treatment.

Acknowledgments

We would like to thank S. De Franceschi, J. König, O. Parcollet, H. Schoeller and A. Shnirman for helpful discussions and especially L. Glazman, who suggested investigating the case of finite B. Part of this work was supported by the Center for Functional Nanostructures and the Emmy-Noether program (A.R.) of the DFG.

References

1. I. Aleiner, P. Brouwer, and L. Glazman, cond-mat/0103008.
2. L. Glazman and M. Raikh, JETP Letters **47**, 452 (1988).
3. T. Ng and P.A. Lee, Phys. Rev. Lett. **61**, 1768 (1988).

4. D. Goldhaber-Gordon, H. Shtrikman, D. Mahalu, D. Abusch-Magder, U. Meirav and M. Kastner, Nature **391**, 156 (1998).
5. S. Cronenwett, T. Oosterkamp and L. Kouwenhoven, Science **281**, 540 (1998).
6. J. Schmid, J. Weis, K. Eberl, and K. v. Klitzing, Physica B **258**, 182 (1998).
7. J. Nygard, D. Cobden and P. Lindelof, Nature **408**, 342 (2000).
8. S. De Franceschi *et al.*, cond-mat/0203146.
9. P. W. Anderson,Phys. Rev. Lett. **18**, 1049 (1967).
10. P. W. Anderson, J. Phys. C **3**, 2436 (1966).
11. K.G. Wilson, Rev. Mod. Phys. **47**, 773 (1975).
12. N. Andrei, Phys. Rev. Lett. **45**, 379 (1980); N. Andrei, K. Furuya and J.H. Lowenstein, Rev. Mod. Phys. **55**, 331 (1983).
13. I. Affleck and A. W. W. Ludwig, Nucl. Phys. B **360**, 641 (1991).
14. J. Kroha and P. Wölfle, in press (Springer), cond-mat/0105491.
15. A. Rosch, J. Kroha and P. Wölfle, Phys. Rev. Lett. **87**, 156802 (2001).
16. M. Pustilnik and L. I. Glazman, Phys. Rev. Lett. **87**, 216601 (2001).
17. P. Coleman, C. Hooley, and P. Parcollet, Phys. Rev. Lett. **86**, 4088 (2001).
18. X.-G. Wen, cond-mat/9812431.
19. J. Appelbaum, Phys. Rev. Lett, **17**, 91 (1966); Phys. Rev. **154**, 633 (1967).
20. J. König, J. Schmid, H. Schoeller, and G. Schön, Phys. Rev. B **54**, 16820 (1996).
21. A.Rosch, J. Paaske, J. Kroha and P. Wölfle, cond-mat/0202404.
22. L.I. Glazman, private communication.
23. A.A. Abrikosov, Physics **2**, 21 (1965.
24. O. Parcollet, and C. Hooley, cond-mat/0202425.
25. A. Kaminski, Yu. V. Nazarov, and L.I. Glazman, Phys. Rev. Lett. **83**, 384 (1999); L. Glazman and A. Kaminski, Phys. Rev. B **62**, 8154 (2000).
26. H. Schoeller and J. König, Phys. Rev. Lett. **84**, 3686 (2000); see also M. Keil, Ph.D. thesis, U. Göttingen (2002).
27. D.C. Ralph and R.A. Buhrman, Phys. Rev. Lett. **72**, 3401 (1994).

Part IV

Thin Films and Surfaces

Electrons, Phonons and Excitons at Semiconductor Surfaces

Johannes Pollmann, Peter Krüger, Albert Mazur, and Michael Rohlfing

Institut für Festkörpertheorie, Universität Münster,
Wilhelm-Klemm-Strasse 10, 48149 Münster

Abstract. We briefly address 'state-of-the-art' *ab-initio* calculations of basic properties of semiconductor surfaces such as their atomic configuration, electronic structure and surface vibrations, as well as, their optical properties and compare exemplary results with experimental data. The surface structure and the electronic ground state are described within the local density approximation (LDA) of density functional theory (DFT). The description of excited electronic states requires to take dynamical correlations in the many electron system into account, which is achieved to a considerable extent within the *GW* approximation (GWA) leading to the concept of the quasiparticle bandstructure. Surface phonons are treated from first principles within density functional perturbation theory (DFPT). The theory of optical surface properties and surface excitons, in particular, requires to account for the electron-hole Coulomb correlation which is done in the framework of the Bethe-Salpeter equation (BSE). These methods yield results which are in good agreement with experiment and can significantly contribute to an interpretation of experimental data from high-resolution surface microscopy and spectroscopy thus enhancing our current understanding of semiconductor surfaces.

1 Introduction

Semiconductor surfaces are intensively studied worldwide because of their large potential for technological applications and their basic importance in fundamental research. Their very intriguing physical properties have been an ever increasing stimulus to fully understand their atomic and electronic structure, as well as, their vibronic and optical properties on a microscopic level. The field of atomic and electronic structure theory of semiconductor surfaces has matured within the last decade. It is nowadays possible to theoretically determine optimal surface structures with a good level of confidence by employing total energy minimization techniques and to self-consistently evaluate the respective surface electronic structure, in particular charge densities, energy bands and wave-vector resolved layer densities of states (LDOS) with high precision. More recently, it has become possible to carry out very accurate *ab-initio* calculations of surface vibronic and optical properties, as well. Here we concentrate on a discussion of *ab-initio* approaches and respective results, therefore.

In the discussion of examples, we focus on a few prototypical systems which we have studied intensively in the last several years. In particular, we relate specific electronic, vibronic and optical properties of exemplary surfaces to their atomic structure. Particular surfaces of a given bulk solid and their structural variants often constitute largely different systems because most physical properties of a surface sensitively depend on its specific atomic configuration. In this respect, e. g., bulk Si and Ge have probably more in common than the Si(001)-(2×1) and the Si(111)-(7×7) surface. The investigation of any physical surface property, therefore, is to a certain extent also a determination of its structural properties. Discussing electrons, phonons or excitons at surfaces, therefore, necessitates in each and every case to address their atomic structure, as well.

2 Electrons at Semiconductor Surfaces

In this Section, we first briefly address current days theoretical LDA calculations of the structure and the electronic ground state of semiconductor surfaces and present some exemplary results in comparison with experimental data. Next we address the calculations of quasiparticle bandstructures and present representative results of GWA calculations in comparison with experiment.

2.1 LDA Calculations of Surface Atomic and Electronic Structure

The calculation of electronic properties of semiconductor surfaces is complicated, in practice, by two obstacles. First, the translational invariance perpendicular to a surface is broken so that Bloch's theorem only allows to classify electronic surface states by a wavevector $\mathbf{k}_\parallel$ that is parallel to the surface. Second, and much more importantly, for many surfaces the actual configuration of the atoms at and near the surface is not very precisely known *a priori*. Since the electronic surface structure is very sensitively dependent on the surface atomic structure, as mentioned above, the calculation of surface electronic properties, in general, constitutes a coupled atomic and electronic structure problem.

Most current days surface electronic structure calculations deal with this situation by referring to density functional theory [1] within local density approximation [2]. Due to its formal and computational simplicity, as well as, due to its very impressive successes in describing ground-state properties of many-electron systems, DFT-LDA has become the dominant approach for calculating structural and electronic properties of bulk semiconductors and their surfaces. Within DFT-LDA the total energy of a surface system is given by:

$$E_{tot}(\rho, \{\mathbf{R}_i\}) = E_{kin} + E_{el-ion} + E_{coul} + E_{xc}^{LDA} + E_{ion-ion}, \quad (1)$$

with

$$E_{kin} = \sum_{sk_\parallel} \int \psi_{sk_\parallel}^*(\mathbf{r})(-\frac{\hbar^2}{2m}\nabla^2)\psi_{sk_\parallel}(\mathbf{r})d^3r \tag{2}$$

$$E_{el-ion} = \int V_{el-ion}(\{\mathbf{R}_j\},\mathbf{r})\rho(\mathbf{r})d^3r \tag{3}$$

$$E_{coul} = \frac{e^2}{2}\int\int \frac{\rho(\mathbf{r})\rho(\mathbf{r}')}{|\mathbf{r}-\mathbf{r}'|}d^3r d^3r' \tag{4}$$

$$E_{xc}^{LDA} = \int \rho(\mathbf{r})f_{xc}^{LDA}(\rho(\mathbf{r}))d^3r \tag{5}$$

$$E_{ion-ion} = \frac{e^2}{2}\sum_{i,j}\frac{Z_i \cdot Z_j}{|\mathbf{R}_i-\mathbf{R}_j|}. \tag{6}$$

Various approximations for $f_{xc}^{LDA}(\rho)$ have been discussed, e.g., by Wimmer and Freemann [3]. The total energy depends parametrically on the positions $\{\mathbf{R}_i\}$ of all atoms in the system and is a functional of the electronic charge density $\rho(\mathbf{r})$ which has to be calculated self-consistently.

Minimizing the total energy with respect to the total valence-charge density ρ under the constraint of orthonormalized wavefunctions yields the Kohn-Sham equations [2]

$$\{-\frac{\hbar^2}{2m}\nabla^2 + V_{eff}^{LDA}(\mathbf{r})\}\psi_{sk_\parallel}(\mathbf{r}) = E_{sk_\parallel}^{LDA}\psi_{sk_\parallel}(\mathbf{r}) \tag{7}$$

$$V_{eff}^{LDA}(\rho(\mathbf{r}),\{\mathbf{R}_i\})\} = V_{ion}(\{\mathbf{R}_i\}) + V_{coul}(\rho(\mathbf{r})) + V_{xc}^{LDA}(\rho(\mathbf{r})) \tag{8}$$

for the one-particle wavefunctions labeled by the quantum numbers s and $\mathbf{k}_\parallel$. The effective one-particle potential in these equations is a sum of an ionic potential V_{ion}, which is most often used in the form of a pseudopotential, the Coulomb potential V_{coul} and the exchange-correlation potential V_{xc} which is most often used in the Ceperley-Alder [4] form as parametrized by Perdew and Zunger [5].

Minimizing the total energy with respect to all structural degrees of freedom $\{\mathbf{R}_i\}$ of a surface system by eliminating the forces

$$\mathbf{F}_i = -\frac{\partial}{\partial \mathbf{R}_i}E_{tot}(\{\mathbf{R}_j\}) \stackrel{!}{=} 0 \quad \forall i \tag{9}$$

yields the optimal surface atomic structure corresponding to a minimum of the total energy in configuration space (see, e.g., Refs. [6,7,8]).

Most current surface electronic structure calculations are carried out using the supercell technique for representing the surface system [9]. Truely

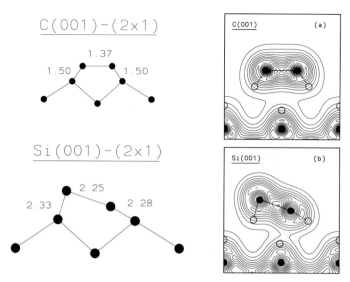

Fig. 1. Structure and valence charge densities of the C(001)-(2×1) and Si(001)-(2×1) surfaces (from Ref. [11])

semi-infinite semiconductors can be described employing Green function techniques [10].

As a first example, we address the intensively studied (001)-(2×1) surfaces of diamond and Si (see e.g. Refs. [11,12]). Fig. reffig1 shows side views of the energy-optimized atomic structures of the (2×1)-reconstructed surfaces of C(001) and Si(001) together with the respective valence charge densities. The C(001) surface shows a symmetric dimer reconstruction while Si(001) shows an asymmetric dimer reconstruction. The asymmetry of the dimers at the Si(001) surface leads to a charge transfer from the *down* to the *up* atoms of the dimers. The charge density of the dimer bond, in consequence, is no longer symmetric, as can be seen in Fig. 1. The *up* atom behaves like an anion and the *down* atom like a cation, so that the Si(001) surface layer is heteropolar or ionic.

The surface electronic structure of these two surfaces, as resulting from LDA calculations for semi-infinite systems, is shown in Fig. 2. In the symmetric dimer models (SDM) of C(001) and Si(001), the π interactions between the two singly occupied dangling bond states give rise to an occupied π band and an empty π^* band. Due to the strong π-interactions in the dimers at the C(001) surface, these two bands are energetically separated and the SDM is semiconducting in agreement with experiment. Any asymmetry of the dimers gives rise to an increase in total energy. Contrary, in the case of Si(001) the π-interactions are considerably weaker so that the π and π^* bands overlap in energy. The surface is metallic in this model, therefore, which is in disagreement with experiment. Allowing for an asymmetry of the dimer recon-

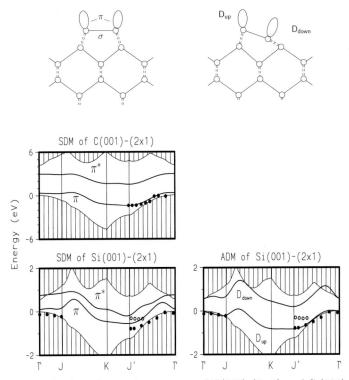

Fig. 2. Surface electronic structure of C(001)-(2×1) and Si(001)-(2×1) in comparison with ARPES data [14,15,16] (from Ref. [11])

struction in the total energy minimization reduces the total energy by some 0.15 eV per surface layer unit cell. The optimized structure in Fig. 1 is in excellent agreement with experimental results from core-level photoelectron diffraction [13]. The two dangling bond bands of this ADM surface, D_{up} and D_{down}, are well separated in energy and the surface is semiconducting.

Experimental angle resolved photoelectron spectroscopy (ARPES) data are shown for comparison in Fig. 2, as well. It is obvious that the data of Graupner et al. [14] for the C(001) surface are in very good agreement with the calculated occupied π band and the ARPES data of Johansson et al. [15] and of Uhrberg et al. [16] for the Si(001) surface are in close agreement with the calculated occupied D_{up} band of the ADM of Si(001)-(2×1). The two latter data sets show a mutual offset of some 0.3 eV which is probably due to differences in the extrinsic Fermi level of the samples used in the two experiments. Apart from this 0.3 eV shift they are in close agreement. The calculated electronic structure of the SDM of the Si(001) surface, shown in Fig. 2, as well, for reference sake, on the contrary, can hardly be reconciled with the data lending further support to the appropriateness of the asymmetric dimer model for the Si(001)-(2×1) surface.

194   Johannes Pollmann et al.

As a second example, we address the Si-terminated SiC(001)-c(4×2) surface. SiC is a wide-band-gap semiconductor material which is of large technological importance. SiC surfaces have been and still are being studied intensively. For a few reviews on experimental and theoretical work see e.g. Refs. [17,18,19].

Among many other reconstructions, the Si-terminated SiC(001) surface shows a c(4×2) reconstruction. The currently most intensively discussed models for this reconstruction are the 'alternate up and down dimer' (AUDD) and the 'missing row asymmetric dimer' (MRAD) models. They have been suggested by Soukiassian and coworkers [20] and by Lu et al. [21], respectively. It should be noted, that the MRAD model has half a Si adlayer on the clean Si-terminated surface while the AUDD model is based on the clean surface without an adlayer.

Figure 3 shows the results of calculations of the formation energy, i.e. the change in grand canonical potential, of various reconstruction models. One is the AUDD model, and another is the 2×1 reconstructed clean surface (labeled (2×1) in the figure) which surves as energy reference. Further models containing one full Si adlayer (labeled Si:(2×1)) or half a Si adlayer with Si adatoms in each second surface row in ideal Si lattice positions (labeled

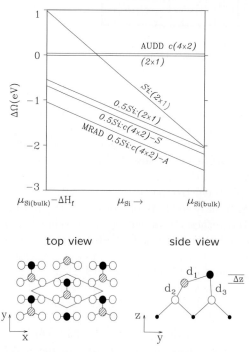

Fig. 3. Formation energy and MRAD model for the Si-terminated SiC(001)-c(4×2) surface (from Ref. [21])

0.5Si:(2×1)) have been studied. When the atoms in the latter model form symmetric or asymmetric dimers we have labeled the resulting configurations as 0.5Si:(2×1)-S and 0.5Si:(2×1)-A, respectively. The latter structure is the MRAD model which is shown in a top and a side view in the bottom panels of Fig. 3. Clearly, on the basis of the formation energy results the MRAD model is most favourable. The MRAD model exhibits *up* and *down* dimer atoms (see the lower right panel of Fig. 3) very much like the Si(001)-(2×1) surface which give rise to respective dangling bond bands.

They are shown in Fig. 4, where we present the surface bandstructure of the Si-terminated SiC(001)-c(4×2) surface. The calculated D_{up} band is found to be in very good agreement with the ARPES data of Duda *et al.* [22]. The occupied dimer bond band around -1 eV is in agreement with experiment, as well. Respective results for the electronic structure of the AUDD model which we have calculated for comparison, as well, do not at all show this kind of good agreement.

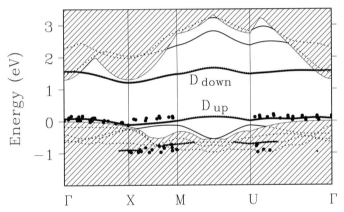

Fig. 4. Electronic band structure of the Si-terminated SiC(001)-c(4×2) surface [21] in comparison with ARPES data [22]

2.2 Quasiparticle Surface Bandstructure Calculations

So far we have discussed DFT-LDA calculations for semiconductor surfaces. But excitation energies do not directly follow from DFT, since the one-particle eigenvalues in LDA are not formally interpretable as quasiparticle energies. For semiconductors, the major difficulty stems from the need to adequately treat dynamical correlations of the electrons in a solid with an energy gap and with a strongly inhomogeneous charge density distribution. To remedy this problem one has to refer to a formalism that allows to include the dynamical correlation between the electrons of the system at least to a certain extent.

Such an approach is the *GW* approximation for the solution of the many-body equations for a correlated many electron system as proposed already some 35 years ago by Hedin and Lundqvist [23].

The basic object of the theory is a nonlocal, non-Hermitian and energy-dependent self-energy operator $\Sigma(\mathbf{r}, \mathbf{r}', E)$. Instead of the Kohn-Sham equation in DFT-LDA (see Eq. 7) one is led to the quasiparticle equation:

$$\left\{ -\frac{\hbar^2}{2m} \nabla^2 + V_{\text{ion}}(\mathbf{r}) + V_{\text{coul}}(\rho(\mathbf{r})) \right\} \psi_{s\mathbf{k}_\parallel}(\mathbf{r})$$

$$+ \int \Sigma(\mathbf{r}, \mathbf{r}', E_{s\mathbf{k}_\parallel}) \psi_{s\mathbf{k}_\parallel}(\mathbf{r}') d^3 r' = E_{s\mathbf{k}_\parallel} \psi_{s\mathbf{k}_\parallel}(\mathbf{r}). \tag{10}$$

As in LDA, the electron-ion interaction V_{ion} is described in many applications by a pseudopotential. Norm-conserving *ab-initio* pseudopotentials are mostly used in these calculations for semiconductors.

The central difficulty connected with Eq. 10 is to find an adequate approximation for the self-energy operator $\Sigma(\mathbf{r}, \mathbf{r}', E)$. In lowest approximation, Σ is given as a product of the Green's function G times the dynamically screened Coulomb interaction W, reading:

$$\Sigma(\mathbf{r}, \mathbf{r}', E) = \frac{i}{2\pi} \int e^{-i\omega 0^+} G(\mathbf{r}, \mathbf{r}', E - \omega) W(\mathbf{r}, \mathbf{r}', \omega) d\omega. \tag{11}$$

The screened interaction W can be written in terms of the inverse dielectric function. The approximation of Σ by Eq. 11 is called the *GW* approximation [23]. Hybertsen and Louie [24], as well as, Godby, Schlüter and Sham [25] developed practicable schemes for evaluating the many body corrections within GWA. They arrived, like Rohlfing *et al.* [26], at theoretical results which are in very good agreement with a whole body of experimental data (for more recent reviews see [27,28]).

To present one result of a quasiparticle bandstructure calculation for a semiconductor surface we have chosen Si(111)-(2×1). This surface is terminated by π-bonded chains of Si atoms in a structure that was suggested by Pandey [29]. Side views of this model are shown in the lower panels of Fig. 5. Different from the originally proposed structure, which was supposed to consist of flat chains, a significant buckling has been observed both in experiment and theory [30,31,32]. By total-energy minimisation within LDA a height difference of 0.51 Å between the up and down atoms of the chains has been found [33].

Figure 5 shows the quasiparticle surface bandstructure, as well as, charge densities and local densities of states. The D_{up} states, giving rise to the D_{up} dangling bond band, are localized at the *up* atoms in the chains and the D_{down} states, giving rise to the D_{down} dangling bond band, are localized at the *down* atoms in the chains. The charge density and LDOS plots clearly highlight the different localization properties of these two states. The D_{up}

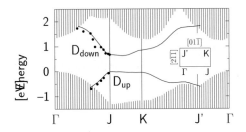

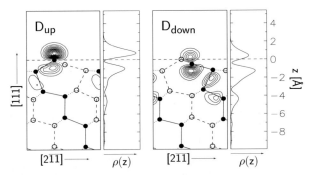

Fig. 5. Quasiparticle surface band structure and charge densities of Si(111)-(2×1) from [33] in comparison with photoemission data [34]

band is fully occupied while the D$_{down}$ band is entirely empty. Therefore, the former can be observed in ARPES measurements while the latter is seen in ARIPES measurements. The quasiparticle bandstructure in Fig. 5 obviously is in excellent agreement with the data confirming both the structure model of the surface and the appropriateness of the GWA to describe excited surface electronic states.

3 Phonons at Semiconductor Surfaces

In order to study surface phonon properties from first principles, one can employ density functional *perturbation* theory (DFPT), as suggested by Baroni and coworkers [35]. Both the optimal atomic structure and the total energy of the ground state as a function of the atomic positions are necessary prerequisites for surface phonon calculations. Therefore, the surface electronic structure and the equilibrum configuration are calculated at the outset within DFT-LDA as described in Section 2. From the known total energy the atomic force constant matrix

$$\Phi_{i\;j}^{\alpha\beta} = \frac{\partial^2 E_{tot}}{\partial u_i^\alpha \, \partial u_j^\beta}\bigg|_{\{\mathbf{u}\}=\{\mathbf{0}\}} \tag{12}$$

is calculated. The dynamical matrix is given as usual by

$$D_{\mu\nu}^{\alpha\beta}(\mathbf{q}) = \frac{1}{\sqrt{M_\mu M_\nu}} \sum_m e^{-i\mathbf{q}\mathbf{R}_m} \Phi_{m\mu,0\nu}^{\alpha\ \beta} \qquad i = m, \mu;\ j = l, \nu. \qquad (13)$$

Here i = m, μ and j = l, ν specify the atomic positions within the unit cell (μ and ν), as well as, in the Bravais lattice (m and l). The indices α and β refer to the cartesian directions. The displacements of the atoms from their equilibrium positions are labeled by $\mathbf{u}_i$ and $\mathbf{u}_j$, respectively.

In DFPT the dynamical matrix takes the form:

$$D_{\mu\nu}^{\alpha\beta}(\mathbf{q}) = D_{\mu\nu}^{\alpha\beta}(\mathbf{q})^{ion} + D_{\mu\nu}^{\alpha\beta}(\mathbf{q})^{el,1} + D_{\mu\nu}^{\alpha\beta}(\mathbf{q})^{el,2} \qquad (14)$$

It consists of an ionic and an electronic part. The former is calculated from the Madelung energy employing Ewald summation techniques. The latter yields two terms when the Hellmann-Feynman theorem is applied in Eq. 12. The first of these is determined by the electronic wavefunctions of the system together with second derivatives of the perturbation potential which occur due to the atomic displacements involved in vibronic excitations. The second is extremely demanding numerically. It accounts for the screening of the electron system in response to the displacement of the ions. Therefore, it has to be calculated iteratively up to self-consistency, because the screened perturbation potential has to be taken into account [35].

Finally, the calculation of the dispersion of surface phonons is carried out by diagonalizing the dynamical matrix. In the actual calculations again supercell geometries are most often used to represent the surface system. To resolve surface vibrational properties very clearly, it is advantageous to 'intercalate' exact bulk layer contributions in the dynamical matrix by some slab filling procedure that enlarges the effective slab thickness correspondingly. These contributions have to be calculated, anyway, for the determination of the projected bulk-phonon spectrum that serves as a reference for the identification of bound surface phonon modes and of surface resonances (see, e.g., Refs. [36,37]).

We briefly address results of surface phonon calculations for two particular systems. The first is the Si(001)-(1×1) surface passivated by a monolayer of sulfur and the second is the Si(111)-(1×1) surface covered by a monolayer of As. We label these systems as S:Si(001)-(1×1) and As:Si(111)-(1×1), respectively. Both systems have a very simple atomic structure. Due to the adlayer of S or As atoms, respectively, the Si(001) or Si(111) substrate surfaces unreconstruct and recover their ideal 1×1 surface geometry. The S or As adatoms basically are located in Si lattice positions above the respective surface with only a slight relaxation towards (for S) and away from (for As) the substrate. (for details see, e.g., Refs. [37,38]).

Our very recent calculations for the first of these two systems [37] yield the equilibrium geometry and the surface phonon dispersions as shown in Fig. 6. The S adatoms give rise to pronounced surface vibrational modes. Those

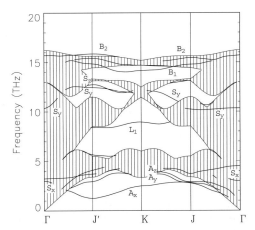

Fig. 6. Optimized surface atomic structure and phonon dispersions of the S-passivated Si(001)-(1×1) surface (from Ref. [37])

labeled A are Rayleigh-type surface modes originating from acoustic bulk phonon modes. Those labeled B are characteristic for the Si bulk substrate. Finally the modes labeled S are highly localized at the surface and they are charactristic surface modes induced by the S adatoms. The mode S_x is a libration mode of the Si-S-Si triangles in the direction perpendicular to the triangles. Actually, only the S adatoms move in this mode and no bond length is altered thereby. In consequence, the S_x mode is a low frequency mode (near 3 THz at Γ), as can be seen in Fig. 6. The S_y mode involves displacements of the S adlayer atoms and the Si substrate surface layer atoms along the direction of the Si-S-Si bonds but in opposite directions. Thus bond lengths are elongated or shortened periodically in this mode. It has a much larger frequency (about 10 THz at Γ) than the S_x mode, therefore. For shortness sake we refrain here from presenting and discussing a considerable number of atomic displacement patterns of the various surface modes at various $\mathbf{q}_\parallel$ wavevectors and refer the interested reader to Ref. [37].

So far, there are no experimental data on surface phonons of this interesting passivating system available. We hope that our recent publication will stimulate experimental interest in the system

To conclude this Section and to compare at least some results of surface phonon calculations with experimental data we finally address surface phonon

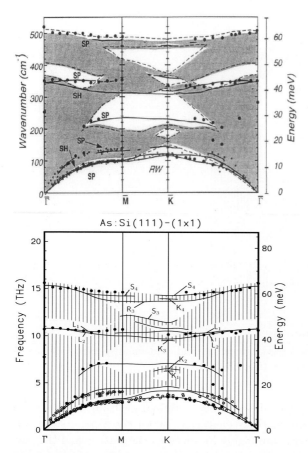

Fig. 7. Comparison of the surface phonon dispersions as calculated within DFPT [39] (upper panel) and ETBM/DFT [38] (lower panel) in comparison with HREELS [40] and HAS [41] data

dispersions of the As:Si(111)-(1×1) system which are shown in Fig. 7. They have been calculated by Honke et al. [39] within DFPT (upper panel) and by Gräschus et al. [38] within a semi-empirical total energy approach (lower panel). The points represent data measured by Schmidt and Ibach [40] using high-resolution electron energy loss spectroscopy (HREELS) and by Doak et al. [41] using helium atom scattering (HAS). The figure clearly reveals that the first-principles DFPT calculations of Honke et al. [39] yield surface phonon dispersions which are basically in quantitative agreement with experiment. The same holds for the results of our semi-empirical calculations [38] which are, however, not free from parameters. The good agreement of the latter calculations is certainly related to the fact, that the parameters of the total energy *ansatz* have been determined from accompanying DFT-LDA calculations [38].

4 Excitons at Semiconductor Surfaces

Optical excitations and excitons at surfaces can be calculated in three successive steps. First, the electronic ground state and the geometric structure is evaluated within DFT-LDA. We have discussed that step in Subsection 2.1. Next, the quasiparticle spectrum is evaluated within the GW approximation, which we have addressed already in Subsection 2.2. In the third step, the Bethe-Salpeter equation for coupled electron-hole excitations [42] is solved. This yields the excitation energies and wave functions and allows to calculate the optical transition matrix elements for bound and continuum exciton states. For a detailed discussion of the theoretical concepts we refer to Refs. [43,44].

Based on the quasiparticle (QP) states and energies coherently coupled electron-hole excitations $|S\rangle$ are addressed. They can be expanded in the basis given by the QP states, i.e. by the ansatz [43]

$$|S\rangle = \sum_{\mathbf{k}} \sum_{v}^{\text{hole}} \sum_{c}^{\text{elec}} A^S_{vc\mathbf{k}} \hat{a}^\dagger_{v\mathbf{k}} \hat{b}^\dagger_{c,\mathbf{k}+\mathbf{Q}} |N,0\rangle$$

$$=: \sum_{\mathbf{k}} \sum_{v}^{\text{hole}} \sum_{c}^{\text{elec}} A^S_{vc\mathbf{k}} |vc\mathbf{k}\rangle \quad . \tag{15}$$

where $\hat{a}^\dagger_{v\mathbf{k}}$ and $\hat{b}^\dagger_{c,\mathbf{k}+\mathbf{Q}}$ simultaneously create a hole (v) or an electron (c), respectively, at wave vector $\mathbf{k}$ (or $\mathbf{k}+\mathbf{Q}$, respectively) to the many-body ground state $|N,0\rangle$ which is given by the fully occupied quasiparticle valence bands. The coefficients $A^S_{vc\mathbf{k}}$ describe the coherent coupling among the independent-particle pair configurations $|vc\mathbf{k}\rangle$.

From the equation of motion of the two-particle Green function one can derive a Bethe-Salpeter equation (BSE) of the states defined by Eq. (15) [42]:

$$(E_{c,\mathbf{k}+\mathbf{Q}} - E_{v\mathbf{k}})A^S_{vc\mathbf{k}} + \sum_{\mathbf{k}'} \sum_{v',c'} \langle vc\mathbf{k}|K^{eh}|v'c'\mathbf{k}'\rangle A^S_{v'c'\mathbf{k}'} = \Omega_S A^S_{vc\mathbf{k}} . \tag{16}$$

In this equation, $E_{c,\mathbf{k}+\mathbf{Q}}$ and $E_{v\mathbf{k}}$ are the QP bandstructure energies of the relevant electron and hole states. The electron-hole interaction $\langle vc\mathbf{k}|K^{eh}|v'c'\mathbf{k}'\rangle$ is responsible for the coupling among the independent electron-hole pair configurations. The electron-hole interaction kernel K^{eh} results from the electron self-energy operator Σ, which is again evaluated in the GWA, and the classical Coulomb interaction V [42]. The kernel K^{eh} consists of two terms, a direct, attractive interaction term $K^{eh,d}$, which is responsible for the binding between electron and hole, and a repulsive exchange term $K^{eh,x}$. They are given by

$$\langle vc\mathbf{k}|K^{eh,d}|v'c'\mathbf{k}'\rangle = \\ -\int d\mathbf{x}d\mathbf{x}'\psi^*_{c,\mathbf{k}+\mathbf{Q}}(\mathbf{x})\psi_{c',\mathbf{k}'+\mathbf{Q}}(\mathbf{x})W(\mathbf{r},\mathbf{r}')\psi_{v\mathbf{k}}(\mathbf{x}')\psi^*_{v'\mathbf{k}'}(\mathbf{x}') \qquad (17)$$

$$\langle vc\mathbf{k}|K^{eh,x}|v'c'\mathbf{k}'\rangle = \\ \int d\mathbf{x}d\mathbf{x}'\psi^*_{c,\mathbf{k}+\mathbf{Q}}(\mathbf{x})\psi_{v\mathbf{k}}(\mathbf{x})V(\mathbf{r},\mathbf{r}')\psi_{c',\mathbf{k}'+\mathbf{Q}}(\mathbf{x}')\psi^*_{v'\mathbf{k}'}(\mathbf{x}') \quad . \qquad (18)$$

The direct interaction term resulting from the (screened-exchange) self-energy operator, is screened by the dielectric function (contained in W), while the exchange interaction term, which results from the unscreened Coulomb potential V, is not.

It goes without saying that the algorithmic realization and the numerical evaluation of this approach is an extremely formidable and demanding task [43]. Applications of this formalism to the calculation of optical properties of bulk semiconductors such as Si [43], GaAs [44] or cubic SiC [45] have been carried out with great success. Very good descriptions of the real and imaginary parts of the bulk dielectric function have been achieved.

For shortness sake, here we address only one example, the results of which are of very basic significance in the context of semiconductor surfaces and the question as to which extent different physical surface properties are sensitive to the surface structure.

The top panels of Fig. 8 show side views of two possible variants of Pandey's chain model of the Ge(111)-(2×1) surface. In the case of the so-called 'positive buckling' the *down* atoms of the chains are directly above atoms in the fourth layer. In the case of 'negative buckling' the *up* atoms of the chains are vertically above these atoms. This seemingly minor structural difference has drastic consequences for various physical properties as we will see. The 'negative buckling' structure is 26 meV lower in energy than the 'positive buckling' model [46].

The middle panels of Fig. 8 show the quasiparticle bandstructures for the two respective models in comparison with ARPES [47] and ARIPES [48] data. Though not very noticeable at a first glance, the calculated dangling bond bands of the 'negative buckling' surface agree much better with experiment than those for the 'positive buckling' case. This holds in particular for the calculated surface band gap, which is in perfect agreement with the data for 'negative buckling' while it deviates considerably for the opposite case. We note in passing that these results (for the 'negative buckling' case) again reveal that the GWA yields an excellent description of both the occupied and the empty states.

The lower panels of Fig. 8 finally address optical spectra. Since the optical wavelength is very long compared to typical surface perpendicular dimensions standard absorption or reflectance spectroscopy do not allow for reasonable surface resolution. If, however, the *change in reflectance* at a clean surface as

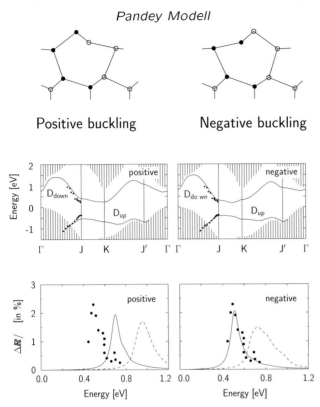

Fig. 8. This figure summarizes results from Ref. [46]. Side views of the energy-optimized Pandey chain model of the Ge(111)-(2×1) surface. Both positive and negative buckling variants are shown (upper panels). Quasiparticle surface bandstructures for the two models in comparison with ARPES [47] and ARIPES [48] data (middle panels). Differential reflectance spectra as calculated for the two models in the GWA and BSE approximations in comparison with experimental data from [49] (bottom panel)

compared to that of an oxide covered surface is measured, i.e. if differential reflectance spectra are taken, a significant surface sensitivity is observed. The two lower panels of Fig. 8 both show the experimental $\Delta R/R$ spectrum [49] in comparison with theoretical results [46]. The experimentally observed peak originates from the surface exciton. The measured excitation energy of the exciton is 0.49 eV. The dashed curves are the results of calculations that neglect electron-hole Coulomb correlations taking only band to band transitions between quasiparticle bands from the GWA into account. Both of these results drastically deviate from experiment and they would not allow to favour one atomic configuration over the other from a comparison with experiment. In the case of the results of full BSE calculations accounting for electron-hole Coulomb interactions appropriately the situation is entirely different. The 'positive buckling' model yields a $\Delta R/R$ spectrum and an excitation energy (0.70 eV) that strongly disagree with experiment. The 'negative buckling' model, on the contrary, yields a $\Delta R/R$ spectrum and an excitation energy (0.51 eV) in excellent agreement with the data allowing for a quantitative interpretation of the experimental results.

5 Conclusion

We have briefly addressed the basic concepts of current days *ab-initio* calculations of basic properties of electrons, phonons and excitons at semiconductor surfaces. A number of theoretical results for exemplary surfaces has been discussed in comparison with high-resolution surface spectroscopy data. The agreement between theoretical and experimental results is good in all cases allowing for an identification of the physical origin and nature of measured data by theory.

Acknowledgments

We would like thank the Deutsche Forschungsgemeinschaft (contracts Po-215/13-1,2 and Ro-1318/1-1, 2-1) and the Bundesministerium für Forschung und Technologie (contract 05 SE8PMA/6) for financial support. Computational resources have been provided by the John von Neumann-Institute for Computing (NIC) of the Forschungszentrum Jülich, the Supercomputing Center Karlsruhe (SCC) and the Bundes-Höchstleistungsrechenzentrum Stuttgart (HLRS).

References

1. P. Hohenberg and W. Kohn: Phys. Rev. B **136**, 864 (1964).
2. W. Kohn and L. J. Sham: Phys. Rev. A **140**, 1133 (1965).
3. E. Wimmer and A.J. Freeman: in *Handbook of Surface Science: Electronic Structure* Vol.**2**, 1 (2000).
4. D. M. Ceperley and B. J. Alder: Phys. Rev. Lett. **45**, 566 (1980).
5. J. P. Perdew and A. Zunger: Phys. Rev. B **23**, 5048 (1981).
6. J. Ihm, A. Zunger, and M.L. Cohen: J. Phys. C **12**, 4409 (1979).
7. M. Scheffler, J.P. Vigneron and G.B.Bachelet: Phys. Rev. B **31**, 6541 (1985).
8. P. Krüger and J. Pollmann: Physica B **172**, 155 (1991).
9. M. Schlüter, J.R. Chelikowsky, S.G. Louie and M.L. Cohen: Phys. Rev. B **12**, 4200 (1975).
10. P. Krüger and J. Pollmann: Phys. Rev. B **38**, 10 578 (1988).
11. P. Krüger and J. Pollmann: Phys. Rev. Lett. **74**, 1155 (1995).
12. P. Krüger and J. Pollmann: in *Handbook of Surface Science: Electronic Structure* Vol.**2**, 93 (2000).
13. E.L. Bullock *et al.*: Phys. Rev. Lett. **74**, 2756 (1995).

14. R. Graupner, M. Hollering, A. Ziegler, J. Ristein, L. Ley and A. Stampfl: Phys. Rev. B **55**, 10 841 (1997).
15. L.S.O. Johansson, R.I.G. Uhrberg, P. Mårtensson and G.V. Hansson: Phys. Rev. B **42**, 1305 (1990).
16. R.I.G. Uhrberg, G.V. Hansson, J.M. Nicholls, and S.A. Flodström: Phys. Rev. B **24**, 4684 (1990).
17. J. Pollmann, P. Krüger and M. Sabisch: phys. stat. sol. (b) **202**, 421 (1997).
18. V. M. Bermudez: phys. stat. sol. (b) **202**, 447 (1997).
19. U. Starke: phys. stat. sol. (b) **202**, 475 (1997).
20. P. Soukiassian et al.: Phys. Rev. Lett. **78**, 907 (1997).
21. W. Lu, P. Krüger and J. Pollmann: Phys. Rev. Lett. **81**, 2292 (1998).
22. L. Duda, L.S.O. Johansson, B. Reihl, H.W. Yeom, S. Hara, and Y. Yoshida: Phys. Rev. B **61**, R2460 (2000).
23. L. Hedin: Phys. Rev. **139**, A796 (1965).
    L. Hedin and S. Lundqvist: Solid State Physics **23**, 1 (1969).
24. M. S. Hybertsen and S. G. Louie: Phys. Rev. Lett. **55**, 1418 (1985); Phys. Rev. B **34**, 5390 (1986).
25. R. W. Godby, M. Schlüter, and L. J. Sham: Phys. Rev. Lett. **56**, 2415 (1986); Phys. Rev. B **37**, 10 159 (1988).
26. M. Rohlfing, P. Krüger, and J. Pollmann: Phys. Rev. Lett. **75**, 3489 (1995); Phys. Rev. B **52**, 1905 (1995).
27. F. Aryasetiawan and O. Gunnarsson: Rep. Prog. Phys. **61**, 237 (1998).
28. W.G. Aulbur, L. Jönsson and J.W. Wilkins: Solid State Physics, Vol. **54**, 2 (2000).
29. K. C. Pandey: Phys. Rev. Lett. **49**, 223 (1982).
30. F. J. Himpsel et al.: Phys. Rev. B **30**, R2257 (1984).
    H. Sakama, A. Kawazu, and K. Ueda: Phys. Rev. B **34**, R1367 (1986).
    R. M. Feenstra, W. A. Thompson, and A. P. Fein: Phys. Rev. Lett. **56**, 608 (1986).
31. J. E. Northrup, M. S. Hybertsen, and S. G. Louie: Phys. Rev. Lett. **66**, 500 (1991).
32. L. Reining and R. Del Sole: Phys. Rev. Lett. **67**, 3816 (1991).
33. M. Rohlfing and S.G. Louie: Phys. Rev. Lett. **83**, 856 (1999).
34. R. I. G. Uhrberg, G. V. Hansson, J. M. Nicholls, and S. A. Flodström: Phys. Rev. Lett. **48**, 1032 (1982).
    P. Perfetti, J. M. Nicholls, and B. Reihl: Phys. Rev. B **36**, 6160 (1987).
35. S. Baroni, P. Giannozzi, and A. Testa: Phys. Rev. Lett. **58**, 1861 (1987).
    P. Giannozzi, S. de Gironcoli, P. Pavone and S. Baroni: Phys. Rev. B **43**, 7231 (1991).
    S. Baroni, S. de Gironcoli, A. Dal Corso, and P. Giannozzi: Rev. Mod. Phys. **73**, 515 (2001).
36. J. Fritsch, C. Eckl, P. Pavone, and U. Schröder: in *Festkörperprobleme/ Advances in Solid State Physics*, ed. by R. Helbig, Vol. **36**, 135 (1996).
37. U. Freking, A. Mazur, and J. Pollmann: Phys. Rev. B **64**, 245341 (2001).
38. V. Gräschus, A. Mazur, P. Krüger and J. Pollmann: Phys. Rev. B **57**, 13 175 (1998).
39. R. Honke, P. Pavone and U. Schröder: Surf. Sci. **367**, 75 (1996).
40. J. Schmidt and H. Ibach: Phys. Rev. B **50**, 14 354 (1994).
41. P. Santini, P. Ruggerone, L. Miglio, and R.B. Doak: Phys. Rev. B **46**, 9865 (1992).

42. L. J. Sham and T. M. Rice: Phys. Rev. **144**, 708 (1966).
    W. Hanke and L. J. Sham: Phys. Rev. Lett. **43**, 387 (1979); Phys. Rev. B **21**, 4656 (1980).
    G. Strinati: Phys. Rev. Lett. **49**, 1519 (1982); Phys. Rev. B **29**, 5718 (1984).
43. M. Rohlfing and S. G. Louie: Phys. Rev. B **62**, 4927 (2000).
44. M. Rohlfing and S. G. Louie: Phys. Rev. Lett. **81**, 2312 (1998).
45. M. Rohlfing and J. Pollmann: Phys. Rev. B **63**, 125201 (2001).
46. M. Rohlfing, M. Palummo, G. Onida, and R. Del Sole: Phys. Rev. Lett. **85**, 5440 (2000).
47. J.M. Nicholls *et al.*: Phys. Rev. B **27**, 2594 (1983).
    J.M. Nicholls *et al.*: Phys. Rev. Lett. **52**, 1555 (1984).
48. J.M. Nicholls and B. Reihl: Surf. Sci. **218**, 237 (1989).
49. S. Nannarone, P. Chiaradia, F. Ciccacci, R. Memeo, P. Sassaroli, S. Selci and G. Chiarotti: Solid State Comm. **33**, 593 (1983).

Diffuse Interface Model for Microstructure Evolution

Britta Nestler

Karlsruhe University of Applied Sciences,
Moltkestrasse 30, 76133 Karlsruhe, Germany
britta.nestler@fh-karlsruhe.de

Abstract. A phase-field model for a general class of multi-phase metallic alloys is proposed which describes both, multi-phase solidification phenomena as well as polycrystalline grain structures. The model serves as a computational method to simulate the motion and kinetics of multiple phase boundaries and enables the visualization of the diffusion processes and of the phase transitions in multi-phase systems. Numerical simulations are presented which illustrate the capability of the phase-field model to recover a variety of complex experimental growth structures. In particular, the phase-field model can be used to simulate microstructure evolutions in eutectic, peritectic and monotectic alloys. In addition, polycrystalline grain structures with effects such as wetting, grain growth, symmetry properties of adjacent triple junctions in thin film samples and stability criteria at multiple junctions are described by phase-field simulations.

1 Introduction

The microstructure formation in metallic alloys involves several different phases and phase transformations. The process is influenced by a great variety of external conditions and physical quantities, e.g. mass and heat diffusion, convection, anisotropy, elasticity etc. and takes place on different time and length scales, see Fig. 1.

Therefore the solidification process of materials yields complex interfacial growth structures and changes in growth topology. The characteristics of the grown microstructure such as the fineness, type of morphology and the spacings themselves determine the physical and mechanical properties of the casting and are hence of great importance with respect to a continuous improvement and optimization of experimental and industrial procedure aiming to produce materials with specific properties. Since solidification in metallic alloys can not in-situ be observed, either experiments on transparent organic substances or modelling methods and numerical simulations are used in order to systematically investigate the influence of the process conditions and of the material properties on the microstructure characteristics.

Traditionally, phase transitions have been expressed mathematically by free boundary problems, where the interface is represented by a sharp surface

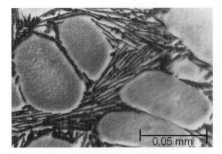 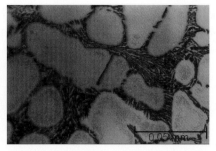

Fig. 1. Examples of the microstructure in a binary Al-Si alloy showing primary Al-dendrites and interdendrititic eutectic regions. Due to the facetted nature of the silicon, the eutectic occurs in an irregular needle-like growth mode

of zero thickness. In these sharp interface formulations the position of the interface has to be calculated explicitly, which leads to great difficulties in the computational treatment of the free boundary problem. In order to handle the moving free boundary numerically, diffuse interface models, e.g. phase-field models have been developed, where the interface is expressed implicitly by a time and space dependent function indicating the phase state and being defined on the whole region. In the last decade, phase-field models have attracted considerable importance as a means of describing and numerically simulating a range of phase transitions and the temporal evolution of complex growth structures that occur during solidification in alloys. Besides the investigation of phenomena on individual length scales, an important challenge for future work on phase-field modelling will be to extend the method so that it can be applied to multiscale problems.

In the standard phase-field description for solid/liquid phase systems, a variable $\phi(\boldsymbol{x}, t)$, called the phase field is introduced whose value characterises the phase of the system in time and space. The interfaces are represented by thin regions in which ϕ varies smoothly between the values of ϕ associated with the adjoining bulk phases. The mathematical model is based on a free energy functional

$$\mathcal{F}(\phi, c) = \int_V \left(\frac{1}{2}\eta^2 |\nabla \phi|^2 + f(\phi, c) \right) dV,$$

where $f(\phi, c; T)$ is typically of the form of a double well potential. A set of governing partial differential equations for the appropriate thermodynamic quantities (e.g. composition $c(\boldsymbol{x}, t)$) with an additional reaction-diffusion equation for the phase field $\phi(\boldsymbol{x}, t)$, often called the phase-field equation can be derived by variational derivatives

$$\frac{\partial \phi}{\partial t} = -\mathbf{M}(\phi) \frac{\delta \mathcal{F}(\phi, c)}{\delta \phi} \quad \text{and} \quad \frac{\partial c}{\partial t} = \nabla \cdot \left\{ \mathcal{D}(\phi) \left[c(1-c) \nabla \left(\frac{\delta \mathcal{F}(\phi, c)}{\delta c} \right) \right] \right\}.$$

This ansatz ensures that the total free energy decreases in time and that the concentration in the system is conserved. The derivation of the governing equations, although originally ad hoc [1], was subsequently placed in the more rigorous framework of irreversible thermodynamics [2,3]. The relationship of the phase-field formulation and the corresponding free-boundary problem (or sharp interface description) may be established by taking the sharp interface limit of the phase-field model, whereby the interface thickness tends to zero and is replaced by interfacial boundary conditions (see e.g. [4] and in the presence of surface energy anisotropy [5]). In [6] Karma and Rappel have developed a framework for second order sharp interface asymptotics which is more appropriate to the simulation of dendritic growth at low undercoolings by the phase-field model. In the field of solidification, phase-field models have been developed to describe both, the solidification of pure materials and binary alloys [7].

For pure materials, phase-field models have been used extensively to simulate numerically dendritic growth into an undercooled liquid, e.g. [6,8,9,10]. These computations exhibit a wide range of realistic phenomena associated with dendritic growth, including side arm production and coarsening. Accurate computations have been conducted at lower undercoolings closer to those encountered in experiments of dendritic growth and have also been used as a means of assessing theories of dendritic growth.

In recent years, the phase-field methodology has been extended to describe the evolution of multiple interfaces, grain boundaries and phase transitions in three phase systems. Grains in a pure material have also been modelled by multi-phase-field models using a vector valued phase field that I denote with $\phi(\boldsymbol{x},t) = (\phi_1(\boldsymbol{x},t), \ldots, \phi_N(\boldsymbol{x},t))$, [11,12,13,14]. The multi-phase-field model discussed in this paper is based on an ad hoc model formulation in which a phase field is associated with each of the different phases, [12]. This model was further developed in [15] to include surface energy anisotropy, to describe and simulate grain structure formation and its sharp interface asymptotic limit was studied. In this limit the classical laws at interfaces (the Gibbs-Thomson condition) and at multi-junctions (the Young's force balance law) were recovered. Based on the same roots, the notion of a generalised ξ-vector formulation was used in order to incorporate anisotropy and the phase-field concept was extended to model the solidification in binary three phase alloy systems including a formulation of convection in the (monotectic) liquid phases, [16].

In this paper a thermodynamically consistent multi-phase-field model is presented which can be used to model and numerically simulate complex microstructure evolution in both, binary three phase alloy systems as well as solid/liquid phase systems with multiple crystals of different crystallographic orientation. The only difference of these two applications is that the components of the phase-field vector either represent the different phases in the alloy system or the different orientational variants of a grain configuration.

The multi-phase-field model discussed here provides a solution of how the occurance of a third/foreign phase contribution at two phase interfaces can be avoided. We present a selection of simulations showing polycrystalline grain structures as well as phase transitions in eutectic, peritectic and monotectic alloy systems.

2 Multi-Phase-Field Model

The vectorial multi-phase-field model is formulated in terms of a generalized Ginzburg–Landau free energy $\mathcal{F}(\boldsymbol{\phi}, c; T)$ which is a functional of the multi-phase-field vector $\boldsymbol{\phi}(\boldsymbol{x},t) = (\phi_1(\boldsymbol{x},t), ..., \phi_N(\boldsymbol{x},t))$ and the concentration variable $c(\boldsymbol{x},t)$

$$\mathcal{F}(\boldsymbol{\phi}, c; T) = \int_V g(\boldsymbol{\phi}, \nabla\boldsymbol{\phi}) + f(\boldsymbol{\phi}, c; T) \, d\boldsymbol{x}.$$

For convenience, the temperature T in this integral expression is treated as an external parameter, so that we discuss isothermal situations. The free energies may be given by

$$g(\boldsymbol{\phi}, \nabla\boldsymbol{\phi}) = \sum_{i<k}^{N,N} \frac{\epsilon_{ik}^2}{2} [\Gamma_{ik}(\phi_i \nabla\phi_k - \phi_k \nabla\phi_i)]^2$$

$$f(\boldsymbol{\phi}, c; T) = \sum_{i<k}^{N,N} \frac{W_{ik}}{4} \phi_i \phi_k + \sum_i^N m_i(c; T)\phi_i + h(c)$$

with e.g. the following choices

$$m_i(c; T) = m_i^B(T)c + m_i^A(T)(1-c) \quad \text{and} \quad h(c) = H(c - c^*)^2.$$

Here, $m_i^A(T)$ and $m_i^B(T)$ are bulk free energies of the pure A and pure B states, respectively. We assume that they have the form

$$m_i^A(T) = L^A \frac{T - T_i^A}{T_i^A} + m_N^A(T) \quad \text{and} \quad m_i^B(T) = L^B \frac{T - T_i^B}{T_i^B} + m_N^B(T).$$

The parameters L^A and T_i^A are the latent heat of fusion per unit volume and the melting temperature of the phases i of pure component A, respectively. A similar interpretation applies to L^B and T_i^B for pure B. The constants ϵ_{ik} are gradient energy coefficients and $\Gamma_{ik}(\phi_i \nabla\phi_k - \phi_k \nabla\phi_i)$ are homogeneous degree one functions of their argument. As shown in [16], the quantity ϵ_{ik} is related to the interface thickness and to the surface energy of the interface between the bulk phases labelled i and k. In particular, the surface energy anisotropy of this interface is described by the dependence of Γ_{ik} on the orientation of its argument. If the interfaces in the system are isotropic Γ_{ik} simplifies to $\Gamma_{ik}(\phi_i \nabla\phi_k - \phi_k \nabla\phi_i) = |\phi_i \nabla\phi_k - \phi_k \nabla\phi_i|$.

In real physical systems, the surface energies of the individual interfaces are in general not equal. In this case, the above phase-field model and also other diffuse multi-order parameter models cited in literature suffer from the difficulty that a third or - more generally - foreign phase contribution occurs at a two phase interface, Fig. 2 a). This effect is to date interpreted to be non-physical and leads to a violation of physical laws of equilibrium thermodynamics such as the Young's force balance law at multiple junctions. Due to this disability, one does e.g. not obtain the correct angle condition at triplejunctions for a system with different surface energies. An explanation in terms of the model formulation is that the surface energies relate to the model parameters according to $\gamma_{ik} \sim \epsilon_{ik}\sqrt{W_{ik}}$, so that for different values of γ_{ik} the model in general involves different values of the W's. As a result the shape of the hyperplane of the multi-obstacle/multi-well potential $\sum_{i<k} W_{ik}/4\phi_i\phi_k$ changes compared to the symmetric case of equal surface energies. The consequence is that the connecting trajectories of two minima, representing stable phase states, do no longer ly along the edges of the Gibbs simplex, but do now have a contribution towards the center involving foreign phases. An interpretation in terms of minimizing energy is that the energy of a pure two phase interface is higher than the energy of an interface involving small amounts of other foreign phases. A possible solution of this difficulty is the introduction of higher order potentials of the form $\sim \phi_i\phi_j\phi_k$ (see [15]) which increase the energy of the center of the Gibbs simplex and which force the third/foreign phase to vanish at a two phase interface, Fig. 2 b). These higher order potentials have no physical meaning and enhance the complexity of the model formulation.

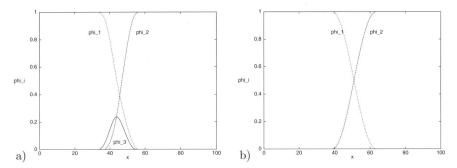

Fig. 2. a) Contour lines of three phase-field variables at a two phase boundary in a three phase system for different surface energies and by using a) the multi-obstacle potential $\frac{W_{ik}}{4}\phi_i\phi_k$ and b) the multi-obstacle potential with additional higher order correction terms $\sim \phi_i\phi_j\phi_k$. The plot in a) illustrates the presence of a third/foreign phase whereas in b) only the two appropriate phases participate in forming the boundary

In the following an alternative solution avoiding the third/foreign phase contribution at two phase interfaces is proposed. We use the notation

$$\nabla_{ik} := \phi_i \nabla \phi_k - \phi_k \nabla \phi_i \quad \text{and} \quad \triangle_{ik} := \phi_i \nabla^2 \phi_k - \phi_k \nabla^2 \phi_i.$$

Next we consider a two phase interface of phases labelled 1 and 2 with $\phi_3 = 0$ in a three phase system. At this interface the evolution equations are

$$\epsilon_{12}^2 (\phi_2 \triangle_{12} + 2\nabla_{12} \cdot \nabla \phi_2) + \frac{1}{4}(W_{12}\phi_2) + m_1(c,T) - \lambda = 0 \tag{1}$$

$$\epsilon_{12}^2 (-\phi_1 \triangle_{12} + 2\nabla_{12} \cdot \nabla \phi_1) + \frac{1}{4}(W_{12}\phi_1) + m_2(c,T) - \lambda = 0 \tag{2}$$

$$\frac{1}{4}(W_{13}\phi_1 + W_{23}\phi_2) - \lambda = 0, \tag{3}$$

where λ is a Lagrangian multiplier. From the third equation, the equality $W_{13} = W_{23}$ follows as a necessary condition for λ to be constant. For an N phase system, we obtain $W_{ik} =: W$ for all $i,k = 1,\ldots,N$. Subsequently, the solution to overcome the difficulties of the third/foreign phase contribution at two phase interfaces is to keep the multi-obstacle potential $\sum_{i<k} \frac{W_{ik}}{4}\phi_i\phi_k$ fixed and independent of the surface energies. As a result, the shape of the hyperplane of $\sum_{i<k} \frac{W_{ik}}{4}\phi_i\phi_k$ is not influenced by any system parameters and does not change. Therefore, the above expression for the free energy contribution $f(\phi,c,T)$ of the multi-phase-field model is replaced by

$$f(\phi,c;T) = \sum_{i<k}^{N,N} \frac{W}{4}\phi_i\phi_k + \sum_i^N m_i(c;T)\phi_i + h(c).$$

Furthermore, the model still consists of enough degrees of freedom to realize the complete set of parameters of a real physical system.

In the following, the relation between the model parameters and the physical parameters will be discussed. From substracting (1) and (2) the evolution of a single two phase interface in one dimension reads $\phi''(x) + \frac{(m_2-m_1)}{\epsilon_{12}^2} = \frac{W}{4\epsilon_{12}^2}(1-2\phi(x))$. The solution of this equation is $\phi(x) = \frac{1}{2}(1+\sin\alpha_{12}x)$ where $\alpha_{12} = \frac{\sqrt{W/2}}{\epsilon_{12}}$. Hence, the interface thickness l_{12} and the surface energy γ_{12} relate to ϵ_{12}

$$l_{12} = \frac{\pi}{2\alpha_{12}} = \frac{\pi\epsilon_{12}}{\sqrt{2W}} \quad \text{and} \quad \gamma_{12} = \frac{1}{2}\int_{l_{12}} \phi(1-\phi)\,dx = \frac{\pi}{8}\sqrt{\frac{W}{2}}\epsilon_{12}$$

The interface mobility μ_{12} is

$$\mu_{12} = \frac{-M_{12}\int_0^1 m(c;T)d\phi}{\int_{l_{12}} \phi'(x)dx} = \frac{-8C}{\pi^2}M_{12}l_{12},$$

with $m(c;T) = m_1(c;T) = m_1^B(T)c + m_1^A(T)(1-c)$ and $m_2(c;T) = 0$ for the two phase interface in equilibrium and where M_{12} is the mobility matrix.

The constant C is derived from the equilibrium concentration equation at the interface

$$\frac{\partial f(\boldsymbol{\phi},c;T)}{\partial c} = (m_1^B(T) - m_1^A(T))\phi + 2H(c-c^*) - \lambda_c = 0,$$

where λ_c is a Lagrange multiplier. It follows that

$$c = \frac{1}{2H}[\lambda_c - (m_1^B - m_1^A)\phi] + c^* \quad \text{and}$$

$$\begin{aligned} m(c;T) &= (m_1^B - m_1^A)c + m_1^A \\ &= -\frac{1}{2H}(m_1^B - m_1^A)^2\phi + (\frac{1}{2H}\lambda_c + c^*)(m_1^B - m_1^A) + m_1^A \\ &=: C_1\phi + C_2. \end{aligned}$$

The definition for C is $C := \frac{1}{2}C_1 + C_2$.

Starting from a real system, the recepy for choosing the model parameters is; first, take suitable values for ϵ_{ik} to represent the surface energies γ_{ik}. This choice of ϵ_{ik} then adjusts certain interface thicknesses l_{ik}. To realize physical mobilities μ_{ik}, it is now necessary to choose the values of the mobility matrix M_{ik} in an appropriate way.

The presented phase-field model is capable to describe both, binary three phase alloy systems such as eutectics, peritectics and monotectics as well as solid/liquid phase systems with an arbitrary number of crystals of different crystallographic orientation. We postulate the set of evolution equations for the phase fields $\phi_i(\boldsymbol{x},t)$ and for the concentration $c(\boldsymbol{x},t)$ from the gradient flow of the energy functional

$$\frac{\partial \phi_i}{\partial t} = -\mathbf{M}(\boldsymbol{\phi})\frac{\delta \mathcal{F}(\boldsymbol{\phi},c;T)}{\delta \phi_i}, \quad i = 1,\ldots,N$$

$$\frac{\partial c}{\partial t} = \nabla \cdot \left\{ \mathcal{D}(\boldsymbol{\phi}) \left[c(1-c)\nabla \left(\frac{\mathcal{F}(\boldsymbol{\phi},c;T)}{\delta c} \right) \right] \right\}$$

with diffusion coeffictions $\mathcal{D}(\boldsymbol{\phi}) = \frac{v_m}{RT}\sum_{i=1}^{N} D_i\phi_i$.

In the binary alloy case $N = 3$ and the functions $m_1(c;T), m_2(c;T)$ are determined by the specific form of the phase diagram choosing $m_3(c,T) = 0$ as a reference frame. Examples can be found in [16]. In the crystal growth case, the number of phase fields is N and the functions $m_i(c;T)$ are $m_1(c;T) = m_2(c;T) = \ldots = m_{N-1}(c)$ with $m_N(c;T) = 0$ representing a lense shape phase diagram of a solid/liquid phase system. In addition the diffusion coefficients of all solid crystals are equal $D_1 = \ldots = D_{N-1}$. They are approximately three orders of magnitude less than in the liquid phase labelled N; $D_i << D_N, i = 1,\ldots,N-1$.

3 Numerical Simulations

The objective of this section is to demonstrate the general utility of the multi-phase-field model as a computational vehicle to simulate a wide variety of grain growth structures and realistic solidification morphologies in peritectic, eutectic and monotectic alloy systems. Pursuing this aim, a selection of numerical solutions of the set of partial differential equations for the phase-field vector $\phi(x,t)$ and for the concentration $c(x,t)$ is presented. A finite difference discretisation on a uniform rectangular mesh allied to an explicit time marching scheme is used for the three phase-fields and for the solute concentration. Effects of fluid flow have not been taken into account. Further details concerning the numerical setup and the system parameters are given in [15].

3.1 Grain Structures

First, we apply the multi-phase-field model to grain growth phenomena. In this context, the components of $\phi(x,t)$ represent grains of the same phase, but of different crystallographic orientations. The simulation depicted in Fig. 3 is performed for a microstructure with four orientational variants and with a sixfold convex crystalline surface energy anisotropy. The incorporation of surface energy anisotropy and explicit expressions for crystalline anisotropy with a typical cusp-like structure are explained in [15]. The evolution of the grain boundaries in Fig. 3 shows the coarsening of the grain structure and the formation of facets in the six preferred directions of the Wulff shape. The anisotropy of the surface energies induces shear forces which change the equilibrium angle condition of 120°.

In isotropic systems, wetting occurs if the surface energies violate the stability condition $\gamma_{ij} \leq \gamma_{ik} + \gamma_{kj}$. Fig. 4 illustrates computations of wetting for a system of four different phase fields. The time sequence in Fig. 4 a) and b) shows the break down of 123 and 234 grain boundary triple junctions accompanied by the building of channels of the grains 1 and 4 between grain

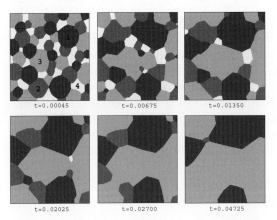

Fig. 3. Growth of grains with a six-fold anisotropy and with relative angles: $\varphi_{12} = 30°, \varphi_{13} = 45°, \varphi_{14} = 90°, \varphi_{23} = 15°, \varphi_{24} = 60°, \varphi_{34} = 45°$

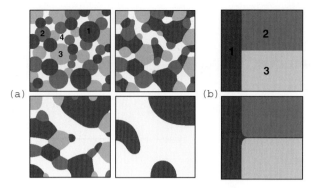

Fig. 4. Wetting in an isotropic system with surface energies $\gamma_{12} = \gamma_{13} = \gamma_{14} = \gamma_{24} = \gamma_{34} = 0.7$ and $\gamma_{23} = 1.8$

2 and grain 3. As a result, the grains remaining last in the coarsening process of Fig. 4 a) are the grains 1 and 4.

3.2 Multiphase Alloy Systems

Next, the phase-field method is used to numerically simulate phase transitions in binary metallic alloys. The following computations of multiphase solidification are carried out under the condition of an isothermally undercooled melt and for isotropic surface energies. Further details concerning the numerical setup and the system parameters are provided in [16].

First, we present a numerical simulation corresponding to peritectic solidification. In a peritectic system, cooled beneath its peritectic temperature, a new solid phase β nucleates, i.e., $L + \alpha \rightarrow \beta$. It grows from the undercooled liquid L and the parent solid phase α. By heterogeneous nucleation, the β phase often occurs at the $\alpha - L$ interface. The β phase then grows around the α phase until, either the α phase is completely melted or it is entirely engulfed within the new β phase. For the computation in Fig. 5, we initially placed circular α solid phase particles of different radii in the liquid region and within the vicinity of a β phase planar front. The solidification velocity of the growing β phase is locally higher in regions close to the melting α particles due to locally different concentration gradients in the liquid phase. As a result, the initially planar β front deformes due to preferential growth towards the solid particles. Triplejunctions of all three phases lead the growth direction, because the α phase provides the supply of solute needed for the growth of the β phase. Depending on the size of the α particles and on the solidification velocity of the β phase, the α particles either dissolve in the liquid or become engulfed in the β phase. Since we assume stoichiometric solid phases, the solid diffusivity is zero and the α particles remain stable in size after engulfment in the β matrix.

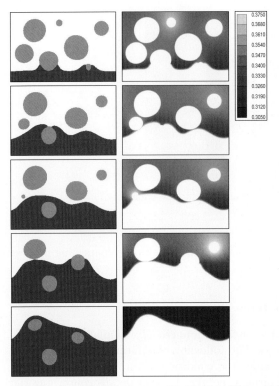

Fig. 5. Engulfment of properitectic α solid particles into a growing β solid front following a peritectic phase transition $L + \alpha \rightarrow \beta$. The plots show the temporal evolution of the phase configuration and the concentration profile in the liquid phase in the left and right column, respectively. In this calculation, diffusion in the solid phases is suppressed, $D_1 = D_2 = 0$

Next, the result of a numerical simulation corresponding to eutectic solidification is reported. In particular, the specific data of the $Al - Si$ phase diagram are used for the computation in Fig. 6. Once the metallic melt in a eutectic system is cooled down below a critical temperature, the liquid phase L transforms into two new solid phases α and β via a eutectic reaction: $L \rightarrow \alpha + \beta$. The black and white regions in the images of Fig. 6 represent the Al- and Si- rich eutectic solid phases α and β of the binary $Al - Si$ alloy system, respectively. The snapshots show the concentration field of Si in the liquid during the growth of a eutectic grain at different time steps. We observe zones of depleted solute ahead of the Si-rich solid phase and concentration enriched regions ahead of the Al-rich solid phase. Due to an increase of the lamellar spacing during growth, deep concave hollows are formed at the solid-liquid interfaces and the phase boundaries of the eutectic microstructure evolve in a disordered manner. The subsequent nucleation of solid particles of the opposite phase within the concave portion of the interfaces stabilizes the growth behaviour and re-establishes lamellar growth. The simulated structure is compared with an experimental photograph in the last image of Fig. 6.

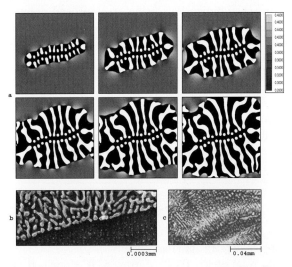

Fig. 6. Phase-field simulation of a eutectic $Al-Si$ grain in comparison with experimental photographs. The snapshots of the calculation display the concentration of silicon and the evolution of the phase boundaries propagating in time

4 Conclusion

The multi-phase-field model discussed here has been derived in a thermodynamically consistent way and shows a way of how a third/foreign phase contribution along a two phase interface can be avoided. Numerical simulations of grain boundary motion and binary three phase solidification are presented which illustrate the general capability of the model to describe realistic microstructure evolutions.

As a next step of progress, we are recently working on an extension of the above multi-phase-field model to multicomponent (ternary) multiphase alloys. To treat the phase transformations in an M-component alloy system with $M+1$ different phases, a vector $\mathbf{c} = (c_i, \ldots, c_M)$ of M concentration fields $c_i, i = 1, \ldots, M$ is introduced, so that the energy functional $\mathcal{F}(\boldsymbol{\phi}, \mathbf{c}, T)$ now also depends on $\mathbf{c}$. The governing equations are complemented by a set of M differential equations for the diffusion fields. This model extension will be used to describe the growth of eutectic colonies in ternary eutectic systems. Another important objective is the investigation of multiscale solidification phenomena such as primary dendrites with interdendritic eutectic regions.

Within this context, the aim is to derive new growth laws on larger length scales by homogenization and averaging the temporal evolution of the structure on finer scales. Furthermore, it is intended to incorporate convection and elasticity into the phase-field model. Based on the promising numerical results in 2 dimensions, simulations in three dimensions using an improved numerical algorithm combined with acceptable computation times are planned in our future work. These developments will allow to reflect and describe effects due to three spatial dimensions.

Acknowledgements

The author B. Nestler is grateful to the German Research Foundation (DFG) for supporting this work through the priority research programs *SPP 1095: Mehrskalenprobleme* and *SPP 1120: Phasenumwandlung in mehrkomponentigen Schmelzen*.

References

1. J. S. Langer, private communication.
2. O. Penrose and P. C. Fife, Physica D **43**, 44 (1990).
3. S-L Wang, R. F. Sekerka, A. A. Wheeler, B. T. Murray, S. R. Coriell, R. J. Braun, G. B. McFadden, Physica D **69**, 189 (1993).
4. G. Caginalp, Phys. Rev. A **39**, 5887 (1989).
5. A. A. Wheeler and G. B. McFadden, Eur. J. Appl. Math. **7**, 369 (1996).
6. A. Karma and W.-J. Rappel, Phys. Rev. E **53**, R3017 (1996).
7. A. A. Wheeler, W. J. Boettinger, G. B. McFadden, Phys. Rev. A **45**, 7424 (1992); A. A. Wheeler, W. J. Boettinger, G. B. McFadden, Phys. Rev. E **47**, 1893 (1993); J. A. Warren and W. J. Boettinger, Acta. metall. mater. **43**, 689 (1994).
8. R. Kobayashi, Bull. Jpn. Soc. Ind. Appl. Math. **1**, 22 (1991); R. Kobayashi, Physica D **63**, 410 (1993); R. Kobayashi, Experimental Math **3**, 60 (1994).
9. S-L. Wang and R. F. Sekerka, J. Comp. Phys. **127**, 110 (1996).
10. N. Provatas, N. Goldenfeld, and J. Dantzig, Phys. Rev. Lett. **80**, 3308 (1998).
11. Long-Qing Chen and Wei Young, Phys. Rev. B **50**, 15752 (1994); Long-Qing Chen, Scr. Metall. Mater. **32**, 115 (1995).
12. I. Steinbach, F. Pezzolla, B. Nestler, J. Rezende, M. Seesselberg and G. J. Schmitz, Physica D **94**, 135 (1996).
13. B. Nestler and A. A. Wheeler, Phys. Rev. E **57**, 2602 (1998).
14. R. Kobayashi, J. A. Warren, and W.C. Carter, Physica D **119**, 415 (1998).
15. H. Garcke, B. Nestler and B. Stoth, Physica D **115**, 87 (1998); H. Garcke, B. Nestler and B. Stoth, SIAM Journal on Applied Mathematics **60**, 295 (1999); H. Garcke, B. Nestler and B. Stoth, J. Interfaces and Free Boundary Problems **1**, 175 (1999); H. Garcke and B. Nestler, Mathematical Models and Methods in Applied Sciences **10**, 895 (2000).
16. B. Nestler and A. A. Wheeler, Phys. Rev. E **57**, 2602 (1998); B. Nestler and A. A. Wheeler, Physica D **138**, 114 (2000); B. Nestler, A. A. Wheeler, L. Ratke, C. Stöcker, Physica D **141**, 133 (2000).

Rapidly Produced Thin Films: Laser-Plasma Induced Surface Reactions

Peter Schaaf, Ettore Carpene, Michael Kahle, and Meng Han

Universität Göttingen, Zweites Physikalisches Institut,
Bunsenstrasse 7/9, 37073 Göttingen, Germany
pschaaf@uni-goettingen.de,
http://www.uni-goettingen.de/~pschaaf

Abstract. Surface coatings and thin films can be easily and rapidly produced by irradiation of materials with short laser pulses in reactive gas atmospheres, where laser-plasma-surface interactions play an important role for the process. As examples for this Laser-Plasma Synthesis, the preparation of Fe-N, Fe-C, AlN, TiN, TiC, SiC, and Si_3N_4 coatings are presented as achieved by employing an Excimer-Laser, a Nd:YAG Laser, a Free Electron Laser and a femtosecond Ti:sapphire laser. This implies laser pulse durations from the nanosecond to the femtosecond regime and wavelengths from UV to IR. The resulting surfaces, thin films, coatings and their properties are investigated by combining Mössbauer Spectroscopy, X-ray diffraction, X-ray absorption spectroscopy, Nanoindentation, Nuclear Reaction Analysis, and Rutherford Backscattering Spectroscopy.

1 Introduction

Thin films are playing a more and more important role in the application of smart materials. This covers the improvement of surface properties like hardness, corrosion resistance, wear resistance, friction, electrical, thermal, magnetic, and optoelectronic properties for specific applications [1,2]. Many methods evolved for the preparation of thin films and coatings. Here, we present a method involving short pulsed laser irradiation for the direct synthesis of surface layers in reactive atmospheres. This Laser Plasma Synthesis, e.g. laser nitriding of aluminum and titanium or laser carburization of iron, is an interesting process both for basic physics as well as industrial technology. The interaction between the molten substrate and the plasma produced by the laser in the ambient atmosphere is the basis for an effective treatment. Nevertheless, the physical mechanisms that lead to the formation of the new phases are not yet fully understood due to the complex phenomena involved [3]. Chen and coworkers [4] investigated the interaction of ambient background gas with the laser plume. Sub-picosecond ablation of metals [5] were also investigated. The treatment performed in nitrogen atmosphere (laser nitriding) has been largely investigated due to the technological importance of nitrogen in metals and alloys [3,6,7,8]. For Ti even the formation of stoichiometric, adherent layers of TiN was reported [9]. The laser nitriding of aluminum and AlSi alloys was investigated by Barnikel and coworkers [10], while Sicard et al. [11] reported the formation of AlN by excimer laser irradiation in nitrogen gas. Although, its efficiency has been demonstrated for nanosecond pulse lasers, the dependence on the pulse duration has never been studied. On this frame, laser pulses ranging from femtosecond to nanosecond have been used to investigate their influence on the Laser-Plasma Synthesis. Some examples are presented in the following.

2 Experimental

Pure materials (Fe, Al, Ti, Si(100)) were used as starting materials for the samples with 0.5-1.5 mm thickness. All materials except the single crystalline Si(100) were mechanically polished with SiC grinding paper followed by 1 μm diamond paste. A Siemens XP2020 XeCl Excimer laser with $\lambda = 308$ nm, 55 ns (FWHM) pulse duration, and 8 Hz repetition rate was used for the Excimer laser treatments. The laser fluence was set to the desired fluence (0.5 to 5 J/cm^2) with a spot size of 5×5 mm^2 employing a focusing beam homogenizer and a variable beam attenuator [3]. The Nd:YAG laser was operated at BIAS Bremen at the wavelength of 1.06 μm and a pulse duration of 8 ns (FWHM) at a repetition rate of 10 Hz. The nearly Gaussian shaped raw beam of the laser had 7 mm diameter and a maximum pulse energy of 0.9 J, corresponding to a mean laser fluence of approximately 2.34 J/cm^2. The Free Electron Laser (FEL) at the Jefferson Lab in Newport News (USA), has been used for the picosecond irradiations at λ=3.1 μm. The micropulse energy is 17 μJ. The time structure of this laser is more complex than for usual lasers. Each micropulse has a duration of 2 ps at a repetition rate of 37 MHz, put together for a total macropulse duration of up to 1 ms and 60 Hz. The Gaussian raw beam of approximately 50 mm diameter was focussed to about 100 μm at the sample surface. The ultra-short femtosecond pulses were delivered by a Ti:sapphire laser with chirped pulse amplifier (CPA) installed at the Laser Zentrum Hannover (LZH). The 150 fs pulses (FWHM) had a broad spectral distribution centered at λ=750 nm, and a maximum single pulse energy of 1 mJ at a repetition rate of 1 kHz. For the irradiations, the samples were mounted in a chamber, which was first evacuated to less than 10^{-3} Pa and then filled with nitrogen gas (99.999%) or methane (99.5%) at a pressure of 0.01-0.15 MPa prior to the treatment. Larger sample areas were treated by meandering (i.e. scanning) the surfaces relative to the laser beam by a motorized and computer-controlled sample stage [3]. The shift of each laser spot was set according to the desired overlap and the number of laser shots. The nitrogen depth profiles have been measured by Resonant Nuclear Reaction Analysis (RNRA) via the reaction ^{15}N(p,$\alpha\gamma$)^{12}C at a resonance energy of 429.6 keV [12]. The carbon profiles have been obtained by Rutherford Backscattering Spectrometry (RBS) with a 900 keV He^{++} beam and a backscattering angle of 165°. The data analysis was performed with the RUMP code [13]. Phase analysis was done by X-ray diffraction in Bragg-Brentano geometry and under 2-5° grazing incidence (GIXRD) employing Cu-K$_\alpha$ radiation. The iron samples have been analyzed by Conversion Electron and Conversion X-ray Mössbauer Spectroscopy (CEMS and CXMS) [3]. It is worth noting that CEMS has an information depth of about 150 nm, whereas CXMS analyses a surface layer of about 20 μm [14]. EXAFS spectra were obtained at the LURE (Orsay, France) synchrotron radiation facility on the beamline SA32 in total electron yield (TEY) mode. The investigated depth was estimated to be about 30 nm. The data analysis was performed with the VIPER software [15]. Further details are given in Ref. [16]. Finally, the surface hardness was determined employing a nanoindenter with Vickers diamond.

3 Results and Discussions

The following results are divided according to the starting material - Fe, Al, Ti, Si - and then subdivided for the atmosphere and the laser type.

3.1 Thin Films on Iron

Iron has been irradiated in nitrogen and methane atmosphere. All lasers gave rise to significant reactions as shown below.

Iron Irradiated in Nitrogen Atmosphere The nitrogen depth profiles for iron irradiated in nitrogen with the various lasers are shown in Fig. 1. The Excimer laser nitrided iron shows a high nitrogen content of about 30 at.% at the surface and 10 at.% in a depth of 400 nm. The nitrogen depth profiles were fitted by the superposition of diffusion profiles according to a simplified one dimensional nitrogen diffusion model [6,7,8]. For single spot XeCl Excimer laser irradiation, the typical nitrogen diffusion length $\sqrt{4Dt}$ deduced

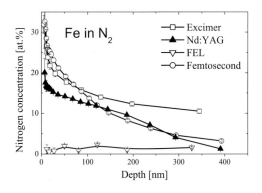

Fig. 1. Nitrogen depth profiles measured by RNRA for iron irradiated in nitrogen with different lasers (Parameters: Excimer, 4 J/cm^2, 64 pulses, 1 bar; Nd:YAG, 2.34 J/cm^2, 100 pulses, 1 bar; FEL, 1 ms/60 Hz, 1 bar, 120 pulses; Ti:sapphire, 4 J/cm^2, 1.5 bar, 400 pulses)

from this approximation is about 100 nm, corresponding to a diffusion time of about 300 ns, in excellent agreement to the calculated melting time [7]. For the Nd:YAG laser nitrided sample the nitrogen concentration is lower, reaching about 20 at.% at the surface and decreases more rapidly with depth (400 nm), due to the shallower melted layer. In fact, decreasing the pulse duration from 55 ns to 8 ns, decreases the melting depth from 800 nm to about 350 nm (at equal fluence) [17]. The most important fact is that there is a nitriding effect at all with the infrared radiation, in contrast to what most investigators claimed [10], that the laser nitriding only works in the UV. The underlying mechanisms appear to be similar for the two types of laser, even if the pulse duration is shortened by a factor of seven, but still remaining in the nanosecond regime.

Shortening the pulse duration to the femtosecond region, the pulses are shorter than the electron collision frequency and a distinct regime of laser material interaction becomes effective. Nevertheless, the nitrogen depth profile shown in Fig. 1 for the Ti:sapphire laser nitrided iron is still comparable with those of the Excimer laser and Nd:YAG laser nitrided iron, having a high concentration of more than 30 at.% at the surface which decreases to about 4 at.% at 400 nm. As there are no significant differences in the depth profiles, the question arises if there are differences in the phase formation, due to non-thermal effects.

In the Excimer laser nitrided iron (also for Nd:YAG laser), the dominant nitride phase is γ-Fe(N) (nitrogen austenite) besides the original α-Fe. This is consistent with the mean nitrogen concentration of about 10 at.% within the first 400 nm deduced from RNRA and the results from the Mössbauer analyses. Surprisingly, the Ti:sapphire laser nitrided iron does not exhibit any other peaks in the diffraction pattern than those of the virgin α-Fe before the laser treatment. This is in contrast to the RNRA result showing a mean nitrogen concentration of nearly 15 at.% within the first 400 nm of the sample. This discrepancy suggests that the nitrogen here is not present as a nitride phase having normal crystalline structure. ^{57}Fe CEMS was used to characterize the iron nitride phases within its sensing depth of about 150 nm [18]. The spectrum of the Ti:sapphire laser nitrided iron exhibits two single lines and a doublet in addition to the virgin α-Fe sextet, the latter having a fraction of 79(3)%. The hyperfine parameters of the paramagnetic phase with a total fraction of 21(3)% coincide very well with those of the γ''/γ'''-FeN phase [19,20]. Assuming that all nitrogen is located inside this paramagnetic Fe-N phase and taking the mean nitrogen concentration within the first 150 nm obtained from RNRA (16.5(7) at.%), this phase should contain 48(7) at.% nitrogen (FeN$_{0.94(8)}$). We suppose that it corresponds to nanocrystalline FeN formed by plasma condensation (recast) of FeN molecules and the following clustering and fast cooling [21], thus being invisible in the GIXRD but appearing in RNRA and CEMS [18]. The picosecond FEL laser is obviously not effective for nitriding iron, irradiations in nitrogen atmosphere lead to an almost negligible nitrogen incorporation, as seen in Fig. 1. Finally, the surface hardness measured by a nanoindenter shows significant differences for the different laser types. Excimer laser nitrided iron achieves a maximum hardness of 5 GPa, the Ti:sapphire laser irradiated sample yields only 0.7 GPa.

Iron Irradiated in Methane Atmosphere By irradiating pure iron substrates in methane atmosphere with a pulsed excimer laser, a rather thick and polycrystalline cementite layer was obtained [22]. Cementite (θ-Fe$_3$C) is of great technological importance for the mechanical properties of steels and iron alloys, however it is hard to obtain as a single phase since it is typically embedded in the steel matrix. Only few investigations on the Fe$_3$C electronic structure [23], its mechanical [24] and thermodynamical [25] properties can be found in the literature, due to the difficulty in obtaining it in its pure state. Recently, a single-phase cementite film has been prepared by a special Physical Vapor Deposition technique [24], and no other method has been successful in synthesizing pure cementite.

The Mössbauer spectra and their analyses are shown in Fig. 2 for iron excimer treated with 4 J/cm^2 in 2 bar CH$_4$ and 33x33 pulses. The CEM spectrum, with an information depth of about 150 nm, shows 75(2)% cementite (Fe$_3$C) and 16(2)% Hägg carbide (Fe$_5$C$_2$), while the remaining part is non-reacted α-Fe. We believe the latter contribution mostly originates from the border of the treated area where the spot overlap is incomplete, but some inhomogeneity of the layer cannot be excluded. The CXMS spectrum, which gives information on a depth of the order of 20 μm, revealed only 15% Fe$_3$C in addition to 3% austenite, indicating that the Fe$_3$C layer is confined to a depth of the order of 3 μm. All hyperfine parameters agree well with those reported in the literature [26,27,28]. This phase analysis and composition coincides with the XRD analysis. The RBS spectra of the original Fe and the

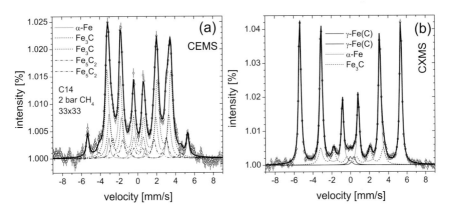

Fig. 2. Mössbauer spectra of excimer laser carburized iron, left CEM spectrum, right CXM spectrum (4 J/cm2, 33x33 pulses, 2 bar CH$_4$)

sample irradiated with 256 pulses show after the laser treatment in methane a reduction of the Fe yield exactly to the cementite stoichiometry (25 at.% C and 75 at.% Fe) to a depth of at least 650 nm.

The microhardness as shown in Fig. 3 revealed that the maximum hardness of the surface layer is about 6.5 GPa, which is much larger than the pure bulk iron. The pressure slightly increases the hardening depth, but the number of pulses does this more effectively from 1 to 3 μm. In conclusion, a method of obtaining a rather thick, polycrystalline cementite layer has been presented.

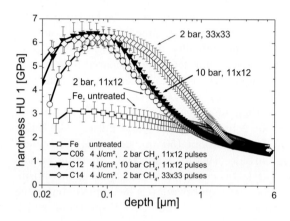

Fig. 3. Nanoindentation measurements of laser carburized iron (4 J/cm2, 11x12 pulses and 33x33 pulses, 2 and 10 bar CH$_4$)

3.2 Thin Films on Aluminum

Surface layers of aluminum nitride were formed by irradiating pure aluminum substrates in nitrogen atmosphere with a pulsed excimer laser. All other lasers are less efficient for the production of AlN films. Irradiations were carried out at various laser fluences, nitrogen gas pressures and numbers of pulses in order to investigate the influence of each parameter on the nitrogen incorporation and the mass transport mechanisms. X-Ray Diffraction showed the formation of polycrystalline AlN phase with the wurtzite structure, and the analysis of the nitrogen depth profiles by means of Resonant Nuclear Reaction Analysis revealed a monotonic increase of the nitrogen concentration with the ambient gas pressure and the number of laser shots. It has been found that the laser fluence directly determines the temperature of the substrate and strongly changes the transport mechanism [29]. The increase of the nitriding efficiency with the gas pressure is seen in Fig. 4.

Diffusion-like profiles with increasing nitrogen content and nitride thickness are formed as the gas pressure increases. A saturation approaching the AlN stoichiometry sets in at about 6 bar. All these profiles can be successfully modeled [29].

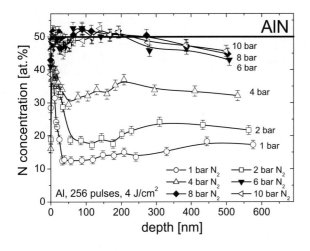

Fig. 4. Nitrogen depth profiles obtained after excimer laser nitriding of Al at gas pressures from 1 to 10 bar (4 J/cm2, 256 pulses)

3.3 Thin Films on Titanium

Laser irradiations on Ti revealed good coating forming efficiency for all pulse duration regimes, even where the irradiation of iron and aluminum are ineffective as shown above. This is shown now for the laser nitriding of titanium to form TiN coatings. TiC films have also been produce by laser carburization but this will be reported elsewhere [30].

Figure 5 shows the nitrogen depth profiles in Ti after ns, ps, and fs laser treatments. In all cases, high amounts of nitrogen are found in the surface (30-40 at.%) and present also deeper than the experimental detection limit of the RNRA method. The fs treatment shows here the highest nitrogen concentration and the FEL and Excimer laser show an almost identical result.

Of course, the results are depending on the actual treatment conditions. Therefore, the FEL treatment was repeated varying the time structure of

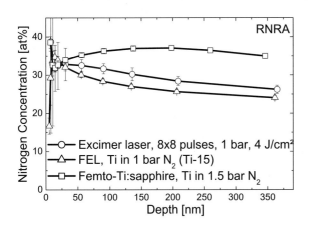

Fig. 5. Nitrogen depth profiles after laser nitriding of Ti with Excimer, FEL, and Femto-Ti:sapphire lasers. The experimental details are given in the graph

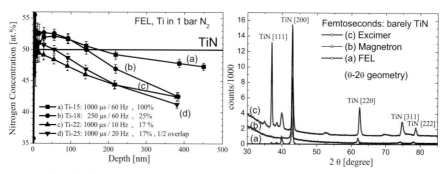

Fig. 6. (a) (Left) Nitrogen depth profiles for FEL nitrided titanium with various time structures (details in graph). (b) (Right) XRD pattern for nitrided Ti samples treated with different lasers)

the FEL. From the nitrogen profiles in Fig. 6(a) it is seen that the influence of the time structure is not very prominent. The titanium seems to have a large parameter window for the nitriding treatment. Nevertheless, additional investigations show that the microstructure and properties of the thin films are heavily depending on the latter [30]. The diffraction pattern in Fig. 6(b) clearly show the predomoninant presence of TiN, but also exhibit a strong (002) texture for the FEL produced TiN. This texture is that strong only for the FEL nitrided sample. The Excimer laser nitrided sample shows no texture at all and the magnetron sputtered TiN film has a much smaller grain size and experiences large stresses as seen from the peak width and a peak shift. The laser nitrided TiN films show sharp and well positioned peaks.

The femtosecond laser treated titanium contains almost 40 at.% nitrogen (Fig. 5) but does not exhibit other phases than α-Ti in the XRD measurement. As discussed above for the iron case, the nitrogen is maybe forming small clusters or nanocrystals of TiN, hardly detected in the diffraction pattern. This is reflected in the microhardness measurements shown in Fig. 7.

The Excimer laser nitrided TiN shows the highest hardness of about 5.5 GPa, but limited to the very surface and rapidly falling to the value of the untreated titanium. The FEL nitrided sample also gives a high hardness of more than 4 GPa, but shows a softer surface and an increasing hardness towards deeper regions (exceeding 4 μm). This may be explained by the different film thicknesses. The femtosecond laser treated sample is very soft (around 0.1 GPa), which results from the loosely sticking TiN clusters and nanocrystals as already concluded from the XRD and RNRA measurements. This cluster formation arises most probably from reactions within the plasma/gaseous state and should be further investigated [18].

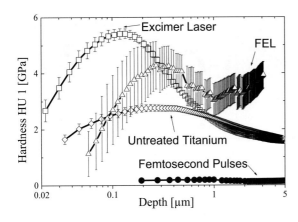

Fig. 7. Nanoindentation microhardness measurements for the laser nitrided samples shown in Fig. 5

3.4 Thin Films on Silicon

The technological importance of silicon carbide (SiC) and silicon nitride (Si_3N_4) has attracted great scientific attention on the synthesis of these materials. The excellent thermal stability, the good chemical resistance at high temperatures and the large band gap of SiC make this material extremely attractive for high power applications [31]. On the other hand, the superior corrosion resistance, the thermal and mechanical properties of Si_3N_4 enhanced the used of this ceramic in many different fields such as automobile engineering, cutting tools, nuclear reactors, etc. Among all polymorphs of Si_3N_4 (α-Si_3N_4, β-Si_3N_4 and c-Si_3N_4 [32]) the amorphous phase (a-Si_3N_4) is employed in the microelectronics industry as gate insulator in transistors [33]. Here, we report preliminary studies on the laser irradiation of B-doped single-crystalline silicon substrate in controlled nitrogen and methane gaseous environment, revealing the formation of carbide and nitride as function of the experimental parameters.

Figure 8 shows the nitrogen depth profiles for excimer laser nitriding of silicon. The relative amount of the nitride phase saturates after few laser shots for the single spot treatment at values between 40% and 50% and a sharp nitrogen peak can be observed ($\sim$ 20 nm thick) which is attributed to the strong segregation of nitrogen in Si during the cooling of the substrate (with a laser fluence of 5 J/cm^2 the temperature of the silicon substrate can largely exceed the melting point), while the meandering treatment performed under the same experimental conditions revealed the migration of nitrogen to larger depth. Thus, the nitrogen depth profiles revealed the strong influence of the meandering treatment on the mass transport. In order to have more information about the microstructure of the laser produced films, also EXAFS measurements have been performed. The EXAFS spectra of the laser-irradiated samples show peaks belonging to both structures, revealing the superposition of the carbide/nitride phase and non-reacted Si. In order to have a quantitative analysis Back-Fourier transformations were performed

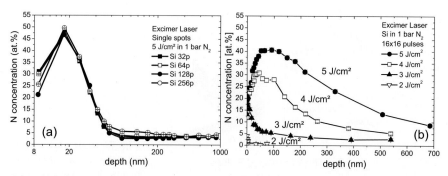

Fig. 8. RNRA depth profiles of excimer laser nitrided and laser carburized silicon ($5\,\mathrm{J/cm^2}$, 1 bar N_2): (a) left: single spot treatment, (b) right: meandering treatment

for SiN and SiC. The results are shown in Fig. 9 and more details are given in Ref. [16].

The FT of the EXAFS oscillations of the single spot and the meandered samples have similar features, seen in Fig. 9a, revealing only the first shell of the Si_3N_4 and all peaks of Si suggesting that the different mechanism of nitrogen migration does not lead to a different crystallization of the nitride phase. It should be noticed that the lower intensity of the nitride peak in the meandered sample is in agreement with the lower nitrogen concentration measured by RNRA, as shown in the inset. The results for SiC shown in Fig. 9b confirm the good crystallinity of the laser-induced SiC. Silicon carbide crystallizes in many polytypes differing from each other only in the stacking sequence. The two most common SiC polytypes are the 3C-SiC (cubic structure) and 6H-

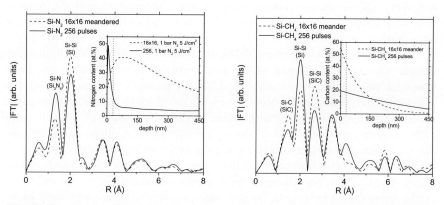

Fig. 9. Fourier transform moduli of the samples irradiated with: (−) single spot with 256 pulses, (···) meandered with 16×16 pulses, **(a)** (Left) in N_2, **(b)** (Right) in CH_4. The insets show the corresponding nitrogen and carbon depth profiles measured with RNRA and RBS. The vertical dotted line is the average information range of the EXAFS in TEY mode (∼30 nm)

SiC (hexagonal structure). The EXAFS oscillations are sensitive to the short range atomic order and all polytypes have the same fine structure pattern. XRD was performed to have information on the crystallographic phase of the laser-induced SiC showing the formation of polycrystalline 3C-SiC (β-SiC).

The EXAFS analysis revealed also the strong influence of the meandering treatment on the mass transport mechanism. As reported in Fig. 9b, the intensities of the carbide peaks of the meandered sample with 16×16 pulses are higher compared to the intensities of the sample treated under identical conditions but as single spot with 256 laser pulses. The carbon depth profiles extracted from the RBS analyses of these two samples are reported in the inset of Fig. 9b, showing that the carbon content on the surface is consistent with the EXAFS observations.

4 Summary and Outlook

The efficiency of Laser Plasma Synthesis, e.g. laser nitriding and laser carburizing in Fe, Al, Ti, Si has been investigated for different types of pulsed lasers. With pulse duration varying from femtosecond to nanosecond regime, the dependence of the laser efficiency on the time scale has been investigated. In Fe and Al the maximum nitrogen incorporation has been found in the nanosecond range, while excellent nitriding efficiency was discovered in the whole time domain for Ti. Due to the strong effect of the wavelength, Si could be efficiently irradiated only with the excimer laser. It is also important to obtain more information on the laser-plasma-substrate interactions, as well as the plasma and melt dynamics.

Acknowledgements

This work is supported by the Deutsche Forschungsgemeinschaft (DFG grant Scha 632). We wish to thank the FEL group at Jefferson Lab, Newport News, USA, especially Drs. Michelle Shinn, Fred Dylla and Gwyn Williams, for their help with the FEL treatments, the LZH Hannover and the BIAS Bremen for the Ti:sapphire and Nd:YAG laser treatments. The EXAFS measurements at LURE were performed within LURE Project PS 016-01.

References

1. D. Bäuerle, *Laser Processing and Chemistry* (Springer Verlag, Berlin, Heidelberg, 2000).
2. R. W. Cahn and P. Haasen, *Physical Metallurgy* (North-Holland Physics Publishing, Amsterdam - Oxford - New York - Tokyo, 1983).
3. P. Schaaf, Progress in Materials Science **47**, 1 (2002).
4. X. Y. Chen, S. B. Xiong, Z. S. Sha, and Z. G. Liu, Appl. Surf. Sci. **115**, 279 (1997).

5. S. Preuss, A. Demchuk, and M. Stuke, Appl. Phys. **A 61**, 33 (1995).
6. E. Carpene, F. Landry, and P. Schaaf, Appl. Phys. Lett. **77**, 2412 (2000).
7. P. Schaaf, F. Landry, and K.-P. Lieb, Appl. Phys. Lett. **74**, 153 (1999).
8. F. Landry, K.-P. Lieb, and P. Schaaf, J. Appl. Phys. **86**, 168 (1999).
9. E. D'Anna, M. L. De Giorgi, G. Leggieri, A. Luches, M. Martino, A. Perrone, I. N. Mihailescu, P. Mengucci, and A. V. Drigo, Thin Solid Films **213**, 197 (1992).
10. J. Barnikel, T. Seefeld, K. Schutte, and H. W. Bergmann, Härterei-Technische Mitteilungen **52**, 94 (1997).
11. E. Sicard, C. Boulmer-Leborgne, and T. Sauvage, Appl. Surf. Sci. **129**, 726 (1998).
12. T. Osipowicz, K. P. Lieb, and S. Brüssermann, Nucl. Instr. and Methods **B 18**, 232 (1987).
13. L. R. Doolittle, Nucl. Instrum. and Methods **B15**, 227 (1986).
14. P. Schaaf, A. Krämer, L. Blaes, G. Wagner, F. Aubertin, and U. Gonser, Nucl. Instrum. and Methods **B 53**, 184 (1991).
15. K. V. Klementev, J. Phys. D: Appl. Phys. **34**, 209 (2001).
16. E. Carpene, A.-M. Flank, A. Traverse, and P. Schaaf, J. Phys. D: Appl. Phys. (2002), submitted.
17. E. Carpene, P. Schaaf, M. Han, K.-P. Lieb, and M. Shinn, Appl. Surf. Sci. **186**, 195 (2002).
18. P. Schaaf, M. Han, K.-P. Lieb, and E. Carpene, Appl. Phys. Lett. **80**, 1091 (2002).
19. H. Nakagawa, S. Nasu, H. Fujii, M. Takahashi, and F. Kanamaru, Hyperfine Interactions **69**, 455 (1991).
20. L. Rissanen, M. Neubauer, K.-P. Lieb, and P. Schaaf, J. Alloys Comp. **274**, 74 (1998).
21. Y. Yamada, H. Shimasaki, Y. Okamura, Y. Ono, and K. Katsumata, Applied Radiation and Isotopes **54**, 21 (2001).
22. E. Carpene and P. Schaaf, Appl. Phys. Lett. **80**, 891 (2002).
23. J. H. Häglung, G. Grimvall, and T. Jarlborg, Phys. Rev. **B 44**, 2914 (1991).
24. S. J. Li, M. Ishihara, H. Yumoto, T. Aizawa, and M. Shimotomai, Thin Solid Films **316**, 100 (1998).
25. J. Kunze, *Nitrogen and carbon in iron and steel* (Akademie Verlag, Berlin, 1990).
26. M. Ron, in *Applications of Mössbauer spectroscopy II*, edited by R. L. Cohen (Academic Press, New York, 1980), pp. 329–392.
27. P. Schaaf, S. Wiesen, and U. Gonser, Acta Metall. **40**, 373 (1992).
28. P. Schaaf, A. Krämer, S. Wiesen, and U. Gonser, Acta Metall. **42**, 3077 (1994).
29. E. Carpene and P. Schaaf, Phys. Rev. **B** (2002), submitted.
30. P. Schaaf and E. Carpene, (2002), in preparation.
31. H. Morkoc, S. Strite, G. B. Gao, M. E. Lin, B. Sverdiov, and M. Burns, J. Appl. Phys. **76**, 1363 (1994).
32. T. Sekine, H. He, T. Kobayashi, M. Zhang, and F. Xu, Appl. Phys. Lett. **76**, 3706 (2000).
33. K. S. Seol, T. Futami, T. Watanabe, and Y. Ohki, J. Appl. Phys. **85**, 6746 (1999).

Nanotechnology – Bottom-up Meets Top-down

O. G. Schmidt, Ch. Deneke, Y. Nakamura, R. Zapf-Gottwick,
C. Müller, and N. Y. Jin-Phillipp

Max-Planck-Institut für Festkörperforschung,
Heisenbergstr. 1, 70569 Stuttgart, Germany
O.Schmidt@fkf.mpg.de

Abstract. Great progress has been made over the last years to elucidate the huge potential of mesoscopic nanostructures such as quantum dots and carbon nanotubes. Nano-circuits have been constructed to carry out simple logic operations, which moves nanoscale electronics from "blue-sky research" to the beginning of a real technology. However, up to now circuits remain rudimentary since the exact and controlled positioning of individual nanostructures requires time consuming and awkward one after the other positioning technologies. Here, we illustrate how the formidable task can be achieved of putting semiconductor quantum dots, nanotubes and other nano-objects into exact and well-defined locations. The undertaking is accomplished by combining self-organization with standard lithographic and processing techniques. In addition, we address the scalability of rolled-up semiconductor nanotubes and present the surprising result that tube diameters from a few nanometers to several hundred nanometers are well-described by a macroscopic continuous mechanical model.

1 Introduction

Nanotechnology is widely accepted as a "key technology" of the 21st century. The importance is drawn from its pronounced interdisciplinary nature, building bridges from and to almost all scientific fields – be it medicine or physics. In spite of its richness and versatility, nanotechnology can be reduced to two fundamental principles.

On the one hand there is traditional processing technology, where existing tools and structures are downscaled into the nanometer region. All of our daily life electronics and opto-electronics components are based on this so-called "top-down" approach. The other approach comes from the bottom. "Let nature do the job" is the credo. Self-formation processes and chemical synthesis create billions and billions of refined and perfect nanometer sized objects within seconds – easy and cheap. However, their distribution on substrate surfaces is governed by statistics – rendering controlled and reproducible use of individual nanostructures impossible.

It needs new and innovative techniques to efficiently combine and integrate the two fundamental approaches of "top-down" and "bottom-up" – to be able to put perfectly formed nanostructures at exactly defined and well-controlled positions in future products and devices. Two prominent types of nanostructures are self-assembled semiconductor quantum dots [1,2,3,4] and chemically synthesized carbon nanotubes [5,6]. Both types inhere the unique feature of perfect crystallinity – a property not achievable by any kind of processing technique. Efforts are huge to arrange such nanostructures into ordered two- and three-dimensional arrays on substrate surfaces [7,8,9,10,11,12,13], because it is only then that their superior properties can be exploited in large-scale integration technologies [14,15,16,17,18,19].

2 Two- and Three Dimensional Periodic Arrays of Self-Assembled Semiconductor Quantum Dots

Figure 1 shows how lithography and self-formation are successfully combined to render self-assembled Ge islands into alignment on a flat Si (001) substrate [9]. The initial Si substrate surface was pre-patterned by e-beam lithography and reactive ion etching. By this technique a one-dimensional pattern of straight grooves was processed into the substrate surface [9]. The grooves had a width and depth of 100 nm and about 10 nm, respectively, with a periodicity of 250 nm. This surface was subsequently overgrown with a Si/SiGe superlattice that transformed the surface modulation into a strain-field modulation at a planar growth front. The strain-field modulation finally induced the ordered formation of self-assembled Ge islands in the final layer on the surface. The virtues of both approaches – "top-down" and "bottom-up" – are preserved by this technique. The exact position is accomplished by the initial pre-patterning and the perfect crystal quality of the structures is guaranteed by the final self-assembling process – far away from the contaminated pre-patterned interface. The technique is universal and can be applied to many other strained material systems.

Another example is the well-established strained InAs/GaAs heterostructure system [7]. Figure 2(a) shows the lateral alignment of self-assembled InGaAs quantum dots grown onto a pre-patterned GaAs (001) substrate. Similar to the Si/SiGe case, the substrate surface was decorated with a square array of shallow 80 nm wide and 25 nm deep holes with a periodicity of 200 nm. The sample was then overgrown with a 13 monolayer (ML) and a 5 ML thick InGaAs layer separated by a 20 nm GaAs-AlGaAs-GaAs spacer layer. With this technique well-developed InGaAs quantum dots form into a regular square array on the surface with the same periodicity as the pre-patterned hole array.

We stacked several layers of In(Ga)As quantum dots on the pre-patterned surface to fabricate a three dimensional quantum dot crystal [7]. The spacer layers between the In(Ga)As layers were chosen thin enough such that the quantum dots reproduce themselves in vertical direction via their strain fields [20,21,22]. The stack consists of four $In_{0.4}Ga_{0.6}As$ layers and two nominally

Fig. 1. Ordered array of self-assembled Ge islands on Si (001). Upper part shows 1.7 x 1.7 mm² AFM scan and lower part shows corresponding autocorrelation. The autocorrelation reveals a two-dimensional hexagonal pattern although the initial Si surface was pre-patterned with one-dimensional trench lines, only

pure InAs quantum dot layers, where all layers are separated by 20nm-thick GaAs-AlGaAs-GaAs spacer layers. Figure 2(b) shows a cross-section transmission electron microscopy (TEM) image of a single column of the quantum dots. The dots are free of any extended crystal defects and the vertical alignment is initiated by the first deposited quantum dot layer, which was grown onto the patterned surface. The InGaAs of the first layer fills in the prepatterned holes. After that the AlGaAs marker layers (bright contrast) allow us to follow the stacking process step by step. After the first InGaAs layer has been deposited the surface becomes almost completely flat again. Nevertheless, the quantum dots of the second layer nucleate exactly on top of the buried ones, which is due to the strain field created by buried InGaAs layer. The first two layers again demonstrate the possibility to transform a periodic structural modulation of a substrate surface into a periodic strain field modulation at the growth front, which constitutes the platform to fabricate a perfect and defect-free quantum dot crystal.

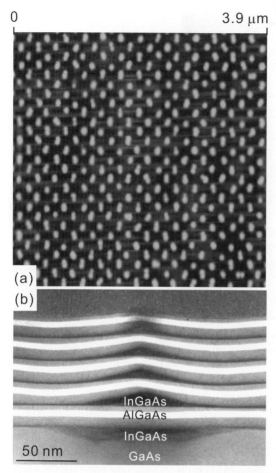

Fig. 2. (a) Two-dimensional square pattern of InGaAs quantum dots on a GaAs (001) surface. (b) Cross-section TEM of vertically stacked In(Ga)As quantum dots. The position of the vertically aligned QDs is defined and induced by the initial pre-patterning of the hole, which is completely filled with InGaAs from the first deposited layer

3 Semiconductor Nanotubes

Recently, it was shown that strained semiconductor epitaxial layers roll up or fold into nanotubes once they are released from their substrate [23,24,25,26]. This technique was generalized to a new kind of nanotechnology, in which 3-dimensional nano-objects of almost arbitrary shape and arbitrary material combination can be placed at almost any arbitrary position on a substrate surface [25,26] – thus making this technology a kind of Origami on the micro and nanometer scale [27]. The technology combines "top-down" and "bottom-up" approach, since on the one hand self-organization processes are utilized to create the nanostructures and on the other hand lithographic techniques are exploited to exactly position the objects. By now a rich variety of differently shaped micro- and nanostructures has been created by this technique – ranging from tubes [23,24,25,26,27,28,29,30], pipelines [25,28] and rings [25,29,3] to coils [23,24], membranes [25], loops [27] and mirrors [31]. Arsenic-based [23,25,3,30] and P-based [27,30] compounds as well as highly doped SiGe [24] and intrinsic SiGe [25,26,27,28,29,30] have been used to form these nano-objects. Nanotubes can be manipulated and cut without destroying the underlying substrate by applying focused laser light [27]. Nanotubes can perform up to about 30 rotations [25,29] on a substrate surface and other materials can be rolled into such nanotubes [25]. If the inner diameter is kept small and the number of rotations is high these nanotubes transform into nanorods [30]. Tubes made up of intrinsic InGaAs, SiGe, and InGaP are presented in Fig. 3. In Fig. 3(a) an InGaAs/GaAs bilayer has rolled-up over a distance of 6 μm and the diameter of the created tube is 120 nm. Fig. 3(b) shows a SiGe tube that has rolled into the [010] direction on a Si (001) surface while in Fig. 3(c) the 50 nm wide opening of an InGaP nanotube is presented.

At first sight, the diameter of the nanotubes can be scaled by simply changing the layer thicknesses and layer composition [23,24,25,26,29,30] and an example of the scaling behavior is given in Fig. 4. Both tubes in Fig. 4 have roughly performed 30 rotations on the surface. For the left tube, the

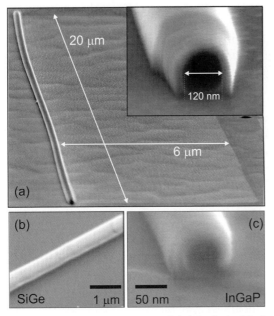

Fig. 3. Nanotubes made up of (**a**) InGaAs/GaAs, (**b**) SiGe, and (**c**) InGaP. (**a**) demonstrates that the tube position is determined by the initial starting point of the roll-up process and the duration of etching

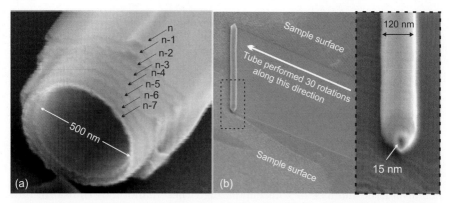

Fig. 4. Multi-wall InGaAs/GaAs nanotubes with different diameters. In (**a**) the outer eight rolled up layers are well-resolved. In (**b**) the wall thickness is 53 nm, resulting in an inner to outer diameter ratio of only 0.125. The structure is called a nanorod

outer eight rolled-up layers are well-resolved. The as-grown bilayer consists of 6.6 ML $In_{0.3}Ga_{0.7}As/11.7$ ML GaAs, whereas the as-grown bilayer for the nanotube on the right incorporates 1.4 ML $In_{0.3}Ga_{0.7}As/4.4$ ML GaAs. Since the layers of both tubes have the same In concentration but the bilayer in the right case is much thinner than the one for the left tube, the tube on the right hand side experiences a much smaller diameter. Recently, we proposed that nanorods could be created if tube diameters were kept small and the number of rotations high [26]. Figure 4(b) represents exactly such a case: The wall-thickness of the tube is 53 nm and the inner to outer diameter ratio amounts to 0.125, thus making this nanotube a so-called nanorod.

Recently, we pointed out that the nominal GaAs epitaxial layer thickness, deposited by e.g. molecular beam epitaxy, changes during the nanotube fabrication (underetching) process [29]. This is caused by oxide formation that consumes a certain thickness of the epitaxial topmost layer as soon as the sample is removed from the ultra-high vacuum chamber into ambient conditions. The oxide is subsequently removed during the selective etching process [32,33] and thus reduces the effective layer thickness relevant for the tube diameter. In References [25,29,3] we assumed a consumption thickness of about 5 ML. However, the thickness of the oxide changes with time. For GaAs the oxide formation was studied in some detail by Y. Mizokawa et al. [33], revealing a stable oxide of 0.8-0.9 nm after the first day under ambient conditions and a steady increase of the oxide thickness after one month. No value for the consumed GaAs is available in literature, but from the oxide thickness we assume the loss of the first 2-3 ML GaAs. Figure 5 shows a cross-section TEM image of an InGaAs/GaAs bilayer and its oxide after the sample was exposed to air for 30 days. From the TEM image we deduce that the bilayer has a thickness of 4.8 nm, although nominally 5.3 nm have been

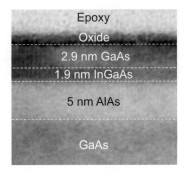

Fig. 5. Cross-section TEM image of an InGaAs/GaAs bilayer grown on an AlAs buffer layer. The top section of the GaAs layer is partially consumed by oxide formation

grown and we conclude that 0.5 nm, i.e. 2 ML GaAs have been consumed by the oxide. Furthermore the top layer is expected to be sightly reduced during the etching process, so that a reasonable assumption would be a total loss of 2-4 ML of the top GaAs layer.

Figure 6 shows the diameter of various $In_{0.3}Ga_{0.7}As/GaAs$ nanotubes as a function of bilayer thickness. We assume that the bilayer thickness was reduced by three monolayers compared to the original as-grown layer thickness due to oxide formation. The specific composition of 30% Indium was chosen since tube diameters can be tuned from a few nanometers to several hundred nanometers by just changing the layer thickness and leaving the composition of the layers constant. The growth of the layers was monitored with reflection high energy electron diffraction, which showed a streaky c(4×4) reconstruction pattern during all growth runs. The growth temperature was kept relatively low at 430°C to avoid In segregation effects [34]. All bilayers were grown onto a 10 ML thick AlAs sacrificial buffer layer.

The diameter D of bilayered nanotubes can be expressed by a continuous mechanical model [31]:

$$D = \frac{d\left[3(1+m)^2 + (1+m\cdot n)\cdot\left[m^2 + (m\cdot n)^{-1}\right]\right]}{3\varepsilon(1+m)^2} \quad (1)$$

where $d = d_1 + d_2$ is the total thickness of the bilayer, ε is the in-plane biaxial strain between the two layers and $n = Y_1/Y_1$ and $m = d_1/d_2$ are the ratio of Young's moduli and the thicknesses of the first and second layer, respectively. Assuming $Y_1 \approx Y_2$ and $d_1 \approx d_2$ we obtain:

$$D \approx \frac{4}{3}\frac{d}{\varepsilon} \quad (2)$$

For the $In_{0.3}Ga_{0.7}As/GaAs$ combination we use the following values: $\varepsilon = 0.023$, $Y_1 = 85.6$ GPa and $Y_2 = 70.53$ GPa [35,36]. The calculated straight line

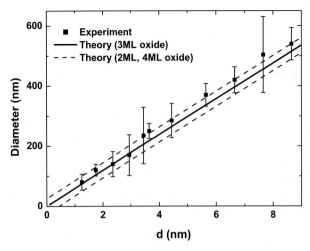

Fig. 6. $In_{0.3}Ga_{0.7}As/GaAs$ tube diameters as a function of bilayer thickness. Experimantal points and calculated values after equation 2 (solid) are given. Because of the uncertainty in oxide formation the theoretical straight lines (dashed) for 2 ML and 4 ML GaAs consumption are inserted, too

is inserted in Fig. 6. Due to the uncertainty in oxide formation we also show the theoretical straight lines (dashed) for 2 ML and 4 ML GaAs consumption. All experimental points are very well described by the macroscopic model in equation 2. This result, by itself, is surprising since the rolled-up layers are only a few monolayers thick and surface stress and reconstruction on the atomic scale were expected to render a macroscopic model in the nanometer region unreliable [28].

4 Conclusion

We have shown that lithographic techniques together with self-organization processes result in well-aligned and well-positioned self-assembled semiconductor nano-islands, quantum dots and nanotubes. The scalability of $In_{0.3}Ga_{0.7}As/GaAs$ nanotubes was experimentally and systematically determined by varying a single parameter – the bilayer thickness. A macroscopic continuous mechanical model is capable to describe the diameters of these nanotubes from several hundred down to only a few nanometers.

Acknowledgements

The authors thank H. Gräbeldinger and H. Schweizer for providing the pre-patterned substrates and K. Eberl for fruitful discussions. The continuous interest and support of K. v. Klitzing is acknowledged.

References

1. Y. W. Mo, D. E. Savage, B. S. Swartzentruber, and M. G. Lagally: Phys. Rev. Lett. **65**, 1020 (1990).
2. D. J. Eaglesham and M. Cerullo: Phys. Rev. Lett. **64**, 1943 (1990).
3. L. Goldstein, F. Glas, J. Y. Marzin, M. N. Charasse, and G. Le Roux: Appl. Phys. Lett. **47**, 1099 (1985).
4. S. Guha, A. Madhukar, and K. C. Rajkumar: Appl. Phys. Lett. **57**, 2110 (1990)
5. S. Iijima: Nature **354**, 56 (1991).
6. R. Sato, G. Dresselhaus, M. S. Dresselhaus *Physical Properties of Carbon Nanotubes* (Imperial College Press, London 1998)
7. O. G. Schmidt, S. Kiravittaya, Y. Nakamura, H. Heidemeyer, R. Songmuang, C. Müller, N. Y. Jin-Phillipp, K. Eberl, H. Wawra, S. Christiansen, H. Gräbeldinger, and H. Schweizer: Surface Science (in press).
8. G. Jin, J. L. Liu, and K. L. Wang: Appl. Phys. Lett. **76**, 3591 (2000).
9. O. G. Schmidt, N. Y. Jin-Phillipp, C. Lange, U. Denker, K. Eberl, R. Schreiner, H. Gräbeldinger, and H. Schweizer: Appl. Phys. Lett. **77**, 4139 (2000).
10. T. Kitajima, B. Liu, and S. R. Leone: Appl. Phys. Lett. **80**, 497 (2002).
11. H. Omi, D. J. Bottomley, and T. Ogino: Appl. Phys. Lett. **80**, 1073 (2002).
12. E. Kuramochi, J. Temmyo, T. Tamamura, and H. Kamada: Appl. Phys. Lett. **71**, 1655 (1997).
13. Z. F. Ren, Z. P. Huang, J. W. Xu, J. H. Wang, P. Bush, M. P. Siegal, and P. Provencio: Science **282**, 1105 (1998).
14. O. G. Schmidt and K. Eberl: IEEE Trans. Electron Devices **48**, 1175 (2001).
15. O. G. Schmidt: Vakuum in Forschung und Praxis 2, 107 (2001).
16. O. G. Schmidt: Spektrum der Wissenschaft, April 04 (2002).
17. O. G. Schmidt, U. Denker, M. Dashiell, N. Y. Jin-Phillipp, K. Eberl, R. Schreiner, H. Gräbeldinger, H. Schweizer, S. Christiansen, and F. Ernst: Mat. Sci. Eng. B **89**, 101 (2002).
18. A. Bachtold, P. Hadley, T. Nakanishi, and C. Dekker: Science **294**, 1317 (2001).
19. R. Martel, V. Derycke, J. Appenzeller, P. H. S. Wong, and P. Avouris: Abstr. Pap. Am. Chem. S. **222**, 36-Phys Part 2 (2001).
20. Q. Xie, A. Madhukar, P. Chen, and N. P. Kobayashi: Phys. Rev. Lett. **75** (1995) 2542.
21. O. G. Schmidt, O. Kienzle, Y. Hao, K. Eberl, and F. Ernst: Appl. Phys. Lett. **74** (1999) 1272.
22. O. G. Schmidt and K. Eberl: Phys. Rev. B **61** (2000) 13721.
23. V. Ya. Prinz, V. A. Seleznev, A. K. Gutakovsky, A. V. Chehovskiy, V. V. Preobrazenskii, M. A. Putyato, and L. A. Nenasheva: Inst. Phys. Conf. Ser. **166**, 199 (2000).
24. V. Ya. Prinz, S. V. Golod, V. I. Mashanov, and A. K. Gutakovsky: Inst. Phys. Conf. Ser. **166**, 203 (2000).
25. O. G. Schmidt, N. Schmarje, C. Deneke, C. Müller, and N.-Y. Jin-Phillipp: Adv. Mat. **13**, 756 (2001).
26. O. G. Schmidt and K. Eberl: Nature **410**, 168 (2001).
27. O. G. Schmidt, C. Deneke and Y. M. Manz: Proceedings of 8th Annual International Conference on Composites Engineering (ICCE), Tenerife, Spain, August 5-11 2001, p.823.
28. O. G. Schmidt and N.-Y. Jin-Phillipp: Appl. Phys. Lett. **78**, 3310 (2001).

29. O. G. Schmidt, C. Deneke, N. Schmarje, C. Müller, and N. Y. Jin-Phillipp: Mat. Sci. Eng. C **19**, 393 (2002).
30. O. G. Schmidt, C. Deneke, Y. M. Manz, and C. Müller: Physica E (in press).
31. P. O. Vaccaro, K. Kubota, and T. Aida: Appl. Phys. Lett. **78**, 2852 (2001).
32. J. F. Bresse: Applied Surface Science **66**, 1-4 (1993)
33. Y. Mizokawa, O. Komoda, and S. Miyase: Thin Solid Films **156**(1), 127-43 (1988).
34. K. R. Evans, R. Kaspi, J. E. Ehret et al.: Journal of Vacuum Science & Technology B **13** (4), 1820-3 (1995).
35. S. C. Jain, M. Willander, and H. Maes: Semiconductor Science & Technology **11** (5), 641-71 (1996).
36. Landolt-Börnstein: *Numerical data and Functional Relationship* **17** IIIa, (Springer-Verlag, Berlin, Heidelberg, New York 1982).

Local Ordering Processes in Ferroelectric, Glass-like and Modulated phases: An EPR Study

G. Völkel, N. Alsabbagh, J. Banys, H. Bauch, R. Böttcher, M. Gutjahr, D. Michel, and A. Pöppl

Universität Leipzig, Fakultät für Physik und Geowissenschaften,
Linné-Str. 5, 4103 Leipzig, Germany

Abstract. By means of two selected examples it will be shown that electron paramagnetic resonance (EPR) spectroscopy delivers unique microscopic information about crystals with ferroelectric, glass-like and modulated phases. The first example concerns a proton glass. Frustrating interactions between the protons of the hydrogen bond system in mixed crystals of ferroelectric betaine phosphite and antiferroelectric betaine phosphate give rise to a glass-like order of the proton system. Electron nuclear double resonance (ENDOR) and two-dimensional pulsed HYSCORE (hyperfine sublevel correlation spectroscopy) investigations give access to the distribution function of the local proton order parameter and to the Edwards-Anderson glass order parameter. The investigations give evidence for a new cluster-like intermediate phase probably similar to the Griffiths phase in magnetic spin glasses. In the second example a crystal is considered which shows a complex sequence of commensurate and incommensurate phases firstly discovered by EPR investigations. Dimethylammonium gallium sulfate hexahydrate (DMAGaS) proves to be a new model system for a phase sequence paraelectric - ferroelectric - IC/C modulated - antiferroelectric not yet observed for any other crystal. For both examples the experimental results will be discussed in comparison to microscopic model theories.

1 Introduction

Magnetic resonance is known to be a powerful tool for studying local ordering processes in materials showing ferroelectric, antiferroelectric, dipolar glass (proton glass) or commensurably and incommensurably modulated phases [1,2,3,4]. Nuclear magnetic resonance (NMR) and electron paramagnetic resonance (EPR) can sensibly complement each other because of their different time windows of investigation. NMR is the more generally applicable technique for studying such materials because most of them embody nuclei with a non-zero nuclear spin. The application of EPR is much more restricted because dielectric crystals are usually not paramagnetic. The complex EPR spectra can be described by an spin Hamiltonian

$$\mathbf{H} = \beta \mathbf{SgB} + \mathbf{SDS} + \mathbf{SAI} + \mathbf{S}\sum_i \mathbf{A}^i \mathbf{I}^i \qquad (1)$$

where **S** and **I** are the effective electron spin operator and nuclear spin operator, respectively. The first term describes the Zeeman interaction of the effective electron spin S of the paramagnetic center with the applied static magnetic field **B** where **g** is the g-tensor and β the Bohr magneton. The second term with the fine structure tensor **D** characterizes the zero-field or fine structure splitting of the electron spin states for the case of an effective electron spin $S \geqslant 1$. The fourth and sixth order contributions are ignored here. For $S = 1/2$ this contribution disappears. The third and fourth terms specify the hyperfine and superhyperfine interaction of the electron spin S with the nuclear spin I of the own nucleus of the paramagnetic ion and with the nuclear spins I^i of the ligand atoms. The tensors **A** and **A**^{i} are called hyperfine and superhyperfine tensor, respectively. All the tensors introduced are symmetric tensors of second rank and describe the anisotropic behavior of the spectra. After some general remarks to EPR in ferroelectrics in the next chapter, in the following two chapters the power of EPR spectroscopy for studying complex ordering processes shall be demonstrated with two examples. The first example concerns a mixed crystal of ferroelectric betaine phosphite (BPI) and antiferroelectric betaine phosphate (BP) which behaves like a dipolar glass or proton glass [5,6,7]. It will be demonstrated how electron nuclear double resonance (ENDOR) and one- and two-dimensional ESE (electron sin echo) investigations give access to the distribution function of the local proton order parameter and to the Edwards-Anderson glass order parameter. In the second example a crystal of dimethylammonium gallium sulfate hexahydrate (DMAGaS) is considered where EPR studies give evidence of a very complex sequence of commensurate and incommensurate phases [8,9]. For both examples NMR fails. For the first example, the protons do not possess a quadrupole moment and the proton chemical shift is not sensitive enough to indicate the small structural changes related to the proton ordering. This is also the general reason that only deuteron glasses but not proton glasses have been studied by NMR. But even for deuterated BPI or BP, the quadrupole split lines of the two deuterons involved in the ordering process at the ferro- or antiferroelectric transition can not show shifts linear with the order parameter [10] because of their inversion site symmetry in the paraelectric phase. In the second case, the quadrupole splitting of the ^{27}Al and the ^{69,71}Ga nuclei [11] is a few 100 kHz and thus much more smaller than the fine structure splitting of few GHz in the Cr^{3+} EPR experiment with the consequence that the spectrum changes related to the relatively fast dimethylammonium reorientations are averaged out for the NMR but not for the EPR experiment.

2   EPR Investigations on Ferroelectrics

Why can the EPR spectra provide information about the ordering behavior in ferroelectrics and related materials? All the tensors of the spin Hamiltonian mentioned above are sensitive against small structural changes in the environment of the paramagnetic probe. The most striking response gives the fine structure tensor. Therefore, paramagnetic ions with $S \geqslant 1$ are preferentially chosen as paramagnetic probes. Let us now consider what kind of changes one has to expect at a transition from a paraelectric into a ferroelectric phase with a vectorial order parameter $\eta = \eta(T_C - T)$ where we focus our consideration on the fine structure tensor $\mathbf{D}$. In the ferroelectric phase at a temperature $T \leqslant T_C$, the $\mathbf{D}$ tensor can be expressed by a contribution being the extrapolation of the tensor of the paraelectric phase $\mathbf{D}_0(T)$ with respect to thermal expansion to the temperature T, and by an additional contribution $\Delta \mathbf{D}$ caused by the structural changes related to the order parameter of the transition:

$$\mathbf{D}(T) = \mathbf{D}_0(T) + \Delta \mathbf{D}(T_C - T). \tag{2}$$

Expanding the ferroelectric contribution $\Delta \mathbf{D}$ with respect to the vectorial order parameter η up to second order terms one ends with

$$\Delta \mathbf{D}(T_C - T) = \mathbf{K}_1 * \eta + \eta * \mathbf{K}_2 * \eta \tag{3}$$

where $\mathbf{K}_1$ and $\mathbf{K}_2$ are tensors of third and fourth rank, respectively. By an appropriate choice of the magnetic field direction with respect to the $\mathbf{D}$ tensor principal axes one can select such spectra for which the change of line positions linearly corresponds to the change of a given $\mathbf{D}$ tensor element. The symmetry properties of the tensors $\mathbf{K}_1$ and $\mathbf{K}_2$ with respect to the local symmetry of the probe site allow interesting propositions for the line shifts expected at the phase transition. As well known from textbooks on tensor properties, a third rank tensor has non-zero tensor elements only for symmetry groups not containing an inversion center. That means only such probes can respond to the structural transition with $\mathbf{D}$ changes linear with η which do not constitute a center of inversion in the para-phase. A rotation of the tensor as a whole is not subjected to these restrictions. There is a series of examples in literature with this behavior [1]. The Ti and Ba sites of BaTiO$_3$ show inversion symmetry in the paraelectric phase. Therefore, paramagnetic probes as Fe^{3+}_{Ti}, Cr^{3+}_{Ti}, Mn^{2+}_{Ti}, Mn^{4+}_{Ti} or Gd^{3+}_{Ba} can only respond with changes of the fine structure tensor which are proportional to the square of the order parameter if there is no change of the local site symmetry due to charge compensation. Because in the paraelectric cubic phase of BaTiO$_3$ the fine structure tensor must be zero, its change $\Delta \mathbf{D}$ at the transition into the tetragonal phase is the now appearing $\mathbf{D}$ tensor itself. The experimental results for Fe^{3+}_{Ti} and Gd^{3+}_{Ba} in BaTiO$_3$ showed nicely that in the tetragonal phase the axial fine structure parameter $D = 3/2 D_{zz}$ is proportional to the square of the spontaneous polarization. The experimental findings for Mn^{2+}_{Ti} and Mn^{4+}_{Ti} in BaTiO$_3$ [1] demonstrate very impressive the importance of the probe choice for the interpretation of the results. As size and charge of the Mn^{4+} ion are comparable to those ones of the substituted Ti ion, it performs the same off-center motions as the Ti ions in the polar phases and is from this point of view the appropriate probe, but with the strong disadvantage to be only detectable in the rhombohedral low-temperature phase because of motional averaging in the higher phases. In contrast, the Mn^{2+}_{Ti} probe does not accomplish the same motion as the Ti ion and remains in the center position. From this point of view the probe gives only restricted information with respect to the real microscopic processes. But nevertheless it gives a measure for the local order parameter and shows the big advantage to be easily detectable in all the phases. The Mn^{2+}_{Ti} probe has successfully been used to study size effects in nanocrystalline $BaTiO_3$ by X, Q and W band EPR [12]. An example for the fine structure parameter variation ΔD linear with the order parameter η was given with chromium doped triglycine sulfate (TGS) [13] where the formed Cr^{3+} chelate complex shows non-inversion symmetry.

3 The Proton Glass BP_xBPI_{1-x} Studied by ENDOR Spectroscopy

The two isostructural compounds betaine phosphite (BPI) $(CH_3)_3NCH_2COO$ $*H_3PO_3$ and betaine phosphate (BP) $(CH_3)_3NCH_2COO*H_3PO_4$ form solid solutions at any concentration. Like in the KDP family, BP_xBPI_{1-x} mixed crystals of antiferroelectric BP and ferroelectric BPI show an interesting phase diagram with ferroelastic, paraelectric, ferroelectric, and antiferroelectric phases as well as glassy phases in dependence on the composition x and temperature [14,15,16,17,18]. BPI undergoes a phase transition from a paraelectric high-temperature phase with space group $P2_1/m$ into an antiferrodistortive phase with space group $P2_1/c$ at 355 K followed by a transition into a ferroelectric low-temperature phase with space group $P2_1$ at 220 K. BP shows a similar high-temperature transition into the antiferrodistortive phase but then two further transitions follow at 86 K into a intermediate ferroelectric phase of $P2_1$ symmetry [19] and at 81 K into an antiferroelectric low-temperature phase with a doubled unit cell length along the crystallographic **a** direction. The inorganic tetrahedral PO_4 or HPO_3 groups are linked by hydrogen bonds to zig-zag chains along the monoclinic **b** axis. The betaine molecules are arranged almost perpendicular to this chains along the **a** direction and linked by one (BPI) or two (BP) hydrogen bonds to the inorganic group. The temperature dependence of the dielectric permittivity of both compounds showed evidence for a quasi-one dimensional behavior as the coupling between the electric dipolar units within a chain is much stronger than that one between the dipols in neighbored chains [20]. In the mixed crystals the competing ferroelectric and antiferroelectric interactions cause a glass-like order behavior of the protons in the system of hydrogen bonds. Such materials belong to a class of orientational glasses that are called proton glasses [2,21]. The experimental results obtained by means of cw and pulsed ENDOR, ESE and HYSCORE studies allow very detailed comparisons with theoretical models such as the one- and three-dimensional random-bond random-field (RBRF) Ising glass model [22,23]. The attempt to describe the experimental behavior within the framework of the random-bond random-field (RBRF) Ising glass which was worked out for deuteron glasses [22] suffers in conspicuous insufficiency at least at lower temperatures. Taking the influence of proton tunneling into account by using the stochastic model of a quantum RBRF Ising glass [24], the experimental results can sufficiently be described.

3.1 The Quantum RBRF Ising Glass Model

Some results of the theoretical approach for the simplified case of incoherent tunneling [24] are shortly outlined. It is assumed that the proton in an O-H....O hydrogen bond is moving in a double well potential such that it spends most of its time at the bottom of either the left or the right well and very little in between. This is considered to be a two level system described by the pseudo-spin variables $S^z = \frac{1}{2}$ which designate the left or right position of the proton in the double well and must not be confused with the electron spin operator used in Eq. (1). A proton with the nuclear spin $I = \frac{1}{2}$ shall have the nuclear magnetic resonance frequencies ω_L and ω_R when it is on left and right side of the potential, respectively. Including interwell interaction, random electric fields and the possibility of tunneling between the two proton sites in a hydrogen bond, the proton spin Hamiltonian extended by the terms of the interacting pseudo-spin system takes the form

$$\mathbf{H} = \left(\frac{\omega_L + \omega_R}{2}\right) \sum_i I_i^z + (\omega_L - \omega_R) \sum_i I_i^z S_i^z \qquad (4)$$
$$- \frac{1}{2} \sum_{i,j} J_{ij} S_i^z S_j^z - \Omega \sum_i S_i^x - \sum_i f_i S_i^z$$

where Ω is the tunneling frequency, J_{ij} is the infinite-ranged quenched random interaction between the pseudo-spins S_i^z and S_j^z, and f_i represents the random longitudinal electric field at site i. The quantities J_{ij} and f_i are supposed to be independently distributed one from another according to their respective Gaussian probability distributions with the variances J and Δ, respectively. The variance J determines the nominal glass temperature $T_0 \equiv J/4$, and often the variance of the random field is expressed by the temperature $T_f = \sqrt{\Delta}/2 = T_0\sqrt{\tilde{\Delta}}$ with $\tilde{\Delta} = 4\Delta/J^2$. We take into consideration a non-zero mean of the random interactions $\langle J_{ij} \rangle = J_0$. The pseudo-spin system of the electric dipoles is treated within the framework of a mean field theory of a quantum spin glass. Using the effective field at a pseudo-spin in z direction $h(\tilde{z}) = \frac{J\sqrt{q+\tilde{\Delta}}}{2}\tilde{z} + J_0\bar{p}$ with a Gaussian noise field $\tilde{z}$, and the strength of the total effective field $h_0(\tilde{z}) = \sqrt{\Omega^2 + h(\tilde{z})^2}$, the mean-field equations of

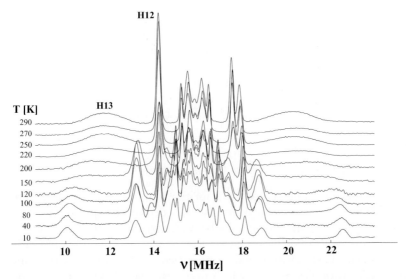

Fig. 1. Pulsed Davis-ENDOR spectra of the PO_3^{2-} probe in $BP_{0.40}BPI_{0.60}$ at various temperatures. The left outer line (H13) is analyzed in Fig. 2

the local polarization p, its nonzero average $\overline{p}$ and the spin-glass order parameter q can then be calculated. The local polarization distribution W(p) defines directly the ENDOR line shape and takes the form

$$W(p) = \frac{4\exp(-z_0^2/2)}{\beta J\sqrt{q+\widetilde{\Delta}}} \left[r(z_0)^2 + \frac{2\Omega^2 p}{\beta h(z_0)h_0(z_0)^2} - p^2 \right]^{-1} \quad (5)$$

where $z_0 = z_0(p)$ has to be calculated from the mean field equation for p for each given p value.

3.2 $BP_{0.15}BPI_{0.85}$ and $BP_{0.40}BPI_{0.60}$

Recently we reported on investigations of the proton glass $BP_{0.15}BPI_{0.85}$ by measuring the proton superhyperfine coupling of a PO_3^{2-} irradiation center [5,6,7]. New results for $BP_{0.40}BPI_{0.60}$ are presented in Figures 1 and 2. Sophisticated methods such as cw and pulsed ENDOR and one- and two-dimensional ESEEM (electron spin echo envelope modulation) techniques like HYSCORE enhance the experimental resolution almost to that one of NMR such that the proton reorientation of the weakly coupled protons becomes detectable. Other ESE techniques give access to the lattice dynamics by measuring the spin-lattice relaxation time. We showed that only the protons in the hydrogen bonds linking the phosphite and phosphate groups to quasi one-dimensional chains along the crystallographic **b** direction show glasslike order. The ENDOR line shape of these protons is a direct measure of the

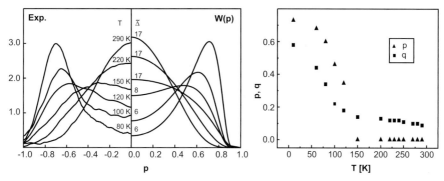

Fig. 2. (a) Experimental ENDOR line shape (left) and simulated polarization distribution function $W(p)$ (right), (b) resulting glass order parameter q and average polarization $\bar{p}$ at various temperatures in $BP_{0.40}BPI_{0.60}$

local polarization distribution $W(p)$. The experimental Edwards-Anderson glass order parameter obtained as the second moment of $W(p)$ shows a temperature dependence that is characteristic for a proton glass with very strong random fields represented by freezing temperatures $T_f = 95$ K and 103 K for $x = 0.15$ and 0.40, respectively. The glass temperature and tunneling energy were obtained to be $T_G = 30$ K and 25 K and $\Omega/k_B = 250$ K and 320 K, respectively. Furthermore, HYSCORE experiments on a partially deuterated $BP_{0.15}BPI_{0.85}$ crystal confirm the different local polarization behavior in the protonated and deuterated hydrogen bonds in agreement with the model expectation. The correspondence of the experimental data with the simulations could only be established by consideration of a non-zero mean value $|J_0/4| = 200$ K and 212 K of the random bond interaction leading to an average polarization $\bar{p}$ below $T_C = 144$ K within the mean-field RBRF model. This intermediate phase transition (see Fig. 3) is also reflected in a temperature anomaly of the spin-lattice relaxation time indicating a singular anomaly in the phonon system. The average proton polarization $\bar{p}$ measured here must have cluster-like nature because no macroscopic polarization does exist for the compositions under study. The local high-resolution EPR measurements show that the system under study is more complex than a usual RBRF proton glass. The new intermediate phase could be an analogy to the Griffith phase in Ising magnets with a non-zero average cluster polarization [25].

4 Modulated Phases in DMAGaS Evidenced by Q Band EPR

DMAGaS and the isomorphous dimethylammonium aluminum sulfate hexahydrate DMAAS are ferroelastic at room temperature. Both they show an order-disorder type transition into a ferroelectric phase at $T_{C1} = 150$ K and 135 K, respectively, but only DMAGaS exhibits a further first-order transition

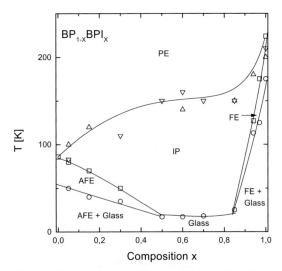

Fig. 3. Extended phase diagram of the solid solution $BP_x BPI_{1-x}$. The open squares and circles are from dielectric measurements, the down and up open triangles from ENDOR and spin-lattice relaxation measurements [10]. The drawn lines are only guides for the eyes. The different phases are abbreviated as follows: PE - paraelectric, FE - ferroelectric, AFE - antiferroelectric, IP - intermediate (cluster), GLASS - dipolar glass, (A)FE + GLASS - coexistence of (anti)ferroelectric and glassy phase

at $T_{C2} = 115$ K into a low temperature non-ferroelectric phase. EPR measurements of chromium doped DMAAS and DMAGaS [8,9,11,26] give insight into the peculiar reorientation order of the polar dimethylammonium (DMA) units on a microscopic level. In the temperature range from room temperature down to liquid helium temperature, cw EPR measurements were performed in Q band at 34 GHz. The measurements proof that in the ferroelastic and ferroelectric phase the Cr^{3+} EPR spectra of both crystals are very similar. Passing in DMAGaS the transition at T_{C2} into the low-temperature phase, striking changes have been observed in form of large line shifts, line broadening and in particular of a multiplied number of lines (Fig. 4). However, at a temperature of about 60 K the spectra show a conspicuous simplification ending up with very sharp lines in the low-temperature region where the number of lines is doubled in comparison to the ferroelastic phase. Taking into consideration the experience gained by the Cr^{3+} EPR investigations in DMAAS [11,26], one may show that the complex line shape results from a modulation of the occupancy of the DMA cation sites [8,9]. Because the order parameter modulation is not influencing the line positions but only the line intensities, another line shape results as such one usually observed in EPR and NMR experiments of incommensurate crystals [4]. As each Cr^{3+} site can be found in one under four configurations formed by the two neighboring DMA cations, the spectrum consists of four lines with intensities given by the statistical weight of

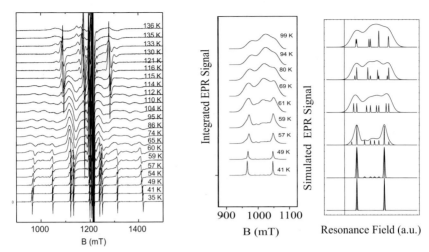

Fig. 4. Q band EPR spectra of chromium doped DMAGaS at various temperatures: recorded spectra (left), integrated part of recorded spectra (middle), simulated spectra (right) using $q = 1/4 - 0.05, 1/4, 1/4 + 0.04, 1/2 - 0.00003, 1/2 - 0.00001, 1/2$

each of the configurations which is assigned by the order parameter of the DMA system [26]. With a doubling of the unit cell along its shortest dimension **b**, the physical properties of two former identical neighboring Ga sites along the **b** direction become different. Therefore, the spectra of two Cr^{3+} probes occupying such former identical sites become different now, and two overlapping quartets result. Because the antiferro-configuration of a Cr^{3+} site corresponds to the low- or high-field line of a quartet [26], after transition into the antiferroelectric state the spectrum of the two former identical Cr^{3+} sites consists of two intense lines at low and high field and six much weaker lines above and below them, respectively. By selecting proper modulation periods b/q of the DMA order parameter along the -DMA-Ga(Cr)-DMA- direction, the EPR line shape below T_{C2} can qualitatively be simulated [9] (Fig. 4). The evidenced phase sequence, with a paraelectric high-temperature phase followed by a ferroelectric phase, by incommensurately and commensurately modulated phases with a pronounced 1/4 phase at medium temperatures and by an antiferroelectric low-temperature phase, is quite unusual and to our knowledge not yet observed in other crystals. However, it can well be explained by means of a Landau approach using a large number of sublattice polarizations [9] and more generally by the extended DIFFOUR model of a linear chain of electric dipoles when a negative E parameter of the fourth order nearest neighbor interaction term is considered [27].

5 Conclusions

As another option to NMR, in special cases EPR investigations can provide unique information about the microscopic ordering processes in ferroelectrics and related dielectric solids like dipole glasses and modulated phases showing more complex ordered states. The very specific results obtained enable a very detailed inspection of the applicability of microscopic models. The appropriate choice of the paramagnetic probe and the methodical techniques available are of vital importance for successful applications. The power of EPR spectroscopy has been demonstrated by two selected examples, the mixed crystals of ferroelectric betaine phosphite and antiferroelectric betaine phosphate representing a proton glass with new properties, and dimethylammonium gallium sulfate hexahydrate that have been proved to be a new model system with a complex sequence of IC/C modulated phases.

Acknowledgements

The financial support of the Deutsche Forschungsgemeinschaft (DFG) is greatly acknowledged. We thank Professor Z. Czapla, Universiy of Wroclaw, and Dr. A. Klöpperpieper, University of the Saarland, for crystal growing and fruitful collaboration.

References

1. K. A. Müller and H. Thomas (Eds.), *Structural Phase Transitions II* (Springer-Verlag, Berlin Heidelberg, 1991).
2. R. Blinc, Z. Naturforsch. **45a**, 313 (1989).
3. R. Blinc, R. Pirc, B. Tadic, B. Zalar, D. Arcon, and J. Dolinsek, Crystallography Reports **44**, 177 (1999).
4. R. Blinc and A. P. Levanyuk (Eds.),*Incommensurate Phases in Dielectrics*, Vols. 1 and 2 (North Holland, Amsterdam, 1986).
5. H. Bauch, G. Völkel, R. Böttcher, A. Pöppl, and H. Schäfer, Phys. Rev. **B 54**, 9162 (1996).
6. G. Völkel, H. Bauch, R. Böttcher, A. Pöppl, H. Schäfer, and A. Klöpperpieper, Phys. Rev. **B 55**, 12151 (1997).
7. R. Böttcher, A. Pöppl, G. Völkel, J. Banys, and A. Klöpperpieper, Ferroelectrics **208**, 105 (1998).
8. G. Völkel, R. Böttcher, Z. Czapla, and D. Michel, phys. stat. sol. (**b**) **223**, R6 (2001).
9. G. Völkel, R. Böttcher, D. Michel, and Z. Czapla, accepted for Ferroelectrics (2001).
10. P. Freude and D. Michel, phys. stat. sol. (**b**) **195**, 297 (1996).
11. N. Alsabbagh, D. Michel, J. Furtak, and Z. Czapla, phys. stat. sol. (a) **167**, 77 (1998).
12. R. Böttcher, C. Klimm, D. Michel, H.-C. Semmelhack, G. Völkel, H.-J. Gläsel, and E. Hartmann, Phys. Rev. **B 62**, 2085 (2000).

13. W. Windsch and G. Völkel, Ferroelectrics **17**, 345 (1975).
14. S.L. Hutton, I. Fehst, R. Böhmer, M. Braune, B. Mertz, P. Lunkenheimer, and A. Loidl, Phys. Rev. Lett. **66**, 1990 (1991).
15. M.L. Santos, J.C. Azevedo, A. Almeida, M.R. Chaves, A.R. Pires, H.E. Müser, and A. Klöpperpieper, Ferroelectrics **108**, 1969 (1990).
16. M.L. Santos, M.R. Chaves, A. Almeida, A. Klöpperpieper, H.E. Müser, and J. Albers, Ferroelectrics Letters **15**, 17 (1993).
17. A. Loidl and R. Böhmer, Glass Transitions and Relaxational Phenomena in Orientational Glasses and Supercooled Crystals. In: Disorder Effects on Relaxational Processes, p. 659, R. Richert, A. Blumen (eds.), (Springer-Verlag, Berlin Heidelberg, 1994).
18. H. Ries, R. Böhmer, I. Fehst, and A, Loidl, Z. Phys. **B 99**, 401 (1996).
19. M. Lopes dos Santos, J.M. Kiat, A. Almeida, M.R. Chaves, A. Klöpperpieper, and J. Albers, phys. stat sol.(b) **189**, 371 (1995).
20. G. Fischer, H.J. Brückner, A. Klöpperpieper, H.G. Unruh, and A. Levstik, Z. Physik **B 79**, 391 (1990).
21. K. Binder and J. D. Reger, Adv. Phys. **41**, 547 (1992).
22. M. Oresic, R. Pirc, Phys. Rev. **B 47**, 2655 (1993).
23. R. Pirc, B. Tadic, R. Blinc, and R. Kind, Phys. Rev. **B 43**, 2501 (1991).
24. S. Dattagupta, B. Tadic, R. Pirc, and R. Blinc, Phys. Rev. **B 44**, 4387 (1991), Phys. Rev. **B 47**, 8801 (1993).
25. V. A. Stephanovich, Eur. Phys. J. **B 18**, 17 (2000).
26. G. Völkel, N. Alsabbagh, R. Böttcher, D. Michel, B. Milsch, Z. Czapla, and J. Furtak, J. Phys.: Condens. Matter **12**, 4553 (2000).
27. G. H. F. van Raaij, K. J. H. van Bemmel, and T. Janssen, Phys. Rev. **B 62**, 3751 (2000).

Part V

Superconducting Systems

Superconductivity and Non-Fermi Liquid Normal State of Itinerant Ferromagnets

Christian Pfleiderer

Physikalisches Institut, Universität Karlsruhe
Wolfgang-Gaede-Str. 1, 76128 Karlsruhe, Germany

Abstract. Itinerant ferromagnetism and superconductivity were traditionally believed to represent incompatible forms of electronic order that are well described by the Fermi liquid model of the metallic state or extensions thereof. The itinerant ferromagnet $ZrZn_2$ exhibits in contrast superconductivity in the milli-Kelvin temperature range that vanishes above the critical pressure p_c where ferromagnetism is suppressed. This suggests that itinerant ferromagnetism may be a precondition for certain forms of superconductivity. The itinerant electron magnet MnSi exhibits, on the other hand, a sharp change from a Fermi liquid T^2 resistivity to a $T^{3/2}$ temperature dependence over an exceptionally large temperature range in the normal metallic state above the critical pressure p_c for which magnetic order is suppressed. The latter property is not consistent with the predictions of the Fermi liquid theory of the normal metallic state. When taken together the superconductivity and non-Fermi liquid normal state may highlight inconsistencies of the current theory of quantum criticality.

The coexistence of magnetism and superconductivity has attracted scientific interest for many decades. Already by the 1960s numerous studies had shown that tiny amounts of magnetic impurities may suppress superconductivity. Extensive investigations of a coexistence of long-range local-moment antiferromagnetic order in the superconducting state of the Chevrel phases were carried out in the 1970s and 80s [1]. Motivated by this work materials were discovered, which even undergo local-moment ferromagnetic order at low temperatures so that superconductivity is suppressed again below the magnetic transition temperature. More recently, the ruthenium-cuprates and the Borocarbides have been identified as new examples of superconductors that undergo magnetic order in the superconducting state. All of these systems have in common that magnetic order and superconductivity arise as microscopically well distinguishable phenomena. This contrasts superconductivity on the border of itinerant antiferromagnetism recently reported for lanthanide and actinide based heavy fermion metals [2]. It has been argued that the latter class of materials includes examples of magnetically mediated superconductivity. A true coexistence of superconductivity and itinerant magnetism were, in contrast, long believed to be impossible as they were regarded competing forms of electronic order.

The focus of the work described here are itinerant ferromagnets that become superconducting at low temperatures, i.e., deep in the ferromagnetic state. As a first example superconductivity was discovered in the sibbling pair of uniaxial ferromagnets UGe$_2$ and URhGe [3]. However, the f-orbitals in U-based compounds overlap little, rendering the magnetic state a difficult problem. This is contrasted by the first d-electron system exhibiting superconductivity in the ferromagnetic state, ZrZn$_2$ [5], which will be reviewed in the first part of this paper. The coexistence of ferromagnetism with superconductivity requires yet consideration of a related important issue, namely itinerant ferromagnetism itself. The many body problem beneath band ferromagnetism has been an extraordinarily demanding task that has attracted scientific interest since the turn of the 20th century. A question that is nevertheless still unresolved concerns the nature of the metallic state above the Curie temperature of itinerant-electron ferromagnets such as iron, cobalt and nickel. To avoid ambiguities related to the high transition temperatures of the elemental ferromagnets, weakly ferromagnetic metals such as ZrZn$_2$, MnSi, YNi$_3$ and Ni$_3$Al are studied in which the ferromagnetic ordering temperature is tuned to zero using high pressure [6,7]. This way a large range of T may be studied above T_c free of the usual ambiguities. In the second part of this paper the emphasis will be on the particularly striking combination of properties of MnSi that question the validity of a description by the standard model of the metallic state, Fermi liquid theory. The paper concludes with a discussion of the possible nature of quantum spin fluctuations as the groundstate evolves with pressure.

1 The Superconducting Ferromagnet ZrZn$_2$

The compound ZrZn$_2$ was first investigated by Matthias and Bozorth in the 1950s [8], who discovered that it is ferromagnetic despite being composed of nonmagnetic, superconducting constituents. ZrZn$_2$ crystallizes in the C15 cubic Laves structure and the magnetic properties derive from the Zr $4d$ orbitals, which have a significant direct overlap [9]. Ferromagnetism develops below the Curie temperature $T_c = 28.5$ K (Fig. 1) with a small ordered moment $\mu_s = 0.17\mu_B$ per formula unit. The normal metallic state above T_c is characterized by a Curie-Weiss temperature dependence of the susceptibility with an effective moment $\mu_{cw} \approx 2\,\mu_B$/f.u.. The low T_c renders ZrZn$_2$ special among stoichiometric ferromagnetic metals and indicates that the compound is close to a ferromagnetic QCP quantum critical point (QCP), defined as the point where ferromagnetism disappears at zero-temperature [10]. This and the prediction [11] that superconductivity is controlled by the QCP led us to revisit this compound. Indeed, at ambient pressure and temperatures below $T_s \approx 0.3$ K we have recently succeeded in observing superconductivity, where T_c still exceeds T_s by two orders of magnitude [5].

The superconductivity in ZrZn$_2$ has a number of remarkable features. First, it only appears to occur in high-purity single-crystal samples. Unconventional or non s-wave forms of superconductivity generally require the su-

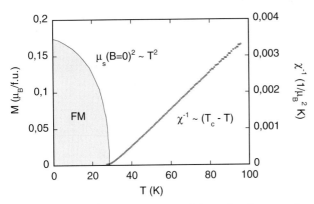

Fig. 1. Temperature dependence of the ordered magnetic moment, μ_s and the paramagnetic susceptibility χ in ZrZn$_2$. The fluctuating Curie-Weiss moment, $\mu_{cw} \approx 2\,\mu_B$ is an order of magnitude larger than the ordered moment $\mu_s \approx 0.17\mu_B$/f.u., where the ferromagnetic transition temperature is $T_c \approx 28.5$ K

perconducting coherence length $\xi_s = 290$ Å, derived from the upper critical field $B_{c2} \approx 0.4$ T, to be somewhat smaller than the electronic mean free path, which is here of the order 1000 Å. In view of the sensitivity to sample quality, the superconductivity is likely to be unconventional in nature [12]. The second interesting feature is the lack of an anomaly in the specific heat at the superconducting transition. If we interpret this literally, it means that the superconducting state is strongly gapless with large portions [13] or even all of the Fermi surface surviving in the superconducting state. The 'zero-field' superconducting transition in ZrZn$_2$ differs fundamentally from that in a conventional superconductor since it occurs in the presence of ferromagnetism. In fact, the transition in ZrZn$_2$ is similar to the transition in a conventional superconductor for applied fields close to B_{c2}, where the superconducting anomaly is suppressed [14]. The third remarkable feature of the superconductivity in ZrZn$_2$ is that it is observed within the ferromagnetic phase, which poses the question [15,16,17] of the microscopic relationship between ferromagnetism and superconductivity. A macroscopic, i.e., uniform co-existence of the two states throughout the sample is consistent with the magnetic response we observe, i.e., flux expulsion upon cooling the sample through T_s in a small magnetic field is almost negligible. In fact, an incomplete Meissner effect, i.e., imperfect flux expulsion is thought to be a signature [18] of the phase coexistence of superconductivity and ferromagnetism.

We can exclude scenarios in which the superconductivity is due to inclusions of a second phase or a surface impurity based on thorough metallurgical tests [19]. As a further important test we have investigated the interplay of superconductivity and ferromagnetism under hydrostatic pressure (Fig. 2). Surprisingly, hydrostatic pressure suppresses both the ferromagnetism and the superconductivity above a critical pressure $p_c = 21$ kbar. Thus, it is not

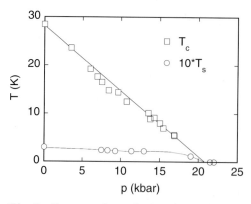

Fig. 2. Pressure dependence of the ferromagnetic ordering temperature T_c and superconducting ordering temperature T_s. The pressure dependence of T_c was determined in d.c. magnetization measurements. The pressure dependence of T_s was determined from the resistivity. Superconductivity disappears for $p > p_c \approx 21$ kbar, in the paramagnetic phase down to the lowest T measured, $T = 15$ mK. Note that, for clarity T_s is magnified by a factor of ten

sufficient to be close to the ferromagnetic quantum critical point for superconductivity to occur in ZrZn$_2$, it must also be in the ferromagnetic state! This may arise naturally in scenarios where the Cooper pairs are in a parallel-spin (triplet) state, which is already favored in the ferromagnetic state. Such behavior could well be universal for itinerant ferromagnets in the limit of small Curie temperature and long electron mean free path.

In order to obtain some insight into the superconductive pairing mechanism it is helpful to emphasize the effect of a magnetic field on the ordered moment. At $T = 1.7$ K a relatively small field $B = 0.05$ T is required to form a single ferromagnetic domain. On further increasing the field, the ordered moment is rapidly increased, with a field of 12 T nearly doubling the moment. M remains unsaturated up to 35 T, the highest field applied [20]. This behavior contrasts strongly with the elemental ferromagnets Fe, Ni and Co where, after a single domain is formed, fields applied parallel to the easy axis only have a small effect on the ordered moment. We note that NMR studies and polarized neutron diffraction are consistent with a purely ferromagnetic state of ZrZn$_2$ [17,22]. Neutron scattering studies moreover show that transverse excitations in the ordered state are well-developed ferromagnetic spin waves [23]. The small ordered moment and the magnetic field induced increase of the magnetization, however, are evidence of an exceptionally large *longitudinal* susceptibility. This makes ZrZn$_2$ an excellent candidate for magnetically mediated pairing.

2 Non-Fermi Liquid Normal State of MnSi

MnSi is perhaps the most extensively studied of the itinerant-electron ferromagnets, apart from the elemental metals Fe, Ni and Co. Though relatively simple, the cubic B20 crystal structure of MnSi is somewhat unusual in that it lacks space-inversion symmetry. This leads to a well-understood long wavelength helical twist in the ferromagnetic polarization [24] that may be instrumental in suppressing superconductivity otherwise expected to arise on the border of itinerant-electron ferromagnetism [3,5].

The ordering temperature of MnSi is $T_c = 29.5$ K at ambient pressure and vanishes above a critical pressure, $p_c = 14.6$ kbar. The transition is second order up to $p^* \approx 12$ kbar and weakly first order in the range $p^* < p < p_c$ [25]. The key result reviewed here is that the resistivity exhibits a temperature dependence of the form $\rho(T) = \rho_0 + AT^{3/2}$ for $p > p_c$, where ρ_0 is the T independent residual resistivity and A a constant. An example of this striking behavior is shown in Fig. 3 for a pressure just above p_c for a pure sample with $\rho_0 \approx 0.33\,\mu\Omega$cm. The conventional T^2 form expected in Fermi liquid theory is in contrast observed only at low T in the ferromagnetic state, where a remarkably abrupt change occurs at p_c (Fig. 4). The $T^{3/2}$ coefficient, A, is

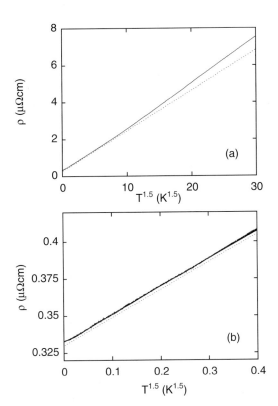

Fig. 3. Temperature dependence of the electrical resistivity of MnSi at a hydrostatic pressure of 14.8 kbar, i.e., above the critical pressure $p_c = 14.6$ kbar, over different T regimes plotted versus $T^{3/2}$. A $T^{3/2}$ dependence is observed between the very low mK regime up to nearly 10 K

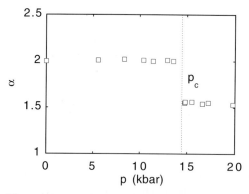

Fig. 4. Pressure dependence of the exponent α derived from $\rho(T) = \rho_0 + AT^\alpha$ at low temperatures. An abrupt change is observed at p_c between a Fermi liquid coefficient $\alpha = 2$ and the experimentally observed normal state dependence $\alpha = 3/2$

only weakly dependent on ρ_0 in the range $0.35 - 4.0\mu\Omega$cm, but is sensitive to p and falls by approximately a factor of 2 from p_c to $2p_c$. The $T^{3/2}$ variations of the resistivity is also observed in two other itinerant-electron ferromagnets, $ZrZn_2$ and Ni_3Al. This suggests that it may have a general origin and is not a consequence of some peculiarity of MnSi alone.

In the search for a possible explanation of the $T^{3/2}$ power law the predictions of the so-called nearly-ferromagnetic-Fermi-liquid (NFFL) model may be considered [26,27,28]. The behavior of the NFFL model depends on the temperature scales $T^* = T_F/(k_F\xi_m S)$, where T_F is the Fermi temperature, k_F is the Fermi wavevector, ξ_m is the magnetic correlation length and S is the Stoner enhancement factor. For $T \ll T^*$, the NFFL model reduces to the Fermi-liquid model which is characterized by a T^2 resistivity. Above T^*, on the other hand, the NFFL model reduces to the so-called Marginal-Fermi-liquid model which is characterized by a $T^{5/3}$ resistivity when T is far below T_F for a 3-dimensional system and for a dominance of small-angle electron scattering typical of electronic transport. When the latter assumption breaks down, $\rho(T)$ is expected to be linear in T. This is also the form predicted by a different kind of Marginal-Fermi-liquid theory introduced to account for the normal-state behavior of the high-temperature superconductors [29].

A comparison of the prediction of the NFFL model with experiment is shown in Fig. 5. At p_c and zero magnetic field, T^* for MnSi is a few K consistent with the first order nature of T_c just below p_c. The resistivity exponent is thus expected to be described by the Marginal-Fermi-liquid model well above 5 K but by Fermi-liquid theory in the milli-Kelvin range. As shown in Fig. 5 the NFFL model does indeed agree with experiment above a few K but is seen to break down dramatically at lower temperatures where the unexpected $T^{3/2}$ power law prevails.

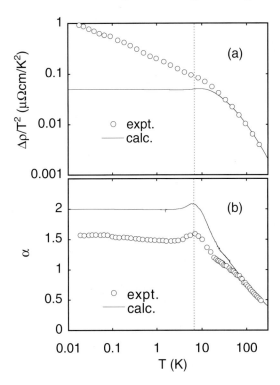

Fig. 5. Comparison of the experimental temperature dependence of the electrical resistivity at $p_c < 14.8$ kbar with the predictions of the model of a nearly ferromagnetic Fermi liquid (NFFL) [26,27,28]. (a) $\Delta\rho/T^2$ vs T. The temperature dependent part divided by T^2 highlights deviations from T^2 behaviour. (b) The exponent α of a power law dependence. The predictions are in agreement with experiment at high T, but below a few K a quadratic T dependence is expected where the $T^{3/2}$ dependence is experimentally observed

It is interesting to note that the $T^{3/2}$ power law qualitatively corresponds to the effects of frozen-in disorder, e.g. observed in amorphous metals and polycrystalline spin glasses [30], which are expected to lead to a diffusive motion of charge carriers on length scales substantially larger than the mean free path l [31,32]. But the single-crystal samples investigated here exhibit quantum oscillations at ambient pressure corresponding to l of the order of 5000 Å in agreement with ρ_0. Thus, only well below a new temperature scale $T_F/(k_F l)$, it is possible that a $T^{3/2}$ term in the resistivity may arise. This scale, however, is at least an order of magnitude too small to explain the range of the $T^{3/2}$ resistivity shown in Figs. 3 and 5. Apart from this, the impurity model would also seem to be at odds with the observed weak dependence of A on ρ_0 and the fact that the first order transition does not exhibit any broadening as function of pressure [25].

The $T^{3/2}$ resistivity in MnSi and related pure systems presently under investigation is not consistent with current models of itinerant-electron ferromagnetism. The origin of this form of $\rho(T)$ may lie in extremely subtle effects of disorder in the highest purity samples, pointing at an entirely unexpected and new aspect of the metallic state. More likely, however, it may lie in a diffusive motion of the electrons resulting from the interactions among the itinerant electrons themselves. This constitutes a drastic breakdown of the Fermi-liquid model long assumed to be valid for the normal state of itinerant-electron ferromagnets.

3 Possible Role of Quantum Criticality

In the following the role of quantum spin fluctuations to the superconducting instability of itinerant ferromagnets and the non-Fermi liquid normal state are discussed further. In $ZrZn_2$ the high sensitivity of superconductivity to sample purity and the absence of a specific heat anomaly are consistent with ferromagnetically mediated pairing. Central to such a pairing potential is the longitudinal susceptibility as opposed to the more familiar transverse susceptibility [11]. The latter describes changes of the orientation of the magnetic moments, which in the ordered state assume the form of spin waves. Transverse fluctuations are generally considered to be pairbreaking but spin waves in the ordered state are only weakly coupled to the single particle excitations forming the Cooper pairs. The longitudinal susceptibility describes in contrast changes of the amplitude of the magnetization, representing a key feature of itinerant magnetism that is not related to a broken symmetry of the ordered state. Longitudinal excitations remain strongly coupled to the spectrum of single particle excitations even in the ordered state and may hence supply an important attractive component in the formation of the Cooper pairs.

Experimentally, tuning the longitudinal susceptibility to become singular thus has long been hoped to generate conditions favorable to ferromagnetically mediated superconductivity [11,33]. This may be readily achieved in materials undergoing a second order phase transition at $T = 0$ between paramagnetism and itinerant ferromagnetism in high pressure experiments. Theoretically, these kinds of quantum critical transitions in itinerant electron systems have been studied in great detail [26,27,28]. The key results may be illustrated with the help of a Ginzburg-Landau free energy.

In Fig. 6 we display schematically the free energy, F, as function of the square of the magnetization, M^2, to illustrate the possible character of the low lying fluctuations near a first and second order phase transition at $T = 0$. Schematic representations of this kind are more familiar for thermally driven phase transitions. A central assumption underlying this picture for $T = 0$ is that single particle fermion excitations may be formally integrated out and that the nature of the collective excitations of the groundstate, described by the magnetization density, do not change when pressure tuning the groundstate [26,27,28]. As the system undergoes a second order phase transition at $T = 0$, the quantum nature of the problem allows a Gaussian description of the fluctuations, and anharmonicities may be accounted for by a renormalization of the linear response only. This may not be the case for weak first order transitions, where fluctuations between order parameter values as indicated by the arrow in Fig. 6(b) remain clearly non-Gaussian.

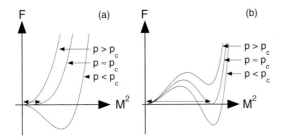

Fig. 6. Form of the free energy, F, as function of magnetization squared, M^2, at various pressures. (a) Fourth order model of a quantum critical point. (b) Sixth order model of a first order transition. In both cases the coefficient of the quadratic term is assumed to vary linearly with p in agreement with present day convention

Shown in Fig. 7 is the evolution of the ordered moment μ_s, ferromagnetic transition temperature T_c, superconducting transition temperature T_s and longitudinal susceptibility χ_L in the vicinity of a QCP as shown in Fig. 6(a). As p_c is approached, μ_s and T_c vanish continuously and the longitudinal susceptibility diverges. T_s, on the other hand, displays maxima to the left and right hand side of the QCP due to pair-breaking effects in the immediate vicinitiy of the QCP [11,34,35,36].

It is interesting to consider the consistency of the experimental properties of the superconducting ferromagnets known to date with a magnetic equation of state and related superconductive pairing described in Figs. 6 and 7. For instance, a surprising property of $ZrZn_2$ is the very weak p dependence of T_s below p_c (Fig. 2) and the absence of superconductivity above p_c. This may be explained, if the longitudinal susceptibility is not pressure dependent, even though the ordered moment vanishes continuously at p_c. This empiri-

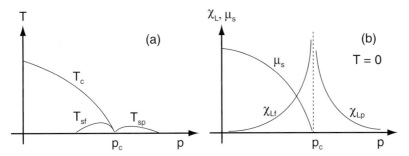

Fig. 7. (a) Qualitative phase diagram of the ferromagnetic transition temperature T_c and superconducting transition temperature T_s theoretically predicted near a ferromagnetic quantum critical point. (b) Pressure dependence of the ordered moment μ_s and longitudinal susceptibility χ_L at $T = 0$ corresponding to the ferromagnetic QCP shown in (a)

cally suggests that $T = 0$ phase transitions are in general not required for superconductivity to coexist with itinerant ferromagnetism. Moreover, a lack of p dependence of the longitudinal susceptibility questions the validity of the current theory [26,27,28], namely that single particle excitations may be integrated out and that the nature of the collective excitations is unchanged as function of p.

Further, even for cubic ferromagnets $T = 0$ phase transitions may be more complex than hitherto anticipated theoretically. The unusual $T^{3/2}$ resistivity observed in MnSi and related systems suggests an unconventional spectral distribution of the spin fluctuations, e.g., that of a diffusive motion of the charge carriers as in strongly disordered materials. From a more general point of view these unusual spin fluctuation spectra must be connected with strongly non-Gaussian behavior, as related to additional structure of the free energy illustrated for the weak first order transition in Fig. 6(b). It is therefore tempting to consider parallels common to all itinerant ferromagnets thus far not considered theoretically. As previously suggested [6] the most prominent microscopic aspect are a dominance of interband transitions, e.g., due to nesting in the complicated band structure. These have long been deemed important in $ZrZn_2$, where the calculated LDA band structure is consistent with de Haas van Alphen spectra [37,38].

Besides the peculiar value of the power law exponent $\alpha = 3/2$, the stability of the $T^{3/2}$ behaviour as function of pressure and magnetic field in MnSi is startling. It is interesting to consider a possible role of small q nesting here. The helical modulation of the ferromagnetic order in MnSi, conventionally associated with the Moriya-Dzyaloshinsky interaction, may in fact be due to nesting representing a short Q spin density wave. The nesting may thus result in long range coherence for the case of MnSi, highlighting the possible role of short q nesting in a particularly pronounced manner. Recently inelastic neutron scattering studies have been reported in MnSi that show chirality of the critical fluctuations at ambient pressure [39]. Thus nesting may modify the spin fluctuations in MnSi in the quantum critical regime, providing a mechanism of coherence thus far not considered for the $T^{3/2}$ resistivity. This is not to be mistaken with large-Q antiferromagnetism for which no experimental evidence exists. Experimental studies of the pressure dependence of the helical state may hence provide a general proof of the role of nesting to itinerant ferromagnetism, as suggested by $T^{3/2}$ resistivities displayed by other weak ferromagnets, namely Ni_3Al and $ZrZn_2$.

The theoretical framework of QCPs described above may be somewhat patched up to account for many-band effects like those alluded to here as follows. The shape of the free energy shown in Fig. 6 derives from the effects of mode-mode coupling, which to date are assumed to be constant in q and ω. Studies of spin density wave antiferromagnetism in Mn_3Si [40] and CuMnSb [41] at high magnetic fields clearly show the need to take into account the detailed form of the q and ω dependence of the mode-mode coupling, thus highlighting an important omission in current theory when nesting is important.

In conclusion the experimental studies described here reveal properties contrasting the conventional wisdom associated with itinerant ferromagnetism. Itinerant ferromagnets may universally become superconducting in the ferromagnetic state if the longitudinal susceptibility is large. The concept of quantum criticality and the nature of the normal metallic state may thereby be ill-described, asking for the inclusion of, for instance, many band effects. These may in turn lead to a phenomenological description of the metallic state that radically differs from Fermi liquid theory over a very large T range.

Acknowledgments

The work presented here is a result of collaborations with A. Faißt, J. Flouquet, A. D. Huxley, S. R. Julian, S. M. Hayden, G. G. Lonzarich, H. v. Löhneysen, H. Stalzer, M. Uhlarz and R. Vollmer as cited throughout the text. I am especially indepted to H. v. Löhneysen for his encouragement and support over the years. Financial assistance by the Deutsche Forschungsgemeinschaft and European Science Foundation under FERLIN are acknowledged.

References

1. *Superconductivity in Ternary Compounds* ed. O. Fischer and M. B. Maple, Springer-Verlag, Berlin (1982).
2. N. Mathur et al., Nature **394**, 39 (1998).
3. S. S. Saxena et al. Nature **406**, 587 (2000); A. D. Huxley et al., Phys. Rev. B **63**, 144519 (2001).
4. D. Aoki et al., Nature **413**, 613 (2001).
5. C. Pfleiderer et al., Nature **412**, 58 (2001); corrigendum ibid 660.
6. C. Pfleiderer, J. Mag. Mag. Mat. 226-230, 26 (2001).
7. C. Pfleiderer, S. R. Julian and G. G. Lonzarich, Nature **414**, 427 (2001).
8. B. T. Matthias and R. M. Bozorth, Phys. Rev. **109**, 604 (1958).
9. T. Jarlborg, A. J. Freeman and D. D. Koelling, J. Mag. Mag. Mat. **23**, 291 (1981).
10. T. F. Smith, J. A. Mydosh and E. P. Wohlfarth, Phys. Rev. Lett. **27**, 1732 (1971).
11. D. Fay and J. Appel, Phys. Rev. B **22**, 3178 (1980).
12. A. P. Mackenzie et al., Phys. Rev Lett. **80**, 161 (1998).
13. N. I. Karchev, K. B. Blagoev, K. S. Bedell and P. B. Littlewood, Phys. Rev Lett. **86**, 846 (2001).
14. D. Sanchez, A. Junod, J. Muller, H. Berger and F. Levy, Physica B **204**, 167 (1995).
15. P. Fulde and R. A. Ferrell, Phys. Rev. **135**, A550 (1964).
16. A. I. Larkin and Y. N. Ovchinnikov, Sov. Phys. JETP **20**, 762 (1975).

17. E. I. Blount and C. M. Varma, Phys. Rev. Lett. **42**, (1979) 1079.
18. E. B. Sonin and I. Felner, Phys. Rev. B **57**, R14000 (1998).
19. S. Z. Huang, M. K. Wu, R. L. Meng, C. W. Chu and J. L. Smith, Sol. St. Comm. **38**, 1151 (1981).
20. A. P. van Deursen et al., J. Mag. Mag. Mat. **54**, 1113 (1986).
21. P. J. Brown, K. R. A. Ziebeck and P. G. Mattocks, J. Mag. Mag. Mat. **42**, 12 (1984).
22. M. Kontani, T. Hioki and Y. Masuda, Physica B **86-88**, 399 (1977).
23. N. R. Bernhoeft, S. A. Law, G. G. Lonzarich and D. McK Paul, Physica Scripta **38**, 191 (1988).
24. Y. Ishikawa, Y. and M. Arai, J. Phys. Soc. Jpn. **53** 2726 (1984).
25. C. Pfleiderer, G. J. McMullan, S. R. Julian and G. G. Lonzarich, Phys. Rev. B **55**, 8330 (1997); C. Thessieu, C. Pfleiderer, A. N. Stepanov and J. Flouquet, J. Phys.: Condens. Matter **9**, 6677 (1997).
26. T. Moriya, *Spin Fluctuations in Itinerant Electron Systems*, Springer, Berlin (1985); G. G. Lonzarich and L. Taillefer, J. Phys. C **18**, 4339 (1985).
27. A. J. Millis, Phys. Rev. B **48**, 7183 (1993); J. Hertz, Phys. Rev. B **14**, 1165 (1976).
28. G. G. Lonzarich in *The Electron*, ed. M. Springford, Cambridge University Press, Cambridge, 1996.
29. C. M. Varma et al., Phys. Rev. Lett. **63**, 1996 (1989).
30. P. Ford and J. A. Mydosh, Phys. Rev. B **14**, 2057 (1976).
31. N. Rivier and A. E. Mensah, Physica B **91**, 85 (1977).
32. K. H. Fischer, Z. Physik B **34**, 45 (1979).
33. F. Pobell, Physica B **109 & 110**, 1485 (1982).
34. R. Roussev and A. J. Millis, Phys. Rev. B **63**, (2001) 140504(R).
35. T. R. Kirkpatrick, D. Belitz, T. Vojta and R. Narayanan, Phys. Rev. Lett. **87**, (2001) 127003.
36. K. B. Blagoev, J. R. Engelbrecht and K. S. Bedell, Phys. Rev. Lett. **82**, 133 (1999).
37. G. Santi, S. B. Dugdale and T. Jarlborg, cond-mat/0107304 (2001).
38. S. Yates and S. M. Hayden, unpublished (2002).
39. B. Roessli et al., cond-mat/0201327 (2002).
40. C. Pfleiderer, J. Bœuf and H. v. Löhneysen, Phys. Rev B in print (2002).
41. J. Bœuf, C. Pfleiderer and H. v. Löhneysen, unpublished (2002).

Infrared Conductivity and Superconducting Energy Gap in MgB$_2$

Andrei Pimenov

Experimentalphysik V, Elektronische Korrelationen und Magnetismus,
Institut für Physik, Universität Augsburg, 86135 Augsburg, Germany

Abstract. Submillimeter-wave and far-infrared conductivities of MgB$_2$ have been investigated using different experimental techniques. The data provide clear experimental evidence for the onset of a superconducting gap at 24 cm^{-1} at $T = 5$ K. On increasing temperature the gap energy increases, contrary to what is expected in isotropic BCS superconductors. The small zero-temperature gap value and its unconventional increase on increasing temperature can be explained by a highly anisotropic or multiple gap function. A peak in the temperature-dependent conductivity is observed at low frequencies, which qualitatively resembles the coherence peak predicted by BCS-theory.

1 Introduction

The investigation of the electrodynamic properties of superconductors has since long been used to understand the mechanism of superconductivity [1]. Direct information about specific parameters like energy gap, quasiparticle scattering rate or spectral weight of the superconducting condensate can be extracted from these experiments. As a classical example, the observation of a gap in the conductivity spectrum of lead, tin and indium by Ginsberg and Tinkham [2] has provided the key information to explain the phenomenon of superconductivity.

The BCS theory of superconductivity [3,4] was able to explain a number of experimental results on conventional superconductors. One of the most striking results of the BCS-theory was the prediction of the coherence peak in the temperature dependence of the conductivity and of the nuclear spin relaxation. The existence of the coherence peak has indeed been observed experimentally in NMR experiments by Hebel and Schlichter [5] and, later on, in the real part of the complex conductivity by the group of G. Grüner [6,7].

The recent discovery of superconductivity at a relatively high temperature ($T_c \approx 39$ K) in a simple binary compound, magnesium boride [8], has stimulated extensive theoretical and experimental studies in this material. Most of these studies have been devoted to find the pairing mechanism responsible for the superconductivity [9]. One of the first indications for phonon-mediated superconductivity in MgB$_2$ comes from the boron-isotope effect [10]. Further hints of the conventional character of superconductivity in MgB$_2$ result from tunneling experiments [11,12], photoemission [13,14], infrared [15,16,17] and Raman [18,19] spectroscopies. Many of these results pointed towards the existence of the BCS-like gap in MgB_2.

The values of the superconducting energy gap in MgB_2 vary substantially from the BCS-estimate $2\Delta \simeq 12$ meV ($T_c = 39$ K) especially when different experimental setups are compared. Techniques like tunneling spectroscopy [11,12], photoemission [13,14] or specific heat [20] reveal significantly different values of the energy gap ranging between $2\Delta = 3 - 16$ meV. These discrepancies together with the observed deviations from the BCS temperature dependence have been partly explained by imperfections of the sample surface, but alternative explanations based e.g. on multiple gaps [14,21] or anisotropy [22] have also been proposed. For a recent review see Ref. [23] and references therein. Different values, observed by various experiments may be partly explained by the preferential sensitivity of particular experiments to different portions of the gap distribution.

2 Infrared Properties of MgB_2

The first results on the optical reflectivity of a bulk MgB_2 have been presented by Gorshunov et al. [24]. The method of grazing incidence [25] has been used in these experiments in order to increase the sensitivity to small changes in reflectance. The normalized reflectivity (R_S/R_N) revealed a gradual increase for decreasing frequency in the range $25 < \nu < 70$ cm^{-1}, which was difficult to interpret within the conventional picture for a superconductor. At the lowest frequencies of these experiment the reflectivity showed a weak maximum around 25-30 cm^{-1}. Taking into account possible anisotropic or multiple character of the gap in MgB_2, the following explanation of the grazing reflectance experiments can be given: probably the gradual change in the reflectivity between 25 cm^{-1} and 70 cm^{-1} corresponds to the gap distribution and the reflectivity maximum around 25-30 cm^{-1} to the lowest gap value. This would produce an estimate of the minimum superconducting gap $2\Delta \sim 3 - 4$ meV.

A different approach to investigate electrodynamics in MgB_2 has been utilized by Jung et al. [26]. In this experiment the transmission of MgB_2 thin films has been measured for frequencies $30 < \nu < 250$ cm^{-1}. According to the BCS-theory of superconductivity, the thin-film transmittance spectra in the superconducting state should reveal a local peak slightly above the gap frequency. In the transmission experiment the properties of the film are averaged along the sample thickness. Therefore, this kind of experiment is less sensitive to the imperfections of the surface compared to the reflection setup. Indeed, the transmittance data of Jung et al. [26] resembled qualitatively the BCS calculations. Comparing their data to the theoretical predictions, Jung et al. obtained the gap value $2\Delta_0 = 42$ cm^{-1} (5.2 meV) [26]. Another parameter, that could be obtained fitting the experimental transmittance to the BCS-theory was the quasiparticle scattering rate, $1/2\pi\tau = 800^{+200}_{-100}\,\mathrm{cm}^{-1}$. This value of the scattering rate puts the sample of Jung et al. into the dirty limit.

Another well established technique to investigate the electrodynamic properties of thin films is the quasioptical spectroscopy in the submillimeter frequency range ($3 < \nu < 40\,\mathrm{cm}^{-1}$) [27,28]. In this method [29] backward-wave oscillators (BWO's) are employed as monochromatic and continuously tunable sources covering the frequency range from $4\,\mathrm{cm}^{-1}$ to $30\,\mathrm{cm}^{-1}$. The Mach-Zehnder interferometer arrangement allows measuring both the intensity and the phase shift of the wave transmitted through the conducting film on a transparent substrate. Using the Fresnel optical formulas for the complex transmission coefficient of the substrate-film system, the complex conductivity of the film can be determined directly from the measured spectra.

The MgB_2 film discussed in this work has been grown by two-beam laser ablation on a plane-parallel sapphire substrate $10 \times 10\,\mathrm{mm}^2$ in size. The crystallographic orientation of the substrate was $[1\bar{1}02]$ with a thickness of about 0.4 mm. Magnetic susceptibility measurements indicated a sharp superconducting transition at 32 K with a width of 1 K. The film has been characterized by four circle X-Ray diffraction. The c-axis of the film is tilted from the substrate normal by an angle of 77°, i.e the film-orientation is close to the a-axis orientation. However, within the experimental accuracy of a few percent the conductivity measurements did not show any substantial anisotropy neither in the normal nor in the superconducting state. The conductivity experiments have been repeated on another film, prepared by the same technique, and similar results have been observed.

Figure 1 represents the example of the transmittance and the phase shift of a MgB_2 film in the normally-conducting and the superconducting state. Both transmittance and phase shift are strongly influenced by the interferences within the substrate. However, this effect does not influence the complex conductivity data, because it is automatically included in the Fresnel equations for the substrate-film system.

Figure 2 shows the complex conductivity $\sigma^* = \sigma_1 + i\sigma_2$ of MgB_2 at various temperatures as obtained from the transmittance and the phase shift spectra. The normal-state conductivity (35 K curve) demonstrates a typical metallic behavior. σ_1 is essentially frequency independent, while σ_2 is almost zero and exhibits a small linear increase for increasing frequencies. This behavior indicates the Drude conductivity with a scattering rate well above the frequency window of the submillimeter experiment. Simultaneous fitting of the σ_1 and σ_2 spectra with a Drude model gives an estimate for the scattering rate at 35 K: $1/2\pi\tau = 150^{+70}_{-50}\,\mathrm{cm}^{-1}$. This value is in a rough agreement with $1/2\pi\tau = 75\,\mathrm{cm}^{-1}$ obtained recently from the analysis of the infrared conductivity of a MgB_2 film [30].

The transition into the superconducting state gives rise to significant changes in the complex conductivity spectra in Fig. 2. σ_2 starts to diverge for

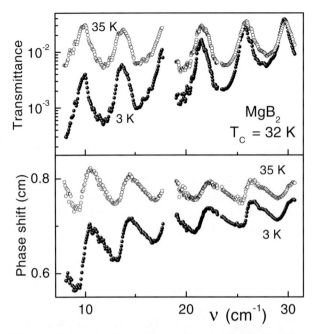

Fig. 1. Frequency dependence of the transmittance (upper panel) and the phase shift (lower panel) of MgB_2 film above and below T_c. The interference patterns are due to the substrate

$\nu \to 0$ (upper frame). This divergence increases with decreasing temperature, reflecting a growth of the spectral weight of the superconducting condensate. The frequency dependence of σ_2 can well be described by a $1/\nu$-dependence, that corresponds to the δ-function in σ_1 at $\nu = 0$ via the Kramers-Kronig relations. In the superconducting state a pronounced frequency dispersion shows up in the σ_1 spectra as well (Fig. 2, lower frame). At low frequencies ($\nu < 9\,\mathrm{cm}^{-1}$) and starting from the normal state, σ_1 initially increases (by approximately a factor of 1.5 at the lowest frequency) and then decreases, while at higher frequencies, $\nu > 9\,\mathrm{cm}^{-1}$, σ_1 monotonically decreases upon cooling.

From the temperature evolution of the conductivity spectra it becomes clear that a maximum in the temperature dependence of σ_1 should exist below T_c and at low frequencies ($\nu < 9\,\mathrm{cm}^{-1}$). The temperature dependencies of σ_1 and σ_2 for several frequencies are shown in Fig. 3. Thick lines in Fig. 3 represent the weak-coupling BCS calculations for the frequencies indicated. The peak in the temperature dependence of σ_1 vanishes at frequencies above $9\,\mathrm{cm}^{-1}$ both in the experiment and in the theory, and the temperature dependence of σ_1 at higher frequencies basically follows the behavior shown in Fig. 3 for $\nu = 16\,\mathrm{cm}^{-1}$. The observed peak in $\sigma_1(T)$ is reminiscent of the coherence peak, predicted for a BCS superconductor and therefore reflects the conventional character of superconductivity in MgB_2.

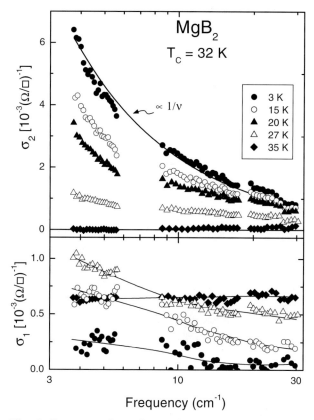

Fig. 2. Frequency dependence of the complex conductivity of MgB_2 film above and below T_c [15]. Upper panel: imaginary part σ_2. Solid line represents the $\sigma_2 \propto 1/\nu$ dependence. Lower panel: real part σ_1. Lines are drawn to guide the eye

The temperature dependence of the imaginary part of the complex conductivity is shown in the lower panel of Fig. 3 in comparison to the BCS-theory. In these calculations [31] the quasiparticle scattering rate $1/2\pi\tau = 150$ cm^{-1} has been taken from the normal-state conductivity. The *only free parameter* that could be changed in this case is the value of the gap energy. A gap value $\Delta_0 = 3.7$ meV was used for the fits, which are shown as solid lines in Fig. 3 and most closely match the low-temperature values of σ_2. In the intermediate temperature range the experimental data deviate significantly from the BCS curves. Qualitatively the same deviations from the BCS calculations can be seen for all frequencies: the slope of the experimentally observed $\sigma_2(T)$ curves is more gradual at temperatures just above T_c, and it is steeper than the BCS prediction for $T \to 0$ K. This disagreement could possibly be improved including the distribution of the superconducting gaps into conductivity calculations. Assuming the anisotropy of the energy gap Haas

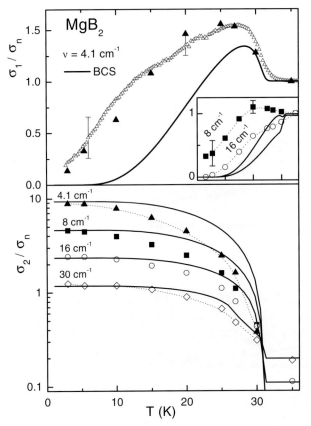

Fig. 3. Temperature dependence of the complex conductivity of a thin film of MgB_2 [15]. Upper panel: σ_1 at $\nu = 4.1$ cm^{-1}. The data for $\nu = 8$ cm^{-1} and 16 cm^{-1} are given in the inset. Solid line represents the BCS model calculation with $\Delta_0 = 3.7$ meV and $1/2\pi\tau = 150$ cm^{-1}. Lower panel: σ_2 at different frequencies. Symbols: experiment, thick lines: BCS calculations with the same parameters as in the upper panel. Thin dotted lines are drawn to guide the eye

and Maki [22] performed the conductivity calculations in the zero-frequency limit. In order to explain the results presented in this work, the extension of these calculations to finite frequencies is highly desired.

Closer analysis of the frequency dependence of σ_1 in Fig. 2 reveals that the conductivity at low temperatures (e.g. $T = 3$ K data) shows a weak minimum around 20 cm^{-1}. This minimum can be interpreted as an indication of the superconducting energy gap in MgB_2. In order to prove this idea, the frequency range of the submillimeter experiment has been increased substantially [16] using the technique of Fourier-transform infrared spectroscopy.

The infrared spectroscopy of thin superconducting films reveals distinct advantages compared to the spectroscopy on bulk samples [32]. The reflectance of bulk superconducting samples rapidly approaches unity both in the normal and in the superconducting state especially at low frequencies. This implies a number of experimental difficulties investigating the changes in reflectance at the superconducting transition. In contrast, the reflectance of thin films can be substantially reduced (compared to unity), which strongly improves the accuracy of the data. In addition, the technique of quasioptical transmission spectroscopy [29] can be applied to the same films at lower frequencies. The transmission technique provides an independent measurement of the complex conductivity at submillimeter frequencies and the reflectance of the sample can be calculated from these data, which expands the low-frequency limit of the available spectrum and considerably improves the quality of the Kramers-Kronig analysis. The combination of these two techniques has been applied previously to thin YBaCuO films [32] and showed the reliability of the method and the advantage of the described procedure compared to conventional far-infrared analyses.

Figure 4 shows the reflectance of the MgB_2 film for different temperatures [16]. The solid lines represent the directly measured reflectance and the symbols correspond to the data as calculated from the submillimeter conductivity (Fig. 2). The most striking feature of the data is the change in slope of the reflectance around 24 cm^{-1}. The reflectance below this frequency is close to one at $T = 5$ K and drops rapidly above 24 cm^{-1}. It is important to note that

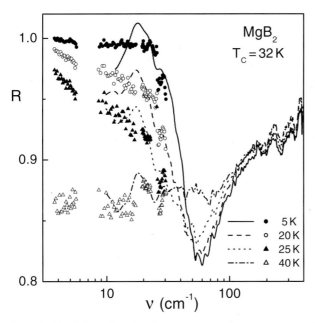

Fig. 4. Far-infrared and submillimeter-wave reflectance of MgB_2 film on an Al_2O_3 substrate. Lines: directly measured reflectance, symbols: data as calculated from the complex conductivity in the submillimeter range (Fig. 2)

this change in slope is seen in the calculated submillimeter data (symbols) as well. Therefore, we interpret this feature as a direct observation of the spectral gap in MgB$_2$. The abrupt change in slope around 24 cm^{-1} clearly indicates the onset of the electromagnetic absorption above this frequency. It therefore resembles the classical experiments on reflectance in conventional superconductors [33], in which the onset absorption across the gap has been observed.

The onset of the absorption in MgB$_2$ corresponds to a superconducting gap $2\Delta \simeq 3$ meV, which is a factor of four smaller than the BCS-estimate and is in the lower limit of the gap values, observed in other experiments ($2\Delta \simeq 3-16$ meV). Remarkably, the energy gaps, estimated from the infrared [17,24,26] and microwave [34] experiments, all reveal quite small values lying in the range $3-5$ meV. We recall further that according to first-principles calculations [35], the Fermi-surface of MgB$_2$ comprises four sheets, each having a distinct energy gap [21]. Indeed, recent tunneling [12] and Raman [18] spectroscopy experiments revealed two different energy gaps in MgB$_2$. It is reasonable to suggest that the experiments which measure the complex conductivity (e.g. microwave or infrared spectroscopy) are sensitive to the lowest gap value. The spectroscopic features due to other gaps are smoothed because they overlap with the absorption across the smallest energy gap. This effect is a possible reason for a relatively low frequency of the "knee" in reflectance observed in Fig. 4.

The complex conductivity of the MgB$_2$ film in the broad frequency range has been calculated from the reflectance data of Fig. 4 [16]. The results are presented in Fig. 5. The real part of the conductivity (upper panel) is frequency independent in the normal conducting state ($T = 40$ K) and below 100 cm^{-1}. The corresponding imaginary part σ_2 (lower panel) is roughly zero for these frequencies. This picture agrees well with the low-frequency limit of the Drude-conductivity of a simple metal. On cooling the sample into the superconducting state, σ_1 becomes suppressed between 5 and 100 cm^{-1} and shows a clear minimum, which corresponds to the superconducting gap. The position of the minimum at $T = 5$ K corresponds well to the change in slope in reflectance ($\nu \approx 24$ cm^{-1}, Fig. 4). The minimum in σ_1 shifts to *higher* frequencies as the temperature increases. This apparently contradicts the expectations for the temperature dependence of the superconducting energy gap. However, as already mentioned discussing the reflectance data of Fig. 4, MgB$_2$ reveals either an anisotropic gap [22] or multiple gaps [21]. At the lowest temperatures the smallest gap dominates and the spectral features at higher frequencies due to other gaps are masked. On increasing temperature the transition across the smallest gap gradually becomes thermally-saturated and the effective "weight" for different gaps is shifted to the higher-energy region. Therefore the conductivity minimum also shifts to higher frequencies, which explains the behavior observed in Fig. 5.

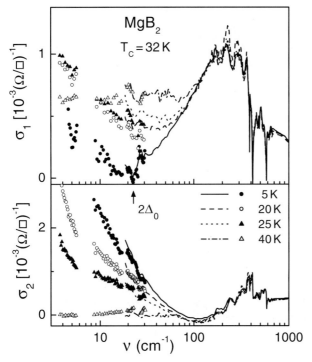

Fig. 5. Complex conductivity of a MgB_2 film at infrared frequencies. Upper panel: real part σ_1. Lower panel: imaginary part σ_2. Lines have been obtained via the Kramers-Kronig analysis of the reflectance data, Fig. 4. Symbols represent the conductivity, which was directly measured by the transmittance technique Fig. 2. Sharp peak-like structures between 400 and 700 cm^{-1} is due to residual influence of phonon contribution of the substrate

Recently, the terahertz time-domain spectroscopy has been applied to the transmission of thin MgB_2 films [17]. Due to phase-sensitivity of the time-domain spectroscopy, the complex conductivity of MgB_2 could be extracted from the spectra without application of the Kramers-Kronig analysis. Similar to the results presented in Fig. 5, the frequency dependence of σ_1 as obtained by the time-domain spectroscopy showed a characteristic minimum in the frequency dependence, which could be attributed to the superconducting gap, $2\Delta \simeq 5$ meV. However, the temperature dependence of the conductivity minimum cannot be described by the BCS theory. Instead, this minimum remained roughly at the same frequency at all temperatures. We believe that similar to the unusual increase of the conductivity minimum in Fig. 5 this behavior points towards the distribution of the energy gaps in MgB_2.

The suppression of σ_1 at infrared frequencies (upper part of Fig. 5) reflects the reduction of the low-frequency spectral weight. At the same time, σ_2 rapidly increases revealing approximately a $1/\omega$ behavior. From the slope of

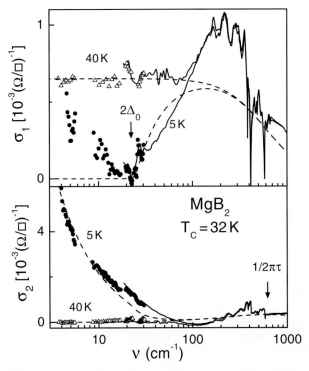

Fig. 6. Comparison between the complex conductivity and the predictions of the weak-coupling BCS-model. Symbols and solid lines represent the experimental data of Fig. 5. Dashed lines: BCS-theory with $2\Delta = 24\,\text{cm}^{-1}$, $1/2\pi\tau = 600\,\text{cm}^{-1}$

$1/\omega$ the spectral weight of the superconducting condensate can be estimated and turns out to be roughly 20% higher than the reduction of the spectral weight in σ_1. This possibly indicates that a substantial extrinsic contribution is present in σ_1 in the superconducting state.

Besides the gap feature, an additional excitation is observed in the conductivity spectra for frequencies above $100\,\text{cm}^{-1}$ (Fig. 5). This feature is seen as a broad maximum around $230\,\text{cm}^{-1}$, which is only weakly temperature dependent. For frequencies higher than $400\,\text{cm}^{-1}$ the conductivity decreases again, which possibly reflects the high-frequency behavior of the Drude-conductivity with a characteristic scattering rate $1/2\pi\tau \approx 400\,\text{cm}^{-1}$. At the same frequency a broad maximum becomes apparent in σ_2 (lower part of Fig. 5), a behavior which is expected within the Drude picture. The scattering rate $1/2\pi\tau \approx 400\,\text{cm}^{-1}$ roughly corresponds to $1/2\pi\tau \approx 150\,\text{cm}^{-1}$, obtained from the low-frequency transmittance of this film (Fig. 2). The discrepancies most probably reflect the experimental uncertainties. The origin of the broad maximum in σ_1 on top of the Drude-conductivity is unclear at present. We cannot attribute it to residual effects due to substrate, because the phonon excitations in Al_2O_3 start around $400\,\text{cm}^{-1}$ and produce sharp anomalies in the spectra for $400 < \nu < 700\,\text{cm}^{-1}$. A possible alternative explanation for the broad maximum could be bound states of localized charge carriers, similar to infrared maxima, observed in Zn-doped $YBa_2Cu_3O_8$ [36].

Finally, Fig. 6 compares the experimental conductivity with the predictions of the weak-coupling BCS-model [31]. We note that the energy gap $2\Delta = 24\,\text{cm}^{-1}$ disagrees with the BCS-value $2\Delta_{BCS} \simeq 80\,\text{cm}^{-1}$ ($T_c = 32\,\text{K}$). In this case the calculations based on the multi-gap conductivity might be more appropriate.

The imaginary part of the conductivity at $T = 5\,\text{K}$ is well reproduced by the BCS-model without additional fitting parameters. For low frequencies and in the superconducting state the real part σ_1 is far above the model calculations, a fact that probably implies an additional absorption in the film. Above the gap frequency, the theoretical curve shows a slightly higher slope compared to the experiment. This again may indicate a distribution of the superconducting gaps in MgB_2.

3 Conclusions

The broadband reflectance of MgB_2 reveals a well-defined change in slope around $\nu \simeq 24\,\text{cm}^{-1}$ and at $T = 5\,\text{K}$, which can be interpreted as a direct evidence of the superconducting gap, $2\Delta \simeq 3\,\text{meV}$. In the real part of the complex conductivity a minimum develops in the superconducting state for $\nu \simeq 24\,\text{cm}^{-1}$, which again corresponds to the superconducting gap. On increasing temperature the gap energy increases, contrary to what is expected in an isotropic BCS superconductor. The small zero-temperature gap value and its unconventional increase on increasing temperature can be explained by a highly anisotropic or multiple gap function.

The results from a variety of different experimental techniques point towards an anisotropic gap or multiple gaps in MgB_2 with characteristic values in the range $2\Delta \simeq 3 - 16\,\text{meV}$. It is a remarkable result, that the analysis of the infrared data leads to comparatively low values of the energy gap: $2\Delta \simeq 3 - 7\,\text{meV}$. However, most experimental techniques, described above, are sensitive to the onset of the electromagnetic absorption and therefore revealed the minimum value of the possible gap distribution. This is most probably the reason why the results from infrared spectroscopies lead to gap values considerably lower than those obtained by other experimental methods.

The quasiparticle scattering rate in MgB_2 is sample-dependent and may be estimated as $1/2\pi\tau \sim 100 - 1000\,\text{cm}^{-1}$ from the complex conductivity data as obtained in different experiments.

The temperature dependence of the real part of the low-frequency conductivity shows a peak below the superconducting transition, which resembles the coherence peak of the BCS-theory. A detailed comparison of the frequency and temperature behavior of the complex conductivity with BCS-model predictions shows significant discrepancies between experiment and theory. The observed deviations may be accounted for including a strongly anisotropic energy gap into the BCS-calculation.

Acknowledgements

I would like to thank A. Loidl, A. V. Pronin, S. I. Krasnosvobodtsev, D. Rainer and A. P. Kampf for valuable discussions. This work has been supported by the BMBF via the contract number 13N6917/A - EKM.

References

1. D. M. Ginsberg, L. C. Hebel, in *Superconductivity*, R. D. Parks (Ed.), p. 193 (Marcel Dekker, New York, 1969).
2. D. M. Ginsberg, M. Tinkham, Phys. Rev. **118**, 990 (1960).
3. J. Bardeen, L. N. Cooper, J. R. Schrieffer, Phys. Rev. **108**, 1175 (1957); *ibid.* **108**, 1175 (1957).
4. D. C. Mattis, J. Bardeen, Phys. Rev. **111**, 412 (1958).
5. L. C. Hebel, C. P. Slichter, Phys. Rev. **113**, 1504 (1959).
6. K. Holczer, O. Klein, G. Grüner, Sol. State Comm. **78**, 875 (1991).
7. O. Klein, E. J. Nicol, K. Holczer, G. Grüner, Phys. Rev. B **78**, 6307 (1994).
8. J. Nagamatsu, N. Nakagawa, T. Muranaka, Y. Zenitani, J. Akimitsu, Nature **410**, 63 (2001).
9. J. E. Hirsch, F. Marsiglio, Phys. Rev. B **64**, 144523 (2001); J. M. An, W. E. Pickett, Phys. Rev. Lett. **86**, 4366 (2001); Y. Kong, O. V. Dolgov, O. Jepsen, O. K. Andersen, Phys. Rev. B **64**, 020501 (2001).
10. S. L. Bud'ko, G. Lapertot, C. Petrovic, C. E. Cunningham, N. Anderson, P. C. Canfield, Phys. Rev. Lett. **86**, 1877 (2001).
11. G. Karapetrov, M. Iavarone, W. K. Kwok, G. W. Crabtree, D. G. Hinks, Phys. Rev. Lett. **86**, 4374 (2001); G. Rubio-Bollinger, H. Suderow, S. Vieira: Phys. Rev. Lett. **86**, 5582 (2001); A. Sharoni, I. Felner, O. Millo: Phys. Rev. B **63**, 220508 (2001); H. Schmidt, J. F. Zasadzinski, K. E. Gray, D. G. Hinks, Phys. Rev. B **63**, 220504 (2001).
12. F. Giubileo, D. Roditchev, W. Sacks, R. Lamy, D. X. Thanh, J. Klein, S. Miraglia, D. Fruchart, J. Marcus, Ph. Monod, Phys. Rev. Lett. **87**, 177008 (2001); P. Szabó, P. Samuely, J. Kačmarčík, T. Klein, J. Marcus, D. Fruchart, S. Miraglia, C. Marcenat, A. G. M. Jansen, Phys. Rev. Lett. **87**, 137005 (2001); F. Laube, G. Goll, J. Hagel, H. v. Löhneysen, D. Ernst, T. Wolf: Europhys. Lett. **56**, 296 (2001).
13. T. Takahashi, T. Sato, S. Souma, T. Muranaka, J. Akimitsu, Phys. Rev. Lett. **86**, 4915 (2001).
14. S. Tsuda, T. Yokoya, T. Kiss, Y. Takano, K. Togano, H. Kitou, H. Ihara, S. Shin, Phys. Rev. Lett. **87**, 177006 (2001).
15. A. V. Pronin, A. Pimenov, A. Loidl, S. I. Krasnosvobodtsev, Phys. Rev. Lett. **87**, 097003 (2001).

16. A. Pimenov, A. Loidl, S. I. Krasnosvobodtsev, cond-mat/0109449 (Phys. Rev. B, in press).
17. R. A. Kaindl, M. A. Carnahan, J. Orenstein, D. S. Chemla, H. M. Christen, H.-Y. Zhai, M. Paranthaman, D. H. Lowndes, Phys. Rev. Lett. **88**, 027003 (2002).
18. X. K. Chen, M. J. Konstantinović, J. C. Irwin, D. D. Lawrie, J. P. Franck, Phys. Rev. Lett. **87**, 157002 (2001).
19. J. W. Quilty, S. Lee, A. Yamamoto, S. Tajima, Phys. Rev. Lett. **88**, 087001 (2002).
20. F. Bouquet, R. A. Fisher, N. E. Phillips, D. G. Hinks, J. D. Jorgensen, Phys. Rev. Lett. **87**, 047001 (2001); Ch. Wälti, E. Felder, C. Degen, G. Wigger, R. Monnier, B. Delley, H. R. Ott, Phys. Rev. B **64**, 172515 (2001); Y. Wang, T. Plackowski, A. Junod, Physica C **355**, 179 (2001).
21. E. Bascones, F. Guinea, Phys. Rev. B **64**, 214508 (2001); A. Y. Liu, I. I. Mazin, J. Kortus, Phys. Rev. Lett. **87**, 087005 (2001).
22. S. Haas, K. Maki, Phys. Rev. B **65**, 020502 (2002).
23. C. Buzea, T. Yamashita, Supercond. Sci. Technol. **14**, R115 (2001).
24. B. Gorshunov, C. A. Kuntscher, P. Haas, M. Dressel, F. P. Mena, A. B. Kuz'menko, D. van der Marel, T. Muranaka, J. Akimitsu, Eur. Phys. J. B **21**, 159 (2001).
25. J. Schützmann, H. S. Somal, A. A. Tsvetkov, D. van der Marel, G. E. Koops, N. Kolesnikov, Z. F. Ren, J. H. Wang, E. Brück, A. A. Menovsky, Phys. Rev. B **55**, 11118 (1997).
26. J. H. Jung, K. W. Kim, H. J. Lee, M. W. Kim, T. W. Noh, W. N. Kang, H.-J. Kim, E.-M. Choi, C. U. Jung, S.-I. Lee, Phys. Rev. B **65**, 052413 (2002).
27. A. V. Pronin, M. Dressel, A. Pimenov, A. Loidl, I. V. Roshchin, L. H. Greene, Phys. Rev. B **57**, 14416 (1998).
28. A. Pimenov, A. Loidl, G. Jakob, H. Adrian, Phys. Rev. B **61**, 7039 (2000); A. Pimenov, A. V. Pronin, A. Loidl, U. Michelucci, A. P. Kampf, S. I. Krasnosvobodtsev, V. S. Nozdrin, D. Rainer, Phys. Rev. B **62**, 9822 (2000).
29. G. V. Kozlov, A. A. Volkov, in *Millimeter and Submillimeter Wave Spectroscopy of Solids*, G. Grüner (Ed.), p. 51 (Springer, Berlin, 1998).
30. J. J. Tu, G. L. Carr, V. Perebeinos, C. C. Homes, M. Strongin, P. B. Allen, W. N. Kang, E.-M. Choi, H.-J. Kim, S.-I. Lee, Phys. Rev. Lett. **87**, 277001 (2001).
31. W. Zimmermann, E.-H. Brandt, M. Bauer, E. Seider, L. Genzel, Physica C **183**, 99 (1991).
32. A. Pimenov, Ch. Hartinger, F. Mayr, A. Loidl, G. Jakob, H. Adrian, Ferroelectrics, **249**, 165 (2001).
33. P. L. Richards, M. Tinkham, Phys. Rev. **119**, 575 (1960); S. L. Norman, D. H. Douglass, Jr., Phys. Rev. Lett. **18**, 1967 (1967).
34. N. Klein, B. B. Jin, J. Schubert, M. Schuster, H. R. Yi, A. Pimenov, A. Loidl, S. I. Krasnosvobodtsev, cond-mat/0107259; F. Manzano, A. Carrington, N. E. Hussey, S. Lee, A. Yamamoto, S. Tajima, Phys. Rev. Lett. **88**, 047002 (2002).
35. J. Kortus, I. I. Mazin, K. D. Belashchenko, V. P. Antropov, L. L. Boyer, Phys. Rev. Lett. **86**, 4656 (2001).
36. D. N. Basov, B. Dabrowski, T. Timusk, Phys. Rev. Lett. **81**, 2132 (1998).

MgB_2 Wires and Tapes: Properties and Potential

W. Goldacker, S.I. Schlachter, S. Zimmer, and H. Reiner

Forschungszentrum Karlsruhe, Institut für Technische Physik,
P.O. Box 3640, 76021 Karlsruhe, Germany

Abstract. High current carrying MgB_2 superconducting wires were developed by different routes and analysed in magnetic background fields and at varied temperatures. MgB_2 has anisotropic mechanical and superconducting properties which are very sensitive to strains. This has consequences for the transport critical current densities and irreversibility fields of wire and tape composites, especially with mechanically reinforced sheath. Varying the degree of the sheath reinforcement in the wires, pre-strain induced degradations of the transport critical currents were investigated. Compensating the pre-strain effects with externally applied axial strains, only a limited recovery of the transport currents was observed. A possible explanation for this result is given and consequences for future improved wires are discussed.

1 Introduction

The discovery of superconductivity in MgB_2 [1] with a T_c of 39 K gave much hope for a new cheap superconducting material for a couple of applications, as magnets, fault current limiters and transformers. In a quite short time already all kind of samples, bulk, thin films, wires and tapes, were investigated [2]. Several authors reported on powder-in-tube (PIT) wires and tapes with very high transport critical current densities up to about $2 \cdot 10^5$ Acm^{-2} in self field at 4.2 K [3,4,5,6]. It was recognized soon that a heat treatment of the final conductor with phase decomposition and reformation is beneficial to the filament densification, the support of the poor sintering of MgB_2 and the recovery of strain induced broadened T_c transitions. However, cold worked tapes with high critical current densities were also reported [4]. In comparison to wires, tapes achieved significant higher J_c values, because during rolling not only the filament densification is improved due to a higher deformation pressure but also some phase texture might be favoured [6]. Investigations on thin films with textured MgB_2 [7] found anisotropic superconducting properties of MgB_2 with upper critical fields H_{c2} up to 39 Tesla ($H||a,b$) as expected for the layered crystal structure. Wires and tapes were mainly prepared as monofilamentary composites. Above a certain transport current level of about $2 \cdot 10^5$ Acm^{-2}, which is reached in low fields ($B < 3$ T) or self field at 4.2 K, transport currents are limited due to a thermal instability of the conductors obviously caused by the large filament size, phase impurities, inhomogeneity and non perfect densification. In this range transport current can become up to one order of magnitude smaller than those calculated from magnetic measurements which outline a potential of $J_c = 10^6\,\text{Acm}^{-2}$ (4.2 K, self field) and more. When the wire and tape preparation involves a heat treatment, typically around the decomposition temperature of MgB_2 (875 – 900 °C), the choice of the sheath material in contact with MgB_2 is restricted for chemical reasons, mainly due to the reactivity of volatile Mg. Ta and Nb were in principle suitable but form surface reaction layers, while Fe sheaths react significantly less with decomposed MgB_2 and became the first choice. For mechanical stable wires a stabilising precompression of the filament is required, which was achieved by adding a steel layer to the sheath [6,8,9]. The disadvantage of such reinforced sheaths are strain effects in the filaments which cause degradations of the superconducting properties as it is quite similar known for Nb_3Sn wires.

The goal of this article is to give more insight in the role of stresses and strains in MgB_2 and the consequences for the properties of this superconductor in wire and tape composites. The investigations were focussed on wires, the most suitable geometry for applications in coils and minimized losses with AC currents. These aspects are of most importance for the requirements of possible technical applications, which are high transport current densities, thermal stabilisation and mechanical strength. Depending on the application, superconducting wires have to withstand stresses of 250 MPa and more, caused by coil winding processes, thermal stresses and Lorentz forces in coils with controlled influence on the superconducting properties.

1.1 Crystal Structure, Thermal Expansion and Residual Strain of MgB_2

MgB_2 has a simple hexagonal C32 crystal structure (space group P6/mmm), the most common structure among the diborides, which consists of alternating layers of Mg and B. From the layered nature of the structure one already expects an anisotropy of the physical properties as found in the HTS compounds YBCO and BSCCO.

A strong anisotropy of the thermal expansion and the elastic properties of the a- and c-axis of the hexagonal crystal lattice was measured between 11 and 297 K [10]. The thermal expansion in c-direction, normal to the alternating Boron and metal layers, is about twice as large as in a-direction, which represents the behaviour in the layer plane. This anisotropy is a not unusual property of this class of compounds, but extraordinary strong in MgB_2. In dense non-textured material processed at high temperatures, this anisotropy of the thermal expansion is expected to lead to residual strain load in the grains after cooling. This aspect has not been regarded so far but can be of crucial importance for wire composites, where additional strains are caused by the sheath and may affect the superconducting properties severely. The application of pressure P to MgB$_2$ samples leads to anisotropic strains of the lattice axes due to the different compressibilities of the lattice axes [10] and to a degradation of the critical temperature T_c. For the $T_c(P)$ data a large T_c dependent scattering of the results between -2 and -0.2 K/GPa was reported comparing different authors applying different experimental methods [2,11]. This variance is obviously correlated to the level of T_c and the hydrostatic nature of stress application. For perfect hydrostatic conditions using He gas and MgB$_2$ powder [10,11], the largest anisotropy of the pressure induced lattice strains was observed. Studies of substitutional effects in Mg$_{1-x}$Al$_x$B$_2$ [12] showed a sensitive correlation between increasing Al content, decreasing T_c and decreasing length of the c-axis. So far all substitutional additives as Mn, Co, Zn, Si, C, Al and others led to a more or less pronounced decrease of T_c, indicating that MgB$_2$ is obviously close to a structural instability with a destruction of superconductivity. No T_c increase was observed so far. The highest T_c obviously corresponds to a maximum length of the c-axis. Changes of the c-axis length were analysed to strongly influence the in-plane phonons of the B planes (E$_{2g}$ modes) which are strongly coupled to the B sigma bands contributing to the Fermi-level density of states. A first indication for residual strain effects on the superconductivity was observed comparing MgB$_2$ powder and a fully dense hot isostatic pressed (HIPped) bulk sample [13]. A T_c drop of approximately 0.7 K was observed which can be explained by residual strains in the lattice [14].

From all these results it is well established that structural changes and lattice strains sensitively influence the superconducting properties and must have severe consequences for the situation in wire and tape composites.

1.2 Composition of Wires and Tapes

Wires and tapes consist of two different components, the superconducting core and the sheath and are in most cases processed by applying a final heat treatment to form or condition the superconducting phase. One consequence is an occurring residual stress and strain state at the operation temperature (pre-stress, pre-strain) due to a mismatch of the thermal expansion of the conductor components during cooling. For a mechanically stable technical wire a sufficient compressive pre-strain with minimized influence on the superconducting current is wanted. It was shown earlier for Nb$_3$Sn and Chevrel phase superconductors that this pre-strain is of 3-dimensional nature and has a hydrostatic and deviatoric component [15,16,17]. The measured and calculated stress values were typically in the range of 0.1 - 1 GPa. For MgB$_2$ this can strongly influence the superconducting properties regarding the $T_c(P)$ results being of the order of -0.2 K/GPa to -2 K/GPa. Sheath components used in our investigations were Nb, Ta, in combination with Cu and stainless steel (SS), or Fe with and without steel. In Fig. 1 the thermal expansion of the composite materials over temperature normalized to $T=10$ K is shown, the data for the Chevrel phase PbMo$_6$S$_8$ are added for comparison. For MgB$_2$

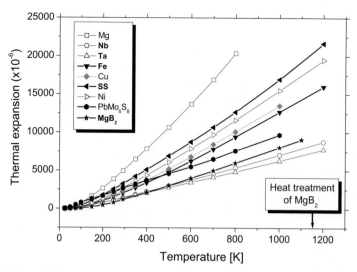

Fig. 1. Thermal expansion of MgB$_2$ (extrapolated from data of ref. [10]), different sheath materials of MgB$_2$ wires and Chevrel phase. All data are normalized to 10 K

the data above room temperature were extrapolated from the crystallographic data below RT [10] using the Einstein equation for a single phonon energy and the cubic root of the unit cell volume data. All sheath materials having a larger thermal expansion than MgB$_2$ can create pre-compression of the filament after cooling. This is not the case for Nb and Ta. Here only a combination with steel can shift the thermal expansion of the composite sheath to higher values than MgB$_2$. The amount of pre-compression of the filament is given through the filling factor, sheath thickness and amount of reinforcement (steel content), the temperature below which all materials become elastic and finally by the elastic properties of the materials, Young's modulus and Poisson ratio. For MgB$_2$ the whole situation becomes more complicated as a result of the expected additional contribution from residual strains from the anisotropic thermal contraction of the lattice axes as already mentioned above.

2 Experimental: Wire Preparation and Characterisation Methods

Two different precursor routes were applied in the preparation of our wires. The two routes are named ex-situ and in-situ method and correspond to the formation of MgB$_2$ before filling and during a heat treatment in the final wire, respectively. In the first concept, commercial MgB$_2$ powders (Alfa Aesar) with about 2 % MgO (from X ray spectrum) secondary phase - obviously varying with production batch - were used (ex-situ samples). In the second

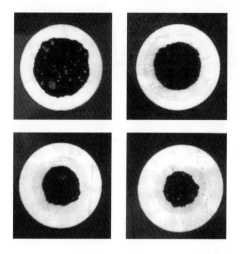

Fig. 2. Cross sections of MgB_2 wires with Fe sheath (top left) and Fe/steel composite sheaths with additional different steel amounts of 37, 54 and 66 %

route, powder mixtures of Mg (< 300 mesh, 99 % purity) and B (< 200 mesh, 99 % purity) were used (in-situ samples) as a promising way to reduce the oxygen ingot which leads to the formation of the main secondary phase MgO and to improve the filament densification supported by melted Mg above 650 °C. The precursor powders were mixed for several hours and not especially ground before filling the sheath tubes. The filling density reached 50 to 60 %.

Ta and Nb tubes (10 mm diam., 1 mm wall thickness) were used in combination with a Cu tube of 12 mm outer diameter or alternatively tubes from low carbon steel (99.5 % Fe). Deformation was made by swaging and drawing to 1.6 mm wire diameter. Then the wire was introduced in a stainless steel tube of 2.5 mm diam., 0.25, 0.35 or 0.45 mm wall thickness, and deformed to final diameters between 0.7 and 1.58 mm. The necessary amount of stainless steel was estimated roughly from earlier experience with Chevrel phase wires in ref. [15] and varied between 36 and 66 % to study pre-strain effects. All wires had quite regular cross sections as shown in Fig. 2. The final heat treatment was performed at 875 to 900 °C in reducing atmosphere of $Ar/H_2(5\%)$, which is slightly above the onset of phase decomposition, followed by a cooling ramp with 20 °C/h to 650 °C with final furnace cooling to RT.

Transport critical currents were measured by means of a conventional 4-probe method in perpendicular background fields up to 12 T. In a special strain rig for high fields [18], axial stresses up to about 700 MPa could be applied. Strains were measured directly over a 14 mm section of the wire via an attached precision strain gage clip. Transport currents were measured for each strain value loading the sample with fast current ramps of typically 0.2 s scan time. Standard criterion for I_c was 1 μV/cm.

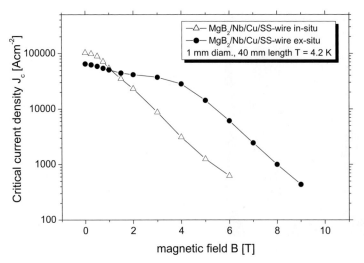

Fig. 3. Critical current densities $J_c(B)$ (4.2 K) of MgB_2 in-situ and ex-situ wires

3 Transport Critical Currents

For the two kinds of conductors fabricated with in-situ and ex-situ technique, a quite different behaviour of the transport currents with field was observed as depicted in Fig. 3. Ex-situ wires from pre-reacted powders reached $65000\,Acm^{-2}$ at 4.2 K, 0 T and possess still relative high transport currents at high fields, corresponding to an irreversibility field of approximately 10 T, estimated from Kramer plots. SEM investigations of the microstructure in the filaments indicate a quite homogeneous grain structure in the major part of the cross section. In-situ wires reached higher current densities of $10^6\,Acm^{-2}$ at 4.2 K in self field, but with a much larger degradation of J_c in increasing fields and H* was extrapolated to only 7 T. The reason for this has mainly to be attributed to the different microstructure as a consequence of the phase formation mechanism. The in-situ reaction of the phase begins with melting of Mg above 650 °C. At temperatures above 850 °C Boron grains react from the surface to the inner part forming MgB_2. Therefore, the grain size of the Boron particles significantly determine the microstructure resulting in much larger and more irregular MgB_2 grains and more secondary phase compared to ex-situ wires. Since flux pinning obviously occurs at grain boundaries, the microstructure has a crucial influence on the superconducting properties and is one reason for the limited currents at high fields and the reduced irreversibility fields.

At fields below approximately 3 T the $J_c(B)$ curves flatten in contrast to $J_c(B)$ curves determined in magnetic measurements which show a mostly linear characteristics of $\ln J_c(B)$ [19]. The reason for this is the poor thermal stabilisation of the wires at high current levels. A non-dense filament material

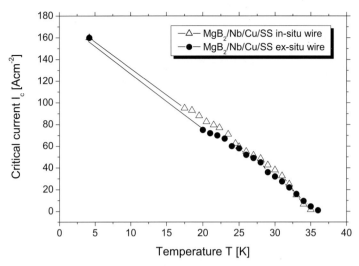

Fig. 4. Temperature dependence of the critical current densities J_c of MgB$_2$ in-situ and ex-situ wires in self field

and the presence of inclusions of secondary phases as MgO are the most probable reasons for occurring "hot spots" when the current locally surpasses I_c. In this field regime many samples heat up very fast in the normal conducting state and even burn through. The $U(I)$ characteristics of the I_c transitions represents this sudden transition. An improvement of phase purity and homogeneity of the microstructure and a multifilamentary arrangement of the superconductor is therefore required to realize thermally stable wires.

The temperature dependence of the critical currents is linear with temperature for both kind of samples as shown in Fig. 4. For the ex-situ wire 52 kAcm^{-2} (20 K), 37 kAcm^{-2} (25 K) and 23 kAcm^{-2} (30 K) were reached as best values.

4 Effect of Mechanical Reinforcement

Application of different steel amounts in the sheath as mechanical reinforcement was performed to study the effects of pre-stress in the filaments and the consequences for the transport currents. Systematic $J_c(B)$ investigations were performed on ex-situ wires with Fe and Fe/SS sheaths (see Fig. 5). A raised steel content led to strongly reduced current densities, the reduction depending on the magnetic field. This effect is very well known from reinforced Nb$_3$Sn and Chevrel wires and can be attributed to pre-stress induced degradations of the critical currents and the irreversibility fields. In MgB$_2$/Fe/SS wires (Fig. 5) the addition of steel causes strong current degradations since already a Fe sheath should lead to some pre-compression of the filament. For the highest steel contents the effect saturates. In Fig. 6 Kramer plots were

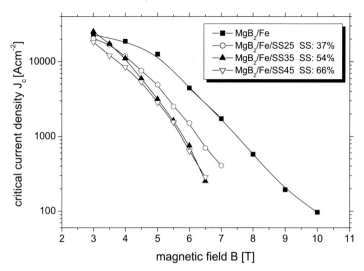

Fig. 5. Critical current densities $J_c(B)$ for MgB$_2$ ex-situ wires with Fe sheath and Fe/steel sheaths with 37, 54, and 66 % steel content

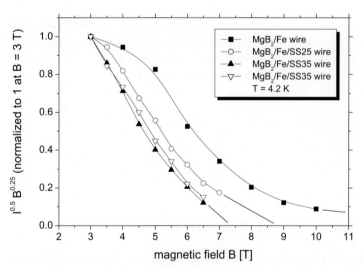

Fig. 6. Kramer plots for MgB$_2$ ex-situ wires with Fe sheath and Fe/steel sheaths with 37, 54, and 66 % steel content

used to perform an extrapolation to the irreversibility field H*. H* changes from about 13 T down to about 7 T with increasing filament pre-compression.

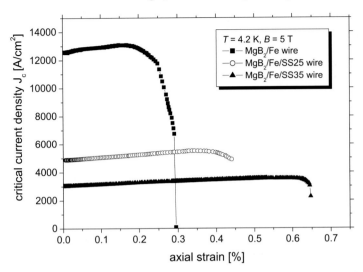

Fig. 7. Critical current density J_c vs. applied tensile axial strain for MgB$_2$ ex-situ wires with Fe sheath (top graph) and Fe/steel sheaths with 37 % (middle graph) and 54 % (lower graph) steel content

5 Critical Currents with Applied Axial Tensile Strain

Axial tensile stress application to the wires with simultaneously measured change of the critical current is a very suitable method to investigate the strain state in superconductors. In these experiments the compressive axial pre-strain of the filaments is compensated applying axial tensile strains and a recovery of the strain induced I_c degradation is expected. For the MgB$_2$/Fe/SS wires the measurements were performed in a background field of 5 T. The strain induced degradation of the critical current density due to the steel reinforcement at 5 T can be obtained from the $J_c(B)$ graphs in Fig. 5. For 37 % steel in the sheath, $J_c(B=5\,\mathrm{T})$ is reduced to 0.4 times the value of the non-reinforced wire, for 54 % steel content $J_c(B=5\,\mathrm{T})$ is even reduced to 0.25 times the value of wire without steel.

For all three investigated wires an increase of the critical current densities with applied strain was observed as shown in Fig. 7. In the regime of increasing critical current densities the changes were reversible. J_c degradations observed above some critical strain limits, 0.2 % for Fe sheath, 0.4 % for Fe/SS25 sheath and 0.6 % for Fe/SS35 sheath, however, were irreversible and have to be attributed to a damage of the superconductor and the formation of cracks in the filament. For the sample with Fe/SS35 sheath an unload of the axial stress in the irreversible regime lead to an increase of the current densities (not shown in the graph). This can be interpreted as closure of cracks in the filaments.

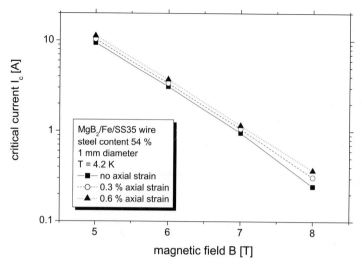

Fig. 8. Field dependence of the critical currents for 0, 0.3 and 0.6 % axial applied strain for a MgB$_2$ ex-situ wire with Fe/steel sheath (54 % steel content)

The increase of the transport current densities in the reversible regime was larger for the steel reinforced wires (see Fig. 7), but only a small part of the J_c difference to the Fe-sheathed wire was recovered. This may be explained by the presence of a still high stress component in the strained samples. 3-dimensional calculations of the stress and strain state during an I_c vs. applied strain experiment for the similar case of Chevrel phase wires showed [17], that an increase of the radial compressive strain components correlates with a compensation of the axial compressive pre-strain component.

For the strain experiment of wire MgB$_2$/Fe/SS35, the $J_c(B)$ dependence was measured at applied strains of 0 %, 0.3 % and 0.6 % to study the influence of the strain effects on the irreversibility fields. The results are depicted in Fig. 8. Only a very small shift to higher current values is observed corresponding to only a slight increase of the irreversibility field with released pre-strain. This results also confirm that inspite of the compensation of the axial pre-strain component, a still strong degradation of the superconducting properties remains. One further aspect was addressed in a former section. The anisotropy of the thermal expansion of the crystal lattice obviously leads to additional residual strain with effects on T_c and propably I_c and therefore a quite complex situation in MgB$_2$.

6 Conclusions and Outlook

The necessary increase of the current carrying capability and the irreversibility fields of MgB$_2$ in wires and tapes and therefore their enhanced chance for an application depends strongly on an improvement of the performance of the wire composites. The superconducting properties of MgB_2 in wires are strongly influenced by strain effects. Conductor reinforcements favour the density of the filament material but lead to strongly field dependent strain induced current degradations at fields $>3\,T$ which can only partly be recovered by compensation with external strains. Future conductor composites need a carefully balanced reinforcement to achieve the necessary strengthening for the specific applications on one hand and to limit the current degradation on the other hand. Transport current densities at low fields are still smaller than those obtained from magnetic measurements. Therefore, optimized precursors and wire preparation techniques are necessary to achieve a microstructure with regular small grains, a completely dense filament and homogeneous element distributions which is a supposition to prepare smaller filament sizes in multifilamentary wires for a sufficient thermal stabilisation of the conductors. A small grained microstructure in the filaments, which is important for the flux pinning, requires new or improved precursor techniques, possibly artificial pinning centres formed by a secondary phase or inclusions. The anisotropic mechanical and superconducting properties ask for quite new conductor geometries and preparation routes to achieve a textured superconductor, which favours tape geometries or even new concepts of coated conductors. From the present stage of the research, applications for MgB_2 wires and tapes are expected in persistent mode MRI coils and magnets for operation at about $20\,K$ and low background fields of $<5\,T$.

References

1. J. Akimitsu, Symposium on Transition Metal Oxides, 10 Jan. 2001, Sendai, Japan.
2. C. Buzea and T. Yamashita, Supercond. Sci. Technol. **14**, R115-R146 (2001).
3. B.A. Glowacki, M. Majoros, M. Vickers, J.E. Evetts, Y. Shi, and I. McDougall, Supercond. Sci. Technol. **14**, 193-199 (2001).
4. G. Grasso, A. Malagoli, C. Ferdeghini, S. Roncallo, V. Braccini, M.R. Cimberle and A.S. Siri, Appl.Phys.Lett. **79**, 230-232 (2001).
5. H.L. Suo, C. Beneduce, M. Dhallé, N. Musolino, J.Y. Genoud, and R. Flükiger, Appl. Phys. Lett. **79**, 3116-3118 (2001).
6. W. Goldacker, S.I. Schlachter, S. Zimmer, and H. Reiner, Supercond. Sci. Technol. **14**, 787-793 (2001).
7. S. Patnaik, L.D. Cooley, A. Gurevich, A.A. Polyanskii, J. Jiang, X.Y. Cai, A.A. Squitieri, M.T. Naus, M.K. Lee, J.H. Choi, L. Belenky, S.D. Bu, J. Letteri, X. Song, D.G. Schlom, E.S. Babcock, C.B. Eom, E.E. Hellstrom, and D. Larbalestier, Supercond. Sci. Technol. **14**, 315-319 (2001).
8. R. Nast, S.I. Schlachter, W. Goldacker, S. Zimmer, and H. Reiner, accept. for publication in Physica C.
9. W.Goldacker and S.I.Schlachter, accept. for publication in Physica C.
10. D.J. Jorgensen, D.G. Hinks, and S.Short, Phys. Rev. B **63**, 224522 (2001).
11. J.S. Schilling, J.D. Jorgensen, D.G. Hinks, S.Deemyad, J.Hamlin, C.W.Looney, and T.Tomita, to appear in: "Studies of High Temperature Superconductors" **38**, edited by A. Narlikar (Nova Science Publishers, N.Y., 2001).

12. J.S. Slusky, N. Rogado, K.A. Regan, M.A. Hayward, P. Khalifah, T. He, K. Inumaru, S.M. Loureiro, M.K. Haas, H.W. Zandbergen, and R.J. Cava, Nature **410**, 343-345 (2001).
13. T.C. Shields, K. Kawano, D. Holdom, and J.S. Abell, Supercond. Sci. Technol. **15**, 202-205 (2002).
14. T.C. Shields, private communication (2002).
15. W. Goldacker, W. Specking, F. Weiss, G. Rimikis, and R. Flükiger, Adv. Cryog. Eng. Plenum Press **36**, 995 (1990).
16. W. Goldacker and R. Flükiger, Adv. Cryog. Eng. Plenum Press **34**, 561 (1987).
17. W. Goldacker, C. Rieger, and W. Maurer, IEEE Trans. Magn. **27** 946 (1991).
18. B. Ullmann, A. Gaebler, M. Quilitz, and W. Goldacker, IEEE Trans. Appl. Supercond. **7**, 2042 (1997).
19. J.K. Song, N.J. Lee, H.M. Jang, H.S. Ha, D.W. Ha, S.S. Oh, M.H. Sohn, Y.K. Kwon, and K.S. Ryu, Cond-Mat/0106124.

Electron-Phonon Coupling and Superconductivity in MgB_2 and Related Diborides

Rolf Heid, Klaus-Peter Bohnen, and Burkhard Renker

Forschungszentrum Karlsruhe, Institut für Festkörperphysik,
P.O.Box 3640, 76021 Karlsruhe, Germany

Abstract. A brief review of the electron-phonon coupling scenario for the superconductor MgB_2 and related diborides on the basis of ab initio density-functional calculations is presented. We show results for phonon dispersions and electron-phonon spectral functions and discuss their implications for superconducting properties in the isotropic limit. The hydrostatic pressure derivative of T_c for MgB_2 is found to be consistent with experiment. Calculations for AlB_2 and a hypothetical superstructure $MgAlB_4$ are performed to study the doping dependence of the electron-phonon coupling. Differences to transition-metal diborides are investigated for NbB_2 and TiB_2. We also include a brief discussion of limitations and extensions of the simple isotropic picture.

1 Introduction

Phonon-mediated superconductivity, often denoted as conventional superconductivity, has long been thought of to be restricted to rather low transition temperatures. High-T_c superconductivity has been observed only in materials with complex crystal structure and was believed to require a more exotic pairing mechanism. Therefore, the discovery of superconductivity in hexagonal MgB_2 with a $T_c \approx 39\,\mathrm{K}$ [1] came as a surprise in view of the simple lattice structure and the apparent absence of strong electron correlations or magnetic interactions. Within the past year, it has initiated a large number of experimental and theoretical investigations of this compound and related diborides dedicated to reveal its main properties and to search for candidates with even improved superconducting properties.

Experimentally, it was soon established that in MgB_2 a spin-singlet pairing state is realized. Although alternative pairing mechanisms have been proposed as well [2,3,4], various experiments point to a phonon-mediated pairing in the medium coupling strength regime (see, e.g., references in [5]). This picture is especially supported by the observation of a significant B-isotope effect [6,7] and of signatures of the phonon spectrum in recent tunneling experiments [8].

The proposed phonon mechanism immediately raises several questions: (i) Is the electron-phonon coupling indeed strong enough to lead to the observed high T_c? (ii) What are the key ingredients? (iii) Is MgB$_2$ unique among the class of diborides, or are there other candidates for even higher T_c's?

On the theoretical side, much insight into these questions has been gained by applications of modern electronic-structure methods based on density-functional theory. On the one hand, they provide an accurate and parameter-free analysis of the electronic structure and of the bonding properties, and have been used extensively in the last year for studies of electronic ground state properties of MgB$_2$ and various other diborides [9,10,11,12]. On the other hand, the same theoretical framework allows to access lattice dynamical properties and quantities related to electron-phonon coupling (EPC). Here, two different approaches have been applied to MgB$_2$. The first utilizes frozen phonon techniques in combination with calculations of deformation potentials to obtain phonon and EPC properties for selected points in the Brillouin zone (BZ), which allows for a crude estimate of average coupling strength only [9,13,14]. The second type of work is based on a perturbational approach and provides information on phonon dispersions and EPC in the whole BZ, thus giving a much more accurate estimate of the coupling. Furthermore, the electron-phonon spectral function $\alpha^2 F$ entering the theory of phonon mediated superconductivity can be calculated. This scheme has been applied to MgB$_2$ by several groups [15,16,17].

In this paper, we review the application of these elaborate electronic-structure techniques to the lattice dynamics and electron-phonon coupling in MgB$_2$ and extend it to other related diborides. We outline the emerging picture of phonon-mediated superconductivity in MgB$_2$ and discuss some open questions. To pinpoint the key features relevant for superconductivity, we present a comparative study with related diborides: (i) the nonsuperconducting reference compound AlB$_2$, (ii) MgAlB$_4$ as a representative of the Mg$_{1-x}$Al$_x$B$_2$-system to study the doping dependence, and (iii) NbB$_2$ and TiB$_2$ as two examples of transition-metal diborides. We will mainly focus on the results obtained by our group and include some material from previous publications [15,18] for sake of comparison.

The paper is organized as follows. In Sect. 2, we give a brief outline of the computational and experimental details, which is followed in Sect. 3 by a discussion of ground-state properties. Section 4 provides a discussion of the lattice dynamical properties with special emphasis on the anomalies related to strong EPC. Calculations of the electron-phonon spectral functions and superconducting properties are presented in Sect. 5. There, first the isotropic limit is considered, which is followed by a discussion of its limitations and possible extensions. The main results are then summarized in Sect. 6.

2 Computational Details

Here we give a brief outline of the computational procedure. All the calculations presented in the following have been performed using the mixed-basis pseudopotential method. Pseudopotentials for Al, Nb, and Ti were constructed according to the description of Bachelet-Hamann-Schlüter [19,20], whereas for Mg a well tested pseudopotential of Martins-Troullier-type has been used [21]. For boron a Vanderbilt-type pseudopotential was created [22]. Non-linear core corrections have been applied for Nb and Ti [23]. The fairly deep p-potential for B and d-potentials for Nb and Ti are efficiently treated by the mixed-basis scheme, which uses a combination of local functions and plane waves for the representation of the valence states. We used p-type local functions for each B site, d-type for each Nb, and s, p, d-type for Ti site, supplemented by plane waves up to a kinetic energy of 16 Ry (24 Ry for TiB_2). All calculations were carried out in the local-density approximation using the Hedin-Lundqvist form of the exchange-correlation functional [24]. Brillouin-zone integration has been performed using Monkhorst-Pack special k-point sets with a Gaussian smearing of 0.2 eV. Phonon dispersions were calculated within the mixed-basis perturbation approach [25]. Complete spectra are obtained from a Fourier interpolation of dynamical matrices calculated on a hexagonal ($6\times6\times6$) q-point mesh. The same method also provides direct access to the screened electron-phonon matrix elements, the key ingredients for the Eliashberg theory of phonon-mediated superconductivity [26,27].

Special care was taken to assure convergence with respect to the number of plane waves and number of k points. For ground state and phonon calculations a ($18\times18\times18$) k-point mesh was sufficient. In the calculation of Eliashberg-type functions, which involve summation over states at the Fermi energy, the δ-functions are replaced by Gaussians of width 0.2 eV, and a much denser ($36\times36\times36$) k-point mesh was utilized. The q-summation was performed by Fourier-interpolation of the coupling matrices obtained for the ($6\times6\times6$) q-point mesh. Finally, estimates of T_c in the dirty limit are obtained by solving the linearized form of the isotropic Eliashberg equations [28].

All phonon calculations are based on a full structural optimization with respect to the total energy. As discussed in [15] for the Γ-point modes of MgB_2 and AlB_2, this is important for accurately determining phonon frequencies.

3 Ground-State Properties

The crystal structure of MB_2, M=Mg, Al, Nb, and Ti, is the hexagonal AlB_2 structure (P6/mmm) consisting of alternating hexagonal M layers and graphite-like B layers. For $MgAlB_4$ we have assumed a superstructure with a regular stacking of Al and Mg layers, which has been recently suggested on the basis of an electron-microscopy study [29]. Table 1 shows the structural parameters for the optimized geometries. For $MgAlB_4$ we find a sizable shift of the B layers in the direction of the Al layers in agreement with recent theoretical studies [30,31].

The calculated bandstructures, shown in Fig. 1 are in good agreement with previous work [9,10,33]. The most important result is related to the σ-

Table 1. Structural parameters of optimized geometries. For MgAlB$_4$, the two c/a-values correspond to the distances between the two B layers surrounding an Mg and Al layer, respectively. Experimental values are given in brackets [32]

	MgB$_2$	AlB$_2$	MgAlB$_4$	NbB$_2$	TiB$_2$
a (Å)	3.056 (3.084)	2.965 (3.009)	3.009	3.093 (3.09)	2.998 (3.03)
c (Å)	3.622 (3.522)	3.232 (3.262)	3.618/3.083	3.337 (3.30)	3.188 (3.23)
c/a	1.153 (1.142)	1.09 (1.084)	1.202/1.024	1.079 (1.068)	1.063 (1.066)

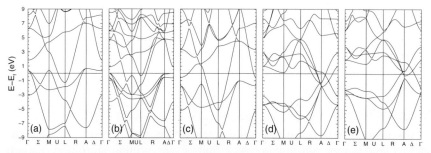

Fig. 1. Calculated band structures of (a) MgB$_2$, (b) MgAlB$_4$, (c) AlB$_2$, (d) NbB$_2$, and (e) (TiB$_2$). The horizontal line denotes the Fermi energy

bands, which form the covalent B-B bonds. For MgB$_2$ they are partially hole-doped, leading to a sizable density-of-states of the p$_{x/y}$-B states at the Fermi energy. The corresponding Fermi surfaces (FS) have a quasi-2D character due to a small dispersion of these bands parallel to the hexagonal axis. The B p$_z$-derived π bands are more 3-dimensional. Doping with Al leads to a gradual filling of the σ bands. For MgAlB$_4$, they are just completely filled, while for AlB$_2$, they are further pushed down in energy with respect to the B-π-states. In the case of the transition-metal diborides, there are additional contributions from the d-bands, while the top of the B-σ-bands is $\approx 3\,\mathrm{eV}$ below E$_f$. As we will see in the following section, this difference is directly reflected in the dynamical properties.

4 Lattice Dynamics

The AlB$_2$-structure possesses four optical modes at the Γ-point. Their calculated frequencies are summarized in Table 2. Due to the light B mass and strong B-B couplings, the two high-frequency modes are of almost pure boron character: an in-plane stretching vibration (E$_{2g}$) and an out-of-plane vibration, where neighboring B atoms move in opposite directions (B$_{1g}$). The

Table 2. Calculated frequencies of the Γ-point modes in meV. For MgAlB$_4$, splittings induced by the superstructure are indicated. Such an assignment is not possible for the low-frequency modes

	MgB$_2$	AlB$_2$	MgAlB$_4$	NbB$_2$	TiB$_2$
E$_{1u}$	40.5	36.6		52.0	65.5
A$_{2u}$	50.2	52.1		60.5	66.4
E$_{2g}$	70.8	125.0	110.9 / 115.2	98.4	112.8
B$_{1g}$	87.0	61.3	78.8 / 80.5	69.8	70.0

lower-frequency modes consist of an in-plane vibration of the metal atoms (E$_{1u}$) and a motion of the metal layer against the B layer (A$_{2u}$).

Because the in-plane B mode probes the strong covalent bonds, one would expect its frequency to be the highest one in the spectrum. This is indeed the case, except for MgB$_2$, where the B$_{1g}$ mode has the highest frequency. The E$_{2g}$ frequency is significantly shifted downwards as compared to the other compounds.

The downward shift of the E$_{2g}$ mode in MgB$_2$ is not a trivial consequence of structural changes, but is a clear signature of a large renormalization due to electron-phonon coupling. This is indicated by the calculated phonon dispersions, shown in Figs. 2 and 3. The branch related to the E$_{2g}$ mode disperses quickly to high energies when going away from the hexagonal axis. In all other cases, it remains high in energy over the whole BZ. The strong softening happens in the vicinity of the ΓA-line.

Experimentally, this picture is supported by two complementary types of experiments: (i) Raman measurements, which are sensitive to the E$_{2g}$ mode at Γ, clearly show a systematic downward shift of its frequency in the series Mg$_{1-x}$Al$_x$B$_2$ with increasing Mg content. At the same time, a significant line broadening is observed [18]. (ii) Inelastic neutron scattering measurements of

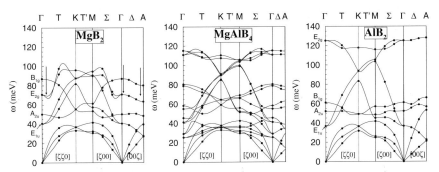

Fig. 2. Calculated phonon dispersions for MgB$_2$, MgAlB$_4$ and AlB$_2$

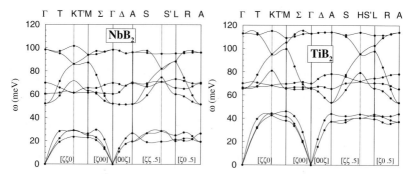

Fig. 3. Calculated phonon dispersions for NbB$_2$ and TiB$_2$

the generalized phonon density-of states (GDOS) access the whole phonon spectrum with enhanced weighting of BZ-boundary modes. There is a good agreement of the experimental GDOS of MgB$_2$ obtained by several groups with our theoretical one [15,34,14]. However, no theoretical evidence exists for the low-energy peak at $\approx 17\,\mathrm{meV}$ observed in two of these experiments [35,36]. Furthermore, the changes in GDOS when partially or totally replacing Mg by Al are well reproduced by the theoretical spectra [18], thus providing further evidence of the reliability of the present calculations.

5 Electron-Phonon Coupling and Superconductivity

This section is devoted to the results obtained for the electron-phonon interaction. Its application to superconducting properties is first discussed in the isotropic or dirty limit, and then followed by a discussion of its validity and possible extensions.

5.1 Isotropic Limit

In this limit, superconducting properties are determined by the isotropic gap equations. Phonon-mediated pairing interaction is described by the isotropic Eliashberg function

$$\alpha^2 F(\omega) = \frac{1}{2\pi N(0)} \sum_{\mathbf{q}\lambda} \frac{\gamma_{\mathbf{q}\lambda}}{\omega_{\mathbf{q}\lambda}} \delta(\omega - \omega_{\mathbf{q}\lambda}), \tag{1}$$

where $\omega_{\mathbf{q}\lambda}$ denotes the frequency of the phonon mode $(\mathbf{q}\lambda)$, $N(0)$ is the electronic density-of-states (per atom and spin) at the Fermi energy, and $\gamma_{\mathbf{q}\lambda}$ is the electronic contribution to the phonon linewidth

$$\gamma_{\mathbf{q}\lambda} = 2\pi \omega_{\mathbf{q}\lambda} \sum_{\mathbf{k}\nu\nu'} |g_{\mathbf{k}+\mathbf{q}\nu',\mathbf{k}\nu}^{\mathbf{q}\lambda}|^2 \delta(\epsilon_{\mathbf{k}\nu} - \epsilon_F) \delta(\epsilon_{\mathbf{k}+\mathbf{q}\nu'} - \epsilon_F). \tag{2}$$

Here, g denotes the screened EPC matrix element. Within the perturbational approach to the lattice dynamics, g is directly accessible from quantities obtained in the calculation of the dynamical matrix.

The only additional material specific quantity entering the gap equations is the effective electron-electron interaction constant μ^*, which is treated as a free parameter in the following. A reasonable approximation to T_c can be obtained from the Allen-Dynes formula [37]

$$T_c = \frac{\omega_{log}}{1.2} \exp\left(\frac{-1.04(1+\lambda)}{\lambda - \mu^*(1+0.62\lambda)}\right), \qquad (3)$$

where the EPC enters only via two moments of $\alpha^2 F$, the average coupling constant

$$\lambda = 2 \int_0^\infty d\omega \frac{\alpha^2 F(\omega)}{\omega} \qquad (4)$$

and an average effective frequency defined as

$$\omega_{log} = \exp\left(\frac{2}{\lambda}\int_0^\infty d\omega \frac{\alpha^2 F(\omega)}{\omega} \ln(\omega)\right). \qquad (5)$$

From these equations it is evident that both a large λ and a large ω_{log} are favorable for high T_c's.

Eliashberg Functions and Estimates of T_c. The results for $\alpha^2 F(\omega)$ are compared with the calculated phonon density-of-states in Fig. 4. Values for the moments of $\alpha^2 F$ are summarized in Table 3.

The most prominent result for MgB$_2$ is a very pronounced coupling of modes in an energy range 60-70 meV. Analysis of the contributions from individual phonons in Eq. (1) shows that this peak results from an unusually strong coupling of the E$_{2g}$-related modes in the vicinity of ΓA, indicated by a very large linewidth $\gamma > 10$ meV. This coupling may be partly responsible for the extreme broadening of the E$_{2g}$ Raman line [15]. This peak contributes about 50% to the total λ of 0.73 (the latter value is in reasonable agreement with other calculations [16,17]). This medium coupling strength combined with a high effective frequency ω_{log} allows to reach the unusual high T_c.

For MgAlB$_4$, we find a rather small coupling. In AlB$_2$, the coupling is sizable due to a larger contribution from the low-energy spectrum, which comes mainly from Al modes.

For NbB$_2$ an average coupling comparable with MgB$_2$ is found. The reason is a large contribution coming from the Nb modes with frequencies below 40 meV, originating from a stronger EPC to the Nb d states. They produce about 75% of the total coupling due to an enhanced weight of the low-frequency part of the spectrum in Eq. 4. But at the same time, ω_{log} is substantially lowered. For TiB$_2$, the overall coupling is very low, consistent with the observed absence of superconductivity.

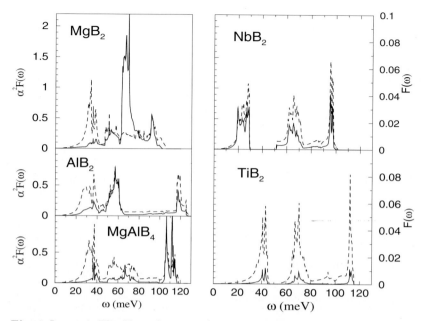

Fig. 4. Isotropic Eliashberg functions (solid lines, left scale) compared with phonon density-of-states (dashed lines, right scale)

Table 3. Average coupling constant λ, effective average phonon frequency ω_{log} (in meV), and estimates of T_c (in K) obtained from the linearized gap equation

	MgB_2	AlB_2	$MgAlB_4$	NbB_2	TiB_2
λ	0.73	0.43	0.25	0.67	0.10
ω_{log}	60.9	49.9	61.9	30.5	52.9
$T_c\ (\mu^*=0)$	54.4	16.3	3.5	23.2	-

Using the calculated spectral functions, estimates for T_c obtained by solving the linearized gap equation are shown in Fig. 5. For MgB_2, the observed T_c of $\approx 40\,K$ is compatible with a μ^* of 0.05, which is a rather small value as compared to 0.1–0.15 usually found for conventional superconductors. However, assuming the same value for the other diborides is at variance with the smallness or absence of T_c for AlB_2 and NbB_2. No superconductivity has yet been observed for AlB_2 above $4\,K$ [18], while for NbB_2 there exist contradicting reports with $T_c = 0.62\,K$ [38] and $T_c = 5.2\,K$ [39]. Two possible explanations could reconcile this contradiction. (i) Screening properties for MgB_2 differ significantly from the other diborides, giving rise to different values for μ^*. (ii) The isotropic limit is not appropriate for MgB_2. This is discussed in the following subsection.

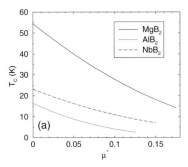 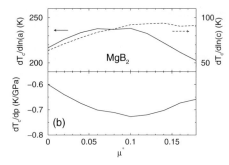

Fig. 5. (a) T_c as function of the parameter μ^* obtained from the linearized gap equation. (b) Derivatives of T_c with respect to lattice parameters and hydrostatic pressure for MgB_2

Pressure Dependence of T_c for MgB_2. An experimental test of the proposed phonon mechanism of superconductivity is provided by the pressure dependence of T_c. The effect of hydrostatic pressure on T_c for polycrystalline samples has been investigated by several groups, which consistently find a drop in T_c with increasing pressure, with values for dT_c/dp of -1.11 K/GPa [40] and -1.6 K/GPa [41].

To see if this behavior is consistent with the EPC picture, we have repeated the calculations of phonon spectra and EPC of MgB_2 for lattice parameters slightly changed from the optimized values. For an increase in the lattice parameter a (c) by 1% (2%), we find an increase in $N(0)$ by 1.2% (1.0%) and an increase of λ by 4.4% (5.0%). Fig. 5b shows the derivatives of T_c with respect to the lattice parameters and with respect to hydrostatic pressure as function of μ^*. The latter quantity is calculated using experimental values for the elastic constants from [42].

These findings can be summarized as follows. (i) The EPC pairing mechanism is consistent with the sign and approximate magnitude of the observed pressure dependence of T_c. (ii) An increase of the electronic DOS is less important than the increase in the effective EPC. (iii) An increase of a or c would be favorable for T_c. A comparison of anisotropic quantities has to await the availability of larger single crystals.

5.2 Beyond the Isotropic Limit

Here we discuss two modifications of the EPC picture which may be of relevance for a proper quantitative description of the superconducting state in MgB_2, and could resolve some of the problems mentioned above.

Anisotropy. Early transport experiments as well as the observed robustness of T_c with respect to sample quality suggested the validity of the dirty limit for which the isotropic description discussed above would be applicable.

With improved sample quality and the availability of single crystals, there is convincing evidence that the clean limit is more appropriate. In view of the very anisotropic nature of the quasi-2D Fermi sheets derived from the σ-bands, it is natural to consider the possibility of anisotropic pairing states. Some popularity has been recently obtained by the multigap scenario, where one assumes gap functions of different magnitudes on the different FS sheets. In its simplest form, one considers two gaps corresponding to the σ and π FS, respectively [43]. Support for this picture comes from the observation of two gaps in photoemission [44] and point-contact [45,46,47] experiments, from low-temperature transport [48] measurements, and from an additional low-temperature contribution to the electronic specific heat [49,50].

A theoretical analysis of the contributions to the total coupling originating from intra- and interband scattering (ν-ν' combinations in Eq. 2) supports the multigap scenario [17]. The intra-band σ-σ coupling accounts to $\approx 60\%$ of the total λ, while π-π and π-σ coupling each contribute $\approx 20\%$ only. As discussed by Lui et al. [17], this anisotropic coupling results in an enhanced effective average coupling constant $\lambda_{eff} \approx 1.0$. Consequently, the observed T_c would be consistent with a $\mu^* \approx 0.13$, which is a value more commonly found for ordinary metals. These results are corroborated by a recent calculation of the gap structure on the basis of the fully anisotropic gap equation [51]. Ab initio derived gap structures provide a good description of the anomalous specific heat data [52,53]. As a similar anisotropy enhancement of λ is not expected for AlB_2 or NbB_2 because of their 3D-like electronic structure, such a large μ^* would reconcile with the absence or smallness of T_c in these compounds despite a sizable EPC.

The multigap theory possesses, however, a conceptual difficulty. The stability of a superconducting state with different gaps requires a very weak interband impurity scattering. This has to be reconciled with the apparent robustness of T_c with respect to sample quality, i.e. impurity concentration. While some optical data support the existence of different impurity scattering rates for the σ and π states [54], no direct experimental confirmation of a weak interband impurity scattering has been achieved.

Anharmonicity. It has been noted from frozen-phonon calculations, that the energy potential of the E_{2g} mode is very anharmonic [14,17]. These calculations predict an effective mode frequency of 17–25% higher than the harmonic one. This effect alone would lead to a *reduction* of the EPC of this mode. An analysis within the multigap picture suggests that this effect reduces λ by 16% [51].

There are, however, some open questions about its actual relevance.

(i) Is the anharmonicity indeed as strong as suggested by these calculations? One argument put forward is the difference between the experimental E_{2g}-frequency and its calculated harmonic value. However, the

latter is quite sensitive to convergence parameters, and is easily underestimated. Our well converged result of 70.8 meV is significantly higher than the value 60.3 meV obtained by Yilderim et al. [14]. As being much closer to the experimental values of 72–77 meV [55], it would imply a significantly smaller anharmonic correction.
(ii) What is the effective phase space of this strong anharmonicity? Calculations for Γ and A suggest a sizable anharmonicity of the E_{2g} mode along the whole ΓA line, but it is not known how fast its strength decays when going away from the hexagonal axis.
(iii) Strong anharmonicity not only leads to a renormalization of the frequencies, but should also give rise to a sizable nonlinear EPC [14]. This should favor pairing and partly compensate for reduction of λ_{eff} discussed above.

6 Summary

We have presented a brief review of the phonon-mediated pairing mechanism for superconductivity in MgB_2 on the basis of first principles density-functional calculations. The same theoretical approach has been applied to several related diborides to clarify the key ingredients responsible for the high T_c of MgB_2. Results for phonon dispersions and electron-phonon spectral functions are shown. They support the picture that the special properties of MgB_2 are related to the existence of boron $p_{x,y}$-derived σ bands at the Fermi surface, which are involved in the strong covalent B-B-bonds. The high T_c results from a strong coupling of these σ states to specific high-frequency in-plane boron vibrations. This feature is missing in AlB_2 due to completely filled σ bands, leading to a significantly reduced average coupling. Within the isotropic limit, the dependence of T_c on pressure and on doping with Al is found to be consistent with this scenario. For NbB_2, despite a sizable average coupling, only a low T_c is expected because the coupling is mainly produced by low-frequency Nb vibrations. In accord with recent experimental and theoretical work, our results lend support to the idea that the isotropic limit is not appropriate for a proper quantitative description of the superconducting state of MgB_2.

References

1. J. Nagamatsu et al: Nature **410**, 63 (2001).
2. J.E. Hirsch: Phys. Lett. A **282**, 392 (2001).
3. J.E. Hirsch, F. Marsiglio: Phys. Rev. B **64**, 144523 (2001).
4. M. Imada: J. Phys. Soc. Jpn. **70**, 1218 (2001).
5. C. Buzea, T. Yamashita: Supercond. Sci. Technol. **14**, R115 (2001).
6. S.L. Bud'ko et al: Phys. Rev. Lett. **86**, 1877 (2001).
7. D.G. Hinks, H. Claus, J.D. Jorgensen: Nature **411**, 457 (2001).

8. A.I. D'yachenko et al: cond-mat/0201200.
9. J. Kortus et al: Phys. Rev. Lett. **86**, 4656 (2001).
10. G. Satta et al: Phys. Rev. B **64**, 104507 (2001).
11. N.I. Medvedeva, A.L. Ivanovskii, J.E. Medvedeva, A.J. Freeman: Phys. Rev. B **64**, 020502 (2001).
12. K.D. Belashchenko, M. van Schilfgaarde, V.P. Antropov: Phys. Rev. B **64**, 092503 (2001).
13. J.M. An, W.E. Pickett: Phys. Rev. Lett. **86**, 4366 (2001).
14. T. Yildirim et al: Phys. Rev. Lett. **8703**, 7001 (2001).
15. K.P. Bohnen, R. Heid, B. Renker: Phys. Rev. Lett. **86**, 5771 (2001).
16. Y. Kong, O.V. Dolgov, O. Jepsen, O.K. Andersen: Phys. Rev. B **64**, 020501 (2001).
17. A.Y. Liu, I.I. Mazin, J. Kortus: Phys. Rev. Lett. **8708**, 7005 (2001).
18. B. Renker et al: Phys. Rev. Lett. **88**, 067001 (2002).
19. G.B. Bachelet, D.R. Hamann, M. Schlüter: Phys. Rev. B **26**, 4199 (1982).
20. K.M. Ho, K.-P. Bohnen: Phys. Rev. B **32**, 3446 (1985).
21. G. Pelg: Dissertation, Universität Regensburg, unpublished.
22. D. Vanderbilt: Phys. Rev. B **32**, 8412 (1985).
23. S.G. Louie, S. Froyen, M.L. Cohen: Phys. Rev. B **26**, 1738 (1982).
24. L. Hedin B.I. Lundqvist: J. Phys. C: Solid St. Phys. **4**, 2064 (1971).
25. R. Heid, K.-P. Bohnen: Phys. Rev. B **60**, R3709 (1999).
26. S.Y. Savrasov, D.Y. Savrasov, O.K. Andersen: Phys. Rev. Lett. **72**, 372 (1994).
27. R. Heid, L. Pintschovius, W. Reichardt, K.-P. Bohnen: Phys. Rev. B **61**, 12059 (2000).
28. G. Bergmann, D. Rainer: Z. Physik **263**, 59 (1973).
29. J.Q. Li et al: cond-mat/0104320.
30. S.V. Barabash, D. Stroud: cond-mat/0111392.
31. P.P. Singh: cond-mat/0201093.
32. R.W.G. Wyckoff: Crystal Structures, Vol. 1 (Wiley & Sons, New York, 1965).
33. I.R. Shein, A.L. Ivanovskii: cond-mat/0109445.
34. R. Osborn, E.A. Goremychkin, A.I. Kolenikov, D.G. Hinks: Phys. Rev. Lett. **8701**, 7005 (2001).
35. T. Muranaka et al: J. Phys. Soc. Jpn. **70**, 1480 (2001).
36. T.J. Sato, K. Shibata, Y. Takano: cond-mat/0102468.
37. P.B. Allen, R.C. Dynes: Phys. Rev. B **12**, 905 (1975).
38. L. Leyarovska etal: J Less-Common Met. **67**, 249 (1979).
39. J. Akimitsu: Annual Meeting Phys. Soc. Japan **3**, 533 (2001).
40. T. Tomita et al: Phys. Rev. B **64**, 092505 (2001).
41. B. Lorenz, R.L. Meng, C.W. Chu: Phys. Rev. B **64**, 012507 (2001).
42. J.D. Jorgensen, D.G. Hinks, S. Short: Phys. Rev. B **63**, 224522 (2001).
43. S.V. Shulga et al: cond-mat/0103154.
44. S. Tsuda et al: Phys. Rev. Lett. **8717**, 7006 (2001).
45. F. Laube et al: Europhys. Lett. **56**, 296 (2001).
46. P. Szabo et al: Phys. Rev. Lett. **8713**, 7005 (2001).
47. N. L. Bobrov et al: cond-mat/0110006.
48. A.V. Sologubenko et al: cond-mat/0111273.
49. F. Bouquet et al: Phys. Rev. Lett. **8704**, 7001 (2001).
50. Y.X. Wang, T. Plackowski, A. Junod: Physica C **355**, 179 (2001).
51. H.J. Choi et al: cond-mat/0111182.

52. A.A. Golubov: cond-mat/0111262.
53. H.J. Choi et al: cond-mat/0111183.
54. A.B. Kuz'menko et al: cond-mat/0107092.
55. K. Kunc et al: J. Phys.-Condes. Matter **13**, 9945 (2001).

Self-Organized Quasi-One Dimensional Structures in High-Temperature Superconductors: the Stripe Phase

Enrico Arrigoni[1], Marc G. Zacher[1], Rober Eder[1], Werner Hanke[1], and Steven A. Kivelson[2]

[1] Institut für Theoretische Physik, Universität Würzburg, 97074 Würzburg, Germany
[2] Department of Physics, University of California, Los Angeles, California 90095, USA

Abstract. Besides superconductivity, high-T_c materials show a number of unconventional phases. In one of these phases doped holes tend to organize themselves in quasi-one-dimensional structures, so-called "stripes". In this paper we analyze some aspects of his phenomenon and its relation with superconductivity. In the first part, we show the important role of the long-range part of the Coulomb interaction, showing that the latter can favor Cooper-pair tunneling between stripes, thus enhancing superconducting correlations. In the second part of the paper, we analyze some features of recent angle-resolved photoemission experiments for LaSrCuO and LaNdSrCuO materials, and show how these can be understood in the light of the stripe picture, supporting the existence of (possibly) dynamic stripes in LaSrCuO. In addition, the analysis allows to distinguish between different stripes geometries and to determine their doping dependence.

1 Introduction

The high-T_c superconducting materials (as well as a number of other so-called strongly-correlated systems) are characterized by the presence of different competing phases in their phase diagram. Besides the well known insulating antiferromagnetic (AF) and superconducting (SC) states, a number of new unconventional phases have been discovered in the past years. Among these, one should mention the spin-glass phase between the AF and the SC regions, the pseudogap phase at higher temperatures [1,2], and different structural phases obtained by decreasing temperatures. Moreover, it has been argued that several competing charge and spin instabilities [3] are present in these materials, such as phase separation [4], incommensurate magnetic phases, and SC states with mixed symmetry [5], which are responsible for their anomalous behavior. Some years ago elastic neutron scattering showed that one of these incommensurate magnetic phase becomes static upon replacing some La with Nd in the "classical" high-T_c superconductor La$_{2-x}$Sr$_x$CuO$_4$ (LSCO) yielding La$_{1.48}$Nd$_{0.4}$Sr$_{0.12}$CuO$_4$ (Nd-LSCO). More interesting, these structures are accompanied by charge-density waves of half the wavelength of the spin-density waves and appear to be one-dimensional [6,7]. The picture is quite appealing: in order to minimize disturbance of the underlying AF structure, doped charges remain essentially confined in one-dimensional "stripes" separating the AF domains [8,9]. These domain walls introduce a phase shift between the AF regions producing a charge density wave with wavelength equal to the distance between them and a spin-density wave of twice this wavelength. While these structures are static in Nd-LSCO, similar diffraction patterns in the *inelastic* neutron-scattering results in the "conventional" high-T_c compound LSCO [10,11] or even in YBa$_2$Cu$_3$O$_{7-x}$ (YBCO) [12], suggesting the presence of fluctuating or dynamic stripes. However, at least in the case of YBCO, it is not yet clear whether the observed pattern is due to one-dimensional spin inhomogeneities (i.e. stripes) [12] or to two-dimensional incommensurable spin waves [13].

In this paper, we discuss the role of stripe fluctuations and the way they affect superconductivity. In particular, in Sec. 2, we show the importance of the long-range part of the Coulomb interaction in increasing stripe fluctuations. In fluctuating stripe systems holes and Cooper pairs can tunnel much better from stripe to stripe strongly *enhancing* long-distance superconducting pair-field correlations, and, thus, SC phase coherence [14,15]. In Sec. 3, we provide numerical arguments showing that angle-resolved photoemission spectroscopy (ARPES) provide a support for the existence of stripes both static (in Nd-LSCO) as well as dynamic (in LSCO) [16,17]. Specifically, we will show that the spectra of LSCO and Nd-LSCO can be described in a very similar way by different types of stripe states (site-centered, bond-centered).

2 Role of Long-Range Coulomb Interaction in the Stripe Phase

We first analyze how stripes can originate spontaneously by simple correlation mechanisms, provided there is a small symmetry-breaking effect that favors their orientation. Specifically, we have studied several $t - J$ clusters with open boundary conditions by means of the numerically accurate density-matrix renormalization-group (DMRG) technique. Our main goal is to investigate the effects of the long-range Coulomb interaction on the stability of stripes and on their physical properties. In order to treat the long-range part, which is difficult to deal with the standard DMRG, we adopt a method for studying the ground-state properties of electrons with strong short- and long-range interactions on fairly large finite systems [15]. The method, which may be termed a "density-functional DMRG", uses numerically very accurate DMRG methods to treat the short-range part of the interactions. The long-range piece is taken into account within the Hartree approximation, which becomes exact in the long-distance limit, and turns out to work well already for intermediate distances. Our Hamiltonian is thus given by

$$H = H_{tJ} + \frac{1}{2}\sum_{\mathbf{r}\neq\mathbf{r}'} V_0 \frac{(n_\mathbf{r} - \bar{n})(n_{\mathbf{r}'} - \bar{n})}{|\mathbf{r} - \mathbf{r}'|}, \qquad (1)$$

where H_{tJ} is the usual $t - J$ Hamiltonian, and $\bar{n}$ is the uniform positive background charge-density. The long-range part of the interaction in (1) is treated in the self-consistent Hartree approximation. The Coulomb prefactor V_0 is given by $V_0 \approx t$ obtained by considering a background dielectric constant $\epsilon \approx 8.5$ [15].

In the figures, we present representative results for the hole ($\rho(x)$) and spin ($S_z(x)$) density *averaged* over the transverse coordinate of the ladder. To probe superconductivity, we have computed the ground-state pair-field correlation function $D(x) = \langle \Delta\left(\frac{N}{2} + \frac{x}{2}\right) \Delta^\dagger\left(\frac{N}{2} - \frac{x}{2}\right)\rangle$. Here $\Delta^\dagger(x)$ creates a $d_{x^2-y^2}$-like pair around the $(x, 2)$ site [18]. In all the figures of this Section, we have adopted a conventional value $J/t = 0.35$.

Consider first the results of the DMRG calculations without the long-range Coulomb potential. Results for the spin and charge density for $n_h = 1/9$ are shown as the dashed lines in Fig. 1 and for the pairing susceptibility $D(l)$ by the dashed line in Fig. 2a. Fig. 1 clearly show the occurrence of charge stripe order. In order to distinguish between stripes and ordinary Friedel oscillations, we have compared the hole-density profile for different length ladders (18×4 and 27×4) at the same doping. As one can see from Fig. 1c and Fig. 1d for the cases with and without Coulomb interaction, respectively, the amplitude in the center of the system is essentially independent of the system

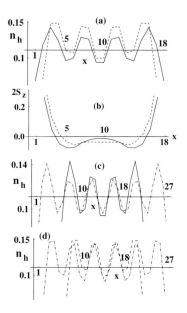

Fig. 1. (a) Hole density $\rho(x)$ and (b) staggered spin density $S_z(x)$ as a function of position x along a 18×4 ladder (*i.e.*, with open boundary conditions in the y direction) with 8 holes ($n_h = 1/9$). The results with nonvanishing Coulomb prefactor ($V_0 = t$, solid lines) are compared with those without Coulomb interaction ($V_0 = 0$, dashed lines). A comparison of $\rho(x)$ in a longer system (12 holes in 27×4) with the same $n_h = 1/9$ (dashed-dotted line in both cases) is shown in (c) and (d) respectively with and without Coulomb interactions

size, while Friedel oscillations should decay as a function of the distance from the boundary [19]. This means that the amplitude oscillations should survive up to long distances in the bulk. There is roughly one hole per two stripe unit cells, which was taken by WS [20] as evidence that the $t - J$ model favors stripes with a minimum energy for a linear charge density of $\lambda \sim 0.5$. This is in agreement with experiments [6,12], which find stripes with $\lambda \approx 0.5$ at hole dopings smaller than $n_{h,c} \approx 1/8$ [21,22,23].

As discussed by WS, the hole clusters locally share a number of features with the two-hole pair state, which accounts for the fact that the energy per hole for a domain wall is close to the energy per hole for a pair [20]. This is suggestive of a competition between stripe stability and superconductivity [24]. Such a competition has already been demonstrated in a model which includes next-nearest-neighbor hoppings t' by WS. For large enough $|t'|$, the domain walls "evaporate" into quasiparticles ($t' < 0$) without significant pairing correlations or into pairs ($t' > 0$) [20]. It has long been clear that stripe formation suppresses long-range superconducting phase coherence [14], as is clear from the rapid falloff with distance of the pair-field correlator in Fig. 2a (dashed line).

We now compare the latter results with the situation with Coulomb interactions. In the lightly doped case, $n_h = 1/9$, shown in Fig. 1, the stripe structure is essentially unchanged, although the amplitude of the charge modulations is suppressed (by roughly a factor of 1.5), and the anti-phase character of the spin correlations is slightly enhanced. We interpret this as meaning that the stripe order is robust, but that the Coulomb interactions enhance the transverse stripe fluctuations. The most dramatic effect of the Coulomb interactions is the strong enhancement of the pair-field correlations shown in Fig. 2. From Fig. 2 one might conclude that the Coulomb interactions primarily increase the *overall* magnitude of the pair correlations. On the other

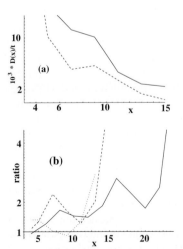

Fig. 2. (a) The long-distance part of the pair-field correlation $D(l)$ for 8 holes in a 18×4 ladder without V_{Coul} (dashed line) and with V_{Coul} (solid line). (b) The ratio between $D(x)$ with V_{Coul} and without V_{Coul} for $N \times 4$ ladders with 8 holes and $N = 18$ (dashed) $N = 16$ (dotted), and $N = 27$ with 12 holes (solid) [18]

hand, from the ratio between the two functions, which is displayed in Fig. 2b over the whole range of x, one can see that it is only the *long-distance* part that is enhanced. Unfortunately, the small system sizes we are restricted due to the limitation of the DMRG do not allow to make any strong statement about the *asymptotic* behavior of the superconducting correlation function. In particular, we don't know whether it saturates at some distance or whether it decays with a power-law behavior. In the latter case, Fig. 2 seems to indicate that the exponent is reduced by the Coulomb interactions. On the other hand, at short distances the effects of the Coulomb interactions are not unique and depend on doping, as it can be seen for the shortest distance ($x = 3$) in Fig. 2.

In order to sort out boundary effects, which are probably responsible for part of the strong increase of the ratio at the longest distances available, we have also treated a larger (27×4) ladder. Although the calculation of $D(x)$ is less accurate for this system [18], we can still draw some conclusions. As one can see from the figure, the ratio of $D(x)$ shows oscillations with twice the stripe periodicity, whose envelope is clearly increasing with distance, even far away from the boundaries. The dependence of $D(x)$ on distance, seen in Fig. 2a, is altered from rapidly decreasing in the absence of Coulomb interactions, to much slower distance dependence in the presence of Coulomb interactions. This result supports the idea that while the longer-range phase coherence (or, in other words, pair delocalization) is inhibited by rigid stripe order, as obtained in the pure $t - J$ model, *stripe fluctuations, induced by the Coulomb interaction, permit the pairs to tunnel from stripe to stripe.*

3 Stripes from Angle-Resolved Photoemission Spectroscopy

In this Section, we want to analyze the effects of stripes on electronic excitations. If static or dynamic stripes are present in HTSC, it is clear that low-energy excitations should be considerably affected. In particular, since the energy scales for stripe fluctuations $\omega \sim 2meV$ are relatively small in comparison with electronic excitations $\omega_{el} \sim 20-500meV$, the corresponding time scales are relatively slow. Therefore, the effects of stripes on electronic spectra should not depend much on whether stripes are static or dynamic. As a matter of fact, indications of such effects on electronic spectra has recently been accumulated by angle-resolved photoemission spectroscopy (ARPES) both on the static stripes in the Nd-LSCO system [17] and on possible dynamic domain walls in the LSCO compound [16]. Here, we want to show that the electronic structure revealed by ARPES, both in Nd-LSCO as well as LSCO contains features which can be explained in terms of a quasi-one dimensional stripe structure.

Our numerical study of the single-electron excitations in a striped phase is carried out via an extension of the cluster-perturbation-theory (CPT)

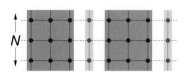

Fig. 3. Visualization of the cluster perturbation approach for stripes: the ground states for the half-filled three-leg ladder $(3 \times N)$ and the quarter-filled 1-leg chain $(1 \times N)$ are calculated exactly via exact diagonalization. The alternating clusters are then coupled via the inter-cluster hopping which is treated perturbatively

method [25,26]. The basic idea is indicated in Fig.(3): it is based on dividing the 2D plane into alternating clusters of metallic (with hole density $n_h = 0.5$) stripes and AF domains. The local many-body physics within the stripes, including the strong correlations, are treated exactly via exact diagonalization (ED). However, the inter-stripe hopping is incorporated perturbatively. Here, the idea is translated to the stripe physics: the theory can be used to incorporate long-distance effects into ED data, which already contain short-distance effects, in particular correlations, exactly. A study of the properties of experimentally observed stripe phases *solely* by ED [27] is precluded by the prohibitively large unit cells. The manageable clusters for ED are simply too small to accommodate even a single such unit cell.

Our main results are: (i) close to $\mathbf{k} = (\pi, 0)$ we see, like in experiments, a two-component electronic feature (see Fig.(4)): a sharp low-energy feature close to E_F and a more broad feature at higher binding energies. Both features can be explained by the mixing of metallic and antiferromagnetic bands at this $\mathbf{k}$-point. (ii) the excitation near $(\pi/2, \pi/2)$ is at higher binding energies than the low-energy excitation at $(\pi, 0)$ and of reduced weight. (iii) the integrated spectral weight of the cluster-stripe calculation resembles the quasi-one-dimensional segments in momentum space (see Fig.(5a)) as seen in the Nd-LSCO experiment. Also in agreement with the Nd-LSCO experiment our calculation finds the low-energy excitations near $(\pm \pi, 0)$ and $(0, \pm \pi)$ (Fig.(5b)). Interestingly, this agreement with experiment occurs only for so called "site-centered" metallic stripes (as shown in Fig.(3)) and not for "bond-centered" metallic stripes. This seems important since it has been argued [28], that, for bond-centered stripes, superconductivity is expected to survive stripe ordering. In the DMRG calculations by WS [20] as well as in dynamical mean-field (DMFT) studies by Fleck *et al.* [29], bond-centered and site-centered domains are very close in energy (ground-state). However, our technique allows to distinguish them dynamically.

3.1 Technique

The computational technique for our calculation of the single-particle spectral weight $A(\mathbf{k}, \omega)$ and the Green's function is illustrated in Fig.(3). We solve the AF cluster and the metallic cluster by ED and combine the individual clusters

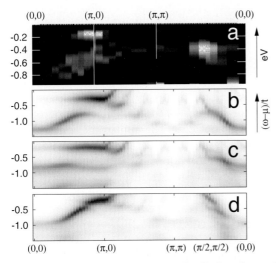

Fig. 4. ARPES results for $La_{2-x}Sr_xCuO_4$, theoretical single particle spectral function $A(\mathbf{k},\omega)$: (a) displays experimental ARPES results by Ino et al. [16]. The gray scale corresponds to the second derivative of the original measured data. Flat regions are black, regions with high curvature (i.e. peaks) are white. (b),(c),(d) show the results of the stripe CPT calculation for $A(\mathbf{k},\omega)$ (b: "3+1" site-centered;$t' = 0$, c: "3 + 1";$t' = -0.2t$, d: "2 + 2" bond-centered;$t' = 0$). Here, $A(\mathbf{k},\omega)$ is plotted directly with maximum intensity corresponding to black

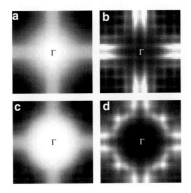

Fig. 5. Integrated spectral weight of site-centered (a,b) and bond-centered (c,d) stripe configurations; (a,c) total integrated weight in photoemission ($n(\mathbf{k})$), (b,d) low energy excitations (integrated weight in $E_F - 0.2t < \omega < E_F$). The data are plotted for the whole Brillouin zone with the Γ-point in the center. The result of the stripe calculations have been symmetrized to account for the differently oriented stripe domains in real materials. Regions of high spectral weight correspond to white areas

to an infinite lattice via CPT as in the homogeneous case. Bond-centered stripes, on the other hand, are modeled by 2-leg ladders with alternating filling (half-filled, $n_h = 0$ and doped, $n_h = 1/4$). In the following we will refer to this bond-centered configuration as "2 + 2".

The intercluster hoppings can be treated within a systematic strong-coupling perturbation expansion [25,26]. We take into account the lowest order contribution,

$$G_\infty(\mathbf{P}, z) = \frac{G_{cluster}(z)}{1 - \epsilon(\mathbf{P})G_{cluster}(z)}. \qquad (2)$$

Here, $\mathbf{P}$ is a superlattice wave vector and G_∞ is the Green's function of the "∞-size" 2D system. For the homogeneous case, the approximation in (2) is exact for vanishing interaction [25]. When the interactions are turned on, (2) is no longer exact but strong interactions are known to be important mainly for short-range correlations. These correlations are incorporated with good accuracy in modest-size clusters and are treated by ED within the cluster.

To allow for larger cluster sizes N, we diagonalize the $t-J$ model and take its spectral (Green's) function as our local $G_{cluster}$ in (2) as an approximation to the Hubbard model's (cluster-) Green's function. A comparison of the Hubbard and $t-J$ model's spectral function on small clusters [30] shows that the strong low energy peaks have similar dispersion and weight in both models, the main difference being a transfer of incoherent high energy spectral weight from momenta near (π, π) to $(0, 0)$. The $t-J$ Hamiltonian is

$$H = -\sum_{ij,\sigma} t_{i,j} \hat{c}^\dagger_{i,\sigma} \hat{c}_{j,\sigma} + J \sum_{<i,j>} (\mathbf{S}_i \mathbf{S}_j - \frac{n_i n_j}{4}). \qquad (3)$$

The hopping matrix element $t_{i,j}$ is nonzero only for nearest (t) and next-nearest neighbors (t'). We have chosen commonly accepted values for the ratio $t'/t = -0.2$, $J/t = 0.4$, where $t \approx 0.5 eV$. In this article, we present calculations for systems with $N = 8$ based on diagonalizations of a $N \times 3 = 24$-site half-filled 3-leg ladder and a quarter-filled 8-site chain. Results for smaller $N = 6$ do not differ much from the results for $N = 8$. However, $N = 6$ is somewhat pathological, since it has an odd number of electrons in the quarter-filled chain.

3.2 Spectral Features

Fig.(4a) shows the experimental ARPES results for LSCO at the superconductor-insulator transition (doping $x = 0.05$) [16]. To enhance the structure in the obtained spectra, the authors of ref. [16] plotted the second derivative of the ARPES spectrum, so areas with high second derivative are marked white and areas with low curvature (i.e. flat intensity) are black. This result is compared with the theoretical CPT calculation for different stripe configurations with overall doping of $x = 1/8$. Fig.(4b,c) are for the "3 + 1" site-centered configuration. Fig.(4b) shows the result for a "3 + 1" configuration with alternating (i.e. π-phase shifted) Néel order between the 3-leg ladders (induced by a staggered magnetic field $B = 0.1t$) without next-nearest neighbor hopping, Fig.(4c) shows the result for the "3 + 1" stripe configuration with next nearest-neighbor hopping $t' = -0.2t$, however without Néel order Fig.(4d) shows the result for a bond-centered "2 + 2" stripe configuration. We observe that the spectra for the site-centered "3 + 1" configuration (Fig.(4b,c)) are in surprisingly good agreement with experiment, and, that the $t' = 0$ calculation (Fig.(4b)) results in much more coherent bands due to the enforced Néel order in the 3-leg ladders. Similar to recent DMFT calculations [29], the sharp excitation near the Fermi surface around $(\pi, 0)$, that has been interpreted by Ino et al. as the quasiparticle peak in the SC state, is visible as well as a dispersive band at higher binding energies which (at least away from $(\pi, 0)$; see discussion below) can be interpreted as remnants of the insulating valence band resulting from the AF domains. Especially in the Néel ordered configuration, we observe a very coherent and pronounced band. This clear dispersion is also visible in the (π, π) direction near $(\pi/2, \pi/2)$. We note that, in agreement with the experimental result (Fig.(4a)), the excitation at $(\pi, 0)$ is at significantly lower binding energy than the excitation at $(\pi/2, \pi/2)$. Neither the 2D $t - J$ model nor a 2D $t - t' - J$ model, with its parameters fitted to the insulating state, can reproduce this result. This is a crucial effect of the stripe assumption: With the stripes oriented along the y-direction, the metallic band is dispersionless in x-direction. Therefore, at $\mathbf{k} = (\pi, 0)$, the minimum of the metallic spinon band (located at $(k_x, 0)$ for any k_x) hybridizes with the top of the insulating valence band resulting in a two-peak structure with one peak pushed to higher and the other pushed to lower binding energies (see below). At $\mathbf{k} = (\pi/2, \pi/2)$, on the other hand, the metallic band has crossed the Fermi surface (its k_F being $\pi/4$) and no mixing takes place. Finally, the "2 + 2" bond-centered stripe configuration (Fig.(4d)) does not show much resemblance to the experimental result. Its main band is much more two-dimensional, normal metal-like, comparable to the dispersion of a 2D tight-binding band.

Fig.(5a) plots the integrated spectral weight $n(\mathbf{k})$ for the "3 + 1" site-centered stripe configuration with $t' = 0$. Although not as clear as in the Nd-LSCO ARPES experiment (from ref. [17]), the "Fermi surface" is rather one-dimensional in structure. Like in Nd-LSCO the low energy excitations (shown in Fig.(5b), calculated by integrating over a $\Delta\omega = 0.2t$ window below the Fermi-energy for each $\mathbf{k}$-point in the Brillouin zone) are located near the $(\pm\pi, 0)$, $(0, \pm\pi)$ points in momentum space. In Fig.(5b) we notice the 8×8 square lattice of bright points. This is the repeated Brillouin zone of the supercell consisting of $(3 + 1 + 3 + 1) \times N = 8 \times 8$ lattice sites (due to the Néel order in x-direction). Clearly, the low energy excitations in the momentum space of the supercell are near the $(\pm\pi, 0)$ and $(0, \pm\pi)$ points as well. Fig.(5c) shows the integrated spectral weight $n(\mathbf{k})$ of the "2 + 2" $t - J$ stripe calculation. Here, the CPT "Fermi surface" is much more rounded, similar to the quasi-2D Fermi surface known from band calculations. The low energy excitations are located isotropically around the "Fermi surface" as well (Fig.(5d)). The loss of one-dimensionality observed for this bond-centered "2 + 2" stripe configuration is accompanied by a substantial enhancement of

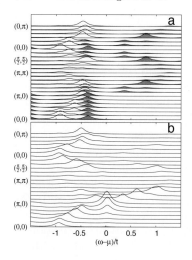

Fig. 6. Single-particle spectral weight prior and after application of the cluster perturbation theory: In (a) the shaded curve gives the spectrum of the 1D metallic chains and the solid line corresponds to the spectrum of the uncoupled AF domains (see Fig.3). In (b) the result of the cluster perturbation is plotted

spectral weight near $(\pi/2, \pi/2)$. From these results for $A(\mathbf{k}, \omega)$ we conclude that, at least in the Nd-LSCO system, the stripes are site-centered and of the type "3 + 1".

With our technique we are able to resolve for each excitation, whether its main origin is from the insulating or the metallic part of the stripe configuration: In Fig.(6) we compare the spectra for the unperturbed stripe configuration with inter-cluster hopping set equal to zero (in Fig.(6a): solid curve for the AF domains, shaded curve for the 1D metal) with the result of our CPT calculation (solid line in Fig.(6b)) with inter-cluster hopping t (for the Néel ordered "3 + 1" configuration). All of the spectral weight in inverse photoemission ($\omega > 0$) naturally stems from the chain since the half-filled 3-leg $t - J$ ladder does not have target states for an inverse photoemission process. The stripes are oriented along the y-direction. Therefore, we can conclude that peaks (in the calculation with $t = 0$), that show a dispersion along $(0, 0)$ to $(\pi, 0)$ direction stem from the AF domains (solid line in (6a)) whereas the metallic excitations prior to the mixing (shaded curve in (6a)) are dispersionless. By comparing the three curves, we therefore conclude that the sharp quasiparticle peak near $(\pi, 0)$ results from the mixing of the (dispersionless) $(k_x, 0)$ minimum of the metallic spinon band and the top of the insulating valence band situated at $(\pi, 0)$. Going from $(\pi, 0)$ to (π, π), the metallic band becomes dispersive and crosses, in agreement with experiment [17,16] the Fermi surface at $k_y = \pi/4$ (since it is quarter-filled). The dispersion of the insulating band, however, is in the opposite direction. For this reason, in the final spectrum, we observe that the sharp quasiparticle peak at $(\pi, 0)$ becomes dispersive going into $(0, \pi)$ direction and eventually crosses the Fermi surface, however, with diminishing weight due to the absence of mixing with the insulating band. This effect is best visible in the gray scale plot of Fig.(4b). This finding may serve to clarify questions raised in the experimental ref. [17] concerning the origin of the quasiparticle peak at $(\pi, 0)$.

4 Summary and Conclusions

Summarizing, in the first part of this paper, we have presented numerical results supporting the view that—while short-range "t-J-like" interactions locally bind holes into pairs—it is the long-range Coulomb interaction which induces their delocalization accompanied with substantial enhancement of superconducting pairing correlations.

In the second part, the single-particle spectral function $A(\mathbf{k}, \omega)$ was calculated for both bond- and site-centered stripe configurations employing an application of the cluster-perturbation technique for inhomogeneous systems. Salient features observed in experiments such as the two-peak structure around $(\pi, 0)$ with a sharp excitation close to the Fermi-energy and a broader feature at higher binding energies, the quasi-1D distribution of spectral weight, and low energy excitations located around the $(\pi, 0)$-points in the Brillouin zone are reproduced by the $A(\mathbf{k}, \omega)$ result for the site-centered stripe-configuration. The origin of these features can be naturally explained by hybridization effects of the metallic and insulating bands coming from the different stripe domains.

Acknowledgments

We are grateful for fruitful discussions with W. Kohn, D. J. Scalapino, Z.-X. Shen and X.J. Zhou, as well as support with the DMRG code by R. M. Noack. We acknowledge financial support from the DFG from the project HA 1537/17-1, and for a Heisenberg fellowship (AR 324/3-1). The calculations were carried out at the high-performance computing centers HLRS (Stuttgart) and LRZ (München). SAK was supported in part by NSF grant number DMR98-08685 at UCLA.

References

1. H. Ding, T. Yokoya, J. C. Campuzano, T. Takahashi, M. Randeria, M. R. Norman, T. Mochiku, K. Kadowaki, and J. Giapintzakis, Nature (London) **382**, 51 (1996).
2. A. G. Loeser, Z.-X. Shen, D. S. Dessau, D. S. Marshall, C. H. Park, P. Fournier, and A. Kapitulnik, Science **273**, 325 (1996).
3. C. Castellani, C. Di Castro, and M. Grilli, Phys. Rev. Lett. **75**, 4650 (1995).
4. S. A. Kivelson and V. J. Emery, Synth. Met. **80**, 151 (1996).
5. M. Vojta, Y. Zhang, and S. Sachdev, cond-mat/0008048.
6. J. M. Tranquada, B. J. Sternlieb, J. D. Axe, Y. Nakamura, and S. Uchida, Nature (London) **375**, 561 (1995).

7. V. J. Emery, S. A. Kivelson, and J. M. Tranquada, Proc. Natl. Acad. Sci. USA **96**, 8814 (1999).
8. J. Zaanen and O. Gunnarsson, Phys. Rev. B **40**, 7391 (1989).
9. J. Zaanen, Science **286**, 251 (1999).
10. K. Yamada, C. H. Lee, K. Kurahashi, J. Wada, S. Wakimoto, S. Ueki, H. Kimura, Y. Endoh, S. Hosoya, G. Shirane, R. J. Birgeneau, M. Greven, M. A. Kastner, and Y. J. Kim, Phys. Rev. B **57**, 6165 (1998).
11. M. Matsuda, M. Fujita, K. Yamada, R. J. Birgeneau, M. A. Kastner, H. Hiraka, Y. Endoh, S. Wakimoto, and G. Shirane, Phys. Rev. B **62**, 9148 (2000).
12. H. A. Mook, P. Dai, F. Dogan, and R. D. Hunt, Nature (London) **404**, 729 (2000).
13. P. Bourges, B. Keimer, L. P. Regnault, and Y. Sidis, J. Supercond. **13**, 735 (2000).
14. S. A. Kivelson, E. Fradkin, and V. J. Emery, Nature (London) **393**, 550 (1998).
15. E. Arrigoni, A. P. Harju, W. Hanke, B. Brendel, and S. A. Kivelson, Phys. Rev. B **65**, 134503 (2002).
16. A. Ino, C. Kim, M. Nakamura, T. Yoshida, T. Mizokawa, Z.-X. Shen, A. Fujimori, T. Kakeshita, H. Eisaki, and S. Uchida, Phys. Rev. B **62**, 4137 (2000).
17. X. J. Zhou, P. Bogdanov, S. A. Kellar, T. Noda, H. Eisaki, S. Uchida, Z. Hussain, and Z.-X. Shen, Science **286**, 268 (1999).
18. The accuracy of the pair-field correlations is especially slowly converging with the number of states kept in the DMRG procedure (see also [20]). However, we have checked that the ratio at large distances plotted in Fig. 2 *increases* with increasing m, i. e. with increasing accuracy. One can, thus, expect that in a more accurate calculation (i. e., with larger m) the Coulomb enhancement effect shown in Fig. 2 would be even more dramatic.
19. C. S. Hellberg and E. Manousakis, Phys. Rev. Lett. **83**, 132 (1999) and S. R. White and D. J. Scalapino, Phys. Rev. Lett. **84**, 3021 (2000).
20. S. R. White and D. J. Scalapino, Phys. Rev. Lett. **80**, 1272 (1998).
21. S.-W. Cheong, G. Aeppli, T. E. Mason, H. Mook, S. M. Hayden, P. C. Canfield, Z. Fisk, K. N. Clausen, and J. L. Martinez, Phys. Rev. Lett. **67**, 1791 (1991).
22. M. G. Zacher, R. Eder, E. Arrigoni, and W. Hanke, Phys. Rev. Lett. **85**, 2585 (2000).
23. J. Orenstein and A. J. Millis, Science **288**, 468 (2000).
24. V. J. Emery, S. A. Kivelson, and O. Zachar, Phys. Rev. B **56**, 6120 (1997).
25. C. Gros and R. Valenti, Phys. Rev. B **48**, 418 (1993).
26. D. Senechal, D. Perez, and M. Pioro-Ladriere, Phys. Rev. Lett. **84**, 522 (2000).
27. T. Tohyama, S. Nagai, Y. Shibata, and S. Maekawa, Phys. Rev. Lett. **82**, 4910 (1999).
28. Y. A. Krotov, D.-H. Lee, and A. V. Balatsky, Phys. Rev. B **56**, 8367 (1997).
29. M. Fleck, A. I. Lichtenstein, E. Pavarini, and A. M. Oles, Phys. Rev. Lett. **84**, 4962 (2000).
30. H. Eskes, R. Eder, Phys. Rev. B **47**, 8929 (1996).

Theory of Superconductivity Due to the Exchange of Spin Fluctuations in Hole- and Electron-Doped Cuprate Superconductors: d-Wave Order Parameter

Dirk Manske and Karl H. Bennemann

Institut für Theoretische Physik, Fachbereich Physik, Freie Universität Berlin, Arnimallee 14, 14195 Berlin

Abstract. Using as a model the Hubbard Hamiltonian we determine various basic properties of hole- and electron-doped cuprate superconductors for a spin-fluctuation-induced pairing mechanism. We treat the corresponding pairing mechanism self-consistently within the framework of the FLuctuation EXchange (FLEX) approximation and study some extensions. Solving the generalized Eliashberg equations for hole- and electron-doped superconductors we obtain both phase diagrams, respectively, and always a d-wave gap function. For hole-doped cuprates we find three characteristic temperature scales which are in qualitatively agreement with the experimental situation. Furthermore, we find that the superconducting transition temperatures $T_c(x)$ for various electron doping concentrations x are calculated to be much smaller than for hole-doped cuprates due to the different energy dispersion and a flat band well below the Fermi level for electron-doped superconductors. Finally, we show how our theory may also explain the neutron scattering data.

1 Introduction

Even 15 years after the discovery by Bednorz and Müller, [1] the nature of the pairing mechanism is still under debate. While hole-doped superconductors have been studied intensively the physics of electron-doped cuprates remained mainly unclear. Our aim here is to show that hole-doped and electron-doped cuprate superconductors can be explained within a unified physical picture, using for example the exchange of antiferromagnetic spin fluctuations as the relevant pairing mechanism. In general, if Cooper-pairing is controlled by antiferromagnetic spin fluctuation, one expects d-wave symmetry pairing also for electron-doped cuprates. Previous experiments did not support this and reported mainly s-wave pairing.[2,3] Maybe as a result of this, so far electron-doped cuprates received much less attention than hole-doped cuprates. However, recently phase sensitive experiments[4] and magnetic penetration depth measurements[5,6] exhibited d-wave symmetry Cooper pairing also for electron-doped cuprates.

In Fig. 1 the general phase diagram of hole-doped cuprates is shown. High-T_c superconductivity occurs in the vicinity of an antiferromagnetic phase transition. Note, the critical temperature T_c for hole-doped cuprates is of the order of 100K and thus much larger than in conventional strong-coupling superconductors like lead ($T_c = 7.2$K) or niobium ($T_c = 9.25$K). This suggests a purely electronic or magnetic mechanism in contrast to the conventional picture of electrons paired through the exchange of phonons.

In this contribution we discuss the underlying physics and calculate the phase diagram of both hole- and electron-doped cuprates within an electronic theory. The first fundamental problem which one has to solve is the theoretical determination of the superconducting transition temperature T_c itself. At around $x = 0.15$ one finds the highest T_c values. This region is called optimal doping. In the overdoped region, i.e. $x > 0.15$, many experimental data suggest that the system is a conventional Fermi liquid. In contrast, on the underdoped side of the phase diagram it is believed that below a mean-field transition temperature T_c^* one find (pre-formed) Cooper-pairs without long-range phase coherence (shaded region in Fig. 1). This part of the phase diagram is sometimes called the 'strong pseudogap' region. Below T_c these

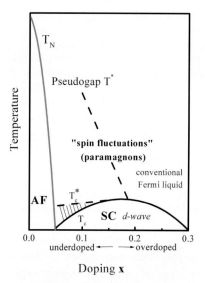

Fig. 1. Schematic phase diagram of hole-doped cuprates. High T_c superconductivity occurs in the vicinity of an antiferromagnetic phase transition. Thus, as we will discuss below Cooper-pairing can mainly be described by the exchange of AF spin fluctuations. In the overdoped region, i.e. $x > 0.15$, the systems behaves like a conventional Fermi liquid, whereas in the underdoped regime below the pseudogap temperature T^* one finds strong antiferromagnetic (AF) correlations. The doping region between T_c and T_c^* (shaded region) is due to local Cooper-pair formation. Below T_c these pairs become phase coherent. The corresponding superconducting order parameter is of $d_{x^2-y^2}$-wave symmetry

pairs become phase coherent and a Meißner effect of the bulk material is observed.

Furthermore, many researchers believe that in order to find the origin of the high-T_c superconductivity in hole-doped cuprates it is necessary to investigate their normal state as a function of the doping concentration. Therefore, phenomenological models like the Marginal-Fermi-liquid (MFL),[9] the Nested-Fermi-liquid (NFL),[10,11] and the Nearly Antiferromagnetic Fermi Liquid theory (NAFL)[12] have been developed in order to understand the unusual non-Fermi liquid properties in the normal-state. At the moment it is not clear whether or not these concepts can also be applied to electron-doped superconductors. We will restrict ourself to the 2D Hubbard Hamiltonian as a minimum model and present an electronic theory which is able to explain both phase diagrams for hole- and electron-doped cuprates as well as the $d_{x^2-y^2}$-wave order parameter which is observed experimentally in both cases.

2 Theory

In order to obtain a unified theory for both hole-doped and electron-doped cuprates we use the same one-band Hubbard Hamiltonian taking into account the different dispersions for the carriers.[27] In the case of electron doping the electrons occupy copper d-like states of the upper Hubbard band while the holes refer to oxygen-like p-states yielding different energy dispersion as used in our calculations. Thus, assuming similar itinerancy of the electrons and holes the mapping on an effective one-band model seems to be justified. We consider U as an effective Coulomb interaction.

We treat the cuprates by using as a model the 2D Hubbard Hamiltonian H which reads in second quantization on a square lattice

$$H = - \sum_{<ij>\sigma} t_{ij} \left(c^+_{i\sigma} c_{j\sigma} + c^+_{j\sigma} c_{i\sigma} \right) + U \sum_i n_{i\uparrow} n_{i\downarrow}$$
$$- \mu t \sum_{i\sigma} n_{i\sigma}, \qquad (1)$$

where $c^+_{i\sigma}$ ($c_{i\sigma}$) creates (annihilates) an electron on site i with spin σ and t_{ij} is a hopping matrix element. The sum performed over nearest neighbors is denoted by $<ij>$. Then, t_{ij} is equal to t. U is the intra-orbital (i.e., on-site) Coulomb repulsion and $n_{i\sigma}$ is equal to $c^+_{i\sigma} c_{i\sigma}$. μ denotes the chemical potential. Therefore, this model can be characterized by two dimensionless parameters, namely U/t and μ.

Using Bloch-wavefunctions we rewrite Eq. (1) as

$$H = \sum_{\mathbf{k}\sigma} \epsilon_{\mathbf{k}} c^+_{\mathbf{k}\sigma} c_{\mathbf{k}\sigma}$$
$$+ \frac{1}{2} \frac{U}{N} \sum_{\mathbf{k},\mathbf{k}',\mathbf{q},\sigma} c^\dagger_{\mathbf{k}\sigma} c^\dagger_{\mathbf{k}',-\sigma} c_{\mathbf{k}'+\mathbf{q},-\sigma} c_{\mathbf{k}-\mathbf{q},\sigma}, \qquad (2)$$

where the one-band electron dispersion in the normal-state ϵ_k reads for nearest neighbor

$$\epsilon_{\mathbf{k}} = -2t\left[\cos k_x - \cos k_y + \mu/2\right] \tag{3}$$

and for next-nearest neighbor hopping

$$\epsilon_{\mathbf{k}} = -2t\left[\cos k_x - \cos k_y - 2B\cos k_x \cos k_y + \mu/2\right], \tag{4}$$

respectively. Here, N is the number of lattice sites.

In Fig. 2 we show energy dispersions for optimally doped $La_{2-x}Sr_xCuO_4$ (LSCO) and $Ne_{1.85}Ce_{0.15}CuO_4$ (NCCO), respectively. Using $t = 250$meV, Eq. (3) describes the Fermi surface of LSCO, whereas $t = 138$meV and $B = 0.3$ in Eq. (4) corresponds to NCCO. One immediately sees the important difference: in the case of NCCO the flat band is approximately 300 meV *below* the Fermi level, whereas for the hole-doped case the flat band lies very close to it. Thus, as will be discussed later, using the resulting ϵ_k in a spin-fluctuation-induced pairing theory we get a smaller T_c for electron-doped cuprates than for the hole-doped ones.

As we have seen earlier, the nearness to a spin-density-wave instability corresponds to a simple physical picture, in which the spins have short-range antiferromagnetic order surviving from the long-range order of the insulating phase. In terms of the Hubbard Hamiltonian we will show that the exchange of longitudinal and transverse spin fluctuations gives rise to an effective electron-electron interaction (originally introduced by Berk and Schrieffer[30]) that provides a pairing interaction leading to d-wave superconductivity near half filling.

From H and the functional differentiation of the free energy F with respect to the Green's function $\mathcal{G}$, $\delta F\{H\}/\delta\mathcal{G} = \Sigma$, one obtains with the help

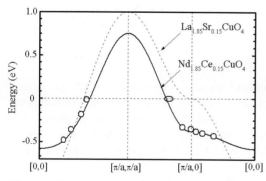

Fig. 2. Results of the energy dispersion ϵ_k of optimally hole-doped $La_{1.85}Sr_{0.15}CuO_4$ (LSCO, dashed curve) and of optimally electron-doped $Ne_{1.85}Ce_{0.15}CuO_4$ (NCCO). The dashed curve corresponds to using $t = 250$ meV and $t' = 0$ and is typical for hole-doped cuprates. The solid curve refers to our tight-binding calculation choosing $t = 138$ meV and $t' = 0.3$. Data (open dots) are taken from Ref. [27]

of the Dyson equation the self-energies. Note, this gives in general coupled equations for the amplitude and phase $\Phi(\mathbf{r})$ of $\mathcal{G}$ [31]. The effective interaction then reads

$$V_{eff}(\mathbf{q}, i\nu_m) = \frac{3}{2} U^2 \frac{\chi_0(\mathbf{q}, i\nu_m)}{1 - U\chi_0(\mathbf{q}, i\nu_m)}$$
$$- \frac{1}{2} U^2 \frac{\chi_0(\mathbf{q}, i\nu_m)}{1 + U\chi_0(\mathbf{q}, i\nu_m)} \quad . \tag{5}$$

In order to solve the generalized Eliashberg equations we will use a self-consistent theory called FLuctuation-EXchange (FLEX) approximation [32,36,37,38]. Remember that the Hubbard Hamiltonian can be rewritten in the form $H = H_0 + H_{int}$ where H_0 describes the one-particle properties and H_{int} denotes a perturbation. After analytical continuation to the real ω-axis, the quasiparticle self-energy components X_ν ($\nu = 0, 3, 1$) with respect to the Pauli matrices τ_ν in the Nambu representation are given by

$$X_\nu(\mathbf{k}, \omega) =$$
$$N^{-1} \sum_{\mathbf{k}'} \int_0^\infty d\Omega \left[P_s(\mathbf{k} - \mathbf{k}', \Omega) \pm P_c(\mathbf{k} - \mathbf{k}', \Omega) \right]$$
$$\times \int_{-\infty}^\infty d\omega'\, I(\omega, \Omega, \omega') A_\nu(\mathbf{k}', \omega'). \tag{6}$$

The spin fluctuation interaction is given by $P_s = (2\pi)^{-1} U^2 \operatorname{Im} (3\chi_s - \chi_{s0})$ with $\chi_s = \chi_{s0} (1 - U\chi_{s0})^{-1}$ and the charge fluctuation interaction is $P_c = (2\pi)^{-1} U^2 \operatorname{Im} (3\chi_c - \chi_{c0})$ with $\chi_c = \chi_{c0} (1 + U\chi_{c0})^{-1}$, where $\operatorname{Im} \chi_{s0,c0}(\mathbf{q}, \omega)$ is given in Ref. [36]. The subtracted terms in P_s and P_c remove a double counting that occurs in second order. In Eq. (6) the plus sign holds for $X_0 = Z$ (quasiparticle renormalization) and $X_3 = \xi$ (energy shift), and the minus sign for $X_1 = \phi$ (gap parameter). The kernel I and the spectral functions A_ν are given in Ref. [36].

For an illustration we show in Fig. 3 how we solve Eq. (6): One starts with a dynamical spin susceptibility $\chi(\mathbf{q}, \omega)$ and constructs the effective pairing interaction using Eq. (5). Then, the strong-coupling gap equation for the superconducting order parameter $\phi(\mathbf{k}, \omega)$ and the corresponding Dyson equation $G^{-1}(\mathbf{k}, \omega) = G_0^{-1}(\mathbf{k}, \omega) - \Sigma(\mathbf{k}, \omega)$ have to be solves, respectively. The full momentum- and frequency dependence of the quantities is kept. Having solved these two equations one has new appropriate starting input values for an electron propagator G which is again used to calculate χ. This procedure is repeated until all equations are solved.

In order to determine the superconducting transition temperature T_c we solve the linearized gap equation. Below T_c we find that the the superconducting gap function has $d_{x^2-y^2}$-wave symmetry. Vertex corrections for the two-particle correlation function (which are not included) have been discussed elsewhere.[41] The doping dependence $n = \frac{1}{N} \sum_{\mathbf{k}} n_{\mathbf{k}} = 1 - x$ is determined

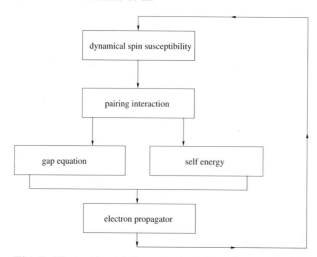

Fig. 3. Illustration of the procedure in order to solve Eq. (6). The full momentum and frequency dependence of the quantities is kept. Thus, our calculation includes pair breaking effects for the Cooper-pairs resulting from lifetime effects of the elementary excitations

with the help of the **k**-dependent occupation number that is calculated self-consistently. $n = 1$ corresponds to half filling.

In order to calculate the phase diagram for hole-doped superconductors we must also calculate the superfluid density $n_s(x,T)/m$ self-consistently from the current-current correlation function and from f-sum rule: the real part of the conductivity $\sigma_1(\omega)$, i.e., $\int_0^\infty \sigma_1(\omega)\,d\omega = \pi e^2 n/2m$ where n is the 3D electron density and m denotes the effective band mass for the tight-binding band considered. $\sigma(\omega)$ is calculated in the normal and superconducting state using the Kubo formula.[43] Vertex corrections have been neglected. Physically speaking, we are looking for the loss of spectral weight of the Drude peak at $\omega = 0$ that corresponds to excited quasiparticles above the superconducting condensate for temperatures $T < T_c^*$. Most importantly, using our results for $n_s(x,T)$, we calculate the doping dependence of the Ginzburg-Landau like free-energy change $\Delta F \equiv F_S - F_N$, where $\Delta F_{cond} \simeq \alpha(n_s/m)\Delta_0(x)$ is the condensation energy due to Cooper-pairing and $\Delta F_{phase} \simeq \hbar^2/2m^* n_s$ the loss in energy due to phase incoherence of the Cooper-pairs. α describes the available phase space for Cooper-pairs (normalized per unit volume) and can be estimated in the strongly overdoped regime. In the BCS-limit one finds $\alpha \simeq 1/400$. Δ_0 is the superconducting order parameter at $T = 0$.

3 Results and Discussion

In Fig. 4 (a) we show the calculated phase diagram for hole-doped superconductors. For a comparison, the solid curve, T_c^{exp}, which describes many classes

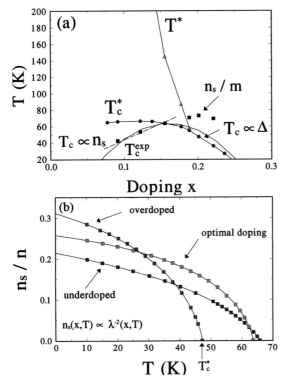

Fig. 4. (a) Phase diagram for hole-doped high-T_c superconductors resulting from a spin fluctuation induced Cooper-pairing including their phase fluctuations. Below T^* (triangles) we get a gap structure in the spectral density. T_c^* denotes the temperature below which Cooper-pairs are formed (circles). Below T_c these pairs become phase coherent. The calculated values for $n_s(T=0)/m$ (squares) are in good agreement with muon-spin rotation experiments (dashed line). (b) Calculated superfluid density n_s/n for three different doping values ($x = 0.12$ (underdoped), $x = 0.15$ (optimally doped), $x = 0.18$ (overdoped)). Note, within London theory $n_s \propto \lambda^{-2}$ which yields good agreement with experimental data on YBCO

of cuprates (taken from Loram and co-workers[44]) is also displayed. Calculating the Ginzburg-Landau free energy, we get for the overdoped cuprates, i.e. $x > 0.15$, mainly BCS-type behavior and consequently $T_c \simeq T_c^* \propto \Delta(T=0)$. In contrast to this, for underdoped superconductors the superfluid density n_s is the relevant energy. In order to demonstrate this we show in Fig. 4 (b) our results for n_s/n versus temperature for three different doping concentrations. We find finite values of n_s/n for temperatures $T < T_c^*$ and a slope close to T_c^* which is in good agreement with experiment.[] Most importantly, $n_s(T=0)/n$ is around $0.25 \ll 1$ which means that only 25 percent of the (dressed) holes become superconducting and is expected for a strongly correlated system. Because also the coherence length of a Cooper-pair is very small, we have no overlap between neighboring Cooper-pair wavefunctions and thus phase fluctuations of Cooper-pairs become important in underdoped cuprates. Consequently, we also calculated the change in the Ginzburg-Landau free energy via n_s/m and then T_c. As a result we find $T_c \propto n_s(T=0)$. Our results for the doping dependence of n_s/m (squares), which are in good agreement with experimental results, are also displayed in Fig. (4). We would like to emphasize that for the underdoped cuprates $T_c \propto n_s$ yields indeed better agreement with experimental results than T_c^* obtained from $\Delta(x,T)=0$ and marking the onset of Cooper-pairing within our mean-field theory. For temperatures $T_c < T < T_c^*$ one finds pre-formed Cooper-pairs. Hence, our electronic theory yields in fair agreement with experiment the non-monotonic doping dependence of $T_c(x)$. Note, we find similar results for the doping dependence of T_c from determining T_c using $n_s(x,T) = 0$. Here, one must include the coupling between Cooper-pairs and their phase fluctuations causing the reduction of $T_c^* \to T_c$ for the underdoped cuprates and $T_c \propto n_s$. Also in Fig. 4 (a) results are given for the characteristic temperature T^* at which a gap appears in the spectral density. Within our FLEX-theory the occurrence of a pseudogap is due to inelastic electron-electron scattering which leads to a loss of spectral weight at the Fermi level which are in qualitatively agreement with experiments. Finally, we have also calculated T_c for the underdoped cuprates calculating $n_s(T)$ and using Kosterlitz-Thouless theory[45] and have found similar T_c values.

We have also calculated the phase diagram $T_c(x)$ and $T_N(x)$ of electron-doped cuprates which is shown in Fig. 5. In order to obtain a unified theory for both hole-doped and electron-doped cuprates it is tempting to use the same Hubbard Hamiltonian taking of course into account the different dispersions for the carriers. Note, in the case of electron doping the electrons occupy copper d-like states of the upper Hubbard band while the holes refer

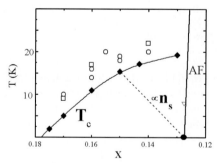

Fig. 5. Phase diagram $T(x)$ for electron-doped cuprates. The AF transition curve is taken from Ref. [46]. The solid curve corresponds to our calculated T_c values obtained from Eq. (6) and $\phi(\mathbf{k},\omega) = 0$. This is in good agreement with experimental data (squares from Ref. [47], circles from Ref. [48], triangle from [49]). The dotted curve refers to $T_s \propto n_s$

to oxygen-like p-states yielding different energy dispersion as used in our calculations. We consider U as an effective Coulomb interaction. Then, assuming similar itinerancy of the electrons and holes the mapping on an effective one-band model is justified.

We find in comparison to hole-doped superconductors smaller T_c values and that superconductivity occurs in a narrower doping range as also observed in experiment. Responsible for this are poorer nesting properties of the Fermi surface and the flat band around $(\pi, 0)$ which lies well below the Fermi level. The narrow doping range for T_c is due to antiferromagnetism up to $x = 0.13$ and rapidly decreasing nesting properties for increasing x.[50] In order to understand the behavior of $T_c(x)$ in underdoped electron-doped cuprates we have calculated the Cooper-pair coherence length ξ_0, i.e. the size of a Cooper-pair, and find similar values for electron-doped and hole-doped superconductors (from 6 Å to 9 Å). If the superfluid density n_s/n becomes small (for example due to strong coupling lifetime effects), the distance d between Cooper pairs increases. If for $0.15 > x > 0.13$ the Cooper-pairs do not overlap significantly, i.e. $d/\xi_0 > 1$, then Cooper-pair phase fluctuations get important. Thus we expect like for hole-doped superconductors $T_c \propto n_s$.

Below T_c we find for all doping concentrations that the gap function has clearly $d_{x^2-y^2}$-wave symmetry which is shown in Fig. 6. This is in agreement with the reported linear and quadratic temperature dependence of the in-plane magnetic penetration depth for low temperatures in the clean and dirty limit, respectively, and with phase-sensitive measurements.[4]

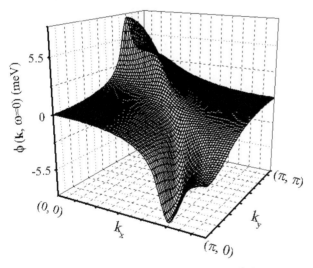

Fig. 6. Calculated $d_{x^2-y^2}$-wave symmetry of the superconducting order parameter $\phi = Z\Delta$ for the electron-doped superconductor NCCO at T/T$_c$=0.8 for x=0.15 in the first square of the BZ

Finally, we discuss the consequences of our electronic theory for inelastic neutron scattering (INS) experiments. In general, if the pairing potential is mainly generated by the spin fluctuations, one expects a strong renormalization of their spectrum below T_c. This is mainly due to opening of the superconducting gap which we calculated self-consistently. For example, for optimally-doped $YBa_2Cu_3O_7$ (YBCO) one finds a resonance peak in INS experiments at 41 meV [51]. Recently, we have shown that the experimentally observed resonance peak and its doping dependence in hole-doped cuprates can be understood as a collective mode which becomes resonant only in the superconducting state[52].

In order to compare the resonance peak in hole-and electron-doped cuprates we show in Fig. 7 the calculated imaginary part of the dynamical spin susceptibility at $\mathbf{Q} = \mathbf{Q}_{AF} = (\pi, \pi)$. One clearly sees that in the case of electron-doped cuprates the rearrangement of the spectral weight is much less than in the case of hole-doped cuprates. This is due to the smaller spectral weight in Im $\chi(\mathbf{Q}, \omega)$ in the normal state at low frequencies and due to a smaller value of the superconducting gap. On the other hand for hole-doped superconductors the characteristic spin fluctuation frequency ω_{sf} (roughly the peak position) is of the order of $2\Delta_0$. Thus one has a strong change in the spectral weight below T_c which is observed as the resonance peak. Therefore, we do not expect a resonance peak in electron-doped NCCO although spin fluctuations provide the pairing interaction also in this material. Further experimental studies should check this.

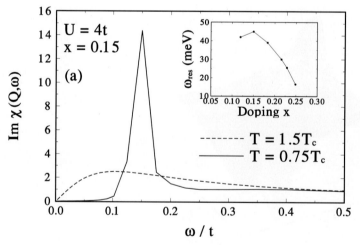

Fig. 7. (a) Calculated resonance peak (solid curve) in Im $\chi(\mathbf{q} = \mathbf{Q} = (\pi, \pi), \omega)$ at $\omega_{res} = 0.15t \approx 41\text{meV}$ below T_c for hole-doped superconductors. The inset shows ω_{res} as a function of the doping concentration x. For a comparison the normal-state result for Im $\chi(\mathbf{Q}, \omega)$ is also displayed. (b) Calculated Im $\chi(\mathbf{Q}, \omega)$ for electron-doped superconductors for the same parameters used in (a). No resonance peak is found

4 Summary

In summary, we have used as a model the Hubbard-Hamiltonian and the self-consistent FLEX-theory to calculate some basic properties of the hole-doped and electron-doped cuprate superconductors. For the hole-doped case we have discussed the superfluid density n_s/m, and the critical temperature T_c as a function of the doping concentration. We found a phase diagram with two different regions: on the overdoped side we obtain a mean-field-like transition and $T_c \propto \Delta(T=0)$, whereas in the underdoped regime we find $T_c \propto n_s(T=0)$. For temperatures $T_c < T < T_c^*$ we find a finite superfluid density but no Meissner effect. This region may be attributed to pre-formed Cooper-pairs without long-range phase coherence. Above T_c^* we find a third energy scale, namely T^* where below we find a gap in the spectral density of states ('pseudogap').

Our unified model for cuprate superconductivity yields for electron-doped cuprates like for hole-doped ones pure $d_{x^2-y^2}$ symmetry pairing in a good agreement with recent experiments.[4] In contrast to hole-doped superconductors we find for electron-doped cuprates smaller T_c values due to a flat dispersion ϵ_k around $(\pi,0)$ well below the Fermi level. Furthermore, superconductivity only occurs for a narrow doping range $0.18 > x > 0.13$, because of the onset of antiferromagnetism and, on the other side, due to poorer nesting conditions. We get $2\Delta/k_B T_c = 5.3$ for $x = 0.15$ for the electron-doped cuprates whereas we obtain much larger values, namely $2\Delta/k_B T_c = 10$ - 12 for the hole-doped ones. Finally, we have discussed the consequences our theory for inelastic neutron scattering experiments. For both hole- and electron-doped superconductors we find a strong feed-back effect of superconductivity on Im $\chi(\mathbf{Q},\omega)$. Furthermore, we only find a resonance peak for the hole-doped case below T_c due to a spin-density wave collective mode. Due to the fact that no resonance feature occurs in our calculations for electron-doped cuprates we conclude that the existence of a resonance peak is no proof for spin-fluctuation-induced pairing. In particular, the overall agreement with experiments on hole- and electron-doped high-T_c superconductors is remarkably good and suggests spin-fluctuation exchange as the dominant pairing mechanism for superconductivity.

Acknowledgments

It's a pleasure to thank I. Eremin, C. Joas, T. Dahm, J. Schmalian, K. Scharnberg, and L. Tewordt for helpful discussions.

References

1. J.G. Bednorz and K.A. Müller, Z. Phys. B **64**, 189 (1986).
2. B. Stadlober et al., Phys. Rev. Lett. **74**, 4911 (1995).

3. S. M. Anlage *et al.*, Phys. Rev. B **50**, 523 (1994).
4. C. C. Tsuei and J. R. Kirtly, Phys. Rev. Lett. **85**, 182 (2000).
5. J. D. Kokales *et al.*, Phys. Rev. Lett. **85** 3696, (2000).
6. R. Prozorov, R. W. Gianetta, P. Furnier, and R. L. Greene, Phys. Rev. Lett. **85** 3700, (2000).
7. J. Bardeen, L. N. Copper, and J. R. Schrieffer, Phys. Rev. B **108**, 1175 (1957).
8. J. Ruvalds *et al.*, Phys. Rev. B **51**, 3797 (1995).
9. C. M. Varma *et al.*, Phys. Rev. Lett. **63**, 1996 (1989).
10. A. Virosztek and J. Ruvalds, Phys. Rev. B **42**, 4064 (1990).
11. J. Ruvalds, C. T. Rieck, J. Zhang, and A. Virosztek, Science **256**, 1664 (1992).
12. A. J. Millis, H. Monien, and D. Pines, Phys. Rev. B **42**, 167 (1990).
13. J. L. Tallon, Phys. Rev. B **51**, 12911 (1995).
14. For a review, see C. P. Slichter in, *Strongly Correlated Electronic Systems.* (Addison Wesley, Reading(MA) 1994).
15. L. P. Regnault *et al.*, Physica B **213-214**, 48 (1995).
16. D. S. Marshall *et al.*, Phys. Rev. Lett. **76**, 4841 (1996).
17. J.M . Harris *et al.*, Phys. Rev. B **54**, R15655 (1996).
18. J. W. Loram *et al.*, Phys. Rev. Lett. **71**, 1740 (1993); J. W. Loram *et al.*, J. Superconductivity **7**, 243 (1994).
19. B. Batlogg *et al.*, Physica C **235-240**, 130 (1994).
20. R. Nemetschek *et al.*, Phys. Rev. Lett. **78**, 4837 (1997); G. Blumberg *et al.*, Science **278**, 1427 (1997).
21. Ch. Renner *et al.*, Phys. Rev. Lett. **80**, 149 (1998).
22. A. K. Gupta and K.-W. Ng, Phys. Rev. B **58**, R8901 (1998).
23. P. C. Hohenberg, Phys. Rev. **158**, 383 (1967).
24. D. Manske and K. H. Bennemann, Physica C **341-348**, 83 (2000).
25. Q. Si, Y. Zha, K. Levin, and J. P. Lu, Phys. Rev. B **47**, 9055 (1993).
26. J. C. Campuzano *et al.*, Phys. Rev. Lett. **64**, 2308 (1990).
27. D. M. King *et al.*, Phys. Rev. Lett. **70**, 3159 (1993).
28. G. M. Eliashberg, Sov.-Phys.-JETP **11**, 696 (1960).
29. P. B. Allen and B. Mitrovic, Solid State Physics **37**, 1 (1982).
30. N. F. Berk and J. R. Schrieffer, Phys. Rev. Lett. **17**, 433 (1966).
31. F. Schäfer, C. Timm, D. Manske, and K. H. Bennemann, J. Low Temp. Phys. **117**, 223 (1999).
32. N. E. Bickers, D. J. Scalapino, and S. R. White, Phys. Rev. Lett. **62** 961 (1989);
33. N. E. Bickers and D. J. Scalapino, Ann. Phys. (N.Y.) **193**, 206 (1989);
34. P. Monthoux and D. J. Scalapino, Phys. Rev. Lett. **72**, 1874 (1994);
35. C.-H. Pao and N.E. Bickers, Phys. Rev. Lett. **72**, 1870 (1994).
36. T. Dahm and L. Tewordt, Phys. Rev. Lett. **74**, 793 (1995); Phys. Rev. B **52**, 1297 (1995).
37. M. Langer, J. Schmalian, S. Grabowski, and K. H. Bennemann, Phys. Rev. Lett. **75**, 4508 (1995).
38. D. J. Scalapino, Phys. Rep. **250**, 329 (1995).
39. A. Abrikosov, L. Gorkov, *Quantum Field Theoretical Methods in Statistical Physics.* (Pergamon Press, Oxford 1965).
40. L. Tewordt, D. Fay, P. Dörre, and D. Einzel, J. Low Temp. Phys. **21**, 645 (1975).
41. D. Manske, *Phonons, Electronic Correlations, and Self-Energy Effects in High-T_c Superconductors.* (PhD thesis, Hamburg, Germany 1997).

42. Z.-X. Shen and D.S. Dessau, Phys. Rep. **253**, 1 (1995).
43. S. Wermbter and L.Tewordt, Physica C **211**, 132 (1993).
44. M. R. Presland *et al.*, *Physica C* **176**, 95 (1991); J. R. Cooper and J. W. Loram, *J. Phys. I* **6**, 1 (1996).
45. J. M. Kosterlitz and D. J. Thouless, J. Phys. C **6**, 1181 (1973); J. M. Kosterlitz, J. Phys. C **7**, 1046 (1974).
46. G. Baumgärtel, J. Schmalian, and K.H. Bennemann, Phys. Rev. B **48**, 3983 (1993).
47. E. F. Paulus *et al.*, Solid State Comm. **73**, 791 (1990).
48. H. Takagi, S. Uchida, and Y. Tokura, Phys. Rev. Lett. **62**, 1197 (1989).
49. G. Liang *et al.*, Phys. Rev. B **40**, 2646 (1989).
50. D. Manske, I. Eremin, and K. H. Bennemann, Phys. Rev. B **62**, 13922 (2000).
51. for a review, see for example L. P. Regnault, Ph. Bourges, and P. Burlet, *Neutron Scattering in Layered Copper-Oxide Superconductors*. (Kluwer Academic Publishers, Dordrecht 1998).
52. D. Manske, I. Eremin, and K. H. Bennemann, Phys. Rev. **B**, Feb. 1 issue (2001).

Part VI

Disordered Systems and Soft Matter

Anomalous Behavior of Insulating Glasses at Ultra-low Temperatures

Christian Enss

Kirchhoff-Institut für Physik, Universität Heidelberg,
Albert-Ueberle-Str. 3-5, 69120 Heidelberg, Germany

Abstract. The low-temperature properties of amorphous solids differ considerably from those of their crystalline counterparts. These differences are caused by the presence of atomic tunneling systems in the irregular structure of glasses. Although the microscopic nature of these tunneling states has not been identified, the so-called tunneling model provides a satisfying description of many properties of amorphous solids below 1 K on a phenomenological basis. For certain properties below about 100 mK, however, serious discrepancies between the predictions of the tunneling model and the experimental results exist. It appears that these discrepancies become more and more pronounced with decreasing temperature. In addition new intriguing phenomena occur that are unexpected on the basis of the tunneling model. These findings might be taken as an indication that elastic and electric interaction between the tunneling systems become important at low temperatures, because in the original formulation of the tunneling model such a mutual interaction was not taken into account. We will discuss a few examples of the shortcoming of the tunneling model.

1 Introduction

Atomic tunneling states give an important contribution to the internal energy of glasses at low temperatures. These degrees of freedom influence drastically many properties of amorphous solids. Prominent examples are the linear specific heat and logarithmic temperature dependence of the sound velocity [1,2]. A phenomenological description of the low-temperature properties of glasses is provided by the so-called tunneling model, that has been proposed independently by Anderson et al. [3] and Phillips [4] in 1972. A central assumption of this model is that single atoms or groups of atoms are not located in well-defined potential minima, but move between two energetically almost equivalent adjacent positions separated by a potential barrier. Such a configuration can be described by a particle moving in a double-well potential. In this approximation the single wells are considered as harmonic and identical, but may are different in depth usually referred to as asymmetry energy Δ. The particle will have vibrational states in each single well separated by an energy $\hbar\Omega$.

At low temperatures, i.e. for $k_\mathrm{B}T \ll \hbar\Omega$, higher vibrational levels are not excited. Therefore only the behavior of the ground state is of interest which is split by tunneling into two levels. Therefore, the tunneling systems are often referred to as *two-level systems*. In addition to the classical potential difference Δ, the quantum mechanical tunnel splitting Δ_0 contributes to the ground state splitting E which is given by

$$E = \sqrt{\Delta^2 + \Delta_0^2} \,. \tag{1}$$

Applying the WKB method the tunnel splitting can be calculated. Neglecting pre-factors of the order of unity one finds

$$\Delta_0 \approx \hbar\Omega e^{-\lambda} \,. \tag{2}$$

Roughly speaking the tunnel splitting Δ_0 is given by the vibrational energy $\hbar\Omega$ of the particle multiplied by the probability $\exp(-\lambda)$ for tunneling. The so-called tunneling parameter $\lambda = d\sqrt{2mV}/2\hbar$ reflects the overlap of the wave functions of the particle of the two sides of the potential barrier.

As a consequence of the irregular structure of glasses, the characteristic parameters of the tunneling states are widely distributed. The tunneling model assumes that the asymmetry energy Δ and the tunnel parameter λ are independent of each other and uniformly distributed as

$$P(\lambda, \Delta)\,\mathrm{d}\lambda\,\mathrm{d}\Delta = \overline{P}\,\mathrm{d}\lambda\,\mathrm{d}\Delta \,, \tag{3}$$

where $\overline{P}$ is a constant. At this point we want to mention, that an equivalent description of the low-temperature properties of glasses is given by the so-called soft potential model [5] which uses a more general form of the atomic potentials. It allows reliable predictions also well above 1 K. In comparison with the tunneling model this leads to a slightly different distribution of the tunneling parameter, but for the properties below 1 K it yields the same predictions as the tunneling model.

Tunneling systems couple to their environment by interaction with phonons and photons. External elastic or electric fields produce changes of the asymmetry energy Δ and lead in turn to relaxation processes. The dominant relaxation mechanism in insulating glasses at temperatures below 1 K is the so-called one phonon process with the rate [6]

$$\tau_1^{-1} = A\left(\frac{\Delta_0}{E}\right)^2 E^3 \coth\left(\frac{E}{2k_\mathrm{B}T}\right) \,, \tag{4}$$

where A is a constant containing mainly the deformation potential of the tunneling systems and the sound velocity of the longitudinal and transversal phonon branches. In glasses the relaxation rates are widely distributed because of the random structure. Even tunneling systems with fixed energy splitting E exhibit a wide range of relaxation rates depending on the ratio Δ_0/E. Note that from experiments it can be concluded that the relaxation rates span at least 15 orders of magnitude. The relaxation time of the fastest relaxing systems with thermal energy is given by $\tau_{\min.} = \tau_1(\Delta_0 = E = k_B T)$.

The coupling to external field also causes resonant processes like resonant absorption and stimulated emission. Together resonant and relaxation processes determine the dynamic response of the two-level systems.

The tunneling model successfully describes many low-temperature properties of glasses [1,2]. However, in the last few years there has been a growing body of evidence that at very low temperatures ($T < 0.1$ K) systematic deviations exist, that cannot be accounted for by simple modifications of the distribution function (3). Therefore the origin of the discrepancies must be something else. An implicit assumption of the tunneling model is that interactions between tunneling systems via their elastic and electric moments are regarded as unimportant. It seems now that this approximation does not hold at temperatures, at which the thermal energy becomes comparable or even smaller than the mean interaction energy of the tunneling systems. Therefore, discrepancies between the experimental observations and the tunneling model are commonly interpreted as a sign that the interaction between tunneling systems is relevant. There is, however, to date no generally accepted model, which describes the properties of glasses on the basis of interacting tunneling system. In this article we will introduce some experimental findings that cannot be explained by the tunneling model and we will discuss their possible origin.

2 Internal Friction

As a first example we consider the results of acoustic experiments on vitreous silica, which provide a clear indication for the shortcoming of the tunneling model at low temperatures. We will restrict our discussion to the internal friction at low frequencies $\omega \ll E/\hbar$. In this case only contributions from relaxation processes have to be taken into account. For single tunneling systems the internal friction at low frequencies should be given by

$$Q_{\rm rel}^{-1} = \frac{\gamma^2}{\varrho v^2}\frac{1}{k_B T}\left(\frac{\Delta}{E}\right)^2 {\rm sech}^2\left(\frac{E}{2k_B T}\right)\frac{\omega\tau_1}{1+(\omega\tau_1)^2} \, , \tag{5}$$

where ϱ represents the mass density and v the sound velocity of the glass. The quantity γ denotes the deformation potential of the tunneling system. To obtain the internal friction as predicted by the tunneling model this expression has to be integrated taking into account the distribution function (3). Although this can be done in general only numerically, analytic expressions can be derived in certain limiting cases. For $\omega\tau_{\min} \gg 1$ one finds $Q^{-1} \propto T^3/\omega$. This means that the internal friction should rise at low temperatures proportional to T^3. At high temperatures ($\omega\tau_{\min} \ll 1$) a temperature and frequency independent internal friction with the magnitude $Q^{-1} = \pi \overline{P} \gamma^2/(2\varrho v^2)$ is expected. This so-called plateau reflects the wide distribution of the parameters λ and Δ.

The high temperature limit has been confirmed in many experiments [1,2]. In contrast, the experimentally observed behavior at low temperatures often differs from the expected T^3 dependence. In general, a weaker temperature dependence is found. As an example, we show in Fig. 1 the internal friction of vitreous silica as a function of temperature for different frequencies. The experiments were carried out with so-called double-paddle oscillators [7,8] of two different sizes cut from a 0.4 mm thick plate of vitreous silica. These mechanical oscillators can be operated in a variety of torsional and bending modes, i.e., at different frequencies and elastic polarizations. A particular advantage of such oscillator is that for many modes the strain at the mounting and in turn the clamping loss are very low. This is important in respect to the experiments here, because any sizable residual absorption may mask the intrinsic loss of the sample at very low temperatures.

To emphasize the deviations a low temperatures we have plotted in Fig. 1 the measured internal friction divided by Q_{STM}^{-1}, the value of the internal friction calculated with the standard tunneling model. Clearly, the discrepancies rise with decreasing temperature following a general trend nearly independent of frequency. The fact, that the measured internal friction is always higher than the theoretically expected value indicates, that at very low temperatures, besides the one-phonon process, additional damping processes become important.

The interaction between tunneling systems may provide such an additional relaxation channel, because it allows to transfer energy within the

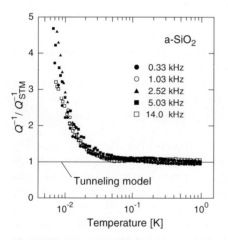

Fig. 1. Temperature dependence of the internal friction of vitreous silica at different frequencies divided by the expected value of the internal friction according to the tunneling model [9]

system of tunneling states. A particular mechanism for such an interaction driven relaxation has been proposed by Burin and Dagan, who considered pairwise interaction between tunneling systems with similar tunnel splittings [10,11]. This mechanism leads to an additional relaxation contribution with the rate [11]

$$\tau_{\text{BK}}^{-1} \simeq \frac{10\, k_B\, C_a^3}{\hbar} \left(\frac{\Delta_0}{E}\right)^2 T \,. \tag{6}$$

Here the so-called macroscopic coupling constant $C_a = \overline{P}\gamma^2/(\varrho v^2)$ enters. Since is numerical value is known there is no free parameter left. Although an additional relaxation contribution of the form $\tau^{-1} \propto (\Delta_0/E)^2 T$ describes in principle the data fairly well, it should be pointed out that the absolute value is several orders of magnitude too small to provide a satisfactory description of the data shown in Fig. 1. However, in a recent publication Burin and coworkers have pointed out that the presence of an alternating field may enhance the relaxation rate (6) considerably [12].

An alternative approach [13] to take into account the interaction between tunneling systems was suggested by adopting a theory worked out by Würger for the tunneling of substitutional defects in alkali halide crystals [14,15]. In this model the interaction leads to an incoherence of the tunneling motion at very low temperatures. As a result, an additional relaxation contribution occurs, and at the same time resonant processes are partially suppressed. Although the basic idea of this approach seems to point in the right direction, the relevance of incoherent tunneling in glasses remains to be seen. To date no developed theory of incoherent tunneling in glasses has been worked out because of conceptional difficulties.

3 Dielectric Constant

Evidence for the importance of interaction between tunneling systems in glasses is also found in dielectric experiments. According to the tunneling model, resonant processes should lead to a logarithmic decrease of the dielectric constant as $\delta\varepsilon/\varepsilon|_{\text{res}} = -C_d \ln(T/T_0)$, where T_0 represents a reference temperature and $C_d = 2\overline{P}p^2/(3\varepsilon_0\varepsilon)$ the macroscopic coupling constant for dielectric experiments. Here p denotes the dipole moment of the tunneling system. Relaxation processes also lead to a logarithmic temperature dependence but of opposite sign $\delta\varepsilon/\varepsilon|_{\text{rel}} = \frac{3}{2}C_d \ln(T/T_0)$. At low temperatures only resonant processes influence the dielectric constant and therefore one expects $\delta\varepsilon/\varepsilon = -C_d \ln(T/T_0)$. With increasing temperature relaxation processes gain importance and eventually become dominant, when $\omega\tau_{\min}$ exceeds unity. In this regime the sum of resonant and relaxation processes determines the dielectric constant and therefore we expect $\delta\varepsilon/\varepsilon = \frac{1}{2}C_d \ln(T/T_0)$. In general, a logarithmic decrease, a minimum and a subsequent logarithmic increase is observed experimentally, but the ratio of the slopes above and below the minimum is roughly $(-1):1$ and not $(-2):1$ as expected from the tunneling model [16,17].

Another remarkable effect at audio frequencies and very low temperatures has been discovered in Stanford [18,19,20]. The dielectric constant jumps rapidly upward after the sudden application of a DC electrical field and begins to relax back with a rate logarithmic in time elapsed after the application of the field. Sweeping the DC field slowly it was found that the dielectric constant exhibits a minimum at the field at which the sample was cooled down from high temperatures. However, if the sample is kept in a different field for several minutes, a new local minimum of ε developed, the depth of which is growing logarithmically with time. This phenomenon has been interpreted as a so-called dipole gap originating from the dipole-dipole interaction between the tunneling systems [21].

A new story in the investigations of the low-temperature properties of glasses opened with the discovery of strong magnetic field effects. Unexpected observations were made in measurements of the dielectric properties of the multi-component glass a-BaO-Al$_2$O$_3$-SiO$_2$ in low-frequency dielectric experiments at ultra-low temperatures. Fig. 2 shows the changes of the dielectric constant at 1.85 mK, caused by small variations of the applied magnetic field.

Experiments at higher temperatures showed that the sensitivity of the dielectric susceptibility to magnetic fields decreases strongly with increasing temperature. In addition, measurements at higher fields showed that the dependence on the magnetic field is non-monotonic [23] and that the magnetic field effect itself depends on the amplitude of the electrical ac-field used to measure the dielectric susceptibility [24,25]. As an example we have plotted in Fig. 3 the magnetic field dependence of the dielectric constant of the

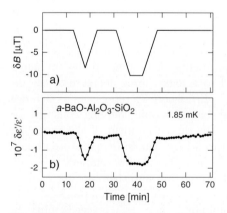

Fig. 2. Influence of small magnetic fields on the dielectric constant of a-BaO-Al$_2$O$_3$-SiO$_2$ [22]. (a) Time variation $\delta B(t)$ of the applied magnetic field. (b) Relative change of the dielectric constant with the variation of the applied magnetic field at 1.85 mK

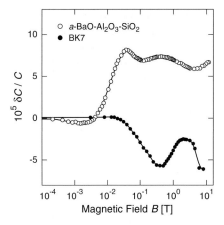

Fig. 3. Magnetic field dependence of the dielectric constant of BK7 (at 15.4 mK) [24] and of a-BaO-Al$_2$O$_3$-SiO$_2$ (at 37 mK) [25]. Both experiments have been performed at 1 kHz and at a field strength of about 10 kV/m. The lines are guides for the eyes

borosilicate glass BK7 in comparison with a-BaO-Al$_2$O$_3$-SiO$_2$. Clearly, the dielectric constant of BK7 is influenced by magnetic fields in a similar way as in case of a-BaO-Al$_2$O$_3$-SiO$_2$. However, interesting differences exist in the course of the magnetic field dependence of the two samples.

Since the temperature dependence of the dielectric constant of glasses is mainly caused by the contribution of atomic tunneling, we might ask, whether a coupling of tunneling states to magnetic fields exists. In several theoretical papers such a coupling mechanism has been suggested [26,27,28,29]. Kettemann, Fulde and Strehlow have considered tunneling systems with a three dimensional path of tunneling [26]. In case the tunneling path is a closed loop, such systems can be described by two currents of opposite direction, which cancel each other exactly in zero magnetic field. If a magnetic flux treads through this loop the symmetry is broken and one expects a periodic variation of the tunnel splitting as a function of the magnetic flux through the loop. With such an approach the non-monotonic magnetic field dependence of the dielectric constant can be explained qualitatively. However, to account quantitatively for the experimentally observed oscillations in the dielectric constant, fields of the order of 10^5 T are necessary assuming tunneling systems with typical atomic dimensions. Therefore, it has been suggested that an amplification of this effect occurs due to a coherent coupling of many tunneling systems. In this case the effective loop might actually be of mesoscopic scale and much lower fields would be necessary to cause a oscillatory behavior. In this context it is interesting to note that indications for collective tunneling at ultra-low temperatures in a-BaO-Al$_2$O$_3$-SiO$_2$ have indeed been reported [22]. In this experiment evidence for a continuous phase transition at 5.84 mK in a-BaO-Al$_2$O$_3$-SiO$_2$ has been observed.

Correlations on long scales are not necessary in the model proposed very recently by Würger [29]. He has investigated the properties of pairwise coupled tunneling systems with almost identical energy splitting. It is conceivable that the tunneling path of two coupled tunneling systems with repulsive (or attractive) interaction gives rise to an effective loop. The tunnel splitting of the pairs would therefore also experience a periodic change with the magnetic flux within the loop. Since the almost degenerate levels of such pairs are much more sensitive to magnetic fields compared to single tunneling systems, the features in the dielectric constants may already appear at fields below 1 T in agreement with the experimental observations.

4 Polarization Echoes

At sufficiently low temperatures the energy relaxation time τ_1 and the phase memory time τ_2 of the tunneling systems become long enough to allow the observation of coherent phenomena, such as polarization echoes. Measurements of the decay of such echoes can be used to deduce τ_1 and τ_2. In addition, echo experiments allow to determine the distribution of the product $p\Delta_0$ at fixed energy. In the following we will concentrate on measurements of so-called spontaneous echoes, which can be generated by applying two microwave excitation pulses. The first pulse, adjusted in amplitude and duration t to fulfill the condition $\Omega_\mathrm{R} t = \pi/2$ leads to a macroscopic polarization which vanishes after a short time because of the distribution of energy splittings of the tunneling states. The quantity Ω_R represents the Rabi frequency which is

$$\Omega_\mathrm{R} = \frac{1}{\hbar} \frac{\Delta_0}{E} \boldsymbol{p} \cdot \boldsymbol{F} , \tag{7}$$

where $\boldsymbol{p}$ stands for the dipole moment of the tunneling systems and $\boldsymbol{F}$ for the amplitude of the driving field. After a short delay time $t = t_{12}$, a second pulse fulfilling the condition $\Omega_\mathrm{R} t = \pi$ is applied, leading to a time reversal in the development of the polarization. Hence a macroscopic polarization, the so-called spontaneous echo, appears at time $2t_{12}$.

4.1 Decay of Spontaneous Echoes

The amplitude of the echo signal is proportional to the number of those tunneling systems which are in resonance with the applied field at the beginning of the experiment and do not lose their phase coherence during the time $2t_{12}$.

In principle, several different mechanism may destroy the phase coherence of the resonant tunneling systems: any energy relaxation process, like the one-phonon process, or processes that involve the interaction between tunneling systems. In the latter case one distinguishes between two different mechanism, namely spin diffusion, that is caused by the interaction among resonant tunneling systems, and spectral diffusion, which results from the interaction of a resonant tunneling system with thermally fluctuating systems with energies of the order of $k_B T$. It is generally expected that spectral diffusion processes dominate the echo decay in glasses at low temperatures. It should be mentioned that the first indications for the relevance of spectral diffusion processes in glasses has been obtained in acoustic hole-burning experiments [30], in which a surprisingly broad hole was observed that is much wider than expected from lifetime broadening or the pulse spectrum.

Theories of spectral diffusion in glasses have been worked out by several groups [31,32,33,34]. Depending on the relative magnitude of the time interval t_{12} and the minimum relaxation time $\tau_{\min}$, different regimes for the decay of the two-pulse echo can be distinguished: The *short time limit* is defined by $t_{12} \ll \tau_{\min}$. The decay is expected to be non-exponential, varying with t_{12} like $A(2t_{12}) \propto \exp[-(2t_{21}/\tau_2)^2]$, where τ_2 denotes the decay time. In the *long time regime*, $t_{12} \gg \tau_{\min}$, the echo decay should vary as $A(2t_{12}) \propto \exp(-2t_{21}/\tau_2)$.

In Fig. 4 the decay of the amplitude of spontaneous echoes in a-BaO-Al_2O_3-SiO_2 is shown as a function of the pulse separation time t_{12} at 6 mK. Clearly the decay pattern is neither exponential nor gaussian. The decay is much faster than expected for spectral diffusion processes involving thermally fluctuating tunneling systems undergoing one-phonon processes. From this we conclude that the decay of spontaneous echoes in this multi-component glass cannot be understood on the basis of spectral diffusion processes involving one-phonon relaxation. We should point out that similar discrepancies are found in many glasses independent of their chemical composition [36,37,38,39]. These observations are unexplained to date. One might speculate that the rapid decay of the echoes is related to the enhanced internal friction at low temperatures (Fig. 1). This could be the case for example if a relaxation mechanism is active in addition at low temperatures besides

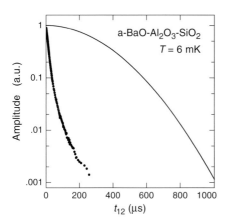

Fig. 4. Decay of the amplitude of the spontaneous echo in a-BaO-Al_2O_3-SiO_2 as a function of the pulse separation t_{12} at 1 GHz and at 6 mK. The solid line represents the theoretical prediction for spectral diffusion

the one-phonon relaxation. As mentioned before, such a mechanism has been suggested by Burin and coworkers involving resonant pairs. However, the predicted decay pattern for two-pulse echo experiments and the temperature dependence of τ_2 are only partially in agreement with the experimental observations [40].

4.2 Magnetic Field Dependence

As a great surprise strong magnetic field effects where recently discovered in polarization echo experiments [41]. As shown in Fig. 5 the amplitude of spontaneous echoes generated in the multi-component glasses a-BaO-Al$_2$O$_3$-SiO$_2$ and BK7 depends strongly on the applied magnetic fields. At small fields the amplitude of the echoes decreases with increasing magnetic field, passes a minimum and a subsequent maximum and finally rises towards higher fields much above its zero field value. One can draw several conclusions from this observation. The fact that the echo amplitude increases in case of a-BaO-Al$_2$O$_3$-SiO$_2$ by more than a factor of three in magnetic fields indicates that the observed phenomenon is not a property of a small sub-group of tunneling systems. The second conclusion we may draw from this experiment is that magnetic impurities are not important for the magnetic field effects in glasses, because in echo experiments only tunneling states contribute to the signal. Therefore any change of the echo signal with magnetic fields indicates that the tunneling states themselves are involved in the magnetic field effects. This clearly supports the theories, that propose a direct coupling of tunneling states to magnetic fields. As stated above, the amplitude of the echo signal is proportional to the number of those tunneling systems, which are in resonance with the applied field at the beginning of the experiment and do not lose their

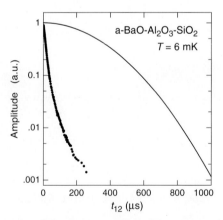

Fig. 5. Magnetic field dependence of the amplitude of spontaneous polarization echoes at 12 mK in a-BaO-Al$_2$O$_3$-SiO$_2$ ($\omega/2\pi = 1.06$ GHz, $F_{el} \approx 1.1$ kV/m) [41] and BK7 ($\omega/2\pi = 0.86$ GHz, $F_{el} \approx 1.8$ kV/m) [42]

phase coherence during the time $2t_{12}$, and is proportional to the magnitude of their dipole moment. Obviously the magnetic field *improves* the coherence in this experiment. Until now the origin of this unexpected phenomenon is totally unclear.

Another, also very surprising aspect of the data shown in Fig. 5 is the non-monotonic dependence of the echo amplitude on the magnetic field seen in both glasses. This reminds of the non-monotonic field dependence in the low-frequency dielectric susceptibility measurements (Fig. 3).

5 Summary

Certain properties of glasses at temperatures below 100 mK cannot be understood in terms of the tunneling model in its present form. As examples we have discussed the excess internal friction, the magnetic field dependence of the dielectric constant, the decay of spontaneous echoes and the magnetic field dependence of the amplitude of such echoes. Some aspects of these findings suggest that interactions between the tunneling systems may be the cause for the anomalous behavior of glasses at very low temperatures. However, the present understanding of these phenomena is incomplete and new experiments and further theoretical work is needed to obtain a satisfactory description.

Acknowledgments

The author would like to acknowledge numerous fruitful discussions with J. Classen, S. Hunklinger, R. Kühn, S. Ludwig, M.v. Schickfus, P. Strehlow, R. Weis and A. Würger. This work has been supported by the DFG under contract Hu359/11.

References

1. Tunneling Systems in Amorphous and Crystalline Solids, ed. P. Esqunazi (Springer, Berlin 1998).
2. S. Hunklinger, and C. Enss, in: Insulating and Semiconducting Glasses, ed. P. Boolchand, Series of Directions in Condensed Matter Physics **17**, 499 (World Scientific 2000).
3. P.W. Anderson, B.I. Halperin, and C.M. Varma, Philos. Mag. **25**, 1 (1972).
4. W.A. Phillips, J. Low Temp. Phys. **7**, 351 (1972).
5. V.G. Karpov, M.I. Klinger, and F.N. Ignat'ev, Sov. Phys. JETP **57**, 439 (1983).
6. J. Jäckle, Z. f. Phys. **257**, 212 (1972)
7. R.N. Kleiman, C.K. Kaminsky, J.D. Reppy, R. Pindak, D.J. Bishop, Rev. Sci. Instrum. **56**, 2088 (1985).
8. C. Spiel, R.O. Pohl, and A.T. Zehnder, Rev. Sci. Instrum. **71**, 1482 (2001).
9. J. Classen, T. Burkert, C. Enss, and S. Hunklinger, Phys. Rev. Lett. **84**, 2176 (2000).

10. A.L. Burin, and Y. Kagan, JETP Lett. **79**, 347 (1994).
11. A.L. Burin, J. Low Temp. Phys. **100**, 309 (1995).
12. A.L. Burin, Y. Kagan, I.Y. Polishuk, Phys. Rev. Lett. **86**, 5616 (2001).
13. C. Enss, and S. Hunklinger, Phys. Rev. Lett. **79**, 2831 (1997).
14. A. Würger, Z. Phys. B **94**, 173 (1994); **98**, 561 (1995).
15. A. Würger, *Springer Tracts in Modern Physics* Vol. 135, (Springer 1997).
16. C. Enss, C. Bechinger, M.v. Schickfus in: Phonons 89 (S. Hunklinger, W. Ludwig, G. Weiss eds.), 474 (World Scientific 1990).
17. J. Classen, C. Enss, C. Bechinger, G. Weiss, and S. Hunklinger, Ann. Phys. **3**, 315 (1994).
18. D.J. Salvino, S. Rogge, B. Tigner, and D.D. Osheroff, Phys. Rev. Lett. **73**, 268 (1994).
19. S. Rogge, D. Natelson, and D.D. Osheroff, Phys. Rev. Lett. **76**, 3136 (1996).
20. D.D. Osheroff, S. Rogge, D. Natelson, Czech. J. Phys. **46**, 3295 (1996).
21. A.L. Burin, D. Natelson, D.D. Osheroff, Y. Kagan,in: Tunneling Systems in Amorphous and Crystalline Solids, ed. P. Esqunazi (Springer, Berlin 1998), p. 223.
22. P. Strehlow, C. Enss, and S. Hunklinger: Phys. Rev. Lett. **80**, 5361 (1998).
23. M. Wohlfahrt, A.G.M. Jansen, R. Haueisen, G. Weiss, C. Enss, and S. Hunklinger, Phys. Rev. Lett. **84**, 1938 (2000).
24. M. Wohlfahrt, C. Enss, S. Hunklinger, and P. Strehlow, Europhys. Lett. **56**, 690 (2001).
25. R. Haueisen, G. Weiss, appears in Physica B.
26. S. Kettemann, P. Fulde, P. Strehlow, Phys. Rev. Lett. **8**, 4325 (1999).
27. P. Fulde, and A. Ovchinnikov, Euro. Phys. J. B**17**, 623 (2000).
28. K.H. Ahn, and P. Fulde, Phys. Rev. B **62**, R4813 (2000).
29. A. Würger, Phys. Rev. Lett **88**, 075502 (2002).
30. W. Arnold, and S. Hunklinger, Solid State Commun. **17**, 833 (1975).
31. J.L. Black and B.I. Halperin, Phys. Rev. B **16**, 2879 (1977).
32. P. Hu and L.R. Walker, Phys. Rev. B **18** 1300 (1978).
33. R. Maynard, R. Rammal, and R. Suchail, J. Phys. Lett. **41**, L291 (1980).
34. B.D. Laikhtman, Phys. Rev. B **31**, 490 (1985).
35. C. Enss, C. Bechinger, and M.v. Schickfus, *Phonons 89*, edited by S. Hunklinger, W. Ludwig, and G. Weiss, 474 (World Scientific 1989).
36. J.E. Graebner and B. Golding, Phys. Rev. B **19**, 964 (1979).
37. L. Bernard, L. Piché, G. Schumacher, and J. Joffrin, J. Low Temp. Phys. **45** (1979) 411.
38. G. Baier and M.v. Schickfus, Phys. Rev. B **38**, 9952 (1988).
39. C. Enss, R. Weis, S. Ludwig, and S. Hunklinger, Czech. J. Phys. **46**, 3287 (1996).
40. A.L. Burin, Y. Kagan, L.A. Maksimov, I.Y. Polishuk, Phys. Rev. Lett. **80**, 2945 (1998); erratum **81**, 2395 (1998).
41. S. Ludwig, C. Enss, P. Strehlow, and S. Hunklinger Phys. Rev. Lett **88**, 075501 (2002).
42. S. Ludwig, C. Enss, and S. Hunklinger, to be published.

Colloidal Suspensions – The Classical Model System of Soft Condensed Matter Physics

Hans-Hennig von Grünberg

Universität Konstanz, 78457 Konstanz, Germany

Abstract. The colloidal suspensions presented here consist of highly-charged mesoscopic particles dispersed in a fluid solvent. They are surrounded by mobile microscopic ions whose distribution in space governs their mutual (effective) interaction. In order to calculate this interaction, one has to integrate out the microionic degrees of freedom, thus arriving at effective pair- and higher-body potentials. Only in some limiting cases, it is sufficient to describe the free energy of the suspension by sums over pair-potentials. Often also many-body forces must be taken into account. This article – rather listing the open questions than presenting the closed chapters of this active field of research – discusses under which condition such many-body effects show up, how they are measured and calculated. All this is essential to gain an understanding of the sometimes complex phase-behavior that charge-stabilized colloidal suspensions can show and that in parts is still unexplained.

1 Introduction – Why it is Worth Studying Colloids

The term "colloid" is used to characterize nothing but a size regime; it is used for particles that are small enough to pass through a membrane with a pore size of a few μm, but which, on the other hand, are substantially larger than the atomic scale. More generally speaking, a colloidal system contains particles of one type of material in a continuous matrix of another type: they can be solid in liquid (paint), liquid in liquid (milk) and even gas in liquid (aerosol). The large class of self-aggregating systems is commonly also regarded as colloidal matter. The system, considered in this article, is thus a rather specific one: it is a colloidal suspension, consisting of solid mesoscopic particles finely dispersed in a liquid solvent, most frequently water.

Because colloids are mesoscopic in size, colloidal suspensions have unusual thermodynamic, rheological and optical properties bridging the gap between the molecular and macroscopic world. Colloids are not visible to the naked eyes, but can easily be observed with modern microscopes. Video-microscopy, for instance, allows the direct observation of a colloidal system: positions of all particles of the suspension are recognized and followed as a function of time. The system can thus be regarded as the realization of a real-time simulation where nature is the computer, generating configuration by configuration. While the systematic exploration of colloidal systems started more than a century ago with investigations of highly diluted suspensions, the interest today is focused rather on highly concentrated suspensions, showing a phase-behavior that is sometimes very complicated. It is found that colloidal systems show analogues of all the states of atomic systems: gas, liquid, crystal, alloy, and glass [1]. Colloidal crystals are grown from readily available dispersions of silica or latex spheres, see Fig. 1, colloidal glasses are also easily produced and much easier to control than glasses made in atomic systems. Colloids can thus be used for detailed studies of fundamental processes of broad relevance such as crystallization and glass formation.

Examples of colloidal systems abound in our everyday life. They occur as dispersion paints, viscosity modifiers, flocculants, superabsorbers, glues, polishes, lubricants, foods and pharmaceuticals, to name but a few technological applications. Colloids play an important role also in biology and life sciences. A great deal of experimental and theoretical research has been devoted to their understanding, and several good textbooks and review articles [2] are available. Two studies, rather randomly chosen from the wealth of available examples, are mentioned in the following to illustrate how useful colloid science can be. Recently, it has been realized that the interstices in a colloidal crystal can be filled with another material and macroporous materials can be made if the templating spheres are afterwards removed without destroying the crystalline order; the filler material then stays behind as a regular 3D air-sphere structure. The 3D regularity of these materials is of great importance, for instance, for photonic applications, but also for processes such as catalysis [3].

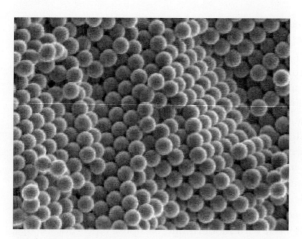

Fig. 1. A scanning electron microscope image of a colloidal crystal grown from a dispersion of silica spheres having a sphere diameter of 300 nm (taken from [3]. With permission.)

That the fundamental physics of model colloids provides a basis on which to build an understanding of more complex systems, is nicely demonstrated by my second example, which stresses at the same time how relevant soft condensed matter can be to biological problems (for other examples in this context see [4]): Protein crystals are usually formed by adding salt or polymers to induce crystallization in a protein solution. This situation can be mimicked by a colloidal suspension in which the interaction between the colloids is varied by adding inert polymers with a radius of gyration much smaller than the colloid sphere radius. It has been found that colloidal crystals form in suspensions that are much more dilute than those without added polymer. The reason is that the polymers add an extra attraction (depletion attraction, to be discussed further below) to the colloid-colloid interaction. Crystallization occurs when this attraction is sufficiently large. In a phase-diagram the strength of attraction is best be quantified by the second virial coefficient B_2. Hence, plotting B_2 over the colloid concentration, one obtains a well-resolved phase-boundary between a single-phase colloid suspension and a crystal-suspension coexistence phase [4]. Surprisingly, over two dozen globular proteins show exactly the same phase-boundary when plotted in this way, thus demonstrating that there is an universal unspecific mechanism responsible for the additive-induced protein crystallization. We see that this mechanism can be studied and understood also in such simple systems as colloid-polymer mixtures which are much easier to produce, to control, to observe and, in addition, much cheaper to produce than the protein solutions.

2 Effective Forces – What Small Particles can do to Large Particles

Colloidal suspensions are complex many-component systems. By definition they consist of at least two components: the solvent molecules of the matrix and the colloidal particles. Usually many other components are present as well (small ions, additional molecules, dust, e.t.c.). Clearly, the interaction between the large colloids is heavily influenced by the presence of the small molecules. The huge number of these small molecules enforces a contracted theoretical description in which the degrees of freedom of the microscopic particles are integrated out, leaving us just with the coordinates of the large colloidal particles and with "effective" forces in between them. By introducing an effective Hamiltonian the initial multi-component system is thus reduced to an effective one-component system. That the inter-colloidal forces are effective forces is at the same time one of the great advantages of soft colloid matter systems over atomic systems: the interaction potential between colloidal particles is not a fixed quantity, but can be manipulated by the experimentalist from the outside world, can be adjusted, tuned or "tailored", for example, by adding salt or small polymers. The tunability of colloidal interaction lets us investigate experimentally how the behavior of an assembly of particles depends on their interactions, something which is absolutely impossible in atomic systems where one is simply stuck with the interaction dictated by electronic structure of the atoms.

The depletion interaction [5] is perhaps the most important non-specific effective interaction, as it can in principle be induced by any solute and solvent. It is based on a simple geometric effect: every solute is surrounded by a region that is inaccessible to the centers of the solvent; if the excluded volumes of neighboring solutes overlap, the total volume available for the solvent increases and the free energy of the system is accordingly lowered. Depletion interactions are almost omnipresent in nature, and can also be used to manipulate biological systems in a controlled way. For example, high-concentrations of polyethylene oxide are used to promote cell-cell aggregation [6]. In colloid-polymer mixture (large hard-sphere like colloids, small polymers of almost spherical shape), the polymers induce an effective attraction between the colloids which increases on increasing the polymer concentration, and can heavily affect the crystallization behavior of the system. An example is the crystallization of protein solutions, mentioned above. The effect of the depletion forces can be nicely seen in Fig. 2, showing a binary hard-sphere system consisting of large and small colloids near a flat glass plate [7]. The volume fraction $\phi_2 = 4\pi a^3 c_2/3$ of the small spheres (radius a, concentration c_2) is increased from the left to the right picture in Fig. 2. If $\phi_2 = 0$, there are no attractive depletion forces, colloids remain in view for a very short time before diffusing back into the bulk. If $\phi_2 = 0.08$ in picture b), the area fraction of colloids increases, since the particles are pressed against the wall through

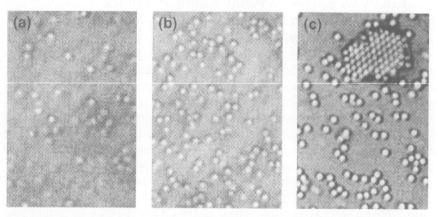

Fig. 2. Optical micrographs of colloids (polystyrene spheres, sphere radius 400 nm) at a flat glass wall. Not visible are additional tiny spheres (radius 35 nm) whose volume-fraction is in a) $\phi_2 = 0$, in b) $\phi_2 = 0.08$ and in c) $\phi_2 = 0.16$. Colloids in c) start crystallizing due to attractive depletion forces. (Taken from [7]. With permission.)

the depletion mechanism. For even larger ϕ_2, lateral depletion forces become strong enough that the 2D colloidal system at the wall starts to crystallize: picture c) shows crystallites that have formed after 40h.

Another important type of effective force in colloidal systems is the electric double-layer force which in the following we want to concentrate on. If a surface covered by some ionizable surface groups is immersed in a polar solvent such as water, the surface releases ions into the solution and becomes charged. These ions are mobile ions, as opposed to the fixed charges remaining at the surface; additional mobile ions are salt ions from the electrolyte solution. The mobile ions experience thermal forces, on the one hand, and the electrical field of both fixed and mobile ions on the other. The Poisson-Boltzmann (PB) equation describes in a mean-field approach how the ions are spatially distributed in response to both forces. An inhomogeneous distribution of mobile ions forms in front of the charged surface, and this layer plus the layer of fixed surface charges constitute the electric double-layer. Every charged surface in solution is surrounded by such a double-layer. Double-layer forces between two charged colloids in solution result from the overlap of the double-layers of the two interacting objects. Being necessarily repulsive, these forces prevent colloids in solution from aggregating and are thus responsible for the stability of a suspension of charged colloids. Charge-stabilized suspensions play a tremendous role in molecular biology, since virtually all proteins in every living cell, as well as the DNA molecule itself, are charged macromolecules dissolved in salty water [8].

3 Many-body Effects – The Whole is More than the Sum of its Parts

As long as only two colloids are involved, the way how to calculate intercolloidal double-layer forces is obvious: fix the macroions at a distance h, calculate the free energy of the whole system, repeat this for varying h, and obtain the effective potential from the free energy as a function of h. This can be done analytically after linearizing the PB equation [9] and leads to the repulsive Yukawa potential which is part of the classical DLVO potential [2], named after Derjaguin, Landau, Vewey and Overbeek. If however a concentrated suspension of charged colloids is considered, this Yukawa interaction potential describes the interaction correctly only in a few limiting cases. Before explaining this is in greater detail let us first take notice of some experimental facts. Fig. 3 shows a 2D colloidal system, where colloids are pressed against a flat glass wall by light forces [10]. To control the particle density the beam of a laser was reflected from a 2D galvanostatic driven mirror and focused into the sample plane. The mirror unit was controlled by a computer in such a way, that the tweezer was repeatedly scanning a rectangular box, resulting in a quasi-static optical trap for the colloidal particles along the contour of the box, see top picture in Fig. 3. The trapped particles then prevent the other particles from leaving the box. By changing the size of the rectangular box, the density of the enclosed particles could be changed in a controlled manner and thus allowed to measure the radial distribution function (RDF) of an identical set of particles at different colloid densities ρ, see lower left plot in Fig. 3. The lower right plot shows the effective pair-potentials obtained by deconvoluting the information contained in the RDFs, using an inversion procedure based, for example, on the Ornstein-Zernicke equation. It is evident from Fig. 3 that the effective colloid-colloid potentials are in fine agreement with the DLVO Yukawa potential (thick dashed line) for the three lower densities, but not for the two higher densities where marked deviations from the Yukawa form is observed at about $2.3\,\sigma$.

This experimental result demonstrates that effective colloid-colloid pair-potentials are density- or 'state'-dependent; this is equivalent to saying that higher-order contributions, such as three-body potentials between the colloids, have to be taken into account which are now partly incorporated in the measured pair-potential. In a charge-stabilized colloidal suspension the many-body character of the interaction clearly results from the fact that the interaction between the colloids is mediated by the screening microions in the interstitial region between the colloids. Important is that the concept of pair-wise additive pair-potentials is applicable only if the mean distance of the colloids in the suspension is much larger than the thickness of the double layer. At higher colloid densities, a third particle surrounded by its double-layer is always close to two interacting colloids and can then severely disturb their mutual interaction. This is actually the explanation for the experimental finding presented in Fig. 3. The fact that strong deviations of the measured potentials take place just at distances comparable to the mean colloid-colloid distance $D = \rho^{-1/2}$, suggests that the interaction of two colloids at a distance $r > D$ is simply blocked by a third macro-ion which on average is located at D and thus somewhere in between the two interacting colloids. This effect is called "macro-ion screening". In a numerical study [11], it has been found that due to macroion-screening the three-body contribution H_3 in the total Hamiltonian $H = H_1 + H_2 + H_3$ of a colloidal system (with H_1 and H_2 being the one- and two-colloid contributions), is always negative. With this in mind, it is clear that when inverting the measured RDFs allowing just for pair-potentials, negative three-body contributions are projected into the pair-potentials, giving thus rise to density dependent pair-potentials that are smaller than the 'real' pair-potentials; for the same reason, even attractive parts in the colloid/colloid pair-potential can be obtained which in the past were erroneously interpreted as "like-charge attraction"[12].

We see that many-body interactions can be directly observed in colloidal systems. Colloidal suspensions are thus an ideal testing ground for concepts of many-body theory, and results obtained here might be useful in a number of related fields as well. An example is the theory of metals where many-body effective forces also play an important role, which however are much more

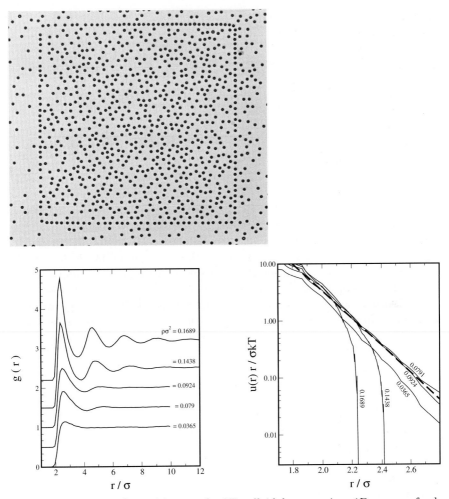

Fig. 3. Top picture: Optical image of a 2D colloidal suspension. 1D arrays of colloids whose positions are held fixed by light forces, are used to realize a 2D colloidal system with a well-defined and controllable colloid density ($300 \times 300\,\mu m^2$). Lower left picture: Radial distribution functions $g(r)$ for the colloidal system inside the rectangular box of the upper picture, for varying box sizes (i.e., colloid densities ρ). Lower right picture: Effective colloid-colloid pair potentials, obtained from inverting the measured $g(r)$. Potentials are multiplied by r and plotted logarithmically (figures at the curves give the reduced density $\rho\sigma^2$ with σ for the colloid sphere diameter). The thick dashed line is a reference Yukawa potential. The data displayed here represent a direct experimental observation of many-body interactions in suspensions of charged colloids

difficult to study in these systems than in colloidal dispersions. It is in particular this many-body aspect dominating the theory of colloidal suspensions that makes the theoretical study of these systems so worthwhile. Reference [13] is a nice example for that, showing how the density dependence of the inter-colloidal potential can lead to a van-der-Waals like instability even in fluids with purely repulsive pair potentials. Fig. 4 gives the results of another experiment where many-body forces seem to be important. We see theoretical and experimental data of structure factors of charge-stabilized colloidal suspensions [14]. Plotted is the peak height of the first peak in $S(k)$ as a function of the colloid volume fraction. This quantity is commonly taken as a measure of the degree of order (or structure) in the system. With decreasing volume fraction starting from very dense suspensions, one first observes a loss of structure, but then a sudden and unexpected re-appearance of structure, with a well-resolved maximum, followed again by a slow decay of structure for extremely diluted suspensions. This complicated behavior, in particular, the observed re-appearance of structure, is due to many-body effects in these suspensions: the poor agreement between two conventional theories and the experimental data in Fig. 4 shows that a description based on pair-wise additive potentials can not capture this interesting feature. This reappearance of structure is in striking contrast to the situation encountered in simple liquids

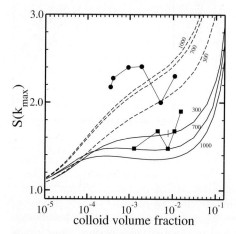

Fig. 4. Structure factors of highly de-ionized suspensions of polystyrene latex particles measured with light scattering [14]. Plotted as symbols is the peak height $S(k_{\mathrm{max}})$ of the first peak in $S(k)$ as a function of volume fraction for two different types of particles (connected by lines as guide to the eye). The curves represent theoretical predictions calculated solving the Ornstein-Zernicke equation with effective pair potentials based on the jellium approximation (solid lines) and on the charge renormalization theory (dashed lines). The curves are labeled with the respective bare colloidal charges Z. Note the non-monotonic behavior of $S(k_{\mathrm{max}})$ which is a fingerprint of many-body effects

where decreasing the packing fraction leads always to a monotonic decrease of inter-particle structure.

4 Phase-Behavior

4.1 Experimental Findings and Current Controversies

One of the major current controversies in the soft colloid matter community centers on the question if there is a fluid-fluid phase separation in charge-stabilized colloidal suspensions, i.e., a separation into one colloid-rich phase ('liquid' phase) and into another colloid-depleted phase ('gas' phase). Why is this an interesting question? In one-component systems phase separation of a homogeneous fluid into a dilute and a dense phase can be explained with intermolecular attractions which can compensate the loss of configurational entropy that a system experiences upon condensation. In other words, in absence of such attractive interactions a fluid-fluid phase separation should be impossible. These ideas have been put forward by van der Waals and represent our classical understanding of gas-liquid phase coexistence. Applying these ideas naively to charge-stabilized colloidal suspensions, one must come to the conclusion that a fluid-fluid phase separation in such suspensions must be ruled out: according to the standard DLVO one-component description of colloidal suspensions, the effective colloid-colloid pair interaction can never be attractive; it is purely repulsive in the salt regime of interest here. Hence, there seems to be no cohesive energy that can stabilize a liquid phase, and a fluid-fluid phase separation in colloidal suspensions should be impossible. And yet it appears to exist, as, for example, the formation of vapor bubbles ('voids') in otherwise homogeneous suspensions seems to suggest. These voids are far from being understood, and the question arises if they possibly result from a fluid-fluid phase separation. What has become clear by now is that two things can have some effect on the stability of colloidal suspensions: (i) density-dependent volume terms in the effective Hamiltonian that include the interaction of double layers with their "own" colloidal particle and (ii) the fact that counterions and macro-ions contribute to screening and that screening thus becomes colloid-density dependent. The effect of the latter has been already discussed in the preceeding section. The first point then is the central element of a theory presented in [16]; it starts from volume terms in the Hamiltonian and finds them to be responsible for large regions of phase instability in the phase diagram, at low volume fractions and ionic strengths, see Fig. 5B. According to this theory, the cohesive energy that stabilizes the dense phases is provided by the Coulomb energy of the relatively compressed double layers in the dense phases at sufficiently low salt concentration. In a recent paper, [17], it has however been shown that the results obtained in [16] are likely to be an artefact that is due to the linearization of the PB equation made in [16]. The question about a possible fluid-fluid phase-coexistence in colloidal suspension is therefore still an open one.

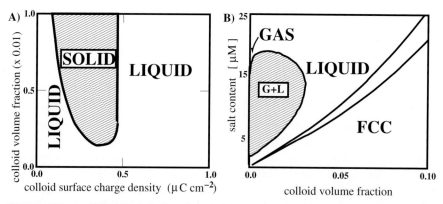

Fig. 5. Phase diagrams of charge-stabilized colloidal suspensions at low ionic strength. A) An experimental phase diagram, reproduced from [15], showing a liquid-solid-liquid re-entrant behavior (salt content: $2\,\mu M$). B) Theoretical phase diagram [16] suggesting a fluid-fluid phase separation in extremely de-ionized suspensions at low volume fraction

Another puzzling observation made in charged colloidal systems at low ionic strength concerns the freezing behavior: Yamanaka et al. [15] found that with increasing colloidal charge and thus with increasing counterionic charge density, the colloidal suspension crystallizes first and, on increasing the density even more, melts again (Fig. 5A). This reentrance behavior is not understandable on the basis of traditional theories on colloidal suspensions which account for the first transition (liquid/solid) but not for the second. Again, one possible conjecture would be that the problem lies in the naive and oversimplifying assumption of pairwise effective interaction potentials. This idea is corroborated by the observation that these results have been found in a parameter regime that is not that much different from the one explored in Fig. 5 where many-body effects lead to the interesting non-monotonic behavior in $S(k_{max})$.

4.2 Pressure and how to Calculate it

Arguably the most fundamental thing to know about colloidal suspensions is their equation of state, i.e., how the (osmotic) pressure depends on other thermodynamic variables like macromolecular charge or concentration. For example, the regions of instability in the phase-diagram of Fig. 5B can be calculated from the equation of state by finding those densities where the derivative of the pressure with respect to the colloid density is negative (negative compressibilities). Fig. 6A shows the pressure of a suspension of charged spherical colloids, as calculated from non-linear and linear PB theory. Within full PB theory the pressure is always positive. Whether or not this also holds in the linearized PB theory depends both on the choice of the linearization

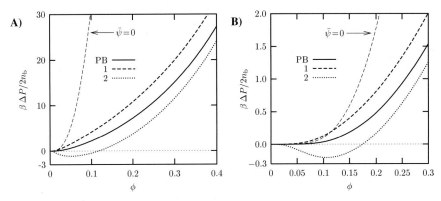

Fig. 6. Pressure as a function of volume fraction ϕ for a suspension of spherical (A) and cylindrical (B) macroions. The solid curves are the result from nonlinear Poisson-Boltzmann (PB) theory, the other curves show analytical results based on the linear PB theory. Compared are three different ways (fine/bold dashed and dotted line) how to incorporate the expansion point $\bar\psi$ for the linearization into the theory, see [18]. The parameters in the calculation in (B) correspond roughly to DNA in an electrolyte of physiological salt concentration, those in (A) to the colloidal system studied in Fig. 5B

point $\bar\psi$ as well as on the precise definition of the pressure itself. It is proven in [18] that the pressure is always positive for symmetric electrolytes if one treats $\bar\psi$ as an independent thermodynamic variable. If one does not, the pressure can become negative at low volume fractions. The implied liquid-gas coexistence (Fig. 5B) – not being present on the nonlinear level – is thus clearly an artifact. As a striking example it has been shown [18] that even a solution of DNA molecules under physiological conditions, see Fig. 6B, would be predicted to phase separate at all relevant densities. If this were true, all DNA in animal cells would tend to aggregate and phase separate!

To conclude: It has not been the purpose of this article to present a final and internally consistent picture of colloidal suspensions, but more to introduce a few of many open questions currently discussed in our community. This list was also intended to prove that colloid science is a rather active field of research. One should not take this for granted considering that theories, for example, on colloidal interaction, date back to the year 1941 (DLVO theory). It is exciting to see that after all the many years that the DLVO concept has ruled over theory and experiment of colloidal interaction – and this means basically all the thermodynamic and dynamic properties of colloidal systems – new experimental observation techniques have made it possible to explore a concentration regime where this celebrated theory ceases to provide an appropriate description for colloidal interaction.

The author gratefully acknowledges many discussions with Rudolf Klein, Matthias Brunner, Clemens Bechinger, Thomas Gisler, Christian Fleck, Emmanuel Mbamala, Jure Dobnikar, Carsten Ruß, Rene van Roij, Markus Deserno, Vladimir Lobaskin, and the financial support from the DFG through SFB 513.

References

1. W. Poon, P. Pusey, H. Lekkerkerker, Physics World, April, 27 (1996).
2. R. J. Hunter, *Introduction to Modern Colloid Science*, Oxford University Press, Oxford (1994); H. Dautzenberg, W. Jaeger, J. Kötz, B. Philipp, Ch. Seidel and D. Stscherbina, *Polyelectrolytes. Formation, Characterization and Application*, Carl Hanser, München (1994). W. B. Russel, D. A. Saville, and W. R. Schowalter, *Colloidal Dispersions*, Cambridge University Press, New York (1989).
3. A. van Blaaderen, Science, **282**, 887 (1998).
4. W.C.K. Poon, T. McLeish, A. Donald, Physics World, May, 34 (2001).
5. S. Asakura and F. Oosawa, J. Polym. Sci. **33**, 183 (1958); B. Götzelmann, R. Evans, and S. Dietrich, Phys. Rev. E **57**, 6785 (1998).
6. W. Norde, in *Physical Chemistry of Biological Interfaces*, edited by A. Baszkin and W. Norde (Marcel Dekker, New York, 2000), p. 115.
7. A.D. Dinsmore, P.B. Warren, W.C.K. Poon, A.G. Yodh: Europhys. Lett. **40**, 337 (1997).
8. H. Lodish, A. Berk, S. L. Zipursky, P. Matsudaira, D. Baltimore, and J. Darnell, *Molecular Cell Biology*, 4th ed., W. H. Freeman, New York (2001).
9. G.M. Bell, S. Levine, and L.N. McCartney, J. Colloid Interface Sci. **33**, 335 (1970).
10. M. Brunner, C. Bechinger, W. Strepp, V. Lobaskin, H.H. von Grünberg, Europhys. Lett., in press, (2002).
11. C. Russ, M. Dijkstra, R. van Roij, H.H. von Grünberg, submitted to Phys. Rev. E, (2002).
12. K.Vondermassen, J.Bongers, A. Müller, and H. Versmold, Langmuir **10**, 1351 (1994); G. M. Kepler and S. Fraden, Phys. Rev. Lett. **73**, 356 (1994); M.D. Carbajal-Tinoco, F. Castro Roman, and J.L. Arauz-Lara, Phys. Rev. E **56**, 3745 (1996); S.H.Behrens and D.G.Grier, Phys. Rev. E **64**, 050401 (2001).
13. M. Dijkstra and R. van Roij, J.Phys.: Condens. Matter **10**, 1219 (1998).
14. L. Rojas, C. Urban, P. Schurtenberger, T. Gisler, H. H. von Grünberg, submitted to Europhys. Lett., (2002).
15. J. Yamanaka, H. Yoshida, T. Koga, N. Ise, T. Hashimoto, Phys. Rev. Lett. **80**, 5806 (1998).
16. R. van Roij, M. Dijkstra, J.P. Hansen, Phys. Rev. E **59**, 2010 (1999).
17. H.H. von Grünberg, R. van Roij, G. Klein, Europhys. Lett. **55**, 580, (2001).
18. M. Deserno, H.H. von Grünberg, Phys. Rev. E, in press, (2002).

Excitable Membranes: Channel Noise, Synchronization, and Stochastic Resonance

Peter Hänggi, Gerhard Schmid, and Igor Goychuk

Dept. Physics, University of Augsburg, 86135 Augsburg, Germany

Abstract. By use of a stochastic generalization of the Hodgkin-Huxley model we investigate the phenomena of Stochastic Resonance (SR) and Coherence Resonance (CR) in assemblies of ion channels. If no stimulus is applied we find the existence of an optimal size of the membrane for which the internal noise alone causes a most regular spiking activity (intrinsic CR). In the presence of an applied stimulus we demonstrate SR vs. decreasing patch size (i. e., vs. increasing internal noise strength). SR with external noise occurs only for large sizes which possess sub-optimal internal noise levels. SR in biology thus seemingly is rooted in the collective properties of large ion channel assemblies. Investigating the signal-to-noise ratio (SNR) for sub-threshold sinusoidal driving vs. the frequency we find a bell-shaped behavior vs. frequency which reflects the existence of a random internal limit cycle. Finally we study the role of synchronization vs. decreasing internal noise intensity (i.e. increasing path area).

1 Introduction

A fundamental question in neurophysiology concerns the limiting factors of the reliability of neuronal responses to a given stimulus. In this article we focus on a particular aspect of this complex issue: the impact of *channel noise*, which is generated by the random gating dynamics of the ion channels, on the reliability of signal transmission. In doing so, we consider the effects of stochastic resonance, coherence resonance and synchronization.

During the last decade, the effect of stochastic resonance (SR) – a cooperative phenomenon wherein the addition of external noise improves the detection and transduction of signals in nonlinear systems (for a comprehensive survey and prominent references, see Ref. [1]) – has been studied experimentally and theoretically in various biological systems [2,3,4,5,6]. For example, SR has been experimentally demonstrated within the mechanoreceptive system in crayfish [2], in the cricket cercal sensory system [3], for human tactile sensation [4], visual perception [5], and response behavior of the arterial baroreflex system of humans [6]. The importance of this SR-phenomenon for sensory biology is by now well established; yet, it is presently not known to which minimal level of the biological organization the stochastic resonance effect can ultimately be traced down. Presumably, SR has its origin in the stochastic properties of the ion channel clusters located in a receptor cell membrane. Indeed, for an artificial model system Bezrukov and

Vodyanoy have demonstrated experimentally that a finite ensemble of the alamethicin ion channels does exhibit stochastic resonance [7]. This in turn provokes the question whether a *single* ion channel is able to exhibit SR, or whether stochastic resonance is the result of a *collective* response from a finite assembly of channels.

Stochastic resonance in single, biological potassium ion channels has also been investigated both theoretically [8] and experimentally [9]. This very experiment [9] did not convincingly exhibit SR in single voltage-sensitive ion channels versus the varying temperature. Nevertheless, the SR phenomenon versus the *externally* added noise can occur in single ion channels if only the parameters are within a regime where the channel is predominantly dwelled in the closed state, as demonstrated within a theoretical modeling for a Shaker potassium channel [8]. The manifestation of SR on the *single*-molecular level, is not only of academic interest, but is also relevant for potential nanotechnological applications, such as the design of single-molecular biosensors. The origin and biological relevance of SR in single ion channels, however, remains still open.

Indeed, biological SR can be a manifestation of *collective* properties of large assemblies of ion channels of different sorts. To display the phenomenon of excitability these assemblies must contain an assemblage of ion channels of at least two different sorts – such as, *e.g.*, potassium and sodium channels. The corresponding mean-field type model has been put forward by Hodgkin and Huxley in 1952 [10] by neglecting the mesoscopic fluctuations which originate from the stochastic opening and closing of channels. SR due to *external* noise in this primary model and related models of excitable dynamics has extensively been addressed [11]. These models further display another interesting effect in the presence of noise, namely so termed coherence resonance (CR) [12,13]: even in absence of an external periodic signal the stochastic dynamics exhibits a surprisingly more regular behavior due to an optimally applied external noise intensity.

Synchronization presents another phenomenon, which is also closely related to SR and CR [14,15]. The mechanisms of synchronization are presently well-established for some chaotic and excitable systems. Depending on the type and strength of coupling, several kinds of synchronization can be distinguished: complete synchronization, generalized synchronization, lag synchronization, phase synchronization, and burst (or train) synchronization [14,15]. In the context of excitable systems, frequency and phase synchronization has been found, for example, in the integrate-and-fire model of neuron driven by white noise and an externally applied stochastic spike train [16]. Namely, for an optimal dose of noise the mean firing rate of the driven neuron becomes locked by the mean frequency of the external spike train [16].

A challenge though still remains: does *internal* noise play a constructive role for SR, CR and synchronization? Internal noise is produced by fluctuation of the number of open channels within the assembly, and diminishes with increasing number of channels. For a large, macroscopic number of channels this noise becomes negligible. Under the realistic biological conditions, however, it may play an important role [17].

2 The Hodgkin-Huxley Model

Our starting point is due to the model of Hodgkin and Huxley [10], *i.e.* the ion current across the biological membrane is carried mainly by the motion of sodium, Na$^+$, and potassium, K$^+$, ions through the selective and voltage-gated ion channels embedded across the membrane. Besides, there is also a leakage current present which is induced by chloride and remaining other ions. The ion channels are formed by special membrane proteins which undergo spontaneous, but voltage-sensitive conformational transitions between open and closed states [18]. Moreover, the conductance of the membrane is directly proportional to the number of the *open* ion channels. This number depends on the potential difference across the membrane, V. The different concentrations of the ions inside and outside the cell are encoded by corresponding reversal potentials $E_{\text{Na}} = 50$ mV, $E_{\text{K}} = -77$ mV and $E_{\text{L}} = -54.4$ mV, respectively, which give the voltage values at which the direction of partial ion currents is reversed [18]. Taking into account that the membrane possesses a capacitance $C = 1$ µF/cm^2, Kirchhoff's first law reads in presence of an *external* current $I_{\text{ext}}(t)$ stimulus:

$$C\frac{d}{dt}V + G_{\text{K}}(n)\,(V - E_{\text{K}}) \\ + G_{\text{Na}}(m,h)\,(V - E_{\text{Na}}) + G_{\text{L}}\,(V - E_L) = I_{\text{ext}}(t). \quad (1)$$

Here, $G_{\text{Na}}(m,h)$, $G_{\text{K}}(n)$ and G_{L} denote the conductances of sodium, potassium, and the remaining other ion channels, respectively. The leakage conductance is assumed to be constant, $G_{\text{L}} = 0.3$mS/cm^2; in contrast, those of sodium and potassium depend on the probability to find the ion channels in their open conformation. To explain the experimental data, Hodgkin and Huxley did assume that the conductance of a potassium channel is gated by four independent and identical gates. Thus, if n is the probability of one gate to be open, the probability of the K$^+$ channel to stay open is $P_{\text{K}} = n^4$. Moreover, the gating dynamics of sodium channel is assumed to be governed by three independent, identical gates with opening probability m and an additional one, being different, possessing the opening probability h. Accordingly, the opening probability of Na$^+$ channel (or the fraction of open channels) reads $P_{\text{Na}} = m^3 h$. The conductances for potassium and sodium thus read

$$G_{\text{K}}(n) = g_{\text{K}}^{\max}\, n^4, \quad G_{\text{Na}}(m,h) = g_{\text{Na}}^{\max}\, m^3 h, \quad (2)$$

where $g_{\text{Na}}^{\max} = 120$ mS/cm^2 and $g_{\text{K}}^{\max} = 36$ mS/cm^2 are the maximal conductances. Furthermore, the gating variables (probabilities) m, h and n obey the two-state, "opening-closing" dynamics,

$$\begin{aligned}\dot{m} &= \alpha_m(V)\,(1-m) - \beta_m(V)\,m\,,\\ \dot{h} &= \alpha_h(V)\,(1-h) - \beta_h(V)\,h\,,\\ \dot{n} &= \alpha_n(V)\,(1-n) - \beta_n(V)\,n\,,\end{aligned} \quad (3)$$

with the experimentally determined voltage-dependent transition rates, reading for a squid giant axon [10,19]:

$$\alpha_m(V) = \frac{0.1(V+40)}{1-\exp[-(V+40)/10]},$$
$$\beta_m(V) = 4\,\exp[-(V+65)/18]\,, \quad (4a)$$

$$\alpha_h(V) = 0.07\,\exp[-(V+65)/20],$$
$$\beta_h(V) = \{1 + \exp[-(V+35)/10]\}^{-1}\,, \quad (4b)$$

$$\alpha_n(V) = \frac{0.01\,(V+55)}{1-\exp[-(V+55)/10]},$$
$$\beta_n(V) = 0.125\,\exp[-(V+65)/80]\,. \quad (4c)$$

The rate constants in (4a)–(4c) are given in ms^{-1} and the voltage in mV. These nonlinear Hodgkin-Huxley equations (1)–(4c) present a cornerstone model in neurophysiology. Within the same line of reasoning this model can be generalized to a mixture of different ion channels with various gating properties [19,20].

3 Stochastic Version of the Hodgkin-Huxley Model

It has been suspected since the time of Hodgkin and Huxley, and known with certainty since the first single-channel recordings of Neher, Sakmann and colleagues, that voltage-gated ion channels are stochastic devices [21]. An essential drawback of the Hodgkin-Huxley model, however, is that it operates with the *average* number of open channels, thereby disregarding the corresponding number fluctuations (or, the so-called *channel noise* [21,17]). These fluctuations, i.e. their strength, scale inversely proportional to the number of ion channels, see eq. (6) below. Thus, the original Hodgkin-Huxley model can be valid, strictly speaking, only within the limit of very large system size. We emphasize, however, that the size of an excitable membrane patch within a neuron is typically finite. As a consequence, the role of internal fluctuations cannot be *a priori* neglected [17]. As a matter of fact, as shown below, they do play a key role for SR, CR, and synchronization.

3.1 Quantifying Channel Noise

The role of channel noise for the neuron firing has been first studied by Lecar and Nossal as early as in 1971 [22]. The corresponding stochastic generaliza-

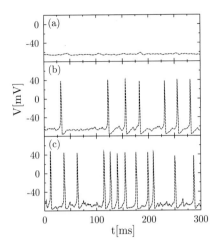

Fig. 1. Numerical simulation of the stochastic Hodgkin-Huxley system (1),(5),(6) with vanishing external stimulus. We computed several realizations of the voltage signal for different numbers of the ion channels: a) $N_{Na} = 6000$, $N_K = 1800$; b) $N_{Na} = 600$, $N_K = 180$; and c) $N_{Na} = 60$, $N_K = 18$. Upon decreasing the system size the influence of channel noise on the spontaneous firing dynamics becomes more and more pronounced. Note that the non-stochastic Hodgkin-Huxley model does not exhibit spikes at all for the parameters given in the text and in the absence of external stimuli

tions of Hodgkin-Huxley model (within a kinetic model which corresponds to the above given description) has been put forward by DeFelice et al. [23] and others; see [17] for a review and further references therein. The main conclusion of these previous studies is that the channel noise can be functionally important for neuron dynamics. In particular, it has been demonstrated that channel noise alone can give rise to a spiking activity even in the absence of any stimulus [17,23,24].

To include the channel noise influence in a theoretical modeling within the stochastic kinetic schemes [17,23], however, necessitates extensive numerical simulations [25]. To aim at a less cumbersome numerical scheme we use a short-cut procedure that starts from Eq. (3) in order to derive a corresponding set of Langevin equations for a stochastic generalization of the Hodgkin-Huxley equations of the type put forward by Fox and Lu [26]. Following their reasoning we substitute the equations (3) with the corresponding Langevin generalization:

$$\begin{aligned} \dot{m} &= \alpha_m(V)\,(1-m) - \beta_m(V)\,m + \xi_m(t)\,, \\ \dot{h} &= \alpha_h(V)\,(1-h) - \beta_h(V)\,h + \xi_h(t)\,, \\ \dot{n} &= \alpha_n(V)\,(1-n) - \beta_n(V)\,n + \xi_n(t)\,, \end{aligned} \qquad (5)$$

with independent Gaussian white noise sources of vanishing mean. The noise autocorrelation functions depend on the stochastic voltage and the corresponding total number of ion channels as follows

$$\langle \xi_m(t)\xi_m(t')\rangle = \frac{2}{N_{\text{Na}}} \frac{\alpha_m \beta_m}{(\alpha_m + \beta_m)} \delta(t-t'),$$

$$\langle \xi_h(t)\xi_h(t')\rangle = \frac{2}{N_{\text{Na}}} \frac{\alpha_h \beta_h}{(\alpha_h + \beta_h)} \delta(t-t'), \quad (6)$$

$$\langle \xi_n(t)\xi_n(t')\rangle = \frac{2}{N_{\text{K}}} \frac{\alpha_n \beta_n}{(\alpha_n + \beta_n)} \delta(t-t').$$

In order to confine the conductances between the physically allowed values between 0 (all channels are closed) and $g^{\max}$ (all channels are open) we have implemented numerically the constraint of reflecting boundaries so that $m(t)$, $h(t)$ and $n(t)$ are always located between zero and one [26].

Moreover, the numbers N_{Na} and N_{K} depend on the actual area S of the membrane patch. With the assumption of homogeneous ion channel densities, $\rho_{\text{Na}} = 60$ μm^{-2} and $\rho_{\text{K}} = 18$ μm^{-2}, one finds the following scaling behavior

$$N_{\text{Na}} = \rho_{\text{Na}} S, \quad N_{\text{K}} = \rho_{\text{K}} S. \quad (7)$$

Upon decreasing the system size S, the fluctuations and, hence, the internal noise increases. Consequently, with abating cell membrane patch the spiking behavior changes dramatically, cf. Fig. 1.

3.2 Numerical Integration

Before integrating the system of stochastic equations (1), (5), (6) numerically, the external stimulus $I_{\text{ext}}(t)$ must be specified. We take a periodic stimulus of the form

$$I_{\text{ext}}(t) = A \sin(\Omega t) + \eta(t), \quad (8)$$

where the sinusoidal signal with amplitude A and angular frequency Ω is contaminated by the Gaussian white noise $\eta(t)$ with the autocorrelation function

$$\langle \eta(t)\eta(t')\rangle = 2D_{\text{ext}} \delta(t-t'), \quad (9)$$

and noise strength D_{ext}.

The numerical integration is carried out by the standard Euler algorithm with the step size $\Delta t \approx 2 \cdot 10^{-3}$ ms. The "Numerical Recipes" routine ran2 is used for the generation of independent random numbers [27] with the Box-Muller algorithm providing the Gaussian distributed random numbers. The total integration time is chosen to be a multiple of the driving period $T_\Omega = 2\pi/\Omega$, as to ensure that the spectral line of the driving signal is centered on a computed value of the power spectral densities. From the stochastic voltage signal $V(t)$ we extract a point process of spike occurrences $\{t_i\}$:

$$u(t) := \sum_{i=1}^{N} \delta(t-t_i), \quad (10)$$

where N is the total number of spikes occurring during the elapsed time interval. The occurrence of a spike in the voltage signal $V(t)$ is obtained by upward-crossing a certain detection threshold value V_0. It turns out that the threshold can be varied over a wide range with no effect on the resulting spike train dynamics. The power spectral density of the spike train (PSD_u), the interspike interval histogram (ISIH) and the coefficient of variation (CV) have been analyzed. The coefficient of variation CV, which presents a measure of the spike coherence, reads:

$$\text{CV} := \frac{\sqrt{\langle T^2 \rangle - \langle T \rangle^2}}{\langle T \rangle}, \tag{11}$$

where $\langle T \rangle := \lim_{N \to \infty} \sum (t_{i+1} - t_i)/N$ and $\langle T^2 \rangle := \lim_{N \to \infty} \sum (t_{i+1} - t_i)^2/N$ are the mean and mean-squared interspike intervals, respectively. From the PSD_u we obtain the height of the spectral line of the periodic stimulus as the difference between the peak value and its background offset. The signal-to-noise ratio (SNR) is then given by the ratio of signal peak height to the background height (in the units of spectral resolution of signals).

3.3 Coherence Resonance and Synchronization

We have analyzed the spike coherence in the autonomous, nondriven regime (*i.e.*, we have used $I_{\text{ext}} = 0$) as a function of the decreasing cluster size. Our simulation reveals, cf. Fig. 2(a), the novel phenomenon of *intrinsic coherence resonance*, where the order in the spike sequence increases; *i.e.* the CV is falling, *solely due to the presence of internal noise*. The fully disordered

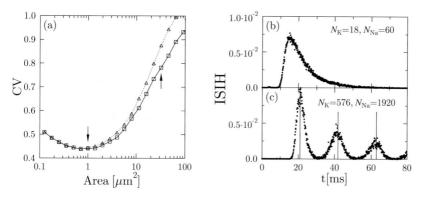

Fig. 2. (a) The coefficient of variation (CV) in (11) is plotted versus the area S in absence of external noise $D_{\text{ext}} = 0$ for periodic sub-threshold driving with $A = 1.0$ µA/cm^2 and $\Omega = 0.3$ ms^{-1} (solid line) and without any stimulus (dotted line). The ISIH are depicted in the presence of the periodic signal for $S = 1$ µm^2 (b), see downward arrow in (a), and $S = 32$ µm^2 (c), see upward arrow in (a). The vertical lines denote the driving period and its first two multiples

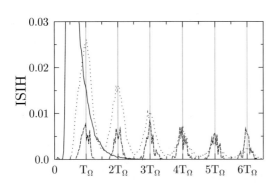

Fig. 3. The interspike interval histogram is plotted for the external stimulus $I_{\text{ext}}(t) = \sin(0.2 \cdot t)$ [μA/cm^2] and different observation areas: $S = 1$ μm^2 (solid line), $S = 32$ μm^2 (dotted line), and $S = 128$ μm^2 (dashed line). For not too small patch sizes the spike occurrences synchronize with the external sinusoidal stimulus at multiples of the driving frequency T_Ω

sequence (which corresponds to a Poissonian spike train) would assume the value CV = 1. We note, however, that near $S = 1$ μm^2 (optimal dose of internal noise), CV ≈ 0.44, *i.e.* the spike train becomes distinctly more ordered! For $S < 1$ μm^2, the internal noise increases further beyond the optimal value and destroys the order in spiking again. It is worth noting that for $S < 1$ μm^2 the model reaches limiting validity; in that regime the kinetic scheme [17,23,24,25] should be used instead. Such a corresponding study, however, has been put forward independently by Jung and Shuai [25]; their results are in qualitative agreement with our findings.

Next we switch on an external sub-threshold sinusoidal driving. Interestingly enough the interspike intervals distribution is not affected for small patch sizes, cf. Fig. 2(b). In this case, the spike-activity possesses an internal rhythm which dominates over the external disturbances. For larger patch sizes the internal noise decreases and the periodic drive induces a reduction of the CV as compared to the undriven case, note the right arrow in Fig. 2(a). In this latter regime the external driving rules the spiking activity as depicted with the characteristic peaks in the ISIH in Fig. 2(c) at multiple driving periods.

In the deterministic limit the Hodgkin-Huxley equations exhibit the phenomenon refractoriness, i.e. no firing event occurs before a minimal time interval of about 15 ms has elapsed since the last firing [10]. In the presence of channel noise, see Fig. 1, the refractory time interval (or dead-time) becomes reduced [24,25]; note the distinct reduction of 12–13 ms in Fig. 2(b) and (c).

For a different sub-threshold driving, with the amplitude $A = 1$ μA/cm^2 and frequency $\Omega = 0.2$ ms^{-1}, the ISIH is plotted for different patch areas in Fig. 3. In the intermediate and small noise regime the spike occurrences are locked to the multiples of driving period. Even though such locking presents clearly a synchronization behavior, the perfect frequency synchronization, characterized by the Rice frequency $\langle \omega \rangle := \lim_{N \to \infty} 2\pi/N \sum_{i=1}^{N} 1/(t_{i+1} - t_i)$, could not be detected. In fact, a frequency mismatch between the driving frequency Ω and the Rice frequency $\langle\omega\rangle$ has been observed (not shown). This frequency mismatch happens due to the multimodal structure of the ISIH, which is caused by the locking of firing occurrences to the external force period in ratios different from 1:1. A similar phenomenon of imperfect synchronization has also been found in human cardiorespiratory activity [28].

3.4 Stochastic Resonance

Next, the focus is on the SNR in absence of external noise, see Fig. 4(a). Here we discover the novel effect of genuine *intrinsic stochastic resonance*, where the response of the system to the sub-threshold external stimulus is optimized *solely* due to internal, ubiquitous noise. For the given parameters it occurs at $S \approx 32$ µm². For $S < 32$ µm² growing internal noise monotonically deteriorates the system response. Under such circumstances, one would predict that the addition of an external noise (which corresponds to the conventional situation in biological SR studies) *cannot* improve SNR further, *i.e.* conventional SR will not be exhibited. Our numerical simulations, Fig. 4(b), fully confirm this prediction. Conventional stochastic resonance therefore occurs only for large membrane patches beyond optimal size, and reaches saturation in the limit $S \to \infty$ (limit of the deterministic Hodgkin-Huxley model). Thus, the observed biological SR [2,3] is rooted in the collective properties of large ion channels arrays, where ion channels are globally coupled via the common membrane potential $V(t)$.

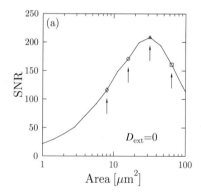
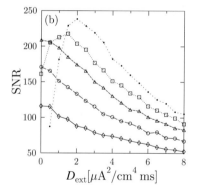

Fig. 4. The signal-to-noise ratio (SNR) for an external sinusoidal stimulus with amplitude $A = 1.0$ µA/cm² and angular frequency $\Omega = 0.3$ ms^{-1} for different observation areas: (a) no external noise is applied; (b) SNR versus the external noise for the system sizes indicated by the arrows in Fig. 4(a): $S = 8$ µm², solid line through the diamonds; $S = 16$ µm², long dashed line connecting the circles; $S = 32$ µm², short dashed line through the triangles; $S = 64$ µm², dotted line connecting the squares. The situation with no internal noise (*i.e.*, formally $S \to \infty$) is depicted by the dotted line connecting the filled dots

In addition, by changing the driving frequency we rediscover the effect of combined stochastic and conventional resonance [13], cf. Fig. 5. In other words, SNR becomes optimized not only versus the patch size, but also versus the driving frequency. Moreover, due to the noisy character of gating variables, the mean frequency of a corresponding random limit cycle in the stochastic Hodgkin-Huxley model (1),(5),(6) depends on the membrane patch area. Thus, the maxima of SNR are located for various system sizes at different driving frequencies.

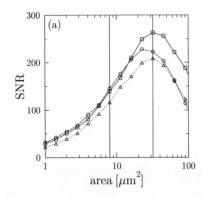
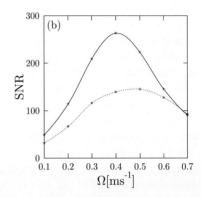

Fig. 5. The signal-to-noise ratio (SNR) for a sub-threshold external stimulus with amplitude $A = 1.0$ µA/cm^2 and different angular frequencies: (a) SNR versus the observation area for $\Omega = 0.3$ ms^{-1} (dotted line through the triangles), $\Omega = 0.4$ ms^{-1} (solid line connecting the squares), and $\Omega = 0.5$ ms^{-1} (dashed line through the circles); (b) SNR versus the driving frequency for two areas ($S = 8$ µm^2, dotted line; $S = 32$ µm^2, solid line), depicted by vertical lines in Fig. 5(a). The curves exhibit clear maxima and, therefore, the effect of double-stochastic resonance

4 Conclusions

In conclusion, we have investigated the stochastic resonance and the coherence resonance [24], as well as the synchronization in a noisy generalization of the Hodgkin-Huxley model. The spontaneous fluctuations of the membrane conductivity due to the individual ion channel dynamics has systematically been taken into account. We have shown that the excitable membrane patches with observation area around $S \approx 1$ µm^2 exhibit a rhythmic spiking activity optimized by omnipresent internal noise. In other words, the collective dynamics of globally coupled ion channels become more ordered solely due to *internal* noise. This new effect can be regarded as the *intrinsic coherence resonance* phenomenon; it presents a first important result of our work. This very finding has also been confirmed independently within a complementary approach by Jung and Shuai [25]. Moreover, for the case of a sub-threshold periodic driving we have shown that for intermediate patch sizes a synchronization of firing events with the multiplies of the driving period occurs. This is reflected by an improper synchronization between the Rice frequency and the frequency of external driving.

The second main result of this study refers to the phenomenon of *intrinsic SR*. Here, the channel noise *alone* gives rise to SR behavior, cf. Fig. 4(a) (see also Ref. [25]). Moreover, such intrinsic SR is optimized versus the driving frequency, cf. Fig. 5. Conventional SR versus the external noise intensity also takes place, but for sufficiently large membrane patches, where the internal noise strength alone is not yet at its optimal value. We thus conclude that the observed biological SR likely is rooted in the *collective* properties of globally coupled ion channel assemblies.

The authors gratefully acknowledge the support of this work by the Deutsche Forschungsgemeischaft, SFB 486 *Manipulation of matter on the nanoscale*, project A10. Moreover, we appreciate most helpful and constructive discussions with Peter Jung and Alexander Neiman.

References

1. L. Gammaitoni, P. Hänggi, P. Jung, and F. Marchesoni, Rev. Mod. Phys. **70**, 223 (1998).
2. J.K. Douglass, L. Wilkens, E. Pantazelou, and F. Moss, Nature (London) **365**, 337 (1993).
3. J.E. Levin, and J.P. Miller, Nature (London) **380**, 165 (1996).
4. J.J. Collins, T.T. Imhoff, and P. Grigg, Nature (London) **383**, 770 (1996).
5. E. Simonotto, M. Riani, C. Seife, M. Roberts, J. Twitty, and F. Moss, Phys. Rev. Lett. **78**, 1186 (1997).
6. I. Hidaka, D. Nozaki, and Y. Yamamoto, Phys. Rev. Lett. **85**, 3740 (2000).
7. S.M. Bezrukov, and I. Vodyanoy, Nature (London) **378**, 362-364 (1995); **385**, 319 (1997).
8. I. Goychuk, and P. Hänggi, Phys. Rev. E **61**, 4272 (2000).
9. D. Petracchi, M. Pellegrini, M. Pellegrino, M. Barbi, and F. Moss, Biophys. J. **66**, 1844 (1994).
10. A.L. Hodgkin, and A.F. Huxley, J. Physiol. (London) **117**, 500 (1952).
11. A. Longtin, J. Stat. Phys. **70**, 309-327(1993); K. Wiesenfeld, D. Pierson, E. Pantazelou, C. Dames, and F. Moss, Phys. Rev. Lett. **72**, 2125 (1994); J.J. Collins, C.C. Chow, A.C. Capela, and T.T. Imhoff, Phys. Rev. E **54**, 5575 (1996); S.-G. Lee, and S. Kim, Phys. Rev. E **60**, 826 (1999).
12. A.S. Pikovsky, and J. Kurths, Phys. Rev. Lett. **78**, 775 (1997); S.-G. Lee, A. Neiman, and S. Kim, Phys. Rev. E **57**, 3292 (1998).
13. B. Lindner, and L. Schimansky-Geier, Phys. Rev. E **61**, 6103 (2000).
14. A. Pikovsky, M. Rosenblum, and J. Kurths, *Synchronization: A Universal Concept in Nonliear Sciences* (Cambridge University Press, Cambridge 2001).
15. Y. Kuramoto, *Chemical Oscillations, Waves and Turbulence* (Springer, Berlin, 1984).
16. A. Neiman, L. Schimansky-Geier, F. Moss, B. Shulgin, and J.J. Collins, Phys. Rev. E **60**, 284 (1999).

17. J.A. White, J.T. Rubinstein, and A.R. Kay, Trends Neurosci. **23**, 131 (2000).
18. B. Hille, *Ionic Channels of Excitable Membranes*, 2nd ed. (Sinauer Associates, Sunderland, MA 1992).
19. R.J. Nossal, and H. Lecar, *Molecular and Cell Biophysics* (Addison-Wesley, Redwood City 1991).
20. S.B. Lowen, L.S. Liebovitch, and J.A. White, Phys. Rev. E **59**, 5970 (1999).
21. B. Sakmann, and E. Neher, *Single-Channel recording* (Plenum Press, 1995).
22. H. Lecar, and R. Nossal, Biophys. J. **11**, 1068 (1971).
23. J.R. Clay, and L.J. DeFelice, Biophys. J. **42**, 151 (1983); A.F. Strassberg, and L.J. DeFelice, Neural Comput. **5**, 843 (1993); L.J. DeFelice, and A. Isaac: Chaotic states in a random world, J. Stat. Phys. **70**, 339 (1993).
24. G. Schmid, and I. Goychuk, P. Hänggi, Europhys. Lett. **56**, 22 (2001).
25. P. Jung, and J. W. Shuai, Europhys. Lett. **56**, 29 (2001).
26. R.F. Fox R, and Y. Lu, Phys. Rev. E **49**, 3421 (1994).
27. W.H. Press, S.A. Teukolsky, W.T. Vetterling, and B.P. Flannery, *Numerical Recipes in C*, 2nd ed. (Cambridge Univ. Press, Cambridge 1992).
28. C. Schäfer, M.G. Rosenblum, J. Kurths, and H.-H. Abel, Nature (London) **392**, 239 (1998); C. Schäfer, M.G. Rosenblum, H.-H. Abel, and J. Kurths, Phys. Rev. E **60**, 857 (1999).

Birth and Sudden Death of a Granular Cluster

Ko van der Weele[1], Devaraj van der Meer[1], and Detlef Lohse[1]

Department of Applied Physics and J.M. Burgers Centre for Fluid Dynamics, University of Twente, P.O. Box 217, 7500 AE Enschede, The Netherlands

Abstract. Granular material is vibro-fluidized in N connected compartments. For sufficiently strong shaking the particles are statistically uniformly distributed over the compartments, but if the shaking intensity is lowered this uniform distribution gives way to a clustered state. The clustering transition is experimentally shown to be of 2nd order for $N = 2$ compartments and of 1st order for $N \geq 3$. In particular, the latter is hysteretic, involves long-lived transient states, and exhibits a striking lack of time reversibility. In the strong shaking regime, a cluster breaks down very abruptly and in its further decay shows anomalous diffusion, with the length scale going as $t^{1/3}$ rather than the standard $t^{1/2}$. We focus upon the self-similar nature of this process. The observed phenomena are all accounted for within a flux model.

1 Introduction

Granular matter is everywhere, yet our understanding of it is still far from complete. At rest, it resembles a solid. When it is flowing, it resembles a fluid. And when it is shaken it resembles a gas. Nevertheless, the dissipative nature of the particle interactions makes it fundamentally different from any standard solid, liquid, or gas, and for this reason it is often called the "fourth state of matter" [1,2,3]. A better understanding is essential not only for physicists, but also for the many industries that handle granular materials (mining, food production, pharmaceutical industry, construction works, chemical reactors, etc.). It has been estimated [1] that no less than 40% of the capacity of these industries is wasted due to problems that have to do with the transportation and sorting of granular matter. In the gas-phase, which is found for instance in gas-fluidized beds and shakers [4], the main problem is the tendency of the particles to spontaneously form clusters.

This clustering effect can be traced back to the fact that the collisions between the granules are inelastic [5]. Some energy is dissipated in every collision, which means that a relatively dense region (where the particles collide more often than elsewhere) will dissipate more energy, and thus become even denser, resulting in a cluster of slow particles. Vice versa, relatively dilute regions will become more dilute. The few particles in these regions are very rapid ones.

A striking illustration of the clustering phenomenon is provided by the so-called Maxwell Demon experiment [6,7], depicted in Fig. 1a. It consists of

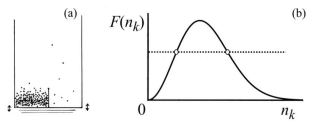

Fig. 1. (a) Clustering in the Maxwell-demon experiment. Below a certain critical shaking level the particles cluster into one compartment, due to the non-elasticity of the collisions. In terms of the granular temperature (i.e., the mean kinetic energy of the particles) the cluster is cold. The few, much faster particles in the other compartment are warm. This separation in a cold and a hot compartment is reminiscent of what Maxwell's Demon was supposed to accomplish [8]. (b) The flux function $F(n_k)$, i.e., the particle flux from compartment k as a function of the fraction n_k within the compartment. As a result of the inelasticity of the collisions between the particles, F does not grow anymore beyond a certain value of n_k, but decreases instead. The horizontal dashed line indicates that the flux from a relatively empty compartment (left intersection point) can be equal to the flux from a well-filled compartment (right intersection point): this is exactly what happens in the clustered state [7]

a box divided into two compartments by a wall of height h, with a few hundred small beads in each compartment. The beads are brought in a gaseous state by shaking the system vertically, with frequency f and amplitude a. If the shaking is vigorous enough, the dissipation from the inelastic collisions is overpowered by the energy input into the system, and the beads divide themselves uniformly over the two compartments, as in any ordinary molecular gas. But if the driving is lowered below a certain level, the beads cluster in one of the two compartments. One ends up with a "cold" compartment containing a lot of beads (moving rather sluggishly, hardly able to jump over the wall anymore) and a "hot" compartment containing only a few (much more lively) beads. Cold and hot here refer to the so-called granular temperature, i.e., the mean kinetic energy of the particles in each compartment.

In this paper we analyze this experiment quantitatively in terms of a dynamical flux model, and we generalize it to an arbitrary number of connected compartments. The flux model is introduced, and tested against experimental data in Section 2. The extension to N compartments is made in Section 3. Already for $N = 3$ it turns out that this is more than just an upscaled version of the 2-box system: the physics changes qualitatively. Whereas the clustering for $N = 2$ takes place via a continuous (2nd order) phase transition, for $N = 3$ we find a discontinuous (1st order) phase transition, including a strong hysteretic effect. In Section 4 two special phenomena associated with this transition for $N = 3$ are discussed. Section 5 is about the continuum limit for $N \to \infty$. We conclude (in Section 6) with a brief overview of further extensions and applications.

2 Flux Model

The clustering can be described quantitatively by means of a flux model [7]. The central idea behind it is that in equilibrium the average particle flux from left to right is equal to the flux from right to left. Hence, the asymmetric equilibrium of the clustered situation can only be explained if the flux F from one compartment to the other does not monotonously increase with the number of particles within that compartment. Rather, it is a function that attains a maximum (at a certain particle number) and thereafter decreases again, as sketched in Fig. 1b. So the flux from a nearly empty box can balance the flux from a well-filled box, and that is exactly what is the case in the clustered situation.

This one-humped form of F can be understood as follows. If there are no particles in the box, it can give nothing to its neighbors: so the flux starts out from $F(0) = 0$. Initially, the flux increases with growing particle number, just as in an ordinary gas. Beyond a certain filling level, however, the increasingly frequent collisions make the particles in the compartment so slow that the flux starts to decrease. Eventually F goes to zero again.

In agreement with this, and based on the kinetic theory for dilute granular gases, Eggers proposed the following analytic approximation for the flux (rewritten here in a form suited for an arbitrary number of N connected compartments) [7,9]:

$$F(n_k) = A n_k^2 e^{-N^2 B n_k^2}. \tag{1}$$

This is indeed a function of the form sketched in Fig. 1b, with a maximum at $n_k = 1/(N\sqrt{B})$. Here n_k is the fraction of particles in the k-th box, normalized to $\sum_k n_k = 1$.

The factors A and B depend on the particle properties (their radius r, and the restitution coefficient e of the collisions between particles), the experimental setup (e.g. the height h of the wall), and the driving parameters a and f. The factor A determines the absolute rate of the flux, and can simply be incorporated in the time scale. The dimensionless factor B is more important: it governs the phase transition towards the clustering state. It is given by

$$B \propto \frac{gh}{(af)^2}(1-e^2)^2 \left(\frac{r^2 P_{tot}}{\Omega N}\right)^2, \tag{2}$$

with P_{tot} the total number of particles, and Ω the ground area of each box. For a given granular material (r and e fixed), B can be raised either by increasing the value of h or by decreasing the value of af, i.e., by reducing the driving. Note that for elastically colliding particles ($e = 1$) the factor B is equal to zero, which means that in that case the flux function F is monotonously increasing, and clustering is impossible.

Let us see how the flux model works out for the 2-box system. The dynamics is given by the balance equation

$$\frac{dn_1}{dt} = -F(n_1) + F(n_2) = -F(n_1) + F(1 - n_1). \tag{3}$$

In equilibrium one has $dn_1/dt = 0$. For $B < 1$ there exists only one equilibrium, namely the stable uniform $\{1/2, 1/2\}$ distribution. For $B > 1$ it becomes unstable and gives way to an asymmetric equilibrium. In other words, there is a symmetry breaking bifurcation at $B_{bif} = 1$. In Fig. 2a we have drawn the corresponding bifurcation diagram [7,9].

For the experimental verification of this diagram we put 600 glass beads ($r = 1.25$ mm, $e = 0.97$) in a perspex tube of inner radius 27.5 mm, divided into two equal compartments by a wall of height $h = 23.0$ mm. The tube was mounted on a shaker with an adjustable frequency and amplitude, so that B can be varied. In the present experiment this was done by means of the frequency f, at a fixed value of $a = 6.5$ mm. The measurements are included as solid dots in Fig. 2a. Clearly, theory and experiment agree on the fact that the clustering transition takes place via a pitchfork bifurcation.

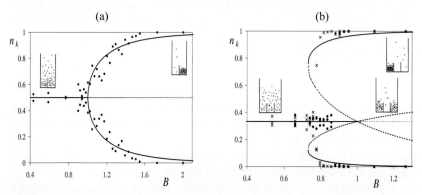

Fig. 2. (a) Bifurcation diagram for the 2-box system ($k = 1, 2$). The solid and dashed lines correspond to the flux model, and the dots are experimental measurements. The transition to the clustering state is seen to take place via a pitchfork bifurcation. (b) Bifurcation diagram for the 3-box system ($k = 1, 2, 3$). The solid and dashed lines correspond to the flux model, and the dots and stars are experimental measurements (dots for measurements that were started from the uniform distribution $\{1/3, 1/3, 1/3\}$, stars for those that were started from a single peaked distribution). Note the hysteresis in the clustering transition

3   More than two Compartments: Hysteresis

The flux function was invented for 2 compartments, but we want to go beyond that and apply it to an arbitrary number of compartments. The dynamics for the general N-box system, assuming only nearest neighbor interactions, is given by [9,10]:

$$\frac{dn_k}{dt} = F(n_{k-1}) - 2F(n_k) + F(n_{k+1}) \qquad \text{for } k = 1, ..., N. \qquad (4)$$

The first three terms on the right hand side denote the flux out of box k into its two neighbors (hence the factor 2), and the flux into the box from its neighbors [12].

The case with $N = 3$ is already considerably richer than the original two-compartment system [9]. Again, we find a uniform distribution $\{1/3, 1/3, 1/3\}$ at high driving levels (small B) and a single-cluster distribution at milder driving levels (large B), but in contrast to the $N = 2$ case the transition between them is hysteretic, see Fig. 2b. That is, the value of B at which the transition occurs when one goes from vigorous to mild driving is different from the B-value when one goes in the opposite direction. The experimental measurements were done with 600 glass beads ($r = 1.25$ mm, $e = 0.97$) in a cyclic 3-compartment tube mounted on a shaker.

The flux confirms these experimental findings. The uniform distribution $\{1/3, 1/3, 1/3\}$ turns unstable at $B_{bif} = 1$. Already at $B_{sn} = 0.73$, however, a stable clustered solution comes into existence by means of a saddle-node bifurcation. The basin of attraction of this clustered state grows as B increases, while its unstable counterpart closes in upon $\{1/3, 1/3, 1/3\}$ and finally destabilizes this uniform solution when it coincides with it at $B_{bif} = 1$. (A more detailed analysis is given in refs. [9,10]). Thus, if one gradually increases the value of B from zero upwards (i.e, if one reduces the driving), the transition from the uniform to the clustered state takes place at $B_{bif} = 1$. If the value of B is then gradually turned down again, the reverse transition occurs at the lower value $B_{sn} = 0.73$ (hysteresis). Physically, this is easily understood: the particles in the cluster keep each other slow, dissipating a lot of energy, and it requires an extra effort in terms of driving power (i.e., a lower value of B) to overcome this.

The unstable counterpart of the clustered state is not only instrumental in destabilizing the uniform solution, but it plays an important role also beyond $B = 1$. If we start out from the (unstable) uniform distribution, the system can go to the clustered state either directly, or - and this is actually more likely - via this counterpart state, which forms a transient situation in which two boxes compete for dominance. This approach towards the 1-cluster state can be interpreted as a coarsening process. We will come back to it in the next section.

4 Coarsening and Sudden Death

What is true for 3 compartments, holds also for $N > 3$ [10], so it is not the 3-box system that is exceptional but the 2-box system. As an example, in Fig. 3a the bifurcation diagram for $N = 5$ compartments is given. The hysteresis is even more pronounced than in the case for $N = 3$ (the first saddle-node bifurcation already takes place at $B_{sn} = 0.34$) and there are also more transient states.

In Fig. 3b we show four stages in the clustering process for the 5-box system, starting out from the uniform distribution, at B slightly larger than 1. The particles do not cluster immediately into one compartment, but form a transient state first, in which two of the boxes (here the leftmost and the center compartment) are in competition. This transient state lives for roughly half a minute, until finally one of the boxes wins. The system always ends up in a one-cluster state, since this is the only truly stable equilibrium (for $B > 1$), but it does not necessarily have to be in the middle box. In fact, the end boxes have a slightly higher probability to draw the cluster towards them, since the extra sidewalls of these boxes form an additional source of energy dissipation.

It is also interesting to have a look at the reverse process (*declustering*), which occurs for $B < B_{sn}$. In fact, in industrial situations such as conveyor belts and sorting machines, declustering is often much more desirable than clustering. It is also very interesting from a purely physical point of view, as Fig. 3c (at $B \approx 0.96 B_{sn}$) shows.

The cluster persists for a remarkably long time. After 42 seconds of shaking it is still clearly visible, and then suddenly (within one second) it vanishes without leaving a trace. At $t = 43$ s it has given way to the uniform distribution. We call this phenomenon "sudden death" [13]. Obviously, declustering is by no means the same as clustering in reverse time order. This lack of reversibility is yet another consequence of the inelastic collisions. It is the rule in dissipative systems, and as such does not come as a surprise, yet the intensity of the effect is striking.

The observed declustering behavior can be understood from the bifurcation diagram. It occurs for B-values just below B_{sn}; i.e., just to the left of $B = 0.73$ for $N = 3$ (Fig. 2b), or $B = 0.34$ for $N = 5$ (Fig. 3a). Initially, all particles are in the middle box; i.e., $n_k = \{0, 1, 0\}$ or $n_k = \{0, 0, 1, 0, 0\}$ respectively. When we start to shake, the material slowly leaks out of the cluster into the surrounding compartments, and the system seems to be heading for a dynamical equilibrium (one well-filled compartment surrounded by nearly empty compartments), just as if the corresponding lines in the bifurcation diagram were extrapolated to the left of B_{sn}. Only when this imaginary line has been passed (and hence the corresponding equilibrium has proven to be false) the system shoots towards the uniform distribution. The time from the start of the experiment until this sudden collapse, the lifetime τ of the cluster, can be calculated from the flux model. The result turns out to agree

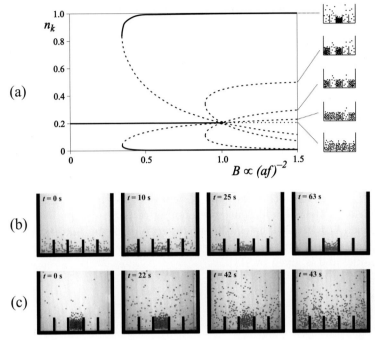

Fig. 3. (a) Bifurcation diagram for the 5-compartment system. The solid lines correspond to stable equilibrium states, the dashed lines to unstable ones. (b) Four snapshots from a clustering experiment. Note that the particles do not immediately cluster into one compartment, but first go through a transient 2-cluster state. (c) Declustering at higher shaking strength. The cluster is clearly present until $t = 42\,\mathrm{s}$, then suddenly collapses, leaving no trace one second later

very well with the experimentally measured lifetime, for all $N \geq 3$. A full account of the sudden death phenomenon is given in ref. [11].

To get an idea how the cluster lifetime is calculated, consider the case of $N = 3$ compartments. Let the initial cluster be positioned in box 2. The material spills symmetrically out of the cluster, so $n_1 = n_3 = \frac{1}{2}(1 - n_2)$, and thus the equation of motion for the cluster fraction is:

$$\frac{dn_2}{dt} = F(n_1) - 2F(n_2) + F(n_3) = 2\left(F(\frac{1}{2}(1-n_2)) - F(n_2)\right), \quad (5)$$

and this is readily integrated to give:

$$\tau = \int_{n_{thr}}^{1} \frac{dn_2}{2\left(F(n_2) - F(\frac{1}{2}(1-n_2))\right)}. \quad (6)$$

Here n_{thr} is the threshold value where the system passes through the imaginary equilibrium line, but its precise value is not too critical. In Fig. 4a the above theoretical expression for τ is compared with the experimental data for the 3-box system, for various values of B, and the agreement is seen to be excellent. Naturally, at $B = B_{sn}$ the lifetime becomes infinite, since the cluster becomes stable there.

In Fig. 4b we present lifetime data for further values of N, based on experiments and numerical simulations. The first thing to note is that the vertical axis (τ) spans no less than 8 orders of magnitude. As explained above, τ goes to infinity at $BN^2 = B_{sn}N^2$, i.e., at the first saddle-node bifurcation for each N. Away from these divergencies, the lifetimes are seen to lie on a universal envelope curve, independent of the value of N. This reflects the fact that for these values of BN^2 the sudden death occurs *before* the particles leaking out of the cluster have had time to fill the outermost boxes of the system to any significant level. Therefore, the behavior does not depend on the value of N. (Or stated otherwise: the system does not feel its finite size during the cluster's lifetime and hence the number of boxes can be taken to be infinite, which brings us to the continuum description of the next section). The form of the envelope curve can be calculated from the continuum version of the flux model and is given by [11]:

$$\tau = K \left[\int_{n_{thr}}^{1} \frac{e^{(3/4)\widetilde{B}n^2} dn}{n^{3/2}(1 - \widetilde{B}n^2 e^{-\widetilde{B}n^2})} \right]^2, \quad \text{with } \widetilde{B} = BN^2. \tag{7}$$

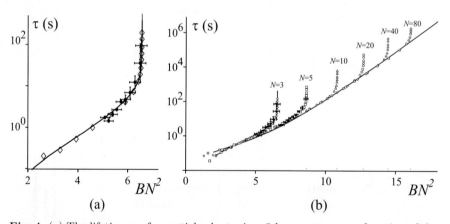

Fig. 4. (a) The lifetime τ of a particle cluster in a 3-box system, as a function of the shaking parameter. The dots with the error bars represent experimental data (based on 15 repetitions of the experiment), the diamonds are numerical data from the flux model, and the solid curve is the theoretical value from Eq. (6). At $B = B_{sn} = 0.73$ (or equivalently, $BN^2 = 6.57$) the cluster becomes stable and hence its lifetime diverges. (b) Lifetime data for various values of N, determined from experiments with $N = 3, 5$ boxes (dots with error bars) and from numerical evaluation of the flux model with $N = 3, 5, 10, 20, 40, 80$ boxes (empty symbols). The solid curves are analytical solutions for $N = 3$ (see also Fig.(a)) and for the envelope curve, which goes roughly as $\exp[(3/2)BN^2]$ (cf. Eq. (7))

The only free parameter in this expression is the constant K, which sets the absolute scale of τ.

All the above effects (hysteresis, coarsening, sudden death) become more pronounced with growing number of compartments. For example, in Fig. 5a the bifurcation diagram for $N = 80$ is depicted. The hysteresis is seen to reach down almost all the way to $B = 0$, and there is a whole forest of transient states, in which the system can easily get stuck. In order to see this happen, we perform a simulation based on the flux model including a noise term. We take $B = 0.9$ and start out from a situation in which the granular material distributed more or less evenly over the boxes. The basin of attraction of the uniform equilibrium is quite narrow at $B = 0.9$, so even a small statistical fluctuation is sufficient to throw the system out of it, and from that moment on the system is heading towards the 1-cluster situation. However, as we know, it does not go there in one stroke but via transient multi-cluster states. In Fig. 5b we show the situation after a few minutes of shaking, when the system happens to be in a transient state consisting of 7 clusters. Next to the clusters there are depletion zones, and in between we see a characteristic hilly pattern.

This 7-cluster transient state may be compared with the two-cluster transient state for the 5-box system in Fig. 3b. And just like there, the clusters will compete each other until eventually we end up in the 1-cluster state, which after all is the only clustered state that is truly stable. But whereas in the 5-box system this coarsening process takes roughly a minute, in the present case it may take years. The competition between clusters is extremely slowed down when the distance between them becomes larger. In practice, the 80-box experiment will be turned off long before the 1-cluster state is reached.

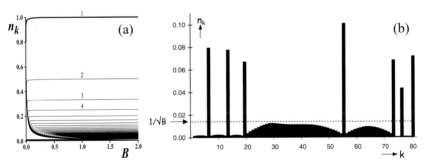

Fig. 5. (a) Bifurcation diagram for the 80-compartment system. The number 1 indicates the single-cluster state (the only clustered state that is mathematically stable); the other numbers indicate transient multi-cluster states. (b) A snapshot from a flux model simulation (including a statistical noise term) at $B = 0.90$. The depicted situation, a transient state with 7 clusters, has arisen from an initial condition in which the 80 boxes were filled more or less uniformly

5 The Limit for $N \to \infty$ Compartments: Anti-Diffusion

Finally, we consider the limit of an infinite number of boxes. In this case the fraction $n_k(t)$ is replaced by its continuum version $n(x,t)$, and the equation of motion takes the following form [11]:

$$\frac{\partial n(x,t)}{\partial t} = (\Delta x)^2 \frac{\partial^2 F(n(x,t))}{\partial x^2} = C(n)\left(\frac{\partial n(x,t)}{\partial x}\right)^2 + D(n)\frac{\partial^2 n(x,t)}{\partial x^2} \quad (8)$$

Here $C(n) = (\Delta x)^2 \partial^2 F/\partial n^2$ and $D(n) = (\Delta x)^2 \partial F/\partial n$, with Δx the (infinitesimal) distance between boxes. The conservation condition now reads $\int n(x,t)dx = 1$.

This partial differential equation looks very similar to the Kardar-Parisi-Zhang equation [13], which is used to describe the roughening of interfaces. In the original KPZ equation C and D are positive constants, and in that case the first term on the right hand side (the nonlinear term) has a steepening effect on the profile of $n(x,t)$, while the second term (which represents diffusion) tends to smoothen it. In the present case this is more intricate: C and D are functions of $n(x,t)$ and can become negative, which means that both terms can either have a steepening or a smoothening effect. For instance, the diffusion coefficient D will be negative on the decreasing right part of the plot in Fig. 2 (i.e., for $n > 1/\sqrt{\widetilde{B}}$, with $\widetilde{B}$ taking the place of $N^2 B$ in this limit). We then get anti-diffusion, local accumulation of granular material, which indeed is the continuum analog to the clustering.

Also on the increasing left part of $F(n)$, where D is positive, the diffusion shows an uncommon feature. The width of a profile in this regime grows as $t^{1/3}$, and its height decreases as $t^{-1/3}$, instead of $t^{1/2}$ and $t^{-1/2}$ as in ordinary diffusion [13]. This anomalous exponent $1/3$ comes out straightforwardly from an analysis of the vigorous-shaking limit $\widetilde{B} = 0$, for which Eq. (8) reduces to

$$\frac{\partial n(x,t)}{\partial t} = 2\widetilde{A}\left(\frac{\partial n(x,t)}{\partial x}\right)^2 + 2\widetilde{A}n\frac{\partial^2 n(x,t)}{\partial x^2} \quad (9)$$

with $\widetilde{A} = A(\Delta x)^2$. The diffusion coefficient ($D = 2\widetilde{A}n$) is invariably positive in this case, so a given profile always spreads out, just as one might expect for very strong shaking. Moreover, in simulations it is observed that the profile diffuses without losing its basic form, indicating that the process is a self-similar one. Therefore, given that $n(x,t)$ has the dimension length^{-1} (as follows directly from the conservation condition) and $\widetilde{A}$ has the dimension length^3/time (as can be seen from Eq. (9)), we try a solution of the following form [11,14]:

$$n(x,t) = \frac{1}{(\widetilde{A}t)^{1/3}}G(\eta), \quad \text{with } \eta = \frac{x}{(\widetilde{A}t)^{1/3}}. \quad (10)$$

It is here that the coefficient $1/3$ comes in. The function G does not depend on x or t separately, but only on the combined dimensionless variable η.

Inserting the ansatz (10) into Eq. (9), the partial differential equation for $n(x,t)$ is transformed into an ordinary differential equation for $G(\eta)$:

$$G + \eta \frac{\partial G}{\partial \eta} + 6 \left(\frac{\partial G}{\partial \eta} \right)^2 + 6G \frac{\partial^2 G}{\partial \eta^2} = 0 \tag{11}$$

It is non-linear, but nevertheless not very difficult to solve: $G(\eta) = G_0 - (1/12)\eta^2$. This inverted parabola represents, in one single curve, the whole evolution of the diffusing profile.

6 Extensions and Applications

There are many interesting generalizations and applications. Presently we are working on a system in which the compartments are not on one level, but form a staircase, resembling an industrial conveyor belt. The central topic here is the competition between clustering and the natural tendency of the particles to stream downwards. The experiment becomes even more interesting when the linear array of boxes is extended to a 2-dimensional grid of compartments. In that case, the clusters are no absolute barriers to the flux anymore (they can be circumvented) and this gives rise to beautiful front propagation patterns.

We also study mixtures consisting of particles of different sizes. This is done with an eye to practical applications, where granular material is rarely mono-disperse. One of the main themes here is the competition between clustering and the tendency of the large and small particles to segregate.

A third application is the traffic jam problem. There exists a strong analogy between granular and traffic flow. In the past decade this has already led to several successful traffic strategies, especially in situations where the cars could still be treated as individual particles [15]. On a much larger scale, e.g. the Dutch network of highways, it is no longer feasible to consider the cars individually and here a coarse-grained approach may help. To this end, we divide the highway in imaginary cells (1 to 5 km per cell) and concentrate on the *density* ρ_k of cars in these cells. The traffic flow from one cell to the next is given by a flux function $F(\rho_k)$. Actually, this function has already been measured a number of times in real-life situations (among traffic analysts it is known as the "fundamental diagram" [16]), and it shows a marked resemblance with the flux function in Fig. 1b. At low densities the cars flow freely from cell to cell and $F(\rho_k)$ is an increasing function of ρ_k, whereas at high densities they hinder each other and $F(\rho_k)$ decreases. As we have seen in the present paper, clustering (traffic jam formation) is adequately described with this type of flux model. It should be possible, given the traffic situation at a certain moment, to predict in which cells traffic jams are likely to be formed within the next hour - and perhaps take measures to prevent them.

In conclusion, the Maxwell Demon experiment is a rich dynamical system full of surprising effects. As for many problems in granular dynamics, the questions are interesting from a fundamental point of view, and at the same time important for practical applications. And the answers are obtained by an approach in which experiment, theory, and numerical simulations go hand in hand.

References

1. H.M. Jaeger, S.R. Nagel, R.P. Behringer: Granular solids, liquids, and gases, Rev. Mod. Phys. **68**, 1259 (1996).
2. L.P. Kadanoff: Built upon sand: theoretical ideas inspired by granular flows, Rev. Mod. Phys. **71**, 435 (1999).
3. P.G. de Gennes: Granular matter: a tentative view, Rev. Mod. Phys. **71**, S374 (1999).
4. J.A.M. Kuipers, W.P.M. van Swaaij: Computational Fluid Dynamics applied to Chemical Reaction Engineering, Adv. Chem. Engineering **24**, 227 (1998).
5. I. Goldhirsch, G. Zanetti: Clustering instability in dissipative gases, Phys. Rev. Lett. **70**, 1619 (1993); A. Kudrolli, M. Wolpert, J.P. Gollub: Cluster formation due to collisions in granular material, Phys. Rev. Lett. **78**, 1383 (1997).
6. H.J. Schlichting, V. Nordmeier: Strukturen im Sand, Math. Naturwiss. Unterr. **49**, 323 (1996).
7. J. Eggers: Sand as Maxwell's demon, Phys. Rev. Lett. **83**, 5322 (1999).
8. A.S. Leff, A.F. Rex: *Maxwell's Demon: Entropy, Information, Computing* (Adam Hilger, Bristol, 1990).
9. K. van der Weele, D. van der Meer, M. Versluis, D. Lohse: Hysteretic clustering in granular gas, Europhys. Lett. **53**, 328 (2001).
10. D. van der Meer, K. van der Weele, D. Lohse: Bifurcation diagram for compartmentalized granular gases, Phys. Rev. E **63**, 061304 (2001).
11. D. van der Meer, K. van der Weele, D. Lohse: Sudden Collapse of a Granular Cluster, Phys. Rev. Lett. **88**, 174302 (2002).
12. Equation (4) is written down for a cyclic array of boxes, for which box N and 1 are (connected) neighbors. If the array is non-cyclic, the equation for the end box 1 reads $dn_1/dt = -F(n_1) + F(n_2)$, and analogously for the end box N. The differences between the two cases are small [10].
13. M. Kardar, G. Parisi, Y.C. Zhang: Dynamic scaling of growing interfaces, Phys. Rev. Lett. **56**, 889 (1986).
14. G.J. Barenblatt: *Scaling, self-similarity, and intermediate asymptotics* (Cambridge Univ. Press, Cambridge, 1996).
15. M. Schreckenberg, D.E. Wolf (eds.): *Traffic and Granular Flow '97* (Springer, Singapore, 1998).
16. D. Helbing: Traffic and related self-driven many-particle systems, Rev. Mod. Phys. **73**, 1067 (2001).

Interacting Neural Networks and Cryptography

Wolfgang Kinzel[1] and Ido Kanter[2]

[1] Institute for Theoretical Physics and Astrophysics, Universität Würzburg, Am Hubland, 97074 Würzburg, Germany
[2] Minerva Center and Department of Physics, Bar-Ilan University, 52100 Ramat-Gan, Israel

Abstract. Two neural networks which are trained on their mutual output bits are analysed using methods of statistical physics. The exact solution of the dynamics of the two weight vectors shows a novel phenomenon: The networks synchronize to a state with identical time dependent weights. Extending the models to multilayer networks with discrete weights, it is shown how synchronization by mutual learning can be applied to secret key exchange over a public channel.

1 Introduction

Neural networks learn from examples. This concept has extensively been investigated using models and methods of statistical mechanics [1,2]. A "teacher" network is presenting input/output pairs of high dimensional data, and a "student" network is being trained on these data. Training means that synaptic weights adopt by simple rules to the input/output pairs.

When the networks — teacher as well as student — have N weights, the training process needs of the order of N examples to obtain generalization abilities. This means that after the training phase the student has achieved some overlap to the teacher, their weight vectors are correlated. As a consequence, the student can classify an input pattern which does not belong to the training set. The average classification error decreases with the number of training examples.

Training can be performed in two different modes: Batch and on-line training. In the first case all examples are stored and used to minimize the total training error. In the second case only one new example is used per time step and then discarded. Therefore on-line training may be considered as a dynamic process: at each time step the teacher creates a new example which the student uses to change its weights by a tiny amount. In fact, for random input vectors and in the limit $N \to \infty$, learning and generalization can be described by ordinary differential equations for a few order parameters [3].

On-line training is a dynamic process where the examples are generated by a static network - the teacher. The student tries to move towards the teacher. However, the student network itself can generate examples on which it is trained. When the output bit is moved to the shifted input sequence, the network generates a complex time series [4]. Such networks are called bit (for binary) or sequence (for continuous numbers) generators and have recently been studied in the context of time series prediction [5].

This work on the dynamics of neural networks - learning from a static teacher or generating time series by self interaction - has motivated us to study the following problem: What happens if two neural networks learn from each other? In the following section an analytic solution is presented [6], which shows a novel phenomenon: synchronization by mutual learning. The biological consequences of this phenomenon are not explored, yet, but we found an interesting application in cryptography: secure generation of a secret key over a public channel.

In the field of cryptography, one is interested in methods to transmit secret messages between two partners A and B. An opponent E who is able to listen to the communication should not be able to recover the secret message.

Before 1976, all cryptographic methods had to rely on secret keys for encryption which were transmitted between A and B over a secret channel not accessible to any opponent. Such a common secret key can be used, for example, as a seed for a random bit generator by which the bit sequence of the message is added (modulo 2).

In 1976, however, Diffie and Hellmann found that a common secret key could be created over a public channel accessible to any opponent. This method is based on number theory: Given limited computer power, it is not possible to calculate the discrete logarithm of sufficiently large numbers [7].

Here we show how neural networks can produce a common secret key by exchanging bits over a public channel and by learning from each other.

2 Dynamic Transition to Synchronization

Here we study mutual learning of neural networks for a simple model system: Two perceptrons receive a common random input vector $\underline{x}$ and change their weights $\underline{w}$ according to their mutual bit σ, as sketched in Fig. 1. The output bit σ of a single perceptron is given by the equation

$$\sigma = \text{sign}(\underline{w} \cdot \underline{x}) \tag{1}$$

$\underline{x}$ is an N-dimensional input vector with components which are drawn from a Gaussian with mean 0 and variance 1. $\underline{w}$ is a N-dimensional weight vector with continuous components which are normalized,

$$\underline{w} \cdot \underline{w} = 1 \tag{2}$$

The initial state is a random choice of the components $w_i^{A/B}, i = 1, ...N$ for the two weight vectors $\underline{w}^A$ and $\underline{w}^B$. At each training step a common random input vector is presented to the two networks which generate two

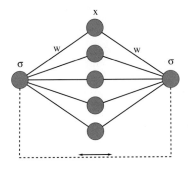

Fig. 1. Two perceptrons receive an identical input $\underline{x}$ and learn their mutual output bits σ

output bits σ^A and σ^B according to (1). Now the weight vectors are updated by the perceptron learning rule [3]:

$$\underline{w}^A(t+1) = \underline{w}^A(t) + \frac{\eta}{N}\underline{x}\sigma^B\,\Theta(-\sigma^A\sigma^B)$$
$$\underline{w}^B(t+1) = \underline{w}^B(t) + \frac{\eta}{N}\underline{x}\sigma^A\,\Theta(-\sigma^A\sigma^B) \qquad (3)$$

$\Theta(x)$ is the step function. Hence, only if the two perceptrons disagree a training step is performed with a learning rate η. After each step (3), the two weight vectors have to be normalized.

In the limit $N \to \infty$, the overlap

$$R(t) = \underline{w}^A(t)\,\underline{w}^B(t) \qquad (4)$$

has been calculated analytically [6]. The number of training steps t is scaled as $\alpha = t/N$, and $R(\alpha)$ follows the equation

$$\frac{dR}{d\alpha} = (R+1)\left(\sqrt{\frac{2}{\pi}}\,\eta(1-R) - \eta^2\frac{\varphi}{\pi}\right) \qquad (5)$$

where φ is the angle between the two weight vectors $\underline{w}^A$ and $\underline{w}^B$, i.e. $R = \cos\varphi$. This equation has fixed points $R = 1, R = -1$, and

$$\frac{\eta}{\sqrt{2\pi}} = \frac{1-\cos\varphi}{\varphi} \qquad (6)$$

Fig. 2 shows the attractive fixed point of (5) as a function of the learning rate η. For small values of η the two networks relax to a state of a mutual agreement, $R \to 1$ for $\eta \to 0$. With increasing learning rate η the angle between the two weight vectors increases up to $\varphi = 133°$ for

$$\eta \to \eta_c \cong 1.816 \qquad (7)$$

Above the critical rate η_c the networks relax to a state of complete disagreement, $\varphi = 180°, R = -1$. The two weight vectors are antiparallel to each other, $\underline{w}^A = -\underline{w}^B$.

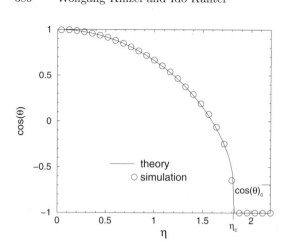

Fig. 2. Final overlap R between two perceptrons as a function of learning rate η. Above a critical rate η_c the time dependent networks are synchronized. From Ref. [6]

As a consequence, the analytic solution shows, well supported by numerical simulations for $N = 100$, that two neural networks can synchronize to each other by mutual learning. Both of the networks are trained to the examples generated by their partner and finally obtain an antiparallel alignment. Even after synchronization the networks keep moving, the motion is a kind of random walk on an N-dimensional hypersphere producing a rather complex bit sequence of output bits $\sigma^A = -\sigma^B$ [8].

3 Random Walk in Weight Space

We want to apply synchronization of neural networks to cryptography. In the previous section we have seen that the weight vectors of two perceptrons learning from each other can synchronize. The new idea is to use the common weights $\underline{w}^A = -\underline{w}^B$ as a key for encryption [9]. But two problems have to be solved yet: (i) Can an external observer, recording the exchange of bits, calculate the final $\underline{w}^A(t)$, (ii) does this phenomenon exist for discrete weights? Point (i) is essential for cryptography, it will be discussed in the following section. Point (ii) is important for practical solutions since discrete weights need less storage and computational efforts. It will be investigated in the following.

Synchronization occurs for normalized weights, unnormalized ones do not synchronize [6]. Therefore, for discrete weights, we introduce a restriction in the space of possible vectors and limit the components $w_i^{A/B}$ to $2L + 1$ different values,

$$w_i^{A/B} \in \{-L, -L+1, ..., L-1, L\} \tag{8}$$

In order to obtain synchronization to a parallel – instead of an antiparallel – state $\underline{w}^A = \underline{w}^B$, we modify the learning rule (3) to:

$$\underline{w}^A(t+1) = \underline{w}^A(t) - \underline{x}\sigma^A \Theta(\sigma^A \sigma^B)$$

$$\underline{w}^B(t+1) = \underline{w}^B(t) - \underline{x}\sigma^B \Theta(\sigma^A \sigma^B) \tag{9}$$

Now the components of the random input vector $\underline{x}$ are binary $x_i \in \{+1, -1\}$. If the two networks produce an identical output bit $\sigma^A = \sigma^B$, then their weights move one step in the direction of $-x_i \sigma^A$. But the weights should remain in the interval (8), therefore if any component moves out of this interval, $|w_i| = L+1$, it is set back to the boundary $w_i = \pm L$.

Each component of the weight vectors performs a kind of random walk with reflecting boundary. Two corresponding components w_i^A and w_i^B receive the same random number ± 1. After each hit at the boundary the distance $|w_i^A - w_i^B|$ is reduced until it has reached zero. For two perceptrons with a N-dimensional weight space we have two ensembles of N random walks on the internal $\{-L, ..., L\}$. If we neglect the global signal $\sigma^A = \sigma^B$ as well as the bias σ^A, we expect that the probability of two random walks being in different states decreases as

$$P(t) \sim P(0) e^{-t/\tau} \tag{10}$$

with some characteristic time scale $\tau = \mathcal{O}(L^2)$

Hence the total synchronization time should be given by $N \cdot P(t) \simeq 1$ which gives

$$t_{\text{sync}} \sim \tau \ln N \tag{11}$$

In fact, our simulations for $N = 100$ show that two perceptrons with $L = 3$ synchronize in about 100 time steps and the synchronization time increases logarithmically with N. However, our simulations also showed that an opponent recording the sequence of $(\sigma^A, \sigma^B, \underline{x})_t$ is able to synchronize, too. Therefore, a single perceptron does not allow a generation of a secret key.

4 Secret Key Generation

Obviously, a single perceptron transmits too much information. An opponent, who knows the set of input/output pairs can derive the weights of the two partners after synchronization. Therefore, one has to hide so much information that the opponent cannot calculate the weights, but on the other side one has to transmit enough information that the two partners can synchronize.

In fact, we found that multilayer networks with hidden units may be candidates for such a task [9]. More precisely, we consider parity machines with three hidden units as shown in Fig. 3. Each hidden unit is a perceptron (1) with discrete weights (8). The output bit τ of the total network is the product of the three bits of the hidden units

$$\begin{aligned} \tau^A &= \sigma_1^A \, \sigma_2^A \, \sigma_3^A \\ \tau^B &= \sigma_1^B \, \sigma_2^B \, \sigma_3^B \end{aligned} \tag{12}$$

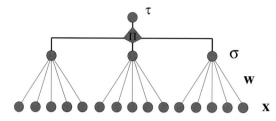

Fig. 3. Parity machine with three hidden units

At each training step the two machines A and B receive identical input vectors $\underline{x}_1, \underline{x}_2, \underline{x}_3$. The training algorithm is the following: Only if the two output bits are identical, $\tau^A = \tau^B$, the weights can be changed. In this case, only the hidden unit σ_i which is identical to τ changes its weights using the Hebbian rule

$$\underline{w}_i^A(t+1) = \underline{w}_i^A(t) - \underline{x}_i \tau^A \qquad (13)$$

For example, if $\tau^A = \tau^B = 1$ there are four possible configurations of the hidden units in each network:

$$(+1, +1, +1), (+1, -1, -1), (-1, +1, +1), (-1, -1, +1)$$

In the first case, all three weight vectors $\underline{w}_1, \underline{w}_2, \underline{w}_3$ are changed, in all other three cases only one weight vector is changed. The partner as well as any opponent does not know which one of the weight vectors is updated.

The partners A and B react to their mutual stop and move signals τ^A and τ^B, whereas an opponent can only receive these signals but not influence the partners with its own output bit. This is the essential mechanism which allows synchronization but prohibits learning. Numerical [9] as well as analytical [10] calculations of the dynamic process show that the partners can synchronize in a short time whereas an opponent needs a much longer time to lock into the partners.

This observation holds for an observer who uses the same algorithm (13) as the two partners A and B. Note that the observer knows 1. the algorithm of A and B, 2. the input vectors $\underline{x}_1, \underline{x}_2, \underline{x}_3$ at each time step and 3. the output bits τ^A and τ^B at each time step. Nevertheless, he does not succeed in synchronizing with A and B within the communication period, it needs a much longer time as shown in Fig. 6

Since for each run the two partners draw random initial weights and since the input vectors are random, one obtains a distribution of synchronization times as shown in Fig. 4 for $N = 100$ and $L = 3$. The average value of this distribution is shown as a function of system size N in Fig. 5. This figure indicates that even an infinitely large network needs only a finite number of exchanged bits - about 400 in this case - to synchronize.

If the communication continues after synchronization, an opponent has a chance to lock into the moving weights of A and B. Fig. 6 shows the distribution of the ratio between the synchronization time of A and B and the learning time of the opponent. In our simulations, for $N = 100$, this

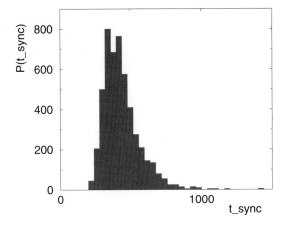

Fig. 4. Distribution of synchronization time for $N = 100, L = 3$

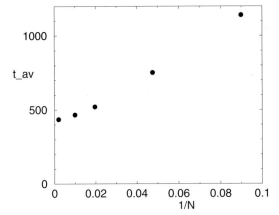

Fig. 5. Average synchronization time as a function of inverse system size

ratio never exceeded the value $r = 0.1$, and the average learning time is about 50000 time steps, much larger than the synchronization time. Hence, the two partners can take their weights $\underline{w}_i^A(t) = \underline{w}_i^B(t)$ at a time step t where synchronization most probably occurred as a common secret key. In real applications, one may include an additional test for synchrony. Thus, synchronization of neural networks can be used as a key exchange protocol over a public channel.

5 Conclusions

Interacting neural networks have been calculated analytically. At each training step two networks receive a common random input vector and learn their mutual output bits. A new phenomenon has been observed: Synchronization by mutual learning. If the learning rate η is large enough, and if the weight

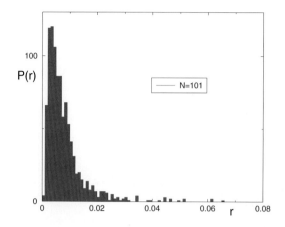

Fig. 6. Distribution of the ratio of synchronization time between networks A and B to the learning time of an attacker E

vectors keep normalized, then the two networks relax to an antiparallel orientation. Their weight vectors still move like a random walk on a hypersphere, but each network has complete knowledge about its partner.

It has been shown how this phenomenon can be used for cryptography. The two partners can agree on a common secret key over a public channel. An opponent who is recording the public exchange of training examples cannot obtain full information about the secrete key used for encryption.

This works if the two partners use multilayer networks, parity machines. The opponent has all the informations (except the initial weight vectors) of the two partners and uses the same algorithms. Nevertheless he does not synchronize.

This phenomenon may be used as a key exchange protocol. The two partners select secret initial weight vectors, agree on a public sequence of input vectors and exchange public bits. After a few steps they have identical weight vectors which are used for a secret encryption key. For each communication they agree on a new secret key, without having stored any secret information before. In contrast to number theoretical methods the networks are very fast; essentially they are linear filters, the complexity to generate a key of length N scales with N (for sequential update of the weights).

Of course, one cannot rule out that algorithms for the opponent may be constructed which find the key in much shorter time. In fact, ensembles of opponents have a better chance to synchronize. In addition, one can show that, given the information of the opponent, the key is uniquely determined, and, given the sequence of inputs, the number of keys is huge but finite, even in the limit $N \to \infty$ [11]. These may be good news for a possible attacker. However, recently we have found advanced algorithms for synchronization, too. Such variations are subjects of active research, and future will show whether the security of neural network cryptography can compete with number theoretical methods.

Acknowledgments

This work profitted from enjoyable collaborations with Richard Metzler and Michal Rosen-Zvi. We thank the German Israel Science Foundation (GIF) and the Minerva Center of the Bar-Ilan University for support.

References

1. J. Hertz, A. Krogh, and R. G. Palmer, *Introduction to the Theory of Neural Computation*, (Addison Wesley, Redwood City, 1991).
2. A. Engel, and C. Van den Broeck, *Statistical Mechanics of Learning*, (Cambridge University Press, 2001).
3. M. Biehl and N. Caticha, Statistical Mechanics of On-line Learning and Generalization, *The Handbook of Brain Theory and Neural Networks*, ed. by M. A. Arbib (MIT Press, Berlin 2001).
4. E. Eisenstein and I. Kanter and D.A. Kessler and W. Kinzel, Phys. Rev. Lett. **74**, 6-9 (1995).
5. I. Kanter, D.A. Kessler, A. Priel and E. Eisenstein, Phys. Rev. Lett. **75**, 2614-2617 (1995); L. Ein-Dor and I. Kanter, Phys. Rev. **E 57**, 6564 (1998); M. Schröder and W. Kinzel, J. Phys. **A 31**, 9131-9147 (1998); A. Priel and I. Kanter, Europhys. Lett. (2000).
6. R. Metzler and W. Kinzel and I. Kanter, Phys. Rev. E **62**, 2555 (2000).
7. D. R. Stinson, *Cryptography: Theory and Practice* (CRC Press 1995).
8. R. Metzler, W. Kinzel, L. Ein-Dor and I. Kanter, Phys. Rev. **E 63**, 056126 (2001).
9. I. Kanter, W. Kinzel and E. Kanter, Europhys. Lett. **57**, 141-147 (2002).
10. M. Rosen-Zvi, I. Kanter and W. Kinzel, cond-mat/0202350 (2002).
11. R. Urbanczik, private communication.

Low Energy Dynamics in Glasses Investigated by Neutron Inelastic Scattering

Jens-Boie Suck

Technical University Chemnitz, Institute of Physics, Materials Research and Liquids, 09107 Chemnitz, Germany

Abstract. Neutron inelastic scattering has been used to investigate the dependence of the low energy atomic dynamics of the bulk glass NiPdP on quench rate and on annealing induced structural relaxation. A decrease of the low energy modes is found with decreasing structural disorder in accordance with expectations from different computer simulations. In spite of this result the calculated contribution of the vibrational entropy to the total entropy change in the transition from the glass to the super-cooled liquid remains small compared to the structural contribution.

1 Introduction

In spite of all efforts made to understand the atomic dynamics of glasses, a surprising number of questions have remained unanswered up to now. Those addressed here concern the dependence of the atomic dynamics on quenched-in disorder and on subsequent structural relaxation, - or expressed differently: the form of the basins in different regions of the n-dimensional energy landscape describing the energy of the disordered system in its phase space [1]. Basins in this energy landscape are related to the energy minima: all points leading to the same minimum by steepest decent define (belong to) the same basin [2]. The expectations of different authors [2,3] concerning the variation of the form of the basins, which determines the vibrational dynamics, differ considerably and thus an experimental investigation is needed.

Glasses are nowadays classified on the basis of the temperature dependence of the viscosity of the super-cooled glass-forming liquid. The viscosity of *strong* glass formers varies with temperature largely following the Arrhenius equation. The "strong" bond network of these glasses, in the sense of the bond network model [4] resists to break up immediately on heating the glass. In contrast to this behavior, the viscosity of *fragile* glass formers change according to the Arrhenius law only at temperatures next to the glass transition temperature T_g and decreases most rapidly at the slightest temperature increase following essentially the Vogel-Fulcher- Tammann relation. Thus the rapidness of the viscosity decrease when going from the glass into the super-cooled liquid decides on the classification of the glasses, and this change again is determined by the rate of entropy change on heating. In this context, the contribution of the vibrational entropy to the entropy change on transforming the glass into the super-cooled liquid remains one of these open questions mentioned above. Computer simulations of Sciortino and collaborators have suggested a nearly complete decoupling of the vibrational entropy from the structural one, leaving a rather small contributions from the vibrations [5]. Considerations of the energy change of the modes, suggested by computer simulations again [6] and experiments on traditional metallic glasses [7], lead Angell and Moynihan to expect a contribution of some importance for this entropy change [3]. Again some more detailed information would be helpful to answer this question at least for the type of glass investigated here, one of the bulk metallic glasses.

2 Inelastic Neutron Scattering

Concerning the questions outlined above the use of neutron inelastic scattering is most likely the best method for obtaining the answers. In all experiments discussed here thermal time-of-flight (t-o-f) spectrometry has been used on a direct geometry spectrometer, i.e. the wave vector of the incident neutrons k_0 was defined by the mono-chromizing of the incident pulse and the wave vector of the scattered neutrons k was determined from the direction of scattering and the time-of-flight of the neutrons between sample and detector. From these wave vectors the energy and momentum transfers, $\hbar\omega = E_0 - E$ and $\hbar Q = k_0 - k$, which have occurred in a single scattering process, are calculated. The intensity of the scattered neutrons is converted to the double differential scattering cross-section (DDSC) applying all necessary corrections and normalizations and from this, for a sample with several different elements, the total dynamic structure factor $S(Q,\omega)$ and the generalized vibrational density of states (GVDOS) $G(\omega)$ are calculated.

The total dynamic structure factor is the weighted sum of the partial dynamic structure factors $S_{ij}(Q,\omega)$ weighted with the coupling of the different elements to the neutrons

$$\sigma_{sc} S(Q,\omega) = 4\pi[\sum \sqrt{c_i c_j} b_i b_j S_{ij}(Q,\omega)] + \sum \sigma_{inc,i} c_i S_{s,i}(Q,\omega) \quad (1)$$

with $\sigma_{sc} = \sum c_i \sigma_{sc,i}$ and $\sigma_{sc,i} = \sigma_{coh,i} + \sigma_{inc,i}$, where $\sigma_{coh,i}$ and $\sigma_{inc,i}$ are the cross-sections for coherent and incoherent scattering of the element i.

In the harmonic approximation, which will be used in the following, the one phonon part of the DDSC of a sample with one incoherently scattering element is directly proportional to the Vibrational Density of States (VDOS), as the incoherent scattering reflects just the single particle motion of the atoms. For a coherently scattering element the VDOS can still be determined within a good approximation, if for each frequency a mean value of the DDSC is determined in a sufficiently large region of reciprocal space [10], as it is the case here

$$g(\omega) = \frac{4\pi}{\sigma_{sc}} \frac{k_0}{k} \hbar\omega(1 - e^{-\beta}) \frac{8Mk_0k}{\hbar^2(Q_{max}^4 - Q_{min}^4)}$$
$$\int e^{2W} \frac{d^2\sigma_{sc}}{d\Omega dE}\bigg|_{1phon} \sin(\Theta)d\Theta . \quad (2)$$

Here $Q_{min} = Q(\Theta_{min})$ and $Q_{max} = Q(\Theta_{max})$ where Θ_{min} and Θ_{max} are the smallest and largest scattering angle used in the experiment. $\beta = \hbar\omega/k_B T$, M is the atomic mass of the scattering unit and 2W is the Debye-Waller coefficient. If one takes into account that the Bose-Einstein occupation factor, in the high temperature approximation, introduces another factor of β in the above equation, one realizes that the measured intensity is essentially proportional to $1/\omega^2$. Thus low energy excitations are measured with good statistical accuracy while the results measured at higher energy transfers suffer from less good statistics.

If the sample contains two or several differently scattering elements, instead of the VDOS

$$F(\omega) = \sum c_i g_i(\omega) \quad (3)$$

the GVDOS is determined, because each element couples differently to the neutrons:

$$G(\omega) = \frac{\sum w_i g_i(\omega)}{\sum w_i} \quad (4)$$

$$w_i = \frac{e^{-2W_i} c_i \sigma_{i,sc}}{M_i} \quad (5)$$

where the $g_i(\omega)$ are the *partial* vibrational density of states of the element i in the sample under investigation. The GVDOS is the quantity , which will be discussed here.

3 Sample Preparation and Experiments

Two samples of $Ni_{40}Pd_{40}P_{20}$ were prepared: one of them as bulk metallic glass consisting of 24 bars with 2mm diameter and 60mm length fixed between two Al- cylinders of 0.4 mm wall thickness and 7.6 and 10 mm inner radius. The other was prepared by melt-spinning of the same pre-alloy. These ribbons were measured in a Al-cylinder of 20 mm inner and 20.4 mm outer diameter. In a second experiment these ribbons were measured after having been annealed for one hour at 520K. The quench rate for the bulk glass was below 100K/s, that of the melt-spun sample above $10^5 K/s$. It is this difference in quench rate and the subsequent annealing, which allows to investigate the atomic dynamics of these samples with different instantaneous structures in different regions of the energy landscape and the rapid quench allows to get some information on the dynamics next to the super-cooled liquid.

The experiments were performed at the thermal neutron time-of-flight spectrometer MARI at the neutron spallation source ISIS (GB) at 200K with an incident energy of 75.4 meV. Scattering angles between 3 and 138 degrees were used, i.e. Q-values between 4 and 110 nm^{-1} were covered at $\hbar\omega = 0$. Of these only intensities measured between 75 and 110 nm^{-1} have been use for the determination of the GVDOS, because of the paramagnetic scattering from the Ni atoms. For Q=75 nm^{-1} the magnetic form factor was so low that the remaining intensity from magnetic scattering could be treated as a background. To compensate for the multi-phonon processes at these high Q-values, which increase proportional to Q^{2n}, with n=2 corresponding to the two-phonon processes etc., the measurements were made at 200K. From the two independent V-calibration measurements (because of the two different sample geometries, hollow cylinder and full cylinder) a resolution of 2.4 meV was determined from the FWHM of the peak of the elastically scattered neutrons in the V-spectra. This good resolution at $\hbar\omega = 0$ improves slightly in the neutron energy loss spectra with increasing energy transfer. Thus the GVDOS discussed in the following were determined from the weighted sum of the neutron energy loss spectra applying all necessary corrections except those for the finite resolution of the spectrometer [8]. Multi-phonon corrections were done in an iterative correction procedure until self-consistency was obtained [9] [10] and all GVDOS shown here were normalized to 1 at a cut-off energy of 60 meV.

4 Results

As here a quantitative comparison of the vibrational and some thermodynamic properties of a rapidly quenched, annealed and bulk glass is aimed for, the discussion will be concentrated on the GVDOS and the quantities calculated using this. Because the differences between the GVDOS is small, the GVDOS of the bulk sample determined from the two independent experiments done at the MARI spectrometer are shown in Fig. 1. The figure proves that the two GVDOS from two independent experiments agree with each other within approximately 3%.

The GVDOS of NiPdP consists of two main bands, of which the one centered at 23.5 meV is by far the dominant one, while the band centered around 45 meV nearly appears as a shoulder at the high energy side of the dominant band. As the weighting factors for the GVDOS of $Ni_{40}Pd_{40}P_{20}$ are 0.768 (Ni), 0.102 (Pd) and 0.13 (P) the Ni vibrations dominate the GVDOS, especially the band at lower energy. The band centered around 45 meV is most likely dominated by the vibration of P atoms.

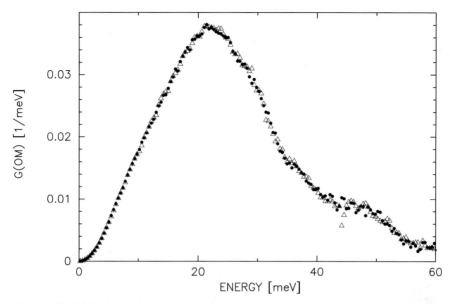

Fig. 1. GVDOS of the bulk metallic glass $Ni_{40}Pd_{40}P_{20}$ determined from two independent experiments. The two low values (open triangles) near 44 meV are caused by electronic errors

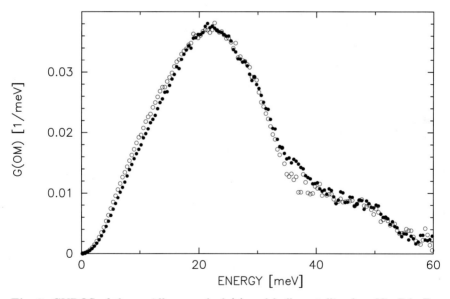

Fig. 2. GVDOS of the rapidly quenched (o) and bulk metallic glass $Ni_{40}Pd_{40}P_{20}$ measured at 200 K. The higher intensity at lower energy transfers is clearly observable

4.1 Rapidly and Slow Quenched NiPdP

Compared to the two GVDOS in Fig. 1 the GVDOS of the rapidly quenched sample clearly has a higher intensity at energies below 18 meV as shown in Fig. 2

This latter remark is qualitatively still true for the comparison of the GVDOS of the bulk sample with that of the melt spun and annealed sample, however quantitatively the differences between the two GVDOS are considerably reduced, as one can see in Fig. 3

In general, the differences observed are restricted to the energy range below 18 meV and near 37 meV.

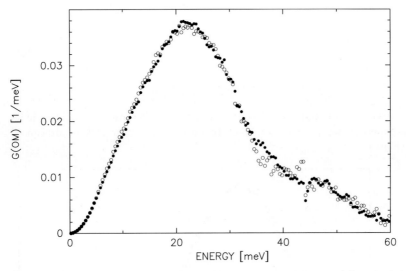

Fig. 3. GVDOS of the rapidly quenched and annealed sample (o) and of the bulk metallic glass $Ni_{40}Pd_{40}P_{20}$ both measured at 200K. The higher intensity in the GVDOS of the former sample is still visible however considerably reduced in comparison with Fig. 2

5 Discussion

The properties of the GVDOS mentioned above can be discussed more clearly with help of the difference of the GVDOS, $\Delta G(\omega) = G(\omega)_{rap.qu.} - G(\omega)_{bulk}$, which is shown in Fig. 4.

Figure 4 demonstrates clearly the higher intensity in the GVDOS of the melt spun sample at energy transfers below 18 meV and the compensation

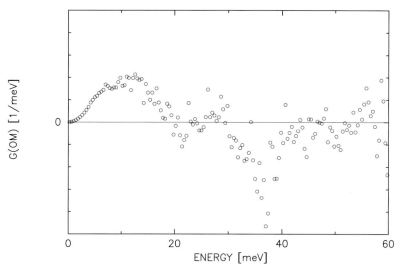

Fig. 4. Difference of the GVDOS of the rapidly quenched and of the bulk metallic glass $Ni_{40}Pd_{40}P_{20}$ both measured at 200K. The higher intensity in the GVDOS of the former sample is significant in the energy region below 18 meV and of the latter sample around 37 meV

of this intensity near 37 meV. 37 meV is just the Debye cut-off energy which one obtains from the second frequency moment of the GVDOS

$$<\omega^{(2)}> = \int_0^{\omega_{max}} \omega^2 G(\omega) d\omega \qquad (6)$$

$$\hbar\omega_D = \sqrt{\frac{7}{3}<\omega^{(2)}>}. \qquad (7)$$

Thus this experimental result proves the correctness of the suggestion [6][3] that the low energy modes expected in the VDOS of the more rapidly quenched sample are compensated at or next to the Debye cut-off energy. The fact that the GVDOS are different after quenches into different energy regions of the energy landscape proves at the same time that the *shape* of the basins must obviously be different in different regions of the energy landscape, as the atomic dynamics is different and this is directly related to the shape of the basins [11].

As here a comparative study is aimed for, the GVDOS can be used in place of the unknown VDOS to calculate thermodynamic and dynamic quantities. From the calculated quantities here the vibrational part of the entropy is of primary interest, because its contribution to the entropy change when going from T_g to T_g+dT is important for modeling the fragility of the glass-forming liquid next to T_g, where all glass-forming liquids are expected to follow an Arrhenius behavior.

$$S_{vib} = -3R \int_0^{\omega_{max}} G(\omega)[\epsilon \coth(\epsilon) - \ln(2\sinh(\epsilon))]d\omega , \qquad (8)$$

$$\epsilon = \frac{\hbar\omega}{2k_B T} \qquad (9)$$

where R is the gas constant and k_B is the Boltzmann constant. Calculating S_{vib} for both samples, which is of the order of 20 [J/mol K], one finds the contribution of the vibrations to the entropy *change* to be approximately 0.67 [J/mol K].

The same quantity can be calculated on the basis of the assumption that the change in the vibrational entropy on going from the slow cooled bulk glass to the higher energy rapidly quenched glass, which should reflect the energy situation of the super-cooled liquid just before reaching the glass transition temperature T_g, is determined by the probability that vibrational energy is transferred between the region of the Debye energy and the low energy region. For this calculation either the equation of reference [3] can be used

$$\Delta S_{vib} = -3R\ln(\frac{\omega_{LEM}}{\omega_D})c \qquad (10)$$

generalizing this by introducing the unknown amount of transferred vibrations c. Here ω_D the Debye cut-off frequency and ω_{LEM} is the center of gravity of the low-energy Modes (LEM), which can be calculated from the low energy region of the difference spectrum shown in Figs. 4 and Fig. 5

$$<\omega_{LEM}> = \int_0^{18} \omega \Delta G(\omega) d\omega . \qquad (11)$$

As one knows the difference spectrum one can also use this directly to calculate the difference of the vibrational part of the entropy on the basis of the same assumption and thus determine the unknown amount c of the transfered spectrum

$$\Delta S_{vib} = -3R \int_0^{18} \Delta G(\omega) \ln(\frac{\omega}{\omega_D}) d\omega . \qquad (12)$$

Using this equation one finds $\Delta S_{vib} = 0.71[J/mol K]$. This value is in very reasonable agreement with the one calculated from the $S_{vib}^{rap.qu.} - S_{vib}^{bulk}$ given above. From the comparison with the value calculated using equation(10) the amount of transfered modes c = 0.022, thus 2.2%. This latter value agrees very well with the direct integration of $\Delta G(\omega)$ shown in Fig. 5, which gives 0.021, i.e.2.1% of the total spectrum. The corresponding value obtained from integrating the difference spectrum of the two GVDOS shown in Fig. 1 in the same energy region is two orders of magnitudes smaller!

Thus the quantitative answer to the question concerning the vibrational contribution to the entropy change is: there is a contribution as suggested by

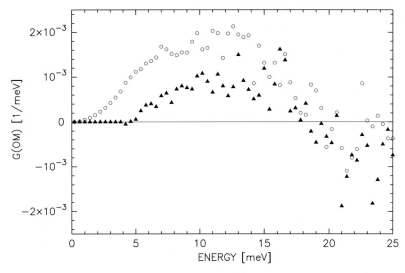

Fig. 5. Difference of the GVDOS of the rapidly quenched and of the bulk metallic glass $Ni_{40}Pd_{40}P_{20}$ (o) and of the rapidly quenched and annealed sample and the bulk sample both measured at 200K. . Structural relaxation obviously relaxes first the most extreme atomic configurations, approaching the situation of the slow cooled glass at lowest frequencies only

Angell and Moynihan [3], but at least for the glass investigated here (and in all computer simulations), $Ni_{80}P_{20}$, it is rather small and therefore much less important than the contribution from structural changes, as it was suggested by the computer simulations by Sciortino and collaborators [5].

Annealing the rapidly quenched sample below T_g will induce structural relaxation for which there was not enough time during the rapid quench in the melt spinning process. The question is here whether or not structural relaxation will lead to a similar situation concerning the atomic dynamics of the glass as does a slow quench. Fig. 5 demonstrates that this is not the case. The low energy intensity in the GVDOS is in fact reduced by more than a factor of 2 compared with the GVDOS of the as quenched sample, but first it is not completely reduced to the level of the slow quenched bulk sample, which - having no real "ideal" limit - we take here as the limiting case as one would do in case of crystals with a "defect free" crystal. And secondly the reduction of the low energy modes is by no means of equal weight in the full energy region shown in Fig. 5 but very strong at lowest energies. This means that annealing of this sample at 520K for one hour anneals out *the most extreme atomic configurations.*

The difference is seen even better when one looks in Fig. 6 at the differences of the vibrational parts of the specific heats calculated using the determined GVDOS of all three glass samples.

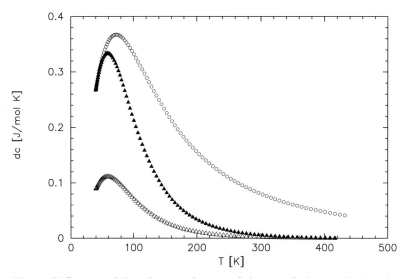

Fig. 6. Difference of the vibrational parts of the specific heats calculated using the GVDOS of the rapidly quenched and of the bulk metallic glass $Ni_{40}Pd_{40}P_{20}$ (o) and of the rapidly quenched and annealed sample and the bulk sample. Structural relaxation obviously narrows considerably the difference specific heat and shifts the maximum to nearly 20% lower temperatures. This is visible more clearly in the upper triangle curve, which is just the lower one multiplied by a factor of 3

When subtracting the calculated vibrational part of the specific heat of the bulk glass from that of the as-quenched sample, the difference has a maximum at 70 K and extends broadly up to 400 K. This is no longer the case when subtracting the calculated specific heat of the bulk glass from that calculated using the GVDOS of the annealed sample. Now the maximum is shifted to about 60K and the intensity reduced by more than a factor of 3. In addition the additional specific heat due to the LEM is now restricted to a much lower temperature range than in the case of the as-quenched sample. This clearly demonstrates the narrowing of the spectrum of additional LEM.

6 Conclusions

From the investigations presented above at least four conclusions can be drawn: Concerning the NiPdP sample used and the differences in quench rate and annealings reached here the atomic dynamics is different in different regions of the energy landscape, i.e. the shape of the basins slightly change.

A transfer of vibrational modes from the low energy region to the energy region of the Debye cut-off energy takes in fact place and can therefore be used to estimate the amount of vibrational entropy, *if* one knows the amount c of modes, which take part in this transfer. For the system studied here and under the conditions of this investigation c was of the order of 2%.

Consequently the contribution of vibrations to the entropy change when heating just above T_g is rather small, approximately 0.7 [J/mol K] in the case investigated here, which corresponds to approximately 3.3% of the vibrational entropy.

Finally short time (1h) structural relaxation of an as-quenched sample *does* reduce considerably the low energy modes, however it does not lead to the same state of relaxation as has the slow quenched bulk metallic glass sample, neither quantitatively nor qualitatively. Obviously the most extreme atomic situations are annealed out first.

Acknowledgments

It is a pleasure to thank H. Teichmann for making my samples. Very fruitful discussions with C.A. Angell and A. Granato are grateful acknowledged.

References

1. M. Goldstein, J. Chem. Phys. **51**, 3728 (1969), ibid.**67**, 2246 (1977).
2. F.A. Stillinger, Science **267**, 1935 (1995).
3. C.A. Angell,C.T. Moynihan, Metall. and Mater. Transact.B **31**, 587 (2000).
4. C.A. Angell, J. Non-Cryst. Solids **73**, 1 (1985).
5. F. Sciortino, W. Kob, P. Tartaglia, Phys. Rev. Lett. **83**, 3214 (1999).
6. W. Kob, F. Sciortino, P. Tartaglia, Europhys. Lett. **49**, 590 (2000).
7. J.- B. Suck, Structural relaxation in a metallic glass, in *Springer Proceedings in Physics 37* , D. Richter, W. Petry, J. Teixeira (Ed.) (Springer, Berlin, Heidelberg 1989) pp. 182–185.
8. J.- B. Suck, H. Rudin, H.- J. Güntherodt, H. Beck, J. Phys.C: Solid State Phys. **14**, 2305 (1981).
9. J.- B. Suck, H. Rudin, Vibrational Dynamics of Metallic Glasses Studied by Neutron Inelastic Scattering, in *Glassy Metals II* ,Topics in Applied Physics **53**, H. Beck, H.-J. Güntherodt (Ed.) (Springer, Berlin, Heidelberg 1983) pp. 217–260.
10. J.- B. Suck, J. Non-Cryst. Solids **153-154**, 573 (1993).
11. D. J. Lacks, Phys. Rev. Lett. **87**, 225502 (2001).

Part VII

Magnetism

Metallic Magnetism

Jürgen Kübler

Institut für Festkörperphysik, Technische Universität,
64289 Darmstadt, Germany

Abstract. Non-collinear moment arrangements in itinerant-electron systems are introduced into density functional theory. Certain ground-state properties of cases like fcc-Fe, invar, and other metallic magnets this way receive a microscopic foundation. Finite-temperature properties are explained by the non-collinearity of the magnetic moments. We discuss how the spin-wave spectrum can be calculated using spiral-spin structures and show how a theory that takes into account longitudinal and transverse spin-fluctuations succeeds in explaining the magnetic phase transition. Numerical results are presented for fcc Fe, invar and Co.

1 Introduction

A great number of ground-state properties of solids are astonishingly well described by the local density functional approximation (LDA) [1]. In particular, a coherent picture of magnetism in metals evolves through electronic structure studies on the basis of the LDA [2]. For the magnetic transition metals and their alloys one discovers that the very electrons that conduct electricity and heat cooperate to form magnetic moments which can, in general, order in non-collinear arrangements.

Excited-state properties of magnets (and other systems), however, are still a great challenge and it was believed until recently that the band-picture, based on the LDA, fails entirely in describing magnetism at elevated temperatures. We emphasize here that this is not so and we show that it is the essential non-collinearity of the magnetic moments at finite temperatures that drives the magnetic phase transition.

In this brief review we demonstrate the power of density-functional-based electronic structure theory by selecting a number of interesting cases pertaining to ground- and excited states properties of metallic magnets.

2 Ground-State Properties

Density functional theory rests on two statements [1,3]: the total ground-state energy of any many electron system in an external potential is a functional of the density, $n(\mathbf{r})$, and a statement that embodies a variational principle, which result in Euler-Lagrange equations; the latter are single-particle

Schrödinger equations where the effective potential depends on the density. Since we want to treat magnetic systems where the spin of the electron is an essential observable, we replace the scalar wave function by two-component spinor functions. The density is thus replaced by the density matrix, $\tilde{n}(\mathbf{r})$, i.e. a two by two matrix defined for every point, $\mathbf{r}$, in space. The density functional theorems still apply and the Euler-Lagrange equations for the spinor wave functions are obtained as (see e.g.[2])

$$\sum_{\beta}[-\delta_{\alpha\beta}\nabla^2 + v_{\alpha\beta}^{\text{eff}}(\mathbf{r}) - \varepsilon_i\,\delta_{\alpha\beta}]\,\psi_{i\beta}(\mathbf{r}) = 0 \qquad (1)$$

where α and β, due to the electron spin, take on the values 1 and 2 and where the effective potential is found to be given by

$$v_{\alpha\beta}^{\text{eff}}(\mathbf{r}) = v_{\alpha\beta}^{\text{ext}}(\mathbf{r}) + 2\,\delta_{\alpha\beta}\int\frac{n(\mathbf{r}')}{|\mathbf{r}-\mathbf{r}'|}\,d\mathbf{r}' + v_{\alpha\beta}^{xc}(\mathbf{r})\,, \qquad (2)$$

with $v_{\alpha\beta}^{xc}(\mathbf{r}) = \delta E_{xc}[\tilde{n}]/\delta\tilde{n}_{\beta\alpha}$. The notation is that usually employed in density functional theory and the last term on the right of eq.(2) is obtained in the LDA from the exchange-correlation energy of the electron gas by transforming locally to a frame of reference where the density matrix is diagonal (see below). The solutions of eq. (1) allow us to write the components of the density matrix as

$$\tilde{n}_{\beta\alpha}(\mathbf{r}) = \sum_{i=1}^{\infty}\psi_{i\beta}(\mathbf{r})\,\psi_{i\alpha}^{*}(\mathbf{r})\,f(\varepsilon_i)\,, \qquad (3)$$

where $f(\varepsilon_i)$ is the Fermi distribution function which here is the unit step function limiting the sum to extend over the lowest occupied states. Knowing the density matrix we write for the density $n(\mathbf{r}) = \text{Tr}\,\tilde{n}(\mathbf{r})$ and the magnetization follows from the expectation value of local spin moments as

$$\mathbf{m}(\mathbf{r}) = \text{Tr}\,\boldsymbol{\sigma}\tilde{n}(\mathbf{r})\,, \qquad (4)$$

where $\boldsymbol{\sigma}$ is given by the Pauli spin matrices.

Turning to the solutions of eq. (1) we assume first the effective potential is diagonal. In this case we face two separate equations for each of the two spin directions, up and down, and the density matrix is diagonal. If the up- and down- effective potentials remain different in the self-consistent-field problem, then we deal with a ferromagnet, for which eq. (4) gives the well-known magnetization $m_z(\mathbf{r}) = \tilde{n}_{\uparrow}(\mathbf{r}) - \tilde{n}_{\downarrow}(\mathbf{r})$, the other components being zero. The conditions for finding this type of solution are controlled by the well-known Stoner exchange integral.

Next we assume the effective potential is not diagonal, a case, which in general leads to non-collinear spin arrangement [4]. In a first step a local coordinate system is chosen such that the density matrix, defined by Eq.

(3), integrated over the atomic sphere is diagonal, i.e. we diagonalize the integrated matrix $n_{\alpha\beta}$ using the well-known spin-1/2 rotation matrix

$$\sum_{\alpha\beta} U_{i\alpha}(\theta,\varphi)\, n_{\alpha\beta}\, U^+_{\beta j}(\theta,\varphi) = n_i\, \delta_{ij} \qquad (5)$$

where the polar angles are given by $\tan\varphi = -\mathrm{Im}\,(n_{12})/\mathrm{Re}\,(n_{12})$ and $\tan\theta = 2\sqrt{(\mathrm{Re}\,(n_{12}))^2 + (\mathrm{Im}\,(n_{12}))^2}/(n_{11} - n_{22})$. A second step consists in expressing the spin dependent part, $v^{xc}_{\alpha\beta}$, in the effective potential by the polar angles as well. For this we use

$$\frac{\delta n_i}{\delta \tilde{n}_{\alpha\beta}} = U_{i\alpha}\, U^+_{\beta i} \qquad (6)$$

to eliminate $\delta\tilde{n}_{\alpha\beta}$ in the expression for the exchange-correlation potential, thus obtaining the desired dependence on the polar angles. Equation (6) is derived from the quadratic equation that is solved in order to diagonalize the integrated density matrix. Furthermore, since the latter is diagonal in the frame of reference defined by the polar angles, the values of θ and ϕ give the direction of the magnetic moment of a given atom in the crystal. Examples of observed and calculated non-collinear ground-states can be found in [2] and [4].

There is a special type of non-collinear state that possesses a parameter, which allows us to continuously tune the system from the ground state to different states having higher or lower energy. The latter case thus leads to replacing what was believed to be the ground-state by a possibly more realistic ground-state. The former case allows a determination of excited states. The non-collinear state in question consists of an incommensurate spin spiral, defined by giving the Cartesian coordinates of the magnetization vector,

$$\mathbf{M}_n = M\, [\cos(\mathbf{q}\cdot\mathbf{R}_n)\sin\theta,\, \sin(\mathbf{q}\cdot\mathbf{R}_n)\sin\theta,\, \cos\theta] \quad . \qquad (7)$$

Here M is the magnitude of the magnetic moment at site $\mathbf{R}_n$, and $(\mathbf{q}\cdot\mathbf{R}_n)$ as well as θ are polar angles. On first sight it appears that the periodicity is lost with respect to lattice translations non-orthogonal to $\mathbf{q}$. One should notice, however, that all atoms of the spiral structure separated by a translation $\mathbf{R}_n$ are equivalent, possessing magnetic moments of equal magnitude. This equivalence leads to an interesting property for the single-particle spinor functions first pointed out by Herring [6] and later by Sandratskii [7], who constructed transformations combining a lattice translation $\mathbf{R}_n$ and a spin rotation about the z-axis by an angle $\mathbf{q}\cdot\mathbf{R}_n$ that leave the spiral structure invariant. The symmetry operators describing this transformation are members of a group whose elements can be shown to commute with the Hamiltonian and whose representation gives a generalized Bloch theorem. This means eigen-spinors $\psi_\mathbf{k}(\mathbf{r})$ are labeled by $\mathbf{k}$-vectors that lie in the first Brillouin zone defined in the usual way. The spin spiral $\mathbf{M}_n$ given by eq. (7) does not break the translational symmetry of the lattice, although, in general, the point-group symmetry may be reduced. This statement is independent of the choice of **q** which, therefore, need not be commensurate with the lattice. For further details see the original literature and Ref. [2,8].

As an example for interesting ground-state properties we first turn to the much dealt with case of fcc or γ-Fe, which has recently been taken up again by Knöpfle et al. [5] using a full-potential calculation and the generalized gradient correction (GGA) [9]. Their results are shown in Fig. 1. They demonstrate that for a certain range of volume, in agreement with experiment, γ-Fe orders in a non-collinear state. Attention is drawn to the cross-over of the two total-energy curves describing high-moment ferromagnetic and low-moment antiferromagnetic γ-Fe, below which the non-collinear total energy curve is found. With some small, but significant changes in the relative positions of the total-energy characteristics this is carried over to the case of NiFe$_3$-invar, our next example.

NiFe$_3$-invar possesses pronounced and unusual thermal expansion and elastic properties [10]. It has received recent theoretical attention by van

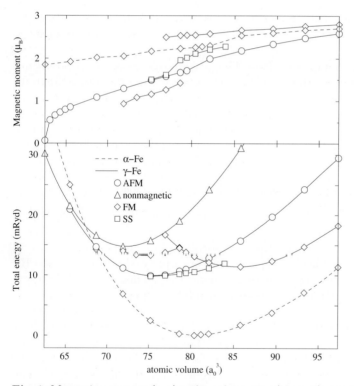

Fig. 1. Magnetic moments (top) and total energies (bottom) counted from the total energy-minimum of bcc Fe (α-Fe) calculated by Knöpfle et al. [5] using the GGA for ferromagnetic (FM), antiferromagnetic (AFM), nonmagnetic, and spiral structures (SS) in cubic (α and γ) iron as a function of atomic volume

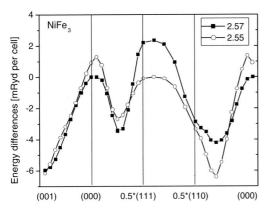

Fig. 2. NiFe$_3$: Total energy differences as a function of the wave vector along symmetry lines in the first Brillouin zone for two different volumes given by the atomic-sphere radii $S = 2.55$ a.u. and $S = 2.57$ a.u.. Energy is counted from the value at the zone center, (000), of the $S = 2.57$ value

Schilfgaarde et al. [11] also establishing with hardly any doubt that the ground-state of NiFe$_3$-invar possesses non-collinear order. In Fig. 2 we graph results of an LDA total-energy calculation for NiFe$_3$-invar. Two volumes given by atomic sphere radii $S = 2.55$ a.u. and $S = 2.57$ a.u. are chosen near the cross-over of a high-moment and a low-moment state that are comparable with the case of γ-Fe. Variable is the value of the **q**-vector chosen along symmetry lines in the Brillouin zone as indicated. One sees that depending on the volume a number of total-energy minima exist, the lowest occurring for $S = 2.55$ a.u. at about (0.35,0.35,0) indicating non-collinear order. This is not necessarily the absolute minimum in the total energy since our scan of the Brillouin zone is incomplete. Furthermore, as Ref. [11] shows, a spiral-structure may only imperfectly model the real state. Still, the multi-minima structure of Fig. 2 leads to the speculation that the real ground state is frustrated resulting in a spin-glass as observed [10]. - Further applications, in particular those involving strong relativistic effects, can be found in the review article [8].

3 Excited States and Thermal Properties

We now solve the Kohn-Sham-Schrödinger equations defined by eq. (1) using particular spin-spiral states as an input. The total energy as a function of the polar angles and the spiral wave vector, **q**, is then obtained. Let us denote the total energy counted from the total energy of the ferromagnetic state by $\Delta E(\mathbf{q}, \theta)$. It can be shown that the magnon frequency of an elementary ferromagnet is given by

$$\omega(\mathbf{q}) = \lim_{\theta \to 0} \frac{4}{M} \frac{\Delta E(\mathbf{q}, \theta)}{\sin^2 \theta} . \tag{8}$$

The quantity M is the magnetic moment (in units of μ_B). The total energy difference $\Delta E(\mathbf{q}, \theta)$ is obtained using the force theorem, i.e. from the change of the band energies when the ferromagnetic ground state is changed to a magnetic spiral characterized by the wave vector $\mathbf{q}$ and the polar angle θ.

This expression has been derived for the magnon spectrum in Ref. [2] and [12] under very general assumptions using the adiabatic approximation. A derivation by Halilov et al. [13], though, rests on assuming the Heisenberg model. Their formula agrees with Eq. (8) and so does that of Rosengaard and Johansson [15] who used similar assumptions as Halilov et al. Another derivation of the same result is due to Antropov et al. [16] who also used an equation of motion method.

3.1 Magnons

Magnon dispersion relations have been calculated by Halilov et al. [13] and, more recently, by Pajda et al.[14]. From the values of the dispersion near the zone origin one determines the spin-wave stiffness constant in very good agreement with measured values. It is to be noted that the Stoner continuum is not obtained with eq. (8).

We now make use of the calculated magnon spectrum to determine the decrease of the magnetization at low temperatures and postulate that thermal equilibrium is properly described by quantizing the magnon spectrum and using the Planck distribution function for calculating the mean occupation number, $\langle n_\mathbf{q} \rangle$, for a magnon of energy $\omega_\mathbf{q}$, i.e.

$$\langle n_\mathbf{q} \rangle = \frac{1}{\exp(\omega_\mathbf{q}/k_\mathrm{B} T) - 1} \,. \tag{9}$$

Assuming at low temperatures the magnons do not interact we add up the single-magnon energies to obtain the total energy as

$$E = \sum_\mathbf{q} \omega_\mathbf{q} \langle n_\mathbf{q} \rangle \,. \tag{10}$$

The total energy change may then be equated with $\omega_\mathbf{q} \langle n_\mathbf{q} \rangle$ and using Eq. (8) for small but finite θ we obtain

$$\langle \theta_\mathbf{q}^2 \rangle = \frac{4}{M} \langle n_\mathbf{q} \rangle \,, \tag{11}$$

where for small θ the mean value of $\sin^2 \theta$ is written as $\langle \theta_\mathbf{q}^2 \rangle$. This equation now enables us to evaluate the decrease of the magnetization at finite temperatures, $\Delta M(T)$:

$$\Delta M(T) = M - M \sum_\mathbf{q} \langle \cos \theta_\mathbf{q} \rangle \simeq M \sum_\mathbf{q} \langle \theta_\mathbf{q}^2 \rangle /2 = 2 \sum_\mathbf{q} \langle n_\mathbf{q} \rangle \,. \tag{12}$$

The same result has been obtained by Halilov et al.[13] who, however, used a different method. In Fig. 3 we show (among things to be discussed below) the magnetization-decrease at low temperatures obtained with eq. (12). At elevated temperatures the occupation numbers $\langle n_{\mathbf{q}} \rangle$ and the amplitudes $\langle \sin \Theta_{\mathbf{q}} \rangle$ become large; the modes begin to interact and the magnon picture is to be replaced by a picture of strong long-wavelength transverse and longitudinal spin fluctuations. In contrast to the pioneering work of Gyorffy et al. [17], who developed a high-temperature theory, we here sketch a method that allows us to treat all temperatures.

3.2 Spin Fluctuations and the Magnetic Phase Transition

We now define spin fluctuations and construct a realistic Hamiltonian that is derived from the energetics of a magnet near its ground-state.

First, the integrated magnetic moment of an atom at the lattice point $\mathbf{R}$ is written as

$$\mathbf{M}(\mathbf{R}) = M\, \mathbf{e}_z + \mathbf{m}(\mathbf{R}) = M\, \mathbf{e}_z + \sum_{j,\mathbf{k}} m_{j\mathbf{k}} \exp(i\mathbf{k} \cdot \mathbf{R})\, \mathbf{e}_j \,. \tag{13}$$

Here we imply that M is the macroscopic magnetization along some direction, say the z-direction, and $\mathbf{m}$ is a local deviation of the magnetization which is expanded in a Fourier series. Since $\mathbf{m}(\mathbf{R})$ is real we require the Fourier coefficients to obey $m_{j-\mathbf{k}} = m_{j\mathbf{k}}^*$. Physically they describe spin fluctuations. The quantities $\mathbf{e}_j$ ($j = 1, 2, 3$) are Cartesian unit vectors.

Second, we require the desired Hamiltonian to consist of two terms, $\mathcal{H} = \mathcal{H}_1 + \mathcal{H}_2$. The first term on the right-hand side is assumed to describe single-site terms, which are written as

$$\mathcal{H}_1 = \frac{1}{N} \sum_i \sum_{n=1}^{2} \alpha_{2n}\, \mathbf{M}(\mathbf{R}_i)^{2n} \,, \tag{14}$$

where N is the number of atoms in the crystal. This Hamiltonian reduces at $T = 0$ to the total energy change $\Delta E_1 = \alpha_2\, M^2 + \alpha_4\, M^4$, which allows us to read off the coefficients α_2 and α_4 from a total-energy calculation. It is possible to use more than two terms in this expansion, but we will not do so here. It should be noted that this part of the total energy is not to be confused with a Landau expansion, since our coefficients α_2 and α_4 are temperature independent. Next we include two-site interaction terms and therefore write

$$\mathcal{H}_2 = \frac{1}{N} \sum_{il} J(\mathbf{R}_i - \mathbf{R}_l)\, \mathbf{M}(\mathbf{R}_i) \cdot \mathbf{M}(\mathbf{R}_l) \,. \tag{15}$$

In keeping with the usual terminology we call the quantities $J(\mathbf{R} - \mathbf{R}')$ exchange constants, which we obtain again from total-energy calculations; here we substitute a spin spiral of the type given in Eq. (7) and obtain the total energy corresponding to Eq. (15), which can be expressed by means of the Fourier-transformed exchange constants as

$$\Delta E_2(M, \mathbf{q}, \theta) = M^2 \, j(\mathbf{q}) \, \sin^2 \theta \, . \tag{16}$$

which defines the second member of the Hamiltonian completely. In practice it is sufficient to use $\theta = 90°$ and the total energy differences are then calculated for a selected set of the spiral wave vectors $\mathbf{q}$.

For the thermodynamics that is next, a manageable method is a variational treatment that defines in a consistent way a mean (or molecular) field theory. We base this on the Bogoliubov-Peierls inequality that states for the free energy $F \leq F_0 + \langle \mathcal{H} - \mathcal{H}_0 \rangle_0$. The quantity $\mathcal{H}_0$ is a model Hamiltonian that should possess a rigorous solution defining the free energy F_0. The right-hand side is made as small as possible by means of variational parameters contained in the definition of the model Hamiltonian $\mathcal{H}_0$. We choose

$$\mathcal{H}_0 = \sum_{j,\mathbf{k}} a_{j\mathbf{k}} \, |m_{j\mathbf{k}}|^2 \, , \tag{17}$$

where $a_{j\mathbf{k}}$ are variational parameters. This choice of $\mathcal{H}_0$ leads to Gaussian statistics which is known to allow all thermal averages and the partition function to be carried out analytically obtaining for the zero-order free energy

$$F_0 = -\frac{1}{2\beta} \sum_{j\,\mathbf{k}} \ln \frac{\pi}{2\beta \, a_{j\,\mathbf{k}}} \quad \text{where} \quad \beta = 1/k_B T \, . \tag{18}$$

Furthermore, to evaluate the thermal average of the real Hamiltonian, $\langle \mathcal{H} \rangle_0$, averages of the type

$$\langle |m_{j\,\mathbf{k}}|^2 \rangle_0 = \frac{1}{2\beta \, a_{j\,\mathbf{k}}} \quad , \quad \langle |m_{j\,\mathbf{k}}|^4 \rangle_0 = 3 \, \langle |m_{j\,\mathbf{k}}|^2 \rangle_0^2 \tag{19}$$

are needed which are easily verified.

In order to write out $\langle \mathcal{H} \rangle_0$ as concisely as possible it is worthwhile to define abbreviations of frequently occurring quantities. There is first the sum $\sum_{\mathbf{k}} \langle |m_{j\,\mathbf{k}}|^2 \rangle_0$ of the Cartesian component j which can be in the direction of the macroscopic magnetization or perpendicular to it. The former we call the longitudinal (l), the latter the transverse (t) fluctuations, i.e. we define

$$t^2 = \sum_{\mathbf{k}} \langle |m_{t\,\mathbf{k}}|^2 \rangle_0 \quad , \quad l^2 = \sum_{\mathbf{k}} \langle |m_{l\,\mathbf{k}}|^2 \rangle_0 \, . \tag{20}$$

The Bogoliubov variational free energy, F_1, can at this stage be expressed in terms of $\langle |m_{j\,\mathbf{k}}|^2 \rangle_0$ which, because of Eq. (19), replaces the variational parameter $a_{j\mathbf{k}}$. We collect

$$F_1 = -\frac{k_B T}{2} \sum_{j\,\mathbf{k}} \left[1 + \ln \left(\pi \, \langle |m_{j\mathbf{k}}|^2 \rangle_0 \right) \right] + \langle \mathcal{H}_1 \rangle_0 + \langle \mathcal{H}_2 \rangle_0 \, . \tag{21}$$

and obtain for the averages on the right-hand side

$$\langle \mathcal{H}_1 \rangle_0 + \langle \mathcal{H}_2 \rangle_0 = \alpha_2 (M^2 + 2t^2 + l^2)$$
$$+ \alpha_4 \left[M^4 + 2M^2(2t^2 + 3l^2) + 8t^4 + 4t^2 l^2 + 3l^4 \right] \quad (22)$$
$$+ 2 \sum_{\mathbf{k}} j(\mathbf{k}) \langle |m_{t\mathbf{k}}|^2 \rangle_0 + \sum_{\mathbf{k}} j(\mathbf{k}) \langle |m_{l\mathbf{k}}|^2 \rangle_0$$

Next there are the variational equations that need be determined, the first being $\partial F_1/\partial M = 0$, the second $\partial F_1/\partial \langle |m_{j\mathbf{q}}|^2 \rangle_0 = 0$ for $j = t$ and $j = l$. The first gives in the ordered state

$$\frac{M^2}{M_s^2} = 1 - \frac{2t^2 + 3l^2}{M_s^2}, \quad (23)$$

where the saturation magnetization is denoted by M_s. This result was apparently first obtained by Moriya [18]. His work and that of Lonzarich and Taillefer [19] as well as Wagner [20] and Uhl [21] has to be credited here who, in one way or the other, developed much of the present theory.

The second optimization condition gives self-consistency equations that follow after some algebra:

$$l^2 = k_B T \sum_{\mathbf{k}} \chi_L(\mathbf{k}), \text{ where } \quad \chi_L^{-1}(\mathbf{k}) = 8\alpha_4 M^2 + 2j(\mathbf{k}) \quad (24)$$

and

$$t^2 = k_B T \sum_{\mathbf{k}} \chi_T(\mathbf{k}), \text{ where } \quad \chi_T^{-1}(\mathbf{k}) = 8\alpha_4(t^2 - l^2) + 2j(\mathbf{k}). \quad (25)$$

In the paramagnetic case we set $M = 0$ in the free energy and differentiate with respect to the paramagnetic mode which we denote by $\langle |m_{p\mathbf{k}}|^2 \rangle_0$ obtaining for the paramagnetic fluctuations, $p^2 = \sum_{\mathbf{k}} \langle |m_{p\mathbf{k}}|^2 \rangle_0$, the relation

$$p^2 = k_B T \sum_{\mathbf{k}} \chi_P(\mathbf{k}), \text{ where } \quad \chi_P^{-1}(\mathbf{k}) = 2\alpha_2 + 20\alpha_4 p^2 + 2j(\mathbf{k}). \quad (26)$$

Finally, we obtain from the free energy for high temperatures the inverse susceptibility $\chi^{-1} = 2\alpha_2 + 20\alpha_4 p^2$ which we compare with the Curie-Weiss law commonly written as

$$\chi = C/(T - T_c), \text{ where } \quad C = q_c(q_c + 2)\mu_B/3k_B. \quad (27)$$

In Fig. 3 the reduced macroscopic magnetization for fcc Co, labeled $M/M_s(0)$, as a function of the temperature is seen to decrease while the reduced transverse, t^2, and longitudinal, l^2, fluctuations increase until solutions to the self-consistency equations, eqs. (23) - (25) cease to exist slightly short of 910 K. The calculated Curie temperature is seen to be smaller than the

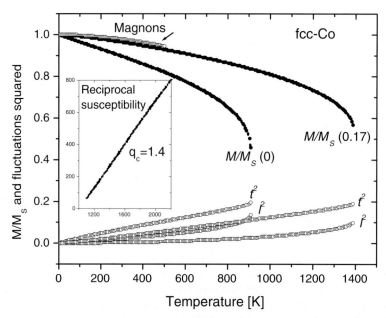

Fig. 3. Calculated magnetization data for fcc Co. Longitudinal, l^2, transverse, t^2, fluctuations and reduced magnetization M/M_s are shown for two different approximations, see text. The curve labeled magnons results from eq. (12). The quantity q_c appearing in the inset for the reciprocal susceptibility is defined by eq. (27)

measured value of $T_c = 1388$ K and the magnetization profile is only qualitatively in agreement with the experimental magnetization. Compared with the magnon-contribution shown, the calculated magnetization decreases too fast at low temperatures, which is due to our treating the fluctuations as classical variables. We briefly will come back to this point below. Furthermore, the magnetization disappears too abruptly at high temperatures resulting in a first order phase transition. We show in the inset the calculated high-temperature inverse susceptibility, which is seen to be approximately linear in temperature, but the slope of χ^{-1} is too high by nearly a factor of two. In fact the calculated value of q_c from eq. (27) shown in the inset of Fig. 3 should be compared with the experimental value of $q_c = 2.29$, which is a measure of the moments above the Curie temperature [22]. The calculations can be improved somewhat if the generalized gradient correction [9] is used instead of the LDA. In this case one obtains a calculated Curie temperature of $T_c = 1070$, but the slope of the Curie-Weiss law is only marginally improved.

In Fig. 4 we show the calculated total energy differences, which describe the hcp-fcc phase transition of Co. The reason for Co to possess the hcp crystal structure at low temperatures can be traced back to the occupation of the minority-bands as pointed out by Söderlind et al. [23] thus explaining

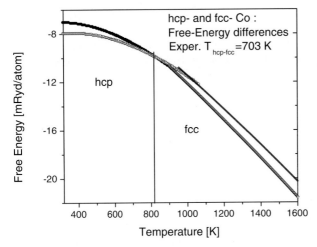

Fig. 4. Calculated free energies for fcc and hcp Co. This is only the contribution from spin fluctuations. The small kinks at high temperatures result from the transition to the paramagnetic state. The crossing of the fcc- and hcp-curves marks the calculated hcp-fcc phase transition, which should be compared with the experimental value given in the figure

the off-set values of the free-energy differences at $T = 0$. The larger curvature of the fcc free energy or the larger rate of change of the fcc entropy as a function of the temperature then overwhelms the hcp-stabilization. The delicate dependence of the crossing of the free energies on the off-set energy results in a somewhat imprecise prediction of the transition temperature. It should be mentioned, however, that the theory is incomplete since effects of lattice vibrations, neglected here, are certainly of importance.

The formalism described above can, with some effort, be generalized to deal with more than one atom per unit cell. Similar to the theory of lattice vibrations, a normal-mode analysis is required, in particular eq. (13) and eq. (17) need be amended by sums over normal modes, which will carry through to eqs.(24)-(26) [2]. Thus hcp-Co (with this Fig. 4) and compounds like CrO_2 and $ZrZn_2$ are well described; furthermore, the Curie temperatures of multilayers have been investigated successfully but cannot be dealt with here.

4 Conclusion

Recognizing the deficiencies of the theory one could begin by improving eqs. (24)-(26), which represent the high-temperature limit of the fluctuation-dissipation theorem, see e.g. Ref. [18]. Utilizing a reasonable approximation for the frequency-dependent susceptibility derived from the Lindhard function as was suggested by Moriya [18] as well as Lonzarich and Taillefer [19] one can, based on the fluctuation-dissipation theorem, derive phenomenological corrections that are characterized by a parameter, which at present cannot be obtained from an *ab initio* calculation, but can be chosen such that the low-temperature magnetization fits the data calculated from the magnon spectrum. Preliminary results of such a treatment are shown in Fig. 3 by the data labeled M/M_s (0.17) the free parameter being 0.17. Unfortunately, the Curie-Weiss law is not significantly improved by this treatment. A unified self-consistent theory as proposed by Moriya [18] seems to be necessary.

References

1. W. Kohn, L.J. Sham: Phys. Rev. A **140**, 1133 (1965).
2. J. Kübler: *Theory of Itinerant Electron Magnetism* (Oxford University Press, Oxford 2000).
3. P. Hohenberg, W. Kohn: Phys. Rev. B **136**, 864 (1964).
4. J. Sticht, K.-H. Höck and J. Kübler: J. Phys.: Condens. Matter **1**, 8155 (1989).
5. K. Knöpfle, L.M. Sandratskii, J. Kübler: Phys. Rev. B **62**, 5564 (2000).
6. C. Herring, in: *Magnetism IV*, Chapter V and XIII, ed. by G. Rado and H. Suhl (Academic Press, New York 1966).
7. L.M. Sandratskii: Phys. Stat. Sol. (b) **135**, 167 (1986) and J. Phys. F: Metal Phys. **16**, L43 (1986).
8. L.M. Sandratskii: Adv. Phys. **47**, 91 (1998).
9. J.P. Perdew, K. Burke, M. Enzerhof: Phys. Rev Lett. **77**, 3865 (1996).
10. E.F. Wassermann: *INVAR: moment-volume instabilities in transition metals and alloys*, in *Ferromagnetic materials*, vol. 5 ed. K.H.J. Buschow and E.P. Wohlfarth (North-Holland, Amsterdam 1990)
11. M. van Schilfgaarde, I.A. Abrikosov, B. Johansson: Nature **400**, 46 (1999).
12. Q. Niu and L. Kleinman: Phys. Rev. Lett. **80**, 2205 (1998).
13. S.V. Halilov, H. Eschrig, A.Y. Perlov, and P.M. Oppeneer: Phys. Rev. B **58**, 293 (1998).
14. M. Pajda, J. Kudrnovsky, I. Turek, V. Drchal, P. Bruno: Phys. Rev. B **64**, 174402 (2001).
15. N.M. Rosengaard and B. Johansson: Phys. Rev. B **55**, 14975 (1997).
16. V.P. Antropov, M.I. Katsnelson, M. van Schilfgaarde, B.N. Harmon, and D. Kuznezov: Phys. Rev. B **54**, 1019 (1996).
17. B.L. Gyorffy, A.J. Pindor, J. Staunton, G.M. Stocks, H. Winter: J. Phys. F: Metal Phys. **15**, 1337 (1985).
18. T. Moriya: *Spin Fluctuations in Itinerant Electron Magnetism* (Springer, Berlin 1985).
19. G.G. Lonzarich, L. Taillefer: J. Phys. C: Sol. State Phys. **18**, 4339 (1985).
20. D. Wagner: J. Phys. Condens. Matter **1**, 4635 (1989).
21. M. Uhl, J. Kübler: Phys. Rev. Lett. **77**, 334 (1996).
22. E.P Wohlfarth: *Iron, Cobalt and Nickel*, in *Ferromagnetic Materials*, vol. 1 ed. E.P. Wohlfarth. (North-Holland, Amsterdam 1980)
23. P. Söderlind, R. Ahuja, O. Eriksson, J.M. Wills, B. Johansson: Phys. Rev. B **50**, 5918 (1994).

Domain State Model for Exchange Bias: Influence of Structural Defects on Exchange Bias in Co/CoO

Bernd Beschoten[1], Andrea Tillmanns[1], Jan Keller[1], Gernot Güntherodt[1], Ulrich Nowak[2], and Klaus D. Usadel[2]

[1] 2. Physikalisches Institut, RWTH Aachen,
   Templergraben 55, 52056 Aachen, Germany
[2] Theoretische Tieftemperaturphysik, Gerhard-Mercator-Universität Duisburg,
   47048 Duisburg, Germany

Abstract. The exchange bias coupling at ferromagnetic/antiferromagnetic interfaces in epitaxially grown Co/CoO bilayers can be intentionally enhanced by a factor of up to 4 if the antiferromagnetic CoO layer is diluted by non-magnetic defects in its volume part away from the interface. Monte Carlo simulations of a simple model consisting of a ferromagnetic layer exchange coupled to a diluted antiferromagnetic layer show exchange bias of the right order of magnitude and qualitatively reproduce the experimentally observed dependence of the exchange bias field on the number of defects. The exchange bias results from a domain state in the antiferromagnet, which is formed during field cooling and carries an irreversible domain state magnetization. Apart from intentionally introduced non-magnetic defects, also structural defects can enhance the exchange bias coupling. Twin boundaries in undiluted CoO increase the exchange bias coupling in Co/CoO by more than a factor of 2 compared to untwined samples. This observation indicates that structural defects in the antiferromagnet, such as twin or grain boundaries, might also stabilize a domain state, suggesting that the domain state model for exchange bias is more generally applicable to understand the origin of the exchange bias phenomenon.

1 Introduction

Direct exchange coupling at the interface between a ferromagnetic (FM) and an antiferromagnetic (AFM) layer may result in exchange biasing, which induces a unidirectional anisotropy of the FM layer. This unidirectional anisotropy causes a shift of the hysteresis loop along the magnetic field axis. The magnitude of the field shift is called the exchange bias (EB) field B_{EB}. Usually, the EB shift occurs after field cooling the system with a saturated FM layer below the Néel temperature T_{N} of the AFM layer or by layer deposition in an external magnetic field. Despite more than four decades of research since its discovery [1,2] and the application in commercially available magnetic sensor devices [3,4], the microscopic understanding of the exchange bias EB effect is still not fully established. For a review of the vast literature on EB the reader is referred to a recent article by Nogués and Schuller [5].

The identification of the microscopic spin structure of the AFM layer close to the interface to the FM layer is of fundamental relevance for the microscopic origin of the EB phenomena. Recently, we studied the EB in ferro-/antiferromagnetic Co/CoO bilayers as a function of volume defects in the antiferromagnet. Of particular importance in this study was the observation that the non-magnetic defects in the volume of the AFM layer can enhance the EB by a factor of up to four [6,7]. The defects were not intentionally placed at the FM/AFM interface but rather throughout the volume part of the AFM layer. Therefore, the observed EB is primarily not due to disorder or defects at the interface. Rather, the strong dependence of the EB field on the dilution of the AFM layer was concluded to have its origin in the formation of a domain state in the volume of the AFM. This domain state gives rise to a small but significant excess of moments at the FM/AFM interface which causes and controls EB. The 'domain state' (DS) model gives a novel microscopic description of exchange bias, which is supported by Monte Carlo (MC) simulations [6,8].

The article is organized as follows. We will give an overview about some basic models for EB in section 2. In section 3 the domain state model for exchange bias is presented and the most important experimental and theoretical results are given including the domain state magnetization in section 4. The relevance of structural defects (twin boundaries) in Co/CoO and their link to the DS model is discussed in section 5. Finally, we conclude in the last section.

2   Models for Exchange Bias

One of the puzzling observations in EB systems is the large variation of the EB coupling strength even between the same type of FM/AFM material systems [5]. In order to better compare the EB coupling strength between various FM/AFM systems independent of the FM material and its thickness, the EB coupling constant or interface energy k_{EB} is often used, which equals the interface energy per unit area

$$k_{EB} = B_{EB} M_{FM} t_{FM}, \tag{1}$$

where M_{FM} and t_{FM} are the saturation magnetization and the thickness of the FM layer. This relation results from a simple energy argument for which the Zeeman energy of the FM in an external magnetic field $E_Z = \mu_0 M_{FM} H$ equals the interface energy normalized to its volume $E_I = k_{EB}/t_{FM}$. Here, the AFM acts as an additional external magnetic field, which induces a shift of the magnetic hysteresis loop by B_{EB}.

In the simplest model for EB, the interface between the FM and the AFM layer is perfectly flat and defect free with the moments of the AFM interface

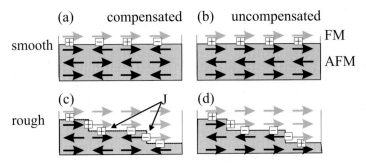

Fig. 1. Schematic structures of magnetic moments at smooth and rough (stepped) FM/AFM interfaces. The AFM interface layer is either compensated (a),(c) or uncompensated (b),(d). The plus and minus signs represent energetically favorable and unfavorable orientations of FM/AFM moment pairs. A ferromagnetic interface coupling (+ sign) is assumed

layer either being compensated or uncompensated. Additionally, the AFM layer is assumed to have an infinite anisotropy, i.e. the AFM spins are either oriented parallel or antiparallel to the FM magnetization. For a compensated moment structure (see. Fig. 1(a)), the exchange couplings between neighboring FM and AFM moments are energetically favorable (+ sign for FM interface coupling J) and unfavorable (− sign) in alternating order. Therefore the macroscopic interface exchange coupling energy vanishes and no EB exists. However, for an uncompensated AFM moment surface, all FM/AFM moment pairs are either favorable (Fig. 1(b)) or unfavorable, which yields different interface exchange coupling energies with $k_{EB} = (NJ_{INT})/F$, where N/F is the number of moment pairs per unit area and J_{INT} is the coupling energy of a FM/AFM moment pair. The resulting EB field is

$$B_{EB} = \frac{NJ_{INT}}{F} \frac{1}{M_{FM} \cdot t_{FM}}. \tag{2}$$

Using reasonable numbers, this equation gives EB fields which are typically two orders of magnitude larger than the experimentally observed values. However, any realistic experimental system has a certain interface roughness. If statistically distributed, these, i.e., stepped interfaces lead to zero interface exchange coupling energy and thus zero EB coupling for both compensated (Fig. 1(c)) and uncompensated (Fig. 1(d)) AFM moment interface layers.

Malozemoff was the first one pointing out the role of AFM domains for the EB effect due to interface roughness [10,11,12]. These domains, which were assumed perpendicular to the FM/AFM interface (see Fig. 2), were supposed to occur during cooling the system below T_N in the presence of the magnetized FM layer and to carry a small net magnetization at the FM/AFM interface. This magnetization was thought to be increasingly stabilized towards low temperatures, consequently shifting the hysteresis loop.

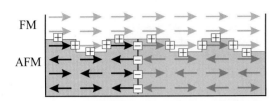

Fig. 2. Schematic structure of magnetic moments at FM/AFM interface. A domain wall in the AFM layer perpendicular to the interface leads to a net interface coupling (plus signs)

However, the formation of domain walls caused only by interface roughness is energetically unfavorable and has not directly been proven.

The magnetic linear dichroism effect and photoelectron emission spectroscopy has recently been used to probe and to image both FM and AFM domains in FM/AFM exchange coupled systems [13,9]. These domains have been shown to be coinciding and to provide evidence of EB coupling on a local scale.

In the micro magnetic model by Schulthess and Butler [15,16] EB is only obtained if uncompensated frozen in moments are assumed at the interface. However their origin and stability during magnetization reversal is not explained within this model, although uncompensated moments have been observed experimentally [17]. Other EB models [18,19] assume the formation of a domain wall in the AFM parallel to the FM/AFM interface.

3 Domain State Model for Exchange Bias

The domain state (DS) model for exchange bias provides a microscopic description for the existence and stabilization of volume domains within the AFM layer carrying a surplus magnetization which causes and controls the EB coupling at the FM/AFM interface [6], which goes beyond the approach by Malozemoff. In the DS model, the stabilization of AFM domains is not due to interfacial roughness, but rather induced and stabilized by the existence of volume defects in the AFM layer. The DS model links the physics of **D**iluted **A**nti**F**erromagnets in an external **F**ield (DAFF) [20,21] to the coupling mechanism of exchange coupled FM/AFM layers.

It is well known that a three-dimensional DAFF develops a domain state when cooled below its Néel temperature [20,21]. The driving force for the domain formation is a statistical imbalance of the number of impurities of the two sub-lattices in a finite region of the DAFF. This leads to a net magnetization which couples to the external field. The necessary energy increase due to domain wall formation can be minimized if the domain walls preferentially pass the non-magnetic defects at no cost of exchange energy [22,23]. Hence, defects substantially favor the formation of domains in the DAFF.

This is illustrated in Fig. 3, where the black dots represent non-magnetic atoms or defects. The dotted line separates a domain inside with a staggered

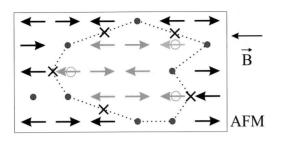

Fig. 3. Schematic structure of magnetic moments in a diluted antiferromagnet. The domain wall (dotted line) preferentially goes through defects (black dots). Frustrated bonds (×) and uncompensated moments (○) are labeled along the domain wall

magnetization reversed compared to the staggered magnetization outside the domain. There are 3 uncompensated moments (○) inside the domain, while 5 frustrated bonds exist along the domain boundary (×). This spin configuration can be stabilized for a cooling field $B > (5/3)|J_{\mathrm{AFM}}|$.

As a result, the DAFF will form a domain state when cooled from its paramagnetic state in an external magnetic field. This domain state will carry a certain surplus magnetization compared to the AFM state. The magnetization of the domain can be determined by field heating from an initially ordered state, which is prepared after first zero-field cooling the system. The difference of the field cooled and the field heated magnetizations corresponds to the irreversible domains state (IDS) magnetization m_{IDS} of the DAFF [8].

In the DS model for EB, the IDS magnetization of diluted AFM layers, which directly controls the EB coupling at the FM/AFM interface, originates from the AFM interface layer. The EB coupling can be established after cooling the system below T_{N} of the AFM layer either in an external magnetic field or with the FM layer remanently magnetized, which acts as an effective interface exchange field. Therefore, the diluted AFM can be expected to form a DS carrying IDS magnetization similar to that of a DAFF after field cooling.

The DS model is based on a combined experimental and theoretical effort. Experimentally, we studied the exchange bias in FM/AFM (111)-oriented Co/CoO bilayers grown by molecular beam epitaxy on (0001)-oriented sapphire substrates as a function of volume defects in the AFM [6,7]. The nonmagnetic defects (dilution) for Co in CoO were realized in two ways: (i) by over oxidizing CoO, leading to Co deficiencies in $Co_{1-y}O$ and (ii) by substituting non-magnetic Mg ions for magnetic Co in $Co_{1-x}Mg_xO$. For all samples investigated a 0.4 nm thick CoO layer with minimum defect concentration was placed at the interface. Therefore, in these systems the observed EB is primarily not due to disorder or defects at the interface.

Theoretically, MC simulations were performed at finite temperatures of a model consisting of a FM monolayer exchange coupled to a diluted AFM film consisting of 9 layers. The FM layer is described by a classical Heisenberg model. The dipolar interaction is approximated by including an anisotropy term leading to an in-plane magnetization. Also, an easy axis in the FM (anisotropy constant 0.1 J_{FM}) was introduced in order to obtain well-defined hysteresis loops. In view of the rather strong anisotropy in CoO an Ising Hamiltonian was assumed for the DAFF with the easy axis is parallel to that of the FM. For the coupling between AFM and FM the same coupling constant ($J_{INT} = J_{AFM}$) as for the AFM. In order to model the same interface structure for all simulations the interface monolayer of the AFM was fixed at a dilution of 50%, while the dilution in the volume of the AFM film (8 layers) was varied, in analogy to what was done in the experiment.

The most striking experimental and theoretical results are summarized in Fig. 4. In Fig. 4(a) the almost unbiased hysteresis loop of Co/CoO ($p(O_2) = 3.3 \times 10^{-7}$ mbar) is shown with the biased loop of Co/Co$_{1-y}$O ($p(O_2) = 5 \times 10^{-6}$ mbar) at T=20K. The dilution dependence of $|B_{EB}|$ for both Mg diluted Co/Co$_{1-x}$Mg$_x$O and Co deficient Co/Co$_{1-y}$O samples are shown in Figs. 4(b) and (c), respectively. For comparison, Fig. 4(d) shows the EB as a function of dilution of the AFM volume as obtained from MC simulations.

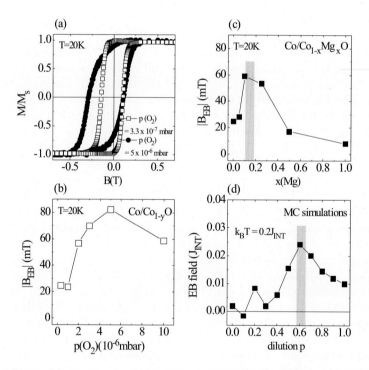

Fig. 4. (a) Hysteresis loops of Co$_{1-y}$O/Co/Al$_2$O$_3$ (0001) at T=20 K with the Co$_{1-y}$O prepared at $p(O_2) = 3 \times 10^{-7}$ mbar and $p(O_2) = 5 \times 10^{-6}$ mbar, respectively. (b) EB field vs. Mg concentration in Co$_{1-x}$Mg$_x$O and (c) vs. oxygen pressure during deposition of Co$_{1-y}$O at 20K. (d) Simulated EB as a function of dilution p of the AFM layer for $k_BT = 0.2 J_{INT}$

The experimental results were obtained after field cooling the samples from above T_N in a magnetic field of $B_{FC} = 5T$ oriented parallel to the plane of the film. EB values were then determined from hysteresis loops measured at $T=20K$ by SQUID magnetometry. As is seen in Fig. 4(b) and (c) the EB can be enhanced by a factor of 3 to 4 for both types of defects in the AFM layer. Maximum enhancement is obtained for $x(Mg)=0.1$ and $p(O_2)=5\times 10^{-6}$mbar. These sample are optimally diluted. Samples with the lowest defect concentration were prepared for $x(Mg)=0.0$ and an oxygen pressure of $p(O_2)=3.3\times 10^{-7}$mbar. As the CoO layer of the latter samples still shows structural defects (see section 5.2), such as twin boundaries, these samples are referred to as unintentionally diluted. Typical hysteresis loops of both unintentionally and optimally diluted $Co/Co_{1-y}O$ samples are shown in Fig. 4(a). Apart from an enhancement of EB, the defects also increase the coercive fields and significantly broaden the transition width of magnetization reversal.

The dilution dependence of the EB as extracted from MC simulations is shown in Fig. 4(d). The overall qualitative agreement with the experimental findings is clearly given. Since dilution favors the formation of domains it leads to an increase of the magnetization in the AFM and thus to a strong increase of the EB upon dilution. The AFM magnetization can directly be measured as will be shown in the next section. For larger dilutions the EB drops again due to a loss of connectivity in the AFM spin lattice.

For small dilutions, the DS model only shows a very small EB (Fig. 4(d)), although the unintentionally diluted samples still have a rather large finite EB (Fig. 4(b),(c)). In addition, the concentration for optimally diluted samples significantly differs between experiment ($x(Mg)\approx 0.15$) and theory ($p \approx 0.6$). This difference might be explained by the presence of twins, grains and grain boundaries which are present in the $CoO/Co/Al_2O_3$ samples or by the existence of other structural defects in the AFM layer. Such defects are presently not included in the DS model, but they are also expected to favor domain wall formation similar to the intentionally introduced defects, thus leading to a finite EB for unintentionally diluted samples.

In section 5 we directly demonstrate that twin boundaries in the AFM CoO enhance the EB with respect to untwined samples.

4 Domain State Magnetization

Like most conventional magnetization probes, SQUID magnetometry is not layer or element specific but rather measures the whole FM/AFM bilayer magnetization. In addition to the magnetization of the FM layer, both interface and volume magnetization of the diluted AFM layer will therefore contribute to the total magnetization. The IDS magnetization of the AFM might be identified by a vertical shift of the total hysteresis loop. This is schematically illustrated in Fig. 5(a). A surplus magnetization of the AFM

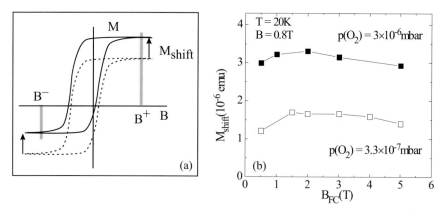

Fig. 5. (a) Schematic illustration of vertical magnetization shift due to irreversible uncompensated magnetic moments or the IDS magnetization of the AFM interface layer. (b) Vertical magnetization shift M_{shift} vs. cooling field for Co/Co$_{1-y}$O samples with Co$_{1-y}$O prepared at different oxygen partial pressures. Data are taken at $B = \pm 0.8$ T and $T = 20$ K and are extracted as described in the text

layer in an exchange coupled FM/AFM bilayer system was first identified in hysteresis loops of Fe/FeF$_2$ and Fe/MnF$_2$ [24]. Both, positive and negative vertical shifts were found and attributed to positive (ferromagnetic) and negative (antiferromagnetic) FM/AFM interface coupling, respectively. However, the origin of the induced uncompensated moments could not directly be assigned to either interface or volume magnetization.

In the DS model, the IDS magnetization of the AFM domain state determines the EB coupling strength. To investigate the change of the AFM magnetization with the number of introduced volume defects in the AFM layer, we performed high accuracy magnetization measurements of the vertical magnetization shift for both, the unintentionally diluted and oxygen diluted samples grown at $p(O_2) = 3.3 \times 10^{-7}$ mbar and $p(O_2) = 3 \times 10^{-6}$ mbar, respectively. The shift was determined at $T = 20$ K and $B = \pm 0.8$T with the FM layer fully saturated and is given by $M_{\text{shift}} = |M(B_+)| - |M(B_-)|$. As is seen in Fig. 5, at cooling fields larger than the saturation field of the FM layer M_{shift} is positive and increases with dilution of the AFM layer at all cooling fields. This increase can directly be linked to the creation of additional volume defects in the AFM layer. It further supports the formation of domains in the AFM carrying surplus magnetization after field cooling. It is important to note that in our experiments we measure the total AFM surplus magnetization, which is not equal to the IDS magnetization. The latter originates from the AFM interface layer. However, we find striking qualitative agreement that the EB field indeed is proportional to the AFM surplus magnetization.

5 Role of Twin Boundaries for Exchange Bias in Co/CoO

In this section we will investigate the relevance of twin boundaries for the EB coupling in Co/CoO layers grown on (111) oriented MgO substrates.

5.1 Sample Preparation

All samples discussed in Fig. 4 were grown on (0001) oriented sapphire substrates with a stacking order CoO/Co/Al$_2$O$_3$. Both, Co and CoO grow epitaxially in (111) orientation with 60° twins in the CoO layer [6] (see also section 5.2). In order to grow epitaxial but untwined CoO, we chose (111) oriented MgO as a new substrate. MgO is iso structural to CoO and has a moderate lattice mismatch of 1.4%. Two sets of samples with different stacking orders were prepared: (I) CoO/Co/MgO and (II) Pt-cap/Co/CoO/MgO. As the EB coupling is well known to depend on both FM and AFM layer thickness, we chose independent of the stacking sequence the same layer thicknesses identical to those previously chosen for the growth on sapphire (6nm for Co and 20nm for CoO). This enables a reliable comparison of the EB coupling between the different samples.

For all samples the oxygen pressure was $p(O_2) = 3.3 \times 10^{-7}$ mbar during the growth of CoO. As it was discussed in Fig. 4 this should lead to unintentionally diluted samples for all cases. Further details of the growth conditions are described elsewhere [6].

5.2 Structural Properties

All samples were characterized *in situ* by reflection high energy electron diffraction (RHEED). First we will focus on the results for the samples grown on the sapphire substrates. RHEED images of the 20nm thick Co$_{1-y}$O layers are shown in Fig. 6 grown at an oxygen pressure of $p(O_2) = 3.3 \times 10^{-7}$ mbar (upper panel) and $p(O_2) = 5 \times 10^{-6}$ mbar (lower panel), respectively. Fig. 6(a) and (e) show diffraction patterns for the electron beam incident parallel to the $[\bar{1}120]$-direction (0°) of the (0001) oriented sapphire substrate, while in Fig. 6(b) and (f) the beam is parallel to the $[\bar{1}010]$-direction (30°).

All RHEED patterns of the CoO layer show transmission images, i. e. diffraction from three-dimensional structures. In order to explain the observed RHEED patterns a (111) orientation of fcc Co$_{1-y}$O is assumed. The calculated diffraction patterns are shown on the right panels in Fig. 6. The full circles represent reciprocal lattice points that fulfill the diffraction condition for an fcc lattice. The pattern of these points does not reproduce the double spot structures as observed for all oxygen concentrations in the 30° direction. In order to explain these diffraction patterns, we furthermore have

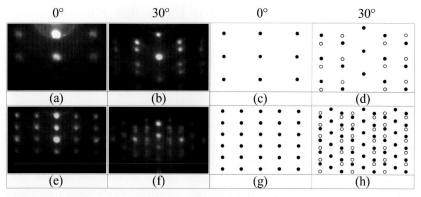

Fig. 6. RHEED images of two 20nm $Co_{1-y}O$ layers prepared at $p(O_2) = 3.3 \times 10^{-7}$ mbar (upper panel (a),(b)) and at $p(O_2) = 5 \times 10^{-6}$ mbar (lower panel (e),(f)). (c), (d) and (g), (h) show simulated diffraction patterns of $Co_{1-y}O$ (111) with 60° in-plane twins. The vertical panels show the patterns for 0° and 30° in plane orientation of the incident electron beam relative to the sapphire $[\overline{11}20]$ axis

to assume that $Co_{1-y}O$ grows in a twined structure where crystallites are oriented 60° relative to each other. The additional spots for the twined structure are included in the calculated pattern by open circles.

The intentionally introduced defects in the CoO layer can structurally directly be identified by the appearance of additional diffraction spots in the RHEED images of diluted AFM layers (Fig. 6(f)). A more detailed analysis about these defects is given in Refs. [6,7].

We now discuss the RHEED patterns of $Co_{1-y}O$ layers prepared on MgO (111), which are shown in Fig. 7 for type I samples (CoO/Co/MgO) in upper panel and type II samples (Co/CoO/MgO) in the lower panel. The RHEED pattern of type I samples (Fig. 7(a),(b)) are almost identical to the RHEED pattern for the CoO grown on sapphire (see Fig. 6(a),(b)). Therefore we conclude that CoO grows in (111) orientation with 60° twins.

However, if the CoO layer is directly grown on MgO such as for type II samples, we obtain diffraction patterns as expected for an fcc (111) oriented surface without twinning (Fig. 7(e),(f)). The RHEED patterns clearly show that both types of samples are unintentionally diluted. We now have a tool to separately control structural defects such as twins and intentionally introduced dilutions (non-magnetic defects) in the AFM layer and can study their relevance for the EB coupling in exchange coupled FM/AFM bilayers.

5.3 Magnetic Properties

To demonstrate the influence of the twinning on the EB coupling, we show in Fig. 8 the temperature dependence of the EB field $|B_{EB}|$ for all undiluted samples as described in the last section. For comparison we include results for

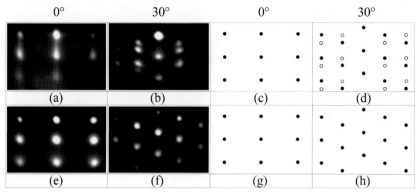

Fig. 7. RHEED images of unintentionally diluted CoO prepared at $p(O_2) = 3.3 \times 10^{-7}$ mbar grown in the stacking sequence CoO/Co/MgO (111) (upper panel, (a),(b)) and Co/CoO/MgO (111) (lower panel, (e),(f)). (c), (d) and (g), (h) show calculated diffraction patterns for twined and untwined CoO, respectively. The vertical panels show the patterns for 0° and 30° in plane orientation of the incident electron beam relative to the MgO [111] axis

an optimally diluted Co/Co$_{1-y}$O sample grown on sapphire (see also Fig. 4). Independent of the particular substrate chosen we find that undiluted and twined Co/Co$_{1-y}$O samples (open and filled squares) show almost identical magnitude and temperature dependence of the EB field. However, the undiluted and untwined sample has the smallest EB of all samples investigated, which is reduced by a factor of more than 2 compared to the twined samples.

These findings strongly suggest that twined AFM layers can also reduce the domain wall energy thus leading to domain wall formation in the volume of the AFM layer and thus to EB even without dilution.

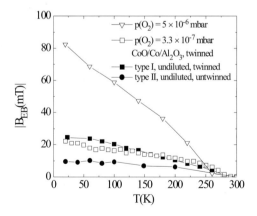

Fig. 8. Temperature dependence of EB field of Co/Co$_{1-y}$O samples grown on Al$_2$O$_3$ (0001) and MgO (111) substrates. Details about the samples are given in the legend and are described in the text. Note that the unintentionally diluted and untwined sample grown on MgO shows the smallest EB of all samples

6 Conclusions

In conclusion, we have shown both theoretically and experimentally that the DS model for EB which is inherently based on disorder in the volume of the AFM layer gives a new and consistent insight into the microscopic mechanism of the EB effect. We could demonstrate how domains in the AFM, which were speculated about previously [10,11,12], can indeed be stabilized. Moreover, we could also identify the origin of the uncompensated moments in the AFM, observed previously [17] and postulated in a previous model [15,16]. We have shown that the EB coupling crucially depends on the density of volume defects (such as cation deficiencies) as well as structural defects (such as grain boundaries, twin boundaries and others) in the AFM layer. This results in different spin configurations and might lead to different IDS magnetization in the AFM interface layer, which can yield a strong variation in the EB coupling strength.

This work has been supported by the DFG through SFB 341 and 491.

References

1. W.H. Meiklejohn and C.P. Bean: Phys. Rev. **102**, 1413 (1956).
2. W.H. Meiklejohn and C.P. Bean: Phys. Rev. **105**, 904 (1957).
3. B. Dieny et al.: Phys. Rev. B **43**, 1297 (1991).
4. C. Tsang: J. Appl. Phys. **55**, 2226 (1984).
5. J. Nogués and Ivan K. Schuller: J. Magn. Magn. Mat. **192**, 203 (1999).
6. P. Miltényi, M. Gierlings, J. Keller, B. Beschoten, G. Güntherodt, U. Nowak, K.D. Usadel: Phys. Rev. Lett. **84**, 4224 (2000).
7. J. Keller, P. Miltényi, B. Beschoten, G. Güntherodt, U. Nowak, and K.D. Usadel: submitted (2002).
8. U. Nowak, K.D. Usadel, J. Keller, P. Miltényi, B. Beschoten, G. Güntherodt: submitted (2002).
9. H. Ohldag, A. Scholl, F. Nolting, S. Anders, F. U. Hillebrecht, and J. Stöhr, Phys. Rev. Lett. **86**, 2878 (2001).
10. A.P. Malozemoff: Phys. Rev. B **35**, 3679 (1987).
11. A.P. Malozemoff: J. Appl. Phys. **63**, 3874 (1988).
12. A.P. Malozemoff: Phys. Rev. B **37**, 7673 (1988).
13. F. Nolting, A. Scholl, J. Stöhr, J.W. Seo, J. Fompeyrine, H. Siegwart, J.-P. Locquet, S. Anders, J. Lüning, E.E. Fullerton, M.F. Toney, M.R. Scheinfein and Padmore, Nature **405**, 767 (2000).
14. H. Ohldag, A. Scholl, F. Nolting, S. Anders, F. U. Hillebrecht, and J. Stöhr, Phys. Rev. Lett. **86**, 2878 (2001).
15. T.C. Schulthess and W.H. Butler: Phys. Rev. Lett. **81**, 4516 (1998).
16. T.C. Schulthess and W.H. Butler: J. Appl. Phys. **85**, 5510 (1999).
17. K. Takano, R.H. Kodama, A.E. Berkowitz, W. Cao, and G. Thomas: Phys. Rev. Lett. **79**, 1130 (1997).
18. D. Mauri, H.C. Siegmann, P.S. Bagus, and E. Kay: J. Appl. Phys. **62**, 3047 (1987).
19. M.D. Stiles and R.D. McMichael: Phys. Rev. B **59**, 3722 (1999).

20. W. Kleemann: Int. J. Mod. Phys. B **7**, 2469 (1993).
21. D.P. Belanger: in *Spin Glasses and Random Fields*, A.P. Young (Ed.) (World Scientific, Singapore 1998).
22. U. Nowak and K. D. Usadel: Phys. Rev. B **46**, 8329 (1992).
23. J. Esser, U. Nowak, and K. D. Usadel: Phys. Rev. B **55**, 5866 (1997).
24. J. Nogues, C. Leighton, and I.K. Schuller: Phys. Rev. B **61**, 1315 (2000).

Itinerant Ferromagnetism and Antiferromagnetism from a Chemical Bonding Perspective

Richard Dronskowski

Institut für Anorganische Chemie, Rheinisch-Westfälische Technische Hochschule, 52056 Aachen, Germany
drons@HAL9000.ac.rwth-aachen.de

Abstract. When quantum-chemical tools such as the Crystal Orbital Hamilton Population (COHP) technique are used for the analysis of nonmagnetic periodic systems, ubiquitous chemical bonding phenomena known from the molecular fields (decrease of total, potential, and single-particle energy as well as increase in kinetic energy) are also found in the solid state; likewise, there is a clear preference for bonding interactions as well as avoidance of antibonding states by means of Jahn–Teller and Peierls distortions. Non-spin-polarized bonding analyses of systems with a tendency for ferromagnetic order (such as the archetypes Fe, Co, and Ni) exhibit *antibonding* interactions at the Fermi level which disappear upon spin polarization; within the elemental metals the onset of magnetism strengthens the metal-metal bond. A typical antiferromagnet (Cr) shows *non-bonding* metal-metal interactions in the highest bands which remain almost unaffected by spin polarization. The bond-theoretical classification of potential ferromagnets and antiferromagnets is easily generalized for intermetallic alloys and can be utilized for rational syntheses. The "chemical" strategy is exemplified for the case of multinary iron/manganese rhodium borides in which various adjustments of the valence electron concentration have been synthetically realized.

Although cooperative magnetic phenomena have made up a rich synthetic and theoretical playground for generations of physicists and chemists [1,2], it may come as a surprise that it remains difficult, even today, to understand why some materials are ferromagnetic (or antiferromagnetic) while others are not. This is especially puzzling considering the paramount importance of magnetic compounds for our information society through data storage and data retrieval. Why, then, are even semiquantitative signpost, to be used in the rational synthesis of new magnetic materials, so very hard to find? This contribution is intended to come up with such theoretical signposts, based on calculational techniques borrowed from solid-state quantum physics and chemical bonding concepts rooted in molecular quantum chemistry.

1   Physics of Cooperative Magnetism in a Nutshell

According to Weiss [3], ferromagnetism depends on the existence of a molecular field aligning neighboring magnetic moments as introduced by

Langevin [4]. The puzzle of this classical field (decaying with the sixth power of the interatomic distance) was independently solved by Heisenberg [5] and Dirac [6] by formulating quantum-mechanical exchange interactions between *localized* electrons. Due to the itinerant nature of the conduction electrons in transition metals, however, a *delocalized* theory was introduced by Stoner [7] which still survives in nowadays solid-state physics textbooks, especially in the form of Stoner's criterion for ferromagnetism [8].

Even now, the localized-delocalized antagonism is omnipresent in the theoretical literature on ferromagnetism. On the one hand, the Hubbard model [9] and related theories for strongly correlated systems have been employed to study ferromagnetism in the rare-earth and transition metals [10]. Typically, a number of simplifications are introduced to make the model Hamiltonians (almost) exactly solvable which, however, renders possible applications difficult for structurally complex systems. On the other hand, density-functional theory (DFT) [11] has matured to a rank which made possible the extraction of Stoner's parameters [12] and a self-consistent description of itinerant magnetism [13]. The theoretical limits of the local spin-density approximation (LSDA) became apparent from Fe phase stability problems [14] and were solved by using gradient corrections [15]; the present status of DFT in the treatment of cooperative magnetism has recently been reviewed [16].

While there seems to be some disagreement between the two theoretical schools (Hubbard theory and DFT) as to the legitimate applicability of either method to cooperative magnetism and, also, conflicting claims of theoretical successes, both sides agree concerning the competing roles of exchange and kinetic energy upon spin polarization. Unfortunately, there haven't been many attempts to translate the theoretical results into a chemical language to be used in the actual synthesis of new magnetic alloys.

2 Chemical Bonds from Band Structure Calculations

Without any disrespect to competing theories it is probably fair to say that DFT has revolutionized the quantum theory of the solid state and, a bit later, large parts of quantum chemistry. Independent from the actual parameterization used, DFT remains an effective one-particle theory involving molecular or crystal orbitals, rendering their chemical interpretation easy since successful concepts from qualitative (molecular-orbital) quantum chemistry can be taken advantage of [17].

Among these, we introduce the Crystal Orbital Hamilton Population (COHP) concept [18] from a simple example, the one-dimensional solid. The left part of Fig. 1 shows the k-dependent wave function of a linear chain of hydrogen atoms with one H atom per unit cell; each H atom carries a single $1s$ atomic orbital. At low energy (zone center), all atomic orbitals have the same plus/minus sign; this is the H–H bonding (in-phase) combina-

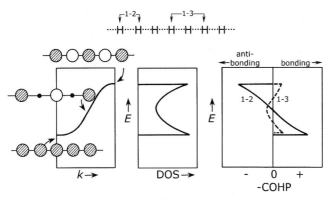

Fig. 1. Electronic bands, densities-of-states (DOS), and Crystal Orbital Hamilton Populations (COHP) of a one-dimensional crystal of hydrogen atoms

tion. At high energies, we find opposite plus/minus signs between neighboring atomic orbitals; this is the H–H antibonding (out-of-phase) combination. The density-of-states (DOS) diagram in the middle results as the inverse slope of the band structure on the left; here, the DOS has been plotted in the (admittedly questionable) "chemical" style in which the band energy is the function of the DOS.

A bonding indicator for the solid can be constructed from a Hamilton population-weighted density-of-states, the above COHP, given in the right part of the figure. The COHP looks similar to the DOS because it results from multiplying the DOS by the corresponding Hamilton (hopping) element, and it adopts negative values (bonding, negative hopping element, gaining energy) and positive values (antibonding, positive hopping element, loosing energy). By comparing the band structure and its orbital icons with the COHP diagram, it is obvious why the nearest-neighbor (1–2) COHP is bonding at low energies and antibonding at high energies. The second-nearest neighbor (1–3) interaction, however, is smaller because of the larger internuclear distance, and it is bonding at low/high energies and antibonding at medium energy. Note that we have effectively partitioned the band structure energy into bonding, non-bonding, and antibonding contributions of orbital (or atomic) pairs. It is customary to plot *negative* COHPs such that bonding interactions are connected with positive values of −COHP.

No matter how complicated the crystal structure, a COHP analysis gives access to the bonding in whatever kind of chemical system as long as a band structure can be computed. If the latter is based on DFT, the COHP carries contributions from all terms of the density-functional electronic energy (kinetic, electron-electron repulsion, electron-nuclear attraction, exchange-correlation), and it is thus intimately connected to the total energy of the system [18,19].

Summarizing two decades of quantum-chemical studies of *non-magnetic* solids, nature successfully tries to energetically optimize the strongest chemical interactions whenever possible by effectively adjusting the Fermi level (stoichiometry, structure) in order to fill up bonding states and empty antibonding states. If antibonding or non-bonding states in the highest occupied bands cannot be avoided for a given structure, there is an extremely high probability for Peierls or Jahn–Teller instabilities which break the structural symmetry to reach an energetically better suited structure [17,18]. Before having a closer look at the bonding of *magnetic* materials, we will first review a few ideas and misconceptions concerning chemical bonding itself.

3 Three Myths of Chemical Bonding

According to common belief, chemical bonding in molecules is due to *spin-pairing* of formerly unpaired electrons. This oversimplification probably stems from valence-bond theory, and the hydrogen molecule often serves as the most prominent "proof". While it is clear that the two unpaired electrons of the isolated H atoms are paired upon bond formation in the H–H single bond, a moment reflection reveals that the pairing of spins is only a side-effect accompanying bond formation in many *but not all* cases. For example, the oxygen molecule in its singlet $^1\Delta_g$ state (with the two energetically upmost electrons spin-paired in an antibonding π^* molecular orbital) is *less* stable (*less* strongly bonded) than $^3\Sigma_g^-$ triplet oxygen by about 0.95 eV [20]. For synthetic reasons, chemists commonly generate spin-paired, diamagnetic, reactive $^1O_2$ by photochemically activating spin-unpaired, paramagnetic $^3O_2$. *Upon spin-pairing, the oxygen molecule becomes less strongly bonded.*

Another myth touches upon the relationship of bond length and bond strength; at first sight, shorter bonds are stronger bonds. This rule of thumb (Pauling's bond length-bond strength method [21]) has been successfully applied to many (mostly organic) molecules, and it is indeed true that most C=C double bonds, to give a prominent example, are shorter and stronger than C–C single bonds. Nonetheless, the C_2 molecule itself shows that above rule of thumb is *principally incorrect*. In its $^1\Sigma_g^+$ ground state, C_2 contains a C–C (single) bond of 124.25 pm but upon excitation into its $^3\Sigma_u^+$ state the bond length *shortens* by almost 3.5 pm although the excited molecule is *less* stable (*less* strongly bonded) by *ca.* 1.1 eV [22]; *this bond weakens upon shortening.* Also, there is growing list of other molecular examples [23] showing that the correlation of bond length and bond strength is weak and cannot be seriously generalized. The problem has certainly been known for decades within the solid-state field: there are numerous metastable crystallographic phases (elements, simple compounds) with higher densities (shorter interatomic distances) than in their ground-state structures.

Third, there is a questionable notion of chemical bonding, still to be found in several textbooks, which erroneously relates chemical bond formation with a decrease in kinetic energy. This flaw stems from a misinterpretation of the pioneering quantum-chemical investigation of the H_2^+ molecule by Hellmann [24], and it goes back to a comparison of non-self-consistent solutions of the Schrödinger equation. When the atomic orbitals of the two H atoms forming the σ_g molecular orbital are not allowed to adapt to the doubled nuclear charge because of *fixed* orbital exponents ($\zeta = 1.000$), the lowered total energy (by -3.40 eV) of H_2^+ compared to H + H^+ is seemingly due to a decrease in kinetic energy (by -6.45 eV); note, however, that the Virial Theorem is violated! When this first approximate wave function is made fully self-consistent by contracting atomic orbitals ($\zeta = 1.228$), the total energy decreases by an *additional* amount (-1.09 eV) in harmony with the variational principle, and the kinetic energy *increases* strongly (by $+4.49$ eV); thus, the lowered potential energy overrides the loss in kinetic energy and drives chemical bond formation.

Summarizing, chemical bonding, following the thermochemical definition, results from a lowered *total* energy which reflects an increased nuclear potential acting on the electrons. Neither spin pairing, nor bond shortening, nor the course of the kinetic energy is a valid measure of bond formation [25].

4 Chemical Bonding and Energetics of α-Fe

Coming back to spin-related phenomena, we start by asking how the chemical bonding of Fe in its body-centered cubic phase (α) will look like [26], and we will first study the band structure of α-Fe by means of linear muffin-tin orbital theory [27,28] *without any spin polarization* (LDA) [29]. It is important to understand that this point of view refers to an electronic structure of an Fe phase which does not exist at all, namely non-magnetic iron (sometimes funnily called "paramagnetic" state in the literature). Thus, it is *not* a model for ferromagnetic Fe above the Curie temperature where there are magnetic moments but no long-ranged order. Fig. 2 offers the bands, the DOS and the Fe–Fe COHP for non-magnetic iron with all spins paired.

The valence bands are mostly $3d$ in character plus a little $4s$, and there results the typical three-peaked DOS of a bcc transition metal. The curious thing about the non-magnetic COHP is that the Fermi level falls in a strong Fe–Fe *antibonding* region. Consequently, one would expect some kind of instability and an associated structural change — which does not take place: bcc-Fe stays bcc-Fe. Nonetheless, non-magnetic iron does undergo a distortion but instead of the atoms rearranging themselves, *the electrons do*. Non-magnetic α-Fe is unstable with respect to an electronic structure distortion, which makes the two spin sublattices inequivalent, thereby reducing the electronic symmetry, lowering the energy, and giving rise to magnetism; for comparison, the spin-polarized DOS and COHP curves are given in Fig. 3.

In the LSDA calculation, the two spin sublattices shift in energy because of the exchange hole; the shapes of the two DOSs do not change very much

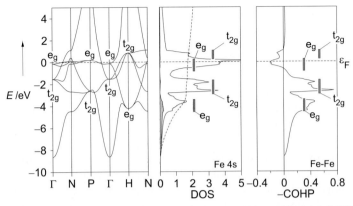

Fig. 2. Non-spin-polarized band structure, DOS, and Fe–Fe COHP curves of α-Fe based on TB-LMTO-ASA calculations (from [19] with kind permission)

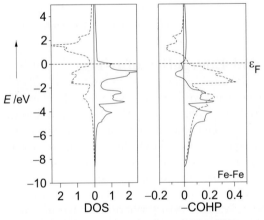

Fig. 3. Spin-polarized DOS and Fe–Fe COHP curves of α-Fe; majority spins in red, minority spins in blue (from [19] with kind permission)

but the majority spins lower in energy while the minority spins increase in energy. The resulting theoretical magnetic moment (2.27 μ_B) is very close to the experimental one (2.21 μ_B). A closer look at the magnetic COHP curve reveals the reason for the lowering of the total energy upon spin polarization: the shifts in the majority (red) and minority (blue) spin sublattices have removed the antibonding states at the Fermi level, thereby maximizing the Fe–Fe bonding as far as possible within this particular structure. A numerical integration reveals that the Fe–Fe bonds become strengthened by roughly 5% upon spin polarization; the bond strength itself decays with the sixth power of the interatomic distance, just like the classic molecular field. A further breakdown of the total energy reveals that, similar to the case of H_2^+, the kinetic energy increases by +4.00 eV upon spin polarization but is overrun by a larger decrease of the potential energy such that the total energy decreases (−0.43 eV), as expected.

The symmetry reduction of the wave function is similar to what happens during a structural Jahn–Teller distortion although only the electronic coordinates are involved in breaking the symmetry. Similar to the case of molecular oxygen (and counterintuitive only at first sight), we reiterate that the strengthening of the chemical bonding in α-Fe results from the *unpairing* of spins which otherwise would be kept in antibonding (crystal) orbitals. In addition, the magnetic crystal structure contains slightly larger interatomic distances than the non-magnetic structure (Fig. 4); this is necessary to optimize bonding between the more diffuse orbitals belonging to the β spins since numerical analyses show that these *minority* spins are much more involved in the bonding than the α spins. As has been demonstrated before, it is only the total energy which allows a proper measure of bond strength, and the total energy is lower for the more strongly bonded magnetic state.

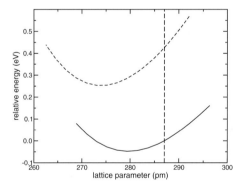

Fig. 4. Relative energies of α-Fe as a function of the lattice parameter within a non-spin-polarized (LDA, dashed curve) and spin-polarized (LSDA, full curve) description and TB-LMTO-ASA theory; note the experimental lattice parameter (vertical dashed line)

5 Magnetic Recipe for Transition Metals and Alloys

When the above analysis is repeated for the other two ferromagnetic transition metals (Co and Ni), the Fermi level is also found in antibonding states. Upon spin polarization, the exchange hole shifts the spin sublattices such that the antibonding states are removed; the bond strengths increase by 4.0 and 0.5%. The theoretical moments are 1.60 and 0.62 μ_B, closely matching the experimental ones [19]. Fig. 5 offers the DOS and COHP figures for the entire series of the $3d$ transition metals. It is obvious that for the early transition

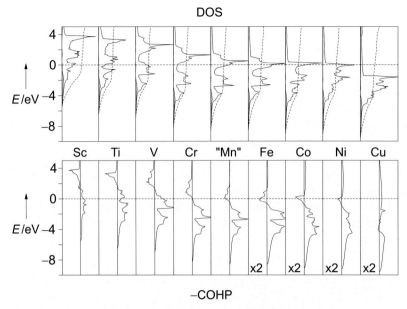

Fig. 5. Non-spin-polarized DOS and M–M COHP curves for the 3d transition metals (from [19] with kind permission)

metals like Ti, the Fermi level lies low in the COHP curve and thereby falls in the metal-metal bonding region: there exists no drive towards ferromagnetism for the early transition metals. For Cr (a typical antiferromagnet), the Fermi level in the COHP curve precisely separates the bonding and antibonding regions, whereas for the metals from "Mn" (hypothetical bcc-Mn) to Ni, the Fermi level lies in the region that is clearly responsible for metal-metal antibonding: while Fe, Co, and Ni are all ferromagnetic, we predict bcc-Mn (with $a = 288$ pm) to be ferromagnetic. For the higher homologues of Fe, Co, and Ni, the valence d orbitals are too well shielded from the nucleus such that sufficiently large energetic changes required for spin polarization are inaccessible [19].

In order to create new metallic ferromagnets and antiferromagnets based on the above COHP reasoning, the synthetic recipe for all transition metal alloys reads as follows: Try to adjust the Fermi level of the starting material in order to position it within the antibonding states (ferromagnets) or nonbonding states (antiferromagnets) between the magnetically active atoms, *i.e.*, those which have narrow band widths in their elemental forms [19,30]. The synthetic tuning of the valence electron concentration may result from either electronic enrichment or depletion.

6 Rational Syntheses of Magnetic Borides

Within the system of quaternary intermetallics $A_2MRh_5B_2$ crystallizing in space group $P4/mbm$, the phase $Mg_2MnRh_5B_2$ is an electron-poor representative of this compound family. Inside the tetragonal unit cell (Fig. 6, left), there are trigonal, tetragonal, and pentagonal rhodium prisms stacked on top of each other along the [001] direction. While the triangular prisms are centered by B atoms, the pentagonal prisms accommodate Mg atoms; the tetragonal prisms contain the magnetically important Mn atoms (Fig. 6, right). We find one-dimensional Mn–Mn chains along the c axis with intra-chain distances of about 291 pm; the inter-chain distances are $ca.$ 660 pm.

$Mg_2MnRh_5B_2$ is characterized by Curie–Weiss behavior above 160 K with a Weiss constant of -130 K, $i.e.$, considerable *antiferromagnetic* Mn–Mn exchange interactions. The Curie constant corresponds to a paramagnetic moment of 3.2 μ_B/Mn; characteristic features for one-dimensional magnetic behavior are not obvious, though. Non-magnetic band structure calculations on $Mg_2MnRh_5B_2$ fully corroborate these findings in that the Fermi level in the COHP analysis (Fig. 7, left) is clearly positioned in the region of *non-bonding* Mn–Mn interactions, the above-mentioned fingerprint for antiferromagnetism. In order to computationally model the (still unknown) antiferromagnetic structure of $Mg_2MnRh_5B_2$, the short tetragonal axis was doubled and the Hamiltonian was switched over to spin-density-functional theory. Starting from a crude guess for the antiparallel moments on the neighboring Mn atoms and a zero net moment, the calculation converges to an antiferromagnetic structure with a self-consistent saturation moment of 3.23 μ_B/Mn; all other atoms are magnetically almost inactive [31]. The stoichiometry of a new *ferromagnet* is easily predicted starting with antiferromagnetic $Mg_2MnRh_5B_2$ and its 62 valence electrons because the experimentalist simply

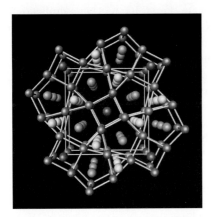 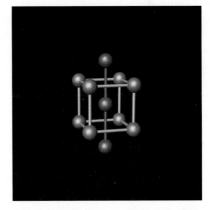

Fig. 6. Perspective view of the $Mg_2MnRh_5B_2$ crystal structure (left) with green Mg atoms, red Mn atoms, blue Rh atoms and yellow B atoms; tetragonal Rh prisms accommodate Mn atoms (right) (from [31] with kind permission)

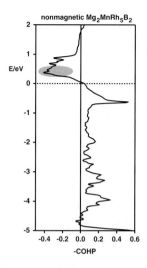 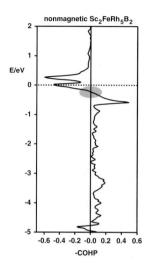

Fig. 7. COHP analysis of Mn–Mn bonding in non-magnetic $Mg_2MnRh_5B_2$ (left) and of Fe–Fe bonding in non-magnetic $Sc_2FeRh_5B_2$ (right) with the Fermi levels set to the energy zero for simplicity; the shaded regions indicate electronic enrichments/depletions by three electrons (from [31] with kind permission)

needs to push the Fermi level up to let it cut through the strongly antibonding states around +0.4 eV as given in the shaded region of Fig. 7 (left). A numerical COHP integration reveals that this corresponds to *three* additional electrons, to be realized by replacing Mg against Sc and Mn against Fe, such that "$Sc_2FeRh_5B_2$" with 65 valence electrons should do.

$Sc_2FeRh_5B_2$ can be synthesized by a classic high-temperature route, and the susceptibility measurement reveals that it is *ferromagnetic*, as predicted, below 450 K with a maximum value of 3.3 μ_B/Fe at 1.7 K; the complicated magnetization curve does not follow the Curie–Weiss law. The shape of the corresponding Fe–Fe COHP analysis within a non-magnetic electronic structure calculation (Fig. 7, right) is in harmony with the ferromagnetic properties of $Sc_2FeRh_5B_2$ since there are strongly *antibonding* Fe–Fe interactions around the Fermi level; these are annihilated in the subsequent spin-polarized band structure calculation. When self-consistent, one finds a theoretical saturation moment of 3.96 μ_B while the individual moments come to -0.05 for Sc, 2.93 for Fe, 0.25 for Rh(1), 0.08 for Rh(2), and 0.02 for B [31].

Additional synthetic steps can also be based on the Fe–Fe COHP curve of ferromagnetic $Sc_2FeRh_5B_2$ (Fig. 7, right). A further electronic *enrichment* will populate even more antibonding states which destabilize the structure and only lead to perovskite-related phases. An electronic *depletion* by three electrons, however, readjusts the Fermi level back into a region of *non-bonding* metal–metal interactions (shaded region), the fingerprint of antiferromagnetism. Thus, starting from $Sc_2FeRh_5B_2$ (65 valence electrons), a lowering by three electrons results in the synthetic target "$Sc_2MnRu_2Rh_3B_2$" (62 valence electrons), very likely to be an antiferromagnet. Alternatively, one might keep to the Fe atom in the 65-electrons alloy $Sc_2FeRh_5B_2$ and lower the valence electron concentration solely through substitutions in the $4d$ elemental substructure. For example, the alternative antiferromagnetic target would then read "$Sc_2FeRu_3Rh_2B_2$", also having 62 valence electrons.

Both $Sc_2MnRu_2Rh_3B_2$ and $Sc_2FeRu_3Rh_2B_2$ can be synthesized in almost quantitative yields. As predicted, the Mn phase exhibits *antiferromagnetic* behavior with a Weiss constant around -300 K and a paramagnetic moment of 2.3 μ_B/Mn. Also, the magnetic properties of the isoelectronic Fe phase characterize it as being an *antiferromagnetic* material, as predicted. Here, the Weiss constant lies at about -90 K while the paramagnetic moment is 4.0 μ_B/Fe. Thus, it is clear that the magnetic type of behavior of both $Sc_2MnRu_2Rh_3B_2$ and $Sc_2FeRu_3Rh_2B_2$ is solely a function of the electron filling; *i.e.*, an electron concentration which positions the Fermi level in non-bonding Mn–Mn or Fe–Fe states.

7 Conclusion

When quantum-chemical bond-detecting tools such as COHP are utilized for the analysis of bonding in the transition metals and their alloys, common bonding phenomena known from molecular chemistry are rediscovered, especially concerning the process of spin-pairing/unpairing and the role of the individual energetics upon bond formation. In addition, potential ferromagnets/antiferromagnets exhibit antibonding/non-bonding M–M interactions at the Fermi level which indicate underlying electronic instabilities. Antibonding interactions are removed by spin polarization in the spirit of electronic distortions taking place solely in the spin degrees of freedom. Given a robust structure type, the semi-quantitative bonding signpost can guide the experimentalist in the search for new ferromagnets and antiferromagnets, as exemplified by the rational syntheses of the magnetic alloys $Mg_2MnRh_5B_2$, $Sc_2FeRh_5B_2$, $Sc_2MnRu_2Rh_3B_2$, and $Sc_2FeRu_3Rh_2B_2$.

Acknowledgments

It is a pleasure to thank my coworkers (G. A. Landrum, A. Decker, Y. Kurtulus, B. Eck) and my colleagues (H. Lueken, W. Jung) for their contributions as well as the Fonds der Chemischen Industrie for their generous support.

References

1. S. Chikazumi, *Physics of Ferromagnetism* (Clarendon Press, Oxford 1997).
2. H. Lueken, *Magnetochemie* (B. G. Teubner, Stuttgart, Leipzig 1999).
3. P. Weiss, J. Phys. **6**, 661 (1907).

4. P. Langevin, Ann. Chim. Phys. **5**, 70 (1905).
5. W. Heisenberg, Z. Phys. **49**, 619 (1928).
6. P. A. M. Dirac, Proc. R. Soc. London A **123**, 714 (1929).
7. E. C. Stoner, Proc. R. Soc. A **165**, 372 (1938);
   E. C. Stoner, Proc. R. Soc. A **169**, 339 (1939).
8. H. Ibach, H. Lüth, *Solid-State Physics* (Springer, Berlin, Heidelberg 1995).
9. J. Hubbard, Proc. R. Soc. London A **276**, 238 (1963).
10. D. Vollhardt, N. Blümer, K. Held, M. Kollar, J. Schlipf, M. Ulmke, J. Wahle, Adv. Solid State Phys. **38**, 383 (1999).
11. P. Hohenberg, W. Kohn, Phys. Rev. B **136**, 864 (1964);
    R. G. Parr, W. Yang, *Density-functional theory of atoms and molecules* (Oxford University Press, New York 1989).
12. J. F. Janak, Phys. Rev. B **16**, 255 (1977).
13. V. L. Moruzzi, P. M. Marcus, K. Schwarz, P. Mohn, Phys. Rev. B **34**, 1784 (1986).
14. J. Kübler, Phys. Lett. A **81**, 81 (1981).
15. P. Bagno, O. Jepsen, O. Gunnarsson, Phys. Rev. B **40**, 1997 (1989).
16. L. Fritsche, B. Weimert, Phys. Status Solidi B **208**, 287 (1998).
17. R. Hoffmann, *Solids and Surfaces: A Chemist's View of Bonding in Extended Structures* (VCH, Weinheim, New York 1988).
18. R. Dronskowski, P. E. Blöchl, J. Phys. Chem. **97**, 8617 (1993); see also website http://www.cohp.de.
19. G. A. Landrum, R. Dronskowski, Angew. Chem. Int. Ed. **39**, 1560 (2000).
20. A. F. Holleman, E. Wiberg, *Lehrbuch der Anorganischen Chemie* (Walter de Gruyter, Berlin, New York 1995).
21. L. Pauling, *The Nature of the Chemical Bond* (Cornell University Press, Ithaca, New York 1960).
22. M. Boggio-Pasqua, A. I. Voronin, Ph. Halvick, J.-C. Rayez, J. Molec. Struct. (Theochem) **531**, 159 (2000).
23. M. Kaupp, B. Metz, H. Stoll, Angew. Chem. Int. Ed. **39**, 4607 (2000).
24. H. Hellmann, *Einführung in die Quantenchemie* (Franz Deuticke, Leipzig, Wien 1937).
25. J. K. Burdett, *Chemical Bonds — A Dialog* (Wiley-VCH, Weinheim, New York 1997).
26. G. A. Landrum, R. Dronskowski, Angew. Chem. Int. Ed. **38**, 1390 (1999).
27. O. K. Andersen, Phys. Rev. B **12**, 3060 (1975);
    O. K. Andersen, O. Jepsen, Phys. Rev. Lett. **53**, 2571 (1984);
    O. K. Andersen, C. Arcangeli, R. W. Tank, T. Saha-Dasgupta, G. Krier, O. Jepsen, I. Dasgupta, in *Tight-Binding Approach to Computational Materials Science*, no. 491 in MRS Symposia Proceedings; MRS, Pittsburgh, 1998.
28. G. Krier, O. Jepsen, A. Burkhardt, O. K. Andersen, The TB-LMTO-ASA program, version 4.7, Stuttgart.
29. U. von Barth, L. Hedin, J. Phys. C **5**, 1629 (1972).
30. A. Decker, G. A. Landrum, R. Dronskowski, Z. Anorg. Allg. Chem. **628**, 303 (2002).
31. R. Dronskowski, K. Korczak, H. Lueken, W. Jung, Angew. Chem. Int. Ed., in press.

Theory of Ferromagnetism in (III,Mn)V Semiconductors

Jürgen König

Institut für Theoretische Festkörperphysik, Universität Karlsruhe,
76128 Karlsruhe, Germany

Abstract. Carrier-induced ferromagnetism has been observed in several (III,Mn)V semiconductors. We review the theoretical picture of these ferromagnetic semiconductors that emerges from a model with kinetic-exchange coupling between localized Mn spins and valence-band carriers. We discuss the applicability of this model, the validity of a mean-field approximation widely used in the literature, and validity limits for the simpler RKKY model. Our conclusions are based in part on our analysis of the system's elementary spin excitations. We use theoretical estimates of spin-wave energies and Monte-Carlo simulations to address the effect of collective fluctuations, neglected by mean-field theory, on limiting the critical temperature. Furthermore, we provide a microscopic theory for phenomenological micromagnetic parameters and evaluate them using either simple or more sophisticated models for the band-structure of the host material. As an application we discuss recently measured domain structures in films of (Ga,Mn)As.

Conventional electronics uses the fact that electrons are charged, i.e., electric currents can be manipulated by electric fields. Electrons, however, also possess a spin degree of freedom yielding a magnetic moment. Spin-electronic effects exploit, per definition, both the electric charge and the magnetic momentum of electrons simultaneously. Spin-electronics applications in metals, such as the use of the giant magnetoresistance and the tunnel magnetoresistance to read magnetically stored information, have already proven tremendous success. An even larger prospect may be anticipated for spin electronics in semiconductor devices. One reason is the compatibility with the present semiconductor technology, another one is due to the fact that electronic transport properties of semiconductors can easily be manipulated by external gates or by doping.

One precondition for spin electronics is the generation of spin-polarized carriers. Spin injection from ferromagnetic metals into semiconductors and spin generation by optical means are two approaches which are presently under investigation. A more direct way is provided by the use of ferromagnetic semiconductors, in which not only charges but also magnetic moments are introduced by doping. In this paper we describe the physics of those ferromagnetic materials.

1 Diluted Magnetic III-V Semiconductors

Semiconductors have rich optical and magnetic properties when Mn or other magnetic ions are doped into the material [1,2]. For a long time, the investigation of these so-called diluted magnetic semiconductors (DMS) concentrated on II-VI compounds such as CdTe or ZnSe. These systems have shown paramagnetic, antiferromagnetic or spin-glass behavior. To achieve ferromagnetism, free carriers had to be introduced by additional doping [3], which is, however, rather difficult experimentally, and only low transition temperatures have been obtained so far.

A more promising approach is to dope Mn into III-V semiconductors. Transition temperatures exceeding 100 K have been realized in (Ga,Mn)As several years ago [4] (very recently, ferromagnetism in (Ga,Mn)N at temperatures close to 1000 K has been reported [5]).

The purpose of the present paper is to describe the origin of ferromagnetism and the ferromagnetic properties of these new materials (for a longer review including a discussion of various different theoretical approaches we refer to Ref. [6].) Our discussion is based on a phenomenological model, which arises from the picture that Mn acts as an acceptor when it substitutes for a cation in a III-V semiconductor lattice, leaving a Mn^{2+} ion which has a half-filled d-shell with angular momentum $L = 0$ and spin $S = 5/2$. The low-energy degrees of freedom are holes in the semiconductor valence band and one local magnetic moment for each Mn ion. The local moments and the itinerant carriers are coupled to each other by an antiferromagnetic kinetic-exchange interaction. Ferromagnetism occurs in these materials because of interactions between Mn local moments that are mediated by holes in the semiconductor valence band. The length scales associated with holes in these compounds are still long enough that a $\boldsymbol{k} \cdot \boldsymbol{p}$, envelope function, description of the semiconductor valence bands is appropriate. The operators in terms of which the phenomenological Hamiltonian is expressed include the spin operator $\boldsymbol{S}_I$ for the $S = 5/2$ local moment on site I and the multi-band envelope function hole-spin-density operator $\boldsymbol{s}(\boldsymbol{r})$. The Hamiltonian is

$$H = H_0 + J_{\rm pd} \int d^3 r \boldsymbol{S}(\boldsymbol{r}) \cdot \boldsymbol{s}(\boldsymbol{r}), \tag{1}$$

where $\boldsymbol{S}(\boldsymbol{r}) = \sum_I \boldsymbol{S}_I \delta(\boldsymbol{r} - \boldsymbol{R}_I)$ is the Mn-spin density, $J_{\rm pd}$ describes the antiferromagnetic exchange coupling, and H_0 models the valence bands of the host semiconductor.

2 Mean-Field Theory

In the following we use a *continuum Mn approximation*, in which we replace the Mn ion distribution by a continuum with the same spin density. This is motivated by the observation that the Fermi wavelength of the valence-band carriers is typically longer than the distance between Mn ions, and seems to be reliable in the limit of main interest, that of high Mn densities and high critical temperatures, where the holes are metallic and their interaction with Mn acceptors will be effectively screened.

Even more simplification is achieved by employing a mean-field picture, in which correlation between local-moment spin configurations and the free-carrier state is neglected: each Mn spin is described as a free spin in an effective external field defined by the average of the free-carrier polarization [7,8,9,10,11]. The free carriers, in turn, are treated as fermions in an effective external field determined by the average of the Mn-spin density. This leads to two coupled equations for the Mn-spin and free-carrier-spin polarization. Both polarizations vanish at the transition temperature T_c^{MF} given by

$$k_B T_c^{\mathrm{MF}} = \frac{N_{\mathrm{Mn}} S(S+1)}{3} \frac{J_{\mathrm{pd}}^2 \chi_f}{(g^* \mu_B)^2} , \qquad (2)$$

where χ_f is the Pauli susceptibility of the itinerant carriers, g^* the g-factor, and μ_B is the Bohr magneton.

The power and the success of the mean-field picture lies in the fact that it is a simple theoretical approach which makes it easy to calculate many observable quantities numerically. Mean-field theory, however, fails to describe the existence of low-energy long-wavelength spin excitations, among other things. Because of its neglect of collective magnetization fluctuations, mean-field theory, e.g., always overestimates the ferromagnetic critical temperature. There are many examples in itinerant electron systems where mean-field theory overestimates ferromagnetic transition temperatures by more than an order of magnitude and it is not *a priori* obvious that mean-field theory will be successful in (III,Mn)V ferromagnets. Indeed, we will find that the multi-band character of the semiconductor valence band plays an essential role in enabling high ferromagnetic transition temperatures in these materials.

3 Collective Spin Excitations

3.1 Beyond Mean-Field Theory and RKKY Interaction

We identify the elementary spin excitations, determine their dispersion [12,13], and discuss implications for the Curie temperature [14,15]. The starting point of our analysis is the itinerant-carrier-mediated ferromagnetic interaction between local magnetic moments. Such an interaction is provided by the familiar Ruderman-Kittel-Kasuya-Yoshida (RKKY) theory. The RKKY picture, however, only applies as long as the perturbation induced by the Mn spins on the itinerant carriers is small. As we will derive below, the proper condition is $\Delta \ll \epsilon_F$ where $\Delta = N_{\mathrm{Mn}} J_{\mathrm{pd}} S$ is the (zero-temperature) spin-splitting gap of the itinerant carriers due to an average effective field induced by the Mn ions, and ϵ_F is the Fermi energy. This is, however, never satisfied in (III,Mn)V ferromagnets, partially because the valence-band carrier concentration p is usually much smaller than the Mn impurity density N_{Mn}. A related drawback of the RKKY picture is that it assumes an instantaneous static interaction between the magnetic ions, i.e. the dynamics of the free carriers are neglected. We will see below that this dynamics is important to obtain all types of elementary spin excitations. As a consequence, *RKKY theory does not provide a proper description of the ordered state in ferromagnetic DMSs.*

3.2 Independent Spin-Wave Theory for Parabolic Bands

The main idea of our theory is to derive an effective description for the Mn spin system by integrating out the valence-band carriers and to look for fluctuations of the Mn spins around their spontaneous mean-field magnetization direction (which we choose as the z-axis). Using the Holstein-Primakoff (HP) representation, we express the Mn spins in terms of bosonic degrees of freedom. We expand the effective action up to quadratic order, i.e., we treat the spin excitations as noninteracting Bose particles.

To keep the discussion transparent we start with a two-band model for the itinerant carriers with quadratic dispersion. Later, in Section 3.5, we extend our theory to a model with a more realistic band structure described by a six-band Kohn-Luttinger Hamiltonian.

For small fluctuations around the mean-field magnetization, we can write the spin operators as $S^+(\boldsymbol{r}) \approx b(\boldsymbol{r})\sqrt{2N_{\text{Mn}}S}$, $S^-(\boldsymbol{r}) \approx b^\dagger(\boldsymbol{r})\sqrt{2N_{\text{Mn}}S}$, and $S^z(\boldsymbol{r}) = N_{\text{Mn}}S - b^\dagger(\boldsymbol{r})b(\boldsymbol{r})$ with bosonic fields $b^\dagger(\boldsymbol{r}), b(\boldsymbol{r})$. The state with fully polarized Mn spins corresponds to the vacuum with no bosons. The creation of a HP boson reduces the magnetic quantum number by one.

The partition function Z can be expressed as a coherent-state path integral in imaginary time over the HP bosons and the valence-band carriers, which are fermions. Since the Hamiltonian is bilinear in the fermionic fields, we can integrate out the itinerant carriers and arrive at an effective description in terms of the impurity spin degree of freedom labeled by the complex number coherent state labels for the boson fields, z and $\bar{z}$. We get $Z = \int \mathcal{D}[\bar{z}z] \exp(-S_{\text{eff}}[\bar{z}z])$ with the effective action

$$S_{\text{eff}}[\bar{z}z] = S_{\text{BP}}[\bar{z}z] - \ln\det\left[(G^{\text{MF}})^{-1} + \delta G^{-1}(\bar{z}z)\right], \tag{3}$$

where $S_{\text{BP}}[\bar{z}z] = \int_0^\beta d\tau \int d^3r\, \bar{z}\partial_\tau z$ is the usual Berry's phase term. In Eq. (3), we have already split the total kernel G^{-1} into a mean-field part $(G^{\text{MF}})^{-1}$ and a fluctuating part δG^{-1},

$$(G^{\text{MF}})^{-1}_{ij} = (\partial_\tau - \mu)\,\delta_{ij} + \langle i|H_0|j\rangle + N_{\text{Mn}}J_{\text{pd}}S s^z_{ij} \tag{4}$$

$$\delta G^{-1}_{ij}(\bar{z}z) = \frac{J_{\text{pd}}}{2}\left[\left(z s^-_{ij} + \bar{z} s^+_{ij}\right)\sqrt{2N_{\text{Mn}}S} - 2\bar{z}z s^z_{ij}\right] \tag{5}$$

where μ denotes the chemical potential, and i and j range over a complete set of hole-band states (i.e., here, for the model with two parabolic bands, i and j

label band wavevectors and spin, ↑,↓), and s^z_{ij} and $s^{\pm}_{ij}$ are matrix elements of the itinerant-carrier spin matrices. The combination $\Delta = N_{\text{Mn}}J_{\text{pd}}S$ defines the mean-field energy to flip the spin of an itinerant carrier.

The independent spin-wave theory is now obtained by expanding Eq. (3) up to quadratic order in z and $\bar{z}$, i.e., spin excitations are treated as noninteracting HP bosons. This is a good approximation at low temperatures, where the number of spin excitations per Mn site is small. We obtain an action that is the sum of the temperature-dependent mean-field contribution and a fluctuation action. The latter is

$$S_{\text{eff}}[\bar{z}z] = \frac{1}{\beta V} \sum_{|\mathbf{k}| \leq k_D, m} \bar{z}(\mathbf{k}, \nu_m) D^{-1}(\mathbf{k}, \nu_m) z(\mathbf{k}, \nu_m), \tag{6}$$

where ν_m are the bosonic Matsubara frequencies. A Debye cutoff k_D with $k_D^3 = 6\pi^2 N_{\text{Mn}}$ ensures that we include the correct number of magnetic-ion degrees of freedom, $|\mathbf{k}| \leq k_D$. The kernel of the quadratic action defines the inverse of the spin-wave propagator,

$$D^{-1}(\mathbf{k}, \nu_m) = -i\nu_m + \frac{J_{\text{pd}}p\xi}{2} + \frac{N_{\text{Mn}}J_{\text{pd}}^2 S}{2V} \sum_{\mathbf{q}} \frac{f[\epsilon_\uparrow(\mathbf{q})] - f[\epsilon_\downarrow(\mathbf{q}+\mathbf{k})]}{i\nu_m + \epsilon_\uparrow(\mathbf{q}) - \epsilon_\downarrow(\mathbf{q}+\mathbf{k})} \tag{7}$$

where $\xi = (p_\downarrow - p_\uparrow)/p$ is the fractional free-carrier spin polarization, $p_{\uparrow,\downarrow}$ denotes the carrier concentration for spin up and down, respectively, p is the total carrier concentration, and $\epsilon_{\uparrow,\downarrow}(\mathbf{q})$ is the energy of spin-up and spin-down valence-band holes, $\epsilon_{\uparrow,\downarrow}(\mathbf{q}) = \epsilon_q \pm \Delta/2$, and $\epsilon_q = \hbar^2 q^2/(2m^*)$. The second term of Eq. (7) is the the energy for a Mn spin excitation in mean-field-theory, $\Omega^{\text{MF}} = J_{\text{pd}}p\xi/2 = x\Delta$. It differs from the itinerant-carrier spin splitting by the ratio of the spin densities $x = p\xi/(2N_{\text{Mn}}S)$, which is always much smaller than 1 in (III,Mn)V ferromagnets. Mean-field theory is, thus, recovered by dropping the last term in Eq. (7). It is this term that describes the response of the free-carrier system to changes in the magnetic-ion configuration.

We mention that the scheme presented above can easily be extended [16] to models beyond the continuum approximation to discuss the role of disorder in the Mn positions close to the metal-insulator regime [16,17].

3.3 Elementary Spin Excitations

We obtain the spectral density of the spin-fluctuation propagator by analytical continuation, $i\nu_m \to \Omega + i0^+$ and $A(\mathbf{k}, \Omega) = \text{Im } D(\mathbf{k}, \Omega)/\pi$. In the following we consider the case of zero temperature, $T = 0$. We find three different types of spin excitations [12], a Goldstone spin-wave mode, a Stoner continuum, and an optical spin-wave branch.

Goldstone-Mode Spin Waves Our model has a gapless Goldstone-mode branch reflecting the spontaneous breaking of spin-rotational symmetry. The dispersion of this low-energy mode for four different valence-band carrier concentrations p is shown in Fig. 1 (solid lines). At large momenta, $k \to \infty$, the spin-wave energy approaches the mean-field result $\Omega_k^{(1)} \to \Omega^{\mathrm{MF}}$ (short-dashed lines in Fig. 1). Expansion of the $T = 0$ propagator for small momenta yields for the collective modes dispersion in the strong and weak-coupling limits, $\Delta \gg \epsilon_F$ and $\Delta \ll \epsilon_F$,

$$\Omega_k^{(1)} = \frac{x}{1-x}\epsilon_k + \mathcal{O}(k^4) \qquad \text{for} \qquad \Delta \gg \epsilon_F, \tag{8}$$

$$\Omega_k^{(1)} = \frac{p}{32 N_{\mathrm{Mn}} S}\epsilon_k \left(\frac{\Delta}{\epsilon_F}\right)^2 + \mathcal{O}(k^4) \qquad \text{for} \qquad \Delta \ll \epsilon_F, \tag{9}$$

where ϵ_F is the Fermi energy of the majority-spin band.

We note that the dependence of the spin-wave energy on the system parameters, namely the exchange interaction strength J_{pd}, hole concentration p, local-impurity density N_{Mn}, and effective mass m^* is different in these two limits, indicating that the microscopic character of the gapless collective excitations differs qualitatively in the two limits. The energy of long-wavelength spin waves is determined by a competition between exchange and kinetic energies: For a given Mn-spin configuration, the valence-band carriers can either follow the spatial dependence of the Mn spin density in order to minimize the exchange energy, as they do in the strong-coupling limit $\Delta \gg \epsilon_F$, or minimize the kinetic energy by forming a state with a homogeneous spin polarization, as they do in the weak-coupling limit $\Delta \ll \epsilon_F$.

Stoner Continuum and Optical Spin Waves We observe that the frequency ν_m is not only present in the first term of Eq. (7), it enters the third term, too. This is the reason why, in addition to the Goldstone mode, other spin excitations can appear in our model. They are absent in a static-limit description, i.e., when the frequency dependence of the third term of Eq. (7) is neglected.

We find a continuum of Stoner spin-flip particle-hole excitations. They correspond to flipping a single spin in the itinerant-carrier system and occur in this simple model at much larger energies near the itinerant-carrier spin-splitting gap Δ. Furthermore, additional collective modes analogous to the optical spin waves in a ferrimagnet appear. Their dispersion lies closely below the Stoner continuum. For a more detailed discussion see Ref. [12]

3.4 Comparison to RKKY and to Mean-Field Picture

For comparison we evaluate the $T = 0$ magnon dispersion assuming an RKKY interaction between magnetic ions. This approximation results from our theory if we neglect spin polarization in the itinerant carriers and evaluate the static limit of the resulting spin-wave propagator defined in Eq. (7). The Stoner excitations and optical spin waves are then not present, and the

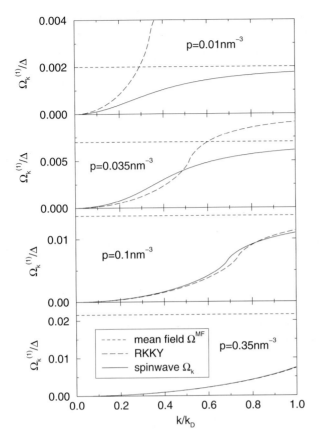

Fig. 1. Spin-wave dispersion (solid lines) for $J_{pd} = 0.06\text{eVnm}^3$, $m^* = 0.5m_e$, $N_{Mn} = 1\text{nm}^{-3}$, and four different itinerant-carrier concentrations p. The short wavelength limit is the mean-field result $\Omega^{MF} = x\Delta$ (short-dashed lines), and the long-dashed lines are the result obtained from an RKKY picture

Goldstone-mode dispersion is incorrect except when $\Delta \ll \epsilon_F$, as depicted in Fig. 1 (long-dashed lines).

The mean-field theory can be obtained in our approach by dropping the last term in Eq. (7). The energy of an impurity-spin excitation is then dispersionless, $\Omega^{MF} = x\Delta$ (short-dashed line in Fig. 1), since correlations among the Mn spins have been neglected, and Ω^{MF} is always larger than the real spin-wave energy.

3.5 Spin-Wave Dispersion for Realistic Bands

For a quantitative analysis [13] of the spin-wave dispersion we extend our parabolic-band model to a six-band Kohn Luttinger Hamiltonian. The effective action for the HP bosons describing the Mn impurity spins is given by the same formal expression Eq. (3) with the contributions Eqs. (4) and (5) to the kernel. The difference is that for each Bloch wavevector i and j now label the states in a six-dimensional Hilbert space (instead of two dimensions for spin up and down), H_0 is the Kohn-Luttinger Hamiltonian, and s_{ij}^z and $s_{ij}^{\pm}$ are 6×6 matrices. The subsequent analysis is straightforward [13], and allows for a numerical evaluation of the spin-wave dispersion.

3.6 Spin Stiffness

As one qualitative difference to the two-band model with isotropic bands, the spin waves in the six-band model are no longer soft modes but acquire a finite gap due to magnetic anisotropy. The quantized energy of a long-wavelength spin wave in a ferromagnet with uniaxial anisotropy can be written as

$$\Omega_k = \frac{2K}{N_{\mathrm{Mn}}S} + \frac{2A}{N_{\mathrm{Mn}}S} k^2 + \mathcal{O}(k^4), \qquad (10)$$

where K is the anisotropy energy constant, and A denotes the spin stiffness or exchange constant. While the anisotropy constant can be obtained from the mean-field energy for different magnetization orientations, the virtue of the spin-wave calculation is to extract the spin stiffness as well.

In Fig. 2 we show the spin stiffness A as a function of the itinerant-carrier density for two values of J_{pd} for both the isotropic two-band and the full six-band model. We find that the spin stiffness is much larger for the six-band

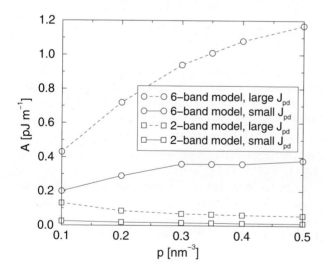

Fig. 2. Exchange constant A as a function of itinerant-carrier density p for the six-band and the two-band model for two different values of $J_{\mathrm{pd}} = 0.068\,\mathrm{eV\,nm}^{-3}$ (solid lines) and $0.136\,\mathrm{eV\,nm}^{-3}$ (dashed lines). The impurity-spin concentration is chosen as $N_{\mathrm{Mn}} = 1.0\,\mathrm{nm}^{-3}$, which yields $\Delta = 0.17\,\mathrm{eV}$ (solid lines) and $\Delta = 0.34\,\mathrm{eV}$ (dashed lines), respectively

calculation than for the two-band model. One reason is that, even in the limit of low carrier concentration, it is not only the (heavy-hole) mass of the lowest band which is important for the spin stiffness. Instead, a collective state in which the spins of the itinerant carriers follow the spatial variation of a Mn spin-wave configuration will involve the light-hole band, too. Our calculations show that accounting for the presence of this second more dispersive band is essential to understanding the success of mean-field theory. Crudely, the large mass heavy hole band dominates the spin-susceptibility and enables local magnetic order at high temperatures, while the dispersive light-hole band dominates the spin stiffness and enables long range magnetic order.

4 Limits on the Curie Temperature

Isotropic ferromagnets have spin-wave Goldstone collective modes whose energies vanish at long wavelengths, $\Omega_k = Dk^2 + \mathcal{O}(k^4)$, where k is the wavevector of the mode. Spin-orbit coupling breaks rotational symmetry which leads to a finite gap, see Eq. (10). According to our numerical studies, though, this gap is negligibly small as far as the suppression of ferromagnetism by collective spin excitations is concerned and can, therefore, be dropped for the present discussion. Each spin-wave excitation reduces the total spin of the ferromagnetic state by 1. The coefficient $D = 2A/(N_{\mathrm{Mn}}S)$ is proportional to the spin stiffness A. If the spin stiffness is small, they will dominate the suppression of the magnetization at all finite temperatures and limit the critical temperature. In this case, the typical local valence-band carrier polarization remains finite above the critical temperature. Ferromagnetism disappears only because of the loss of long-range spatial coherence.

A rough upper bound on the critical temperature T_c^{coll}, which accounts for collective fluctuations, is provided by the temperature at which the number of excited spin waves equals the total spin of the ground state [14]. This yields $k_B T_c^{\mathrm{coll}} = Dk_D^2$ for $S = 5/2$. To get a qualitative but transparent picture we employ the two-band model with parabolic bands, and use Eqs. (8) and (9) for the strong and weak-coupling regime, respectively. For strong coupling, $\Delta/\epsilon_F \gg 1$, the exchange coupling completely polarizes the valence-band electrons, and we find (using $p \ll 2N_{\mathrm{Mn}}S$) the T_c bound

$$T_c^{\mathrm{coll,s}} = \frac{2S+1}{12S}\epsilon_F \left(\frac{p}{N_{\mathrm{Mn}}}\right)^{1/3}. \tag{11}$$

For small Δ/ϵ_F, the weak-coupling or RKKY regime, exchange coupling is a weak perturbation on the band system. In this regime we get

$$T_c^{\mathrm{coll,RKKY}} = T_c^{\mathrm{MF}} \frac{2S+1}{12(S+1)\sqrt[3]{2}} \left(\frac{N_{\mathrm{Mn}}}{p}\right)^{2/3}, \tag{12}$$

where T_c^{MF} is the mean-field Curie temperature, i.e., mean-field theory is reliable only for $p/N_{\mathrm{Mn}} \ll 1$ since in this case the RKKY interaction has a range which is long compared to the distance between Mn spins.

A more quantitative study of Curie temperature trends in (III,Mn)V semiconductors is presented in Ref. [15].

5 Monte-Carlo Approach

The idea, that ferromagnetism may disappear because of loss of long-range spatial coherence, is also underpinned by Monte Carlo simulations, where the collective fluctuations are treated beyond the independent spin-wave theory and disorder associated with the randomness of Mn positions is accounted for [18]. In Fig. 3 we show an example where the local hole polarization is finite even above the Curie temperature, at which the global polarization vanishes. As a consequence, the Curie temperature is reduced as compared to the mean-field-theory result, and in particular shows a different functional dependence on the density of states and the itinerant-carrier concentration. In particular, we find that T_c cannot grow unlimited by increasing the density of states.

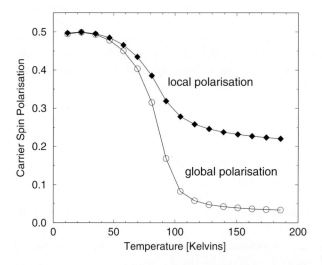

Fig. 3. Monte-Carlo data for the local and the global polarization of the itinerant carriers for a two-band model with $m^* = 0.5 m_e$, $p = 0.1\,\mathrm{nm}^{-3}$, $N_{\mathrm{Mn}} = 1\,\mathrm{nm}^{-3}$, and $J_{\mathrm{pd}} = 0.15\,\mathrm{eV}\,\mathrm{nm}^3$

6 Magnetic Domains

A phenomenological long-wavelength description of ferromagnets usually requires only a small set of parameters such as the saturation magnetization, the anisotropy constant K, and the exchange constant A for a ferromagnet with uniaxial asymmetry [see Eq. (10)]. A whole variety of magnetic properties can be expressed in terms of these parameters. As an example, we study stripe domains which appear in perpendicular films, as measured by Shono et al. [19]. Their width is determined by the ratio λ_c of the Bloch domain-wall energy and the stray-field energy, which in turn depends on K and A [20]. Computing λ_c and, the domain width W for the specific sample we find for low temperature $W = 1.1\,\mu$m, in good agreement with the experimental value $W = 1.5\,\mu$m, as shown in Fig. 4. This is an indication that the (III,Mn)V ferromagnets behave like usual magnets in the sense that they, indeed, can be described within an micromagnetic approach.

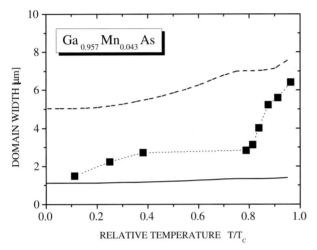

Fig. 4. Temperature dependence of the width of domain stripes as measured for a $Ga_{0.957}Mn_{0.043}As$ film (squares). Solid line: computed domain width for given sample parameters. Dashed line: domain width assuming that λ_c is 1.8 times larger

Summary

In summary, we reviewed the theoretical picture ferromagnetism in (III,Mn)V semiconductors that arises from a kinetic-exchange model between localized Mn spins and valence-band carriers. We discussed collective spin excitations, studied their role on limiting the Curie temperature, and extracted the parameters necessary for a micromagnetic description of magnetic properties from the spin-wave dispersion.

Acknowledgements

Different parts of the presented work has been done in collaboration with T. Dietl, T. Jungwirth, J. Kucera, H.H. Lin, A.H. MacDonald, J. Schliemann, and J. Sinova.

References

1. J.K. Furdyna and J. Kossut, *Diluted Magnetic Semiconductors*, Vol. 25 of *Semiconductor and Semimetals* (Academic Press, New York, 1988).
2. T. Dietl, *Diluted Magnetic Semiconductors*, Vol. 3B of *Handbook of Semiconductors*, (North-Holland, New York, 1994).
3. A. Haury, A. Wasiela, A. Arnout, J. Cibert, S. Tatarenko, T. Dietl, Y. Merle d' Aubigné, Phys. Rev. Lett. **79**, 511 (1997).
4. H. Ohno, Science **281**, 951 (1998); F. Matsukura, H. Ohno, A. Shen, and Y. Sugawara, Phys. Rev. B **57**, R2037 (1998); H. Ohno, J. Magn. Magn. Mater. **200**, 110 (1999).
5. S. Sonoda, S. Shimizu, T. Sasaki, Y. Yamamoto, and H. Hori, cond-mat/0108159.
6. J. König, J. Schliemann, T. Jungwirth, and A.H. MacDonald, to be published in "Electronic Structure and Magnetism of Complex Materials", Eds. D.J. Singh and D.A. Papaconstantopoulos, Springer; cond-mat/0111314.
7. T. Jungwirth, W.A. Atkinson, B.H. Lee, and A.H. MacDonald, Phys. Rev. B **59**, 9818 (1999).
8. T. Dietl, H. Ohno, F. Matsukura, J. Cibert, and D. Ferrand, Science **287**, 1019 (2000).
9. B.H. Lee, T. Jungwirth, and A.H. MacDonald, Phys. Rev. B **61**, 15606 (2000).
10. T. Dietl, H. Ohno, and F. Matsukura, Phys. Rev. B **63**, 195205 (2001).
11. M. Abolfath, T. Jungwirth, J. Brum, and A.H. MacDonald, Phys. Rev. B **63**, 054418 (2001).
12. J. König, H.H. Lin, and A.H. MacDonald, Phys. Rev. Lett. **84**, 5628 (2000); and in *Interacting Electrons in Nanostructures*, Eds. R. Haug and H. Schoeller, Lecture Notes in Physics **579**, p.195-212 (Springer-Verlag, Berlin, 2001);
13. J. König, T. Jungwirth, and A.H. MacDonald, Phys. Rev. B **64**, 184423 (2001).
14. J. Schliemann, J. König, H.H. Lin, and A.H. MacDonald, Appl. Phys. Lett. **78**, 1550 (2001);
15. T. Jungwirth, J. König, J. Sinova, J. Kucera, and A.H. MacDonald, cond-mat/0201157.
16. J. Schliemann and A.H. MacDonald, cond-mat/0107573.
17. S.R. Eric Yang and A.H. MacDonald, cond-mat/0202021.
18. J. Schliemann, J. König, and A.H. MacDonald, Phys. Rev. B **64**, 165201 (2001).
19. T. Shono, T. Hasegawa, T. Fukumura, F. Matsukura, and H. Ohno, Appl. Phys. Lett. **77**, 1363 (2000).
20. T. Dietl, J. König, and A.H. MacDonald, Phys. Rev. B **64**, 241201(R) (2001).

Tetrahedral Quantum Magnets in One and Two Dimensions

Wolfram Brenig[1], Andreas Honecker[1], and Klaus W. Becker[2]

[1] Institut für Theoretische Physik, Technische Universität Braunschweig,
    Mendelssohnstr. 2-3, 38106 Braunschweig, Germany
[2] Institut für Theoretische Physik, Technische Universität Dresden,
    01062 Dresden, Germany

Abstract. We discuss the magnetic properties of two quantum magnets formed from tetrahedral units, i.e. the edge-sharing tetrahedral chain and the corner-sharing two-dimensional tetrahedral checkerboard magnet. Using a combination of exact diagonalization (ED) and bond-operator theory the quantum phase diagram of the former system is shown to incorporate a singlet-product, a dimer, and a Haldane phase. Moreover the one-, and two-triplet excitations in the dimer phase as well as two-triplet bound states are considered. On the two-dimensional spin-1/2 checkerboard lattice and starting from the limit of isolated quadrumers, we perform a complementary analysis of the evolution of the spectrum as a function of the inter quadrumer coupling j using both, ED and series expansion (SE) by continuous unitary transformation. We have computed the ground state energy, the elementary triplet as well as singlet excitations, and find very good agreement between SE and ED. Our results strongly support the notion of a valence-bond crystal ground-state of the checkerboard magnet.

1 Introduction

Low-dimensional quantum-magnetism has received considerable interest recently due to the discovery of numerous novel materials with spin-$\frac{1}{2}$ moments arranged in chain, ladder, and depleted planar structures. On the one hand many of these materials exhibit unconventional magnetic phases without long-range magnetic order (LRO) and gapped elementary excitation spectra due to various types of dimerization and frustration of antiferromagnetic exchange interactions. On the other hand strong geometrical frustration may also induce ground states with near macroscopic degeneracy leading to novel low energy collective spin dynamics and possibly even zero temperature entropy [1,2].

In one dimension complete frustration can occur in two-leg spin-ladders if an additional cross-wise exchange is included as depicted in Fig. 1a) which resembles a chain of edge sharing tetrahedra. Each tetrahedron at $J_1 = J_2$ and $J_3 = 0$ displays a two-fold degenerate singlet ground-state suggesting the existence of low-energy singlet excitations in the case of finite inter-tetrahedral

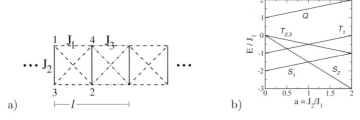

Fig. 1. a) The tetrahedral cluster-chain. l labels the unit cell containing spin-1/2 moments $\mathbf{s}_{i\,l}$ at the vertices $i = 1,\ldots,4$. b) Spectrum of a single tetrahedron

coupling. For $J_1 = J_3$ such tetrahedral chains have been investigated in the past [3,4,5]. Very recently, tellurates of type $Cu_2Te_2O_5X_2$ with X=Cl, Br have been identified as a new class of spin-1/2 tetrahedral-cluster compounds [6]. Bulk thermodynamic data have been analyzed in the limit of isolated tetrahedra [6]. Raman spectroscopy, however indicates a substantial inter-tetrahedral coupling along the c-axis direction of $Cu_2Te_2O_5X_2$ [7]. In this direction the exchange topology is likely to be analogous to that of Fig. 1a) with $J_1 \neq J_3$.

While in principle magnetic LRO is possible above one dimension the impact of frustration on higher dimensional quantum magnets is of great interest. This pertains in particular to the identification of systems with spin-liquid ground states [2]. In this respect quantum spin systems on the kagomé, and more recently on the pyrochlore lattice have been at the center of several investigations. While their classical counterparts have been studied in considerable detail the role of quantum fluctuations in such systems is an open issue. For the spin-1/2 kagomé lattice gap less singlet excitations and a high density of singlet states in the singlet-triplet gap have been established on finite systems by ED [2]. Similar analysis of the pyrochlore quantum-magnet is severely constrained by its three-dimensional structure. Therefore, and as a first step, several investigations [8,9,10,11] have focused on the planar projection of the pyrochlore quantum-magnet, i.e. the spin-1/2 checkerboard lattice of Fig. 2a) for the case of $j = 1$. For this system, ED at $j = 1$ has resulted in a

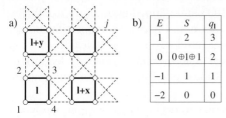

Fig. 2. a) The checkerboard lattice, spin-1/2 moments are located on the open circles. Full (dashed) lines label the quadrumer bonds (bonds corresponding to the expansion parameter j). b) Energy (E), spin (S), and local q_1 quantum-number of a single quadrumer

sizeable spin gap and a large number of in-gap singlet states [8,11]. Moreover, recent ED [11] has given strong evidence in favor of a valence bond crystal (VBC) ground state with long range order in the $S=0$ quadrumers shown in Fig. 2a).

In this note we will summarize several aspects regarding the ground state correlations and the elementary excitations both, for the tetrahedral chain in section two and for the checkerboard magnet in section three.

2 The Tetrahedral Chain

The Hamiltonian of the tetrahedral chain of Fig. 1a) can be written as a 1D chain in terms of the total edge-spin operators $\mathbf{T}_{1(2)\,l} = \mathbf{S}_{1(4)\,l} + \mathbf{S}_{3(2)\,l}$ and the dimensionless couplings $b = J_3/J_1$ and $a = J_2/J_1$

$$\frac{H}{J_1} = \sum_l [\mathbf{T}_{1l}\mathbf{T}_{2l} + b\mathbf{T}_{2l}\mathbf{T}_{1l+1} + \frac{a}{2}(\mathbf{T}_{1l}^2 + \mathbf{T}_{2l}^2) - \frac{3a}{2}] \qquad (1)$$

This model displays infinitely many local conservation laws: $[H, \mathbf{T}_{i(=1,2)\,l}^2] = 0;\ \forall l, i = 1, 2$. The Hilbert space decomposes into sectors of fixed distributions of edge-spin eigenvalues $T_{i\,l} = 1$ or 0, each corresponding to a sequence of dimerized spin-1 chain-segments intermitted by chain-segments of localized singlets. The Hilbert space of a single tetrahedron consists of 16 states, i.e., two singlets $\mathcal{S}_{1,2}$, three triplets $\mathcal{T}_{1,2,3}$ and one quintet $\mathcal{Q}$, where $\mathcal{S}_2$ refers to $T_{1\,l} = T_{2\,l} = 0$ (cf. fig 1b)). For the single tetrahedron Johnsson and collaborators [6] have pointed out that the respective level scheme implies a singlet to reside within the singlet-triplet gap of the tetrahedron for $1/2 < a < 2$. Moreover the ground state switches from $\mathcal{S}_1$ to $\mathcal{S}_2$ at $a = 1$ suggesting a line of quantum phase transitions in the (a, b)-plane for the lattice model.

We now discuss the ground state of the tetrahedral chain. To begin, we note that by a shift of one half of the unit cell, i.e. $\mathbf{T}_{2\,l(1\,l+1)} \to \mathbf{T}_{1\,l(2\,l)}$, model (1) is symmetric under the operation $(J_1, a, b) \to (J_1 b, a/b, 1/b)$. Therefore, in order to cover the *complete* parameter space for $a, b > 0$ it is sufficient to consider the phase diagram in the range of $a \in [0, \infty]$ and $b \in [0, 1]$. Next we note, that the ground state of (1) will be either in the dimerized $S = 1$-chain sector or in a homogeneous product of $\mathcal{S}_2$ states only. Inhomogeneous phases consisting of both, $T_{i\,l} = 0$ and $T_{i\,l} = 1$ sites are not allowed for as ground-states. To see this we fix b and assume $a \to \infty$, in which case the ground state is a pure product of $\mathcal{S}_2$-type singlets: $|\psi_0\rangle = \prod_l |\tilde{s}_l\rangle$. Next we check for the ground-state energy change $\Delta E(a, b, N)$ upon forming a single connected chain-segment of length N composed out of $T_{i\,l} = 1$-sites within the homogeneous state $|\psi_0\rangle$. To be specific we first assume the chain-segment to consist of $D' = N/2$ tetrahedra in which case $\Delta E(a, b, D') = D'[2(a-1) - e(b, D')]$. Here $-e(b, D') < 0$ is the ground-state energy gain *per two sites*. The latter has been checked [12] to be a monotonously increasing function of D'. Therefore the largest critical value $a_c = max\{a_c(D')\}$ at which the formation of tetrahedra in the $S = 1$ sector is favorable, i.e. at which $\Delta E(a_c(D'), b, D')$ turns negative, results for $D' \to \infty$. This implies a single first order quantum phase transition into the infinite-length, dimerized $S = 1$-chain sector as a function of decreasing a. Similar arguments can be pursued for odd N.

From $\Delta E(a, b, D')$ the first order critical line $a_c(b)$ between the infinite-length, dimerized spin-1 chain for $a < a_c(b)$ and the S_2-type singlet product-state for $a > a_c(b)$ is fixed by $a_c(b) = 1 + e(b)/2$, where $e(b) = \lim_{D' \to \infty} e(b, D')$. To determine $e(b)$ we have calculated the ground-state energy of dimerized spin-1 chains using exact diagonalization (ED) with periodic boundary conditions (PBC) on up to $N = 16$ sites and a bond-boson theory detailed below. Figure 3 shows the corresponding quantum phase diagram. Regarding the ED the critical value of $a_c(b = 0) = 1$ agrees with [6], while $a_c^{N=16}(b = 1) \simeq 1.403$ agrees with [13] and is consistent with an extrapolated value of $a_c^{N=\infty}(b = 1) \simeq 1.401$ from Density-Matrix-Renormalization-Group (DMRG) calculations [14,5] and ED on 22 sites [15].

In Fig. 3 MFT and LHP refer to phase boundaries determined from a mapping of the tetrahedral chain onto a system of interacting bosons within the dimerized $S = 1$-chain sector by applying the bond operator method [16,17,18,19,20,21]. There one introduces a set of singlet- $(s_l^\dagger)$, triplet- $(t_{l\alpha}^\dagger)$, and quintet-bosons $(q_{l\alpha}^\dagger)$ for each tetrahedron at site l. These bosons create all states within the multiplets S_1, T_1, and Q. In terms of these bosons one

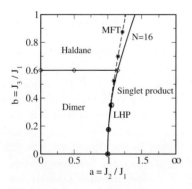

Fig. 3. Quantum phase diagram of the tetrahedral chain. Bare solid line: 1st-oder transition from ED for N=16 sites and PBC at 41 values of $b \in [0, 1]$. The critical value a_c at $b = 1$ from ED is $a_c^{N=16}(b = 1) \simeq 1.40292$. Solid line with diamond markers: 2nd-order Haldane-Dimer transition at $b \simeq 3/5$, extrapolated from ED and ref.[14,15]. Dashed(Solid) line with stared(circled) markers refers to bond-boson mean-field/MFT (Holstein-Primakoff/LHP) approach. LHP terminates at $b = 3/8$

may express the $\alpha = x, y, z$ components of the edge-spins $S^\alpha_{l1,2}$ by [22,23]

$$S^\alpha_{l1,2} \hat{=} \sqrt{\frac{2}{3}}(\pm s^\dagger_l t_{l\alpha} \pm t^\dagger_{l\alpha} s_l) - \frac{i}{2}\varepsilon_{\alpha\beta\gamma} t^\dagger_{l\beta} t_{l\gamma}$$
$$\pm M_{\alpha\hat{\beta}\hat{\gamma}} t^\dagger_{l\hat{\beta}} q_{l\hat{\gamma}} \pm M^*_{\alpha\hat{\beta}\hat{\gamma}} q^\dagger_{l\hat{\gamma}} t_{l\hat{\beta}} + N_{\alpha\hat{\beta}\hat{\gamma}} q^\dagger_{l\hat{\beta}} q_{l\hat{\gamma}} , \qquad (2)$$

To suppress unphysical states the bosons have to fulfill a hard core constraint of no double-occupancy

$$s^\dagger_l s_l + t^\dagger_{l\alpha} t_{l\alpha} + q^\dagger_{l\alpha} q_{l\alpha} = 1 , \qquad (3)$$

where doubly appearing Greek indices are to be summed over their respective ranges. Inserting (2) into (1) an interacting Bose gas constrained by (3) results which allows for an approximate treatment only. In the limit of weak inter-tetrahedral coupling, i.e. $b \ll 1$, the singlet bosons condense [16,18] with $s^{(\dagger)}_l \to s \in \Re$. Focusing on this limit and keeping only terms up to quadratic order in the boson operators and, moreover, satisfying the constraint (3) by a global Lagrange multiplier λ we arrive at a mean-field theory (MFT)

$$H_{MFT} = D(\frac{3}{2} - 2s^2 + \lambda s^2 - \frac{5}{2}\lambda + \frac{a}{2})$$
$$+ \sum_{l\alpha} (\lambda + 1) q^\dagger_{l\alpha} q_{l\alpha} + \sum_{k\alpha} E_k (a^\dagger_{k\alpha} a_{k\alpha} + \frac{1}{2}) , \qquad (4)$$

where the threefold degenerate triplet energy E_k is given by $E_k = [(\lambda - 1)^2(1 + s^2/(\lambda - 1)2\epsilon_k)]^{1/2}$ and $a^{(\dagger)}_{k\alpha}$ are Bogoliubov transforms of the original triplet bosons. The ground state energy E^0_{MFT} is obtained from a self consistent minimization with respect to λ and s^2. In the limit of vanishing inter-tetrahedral coupling, i.e. $b = 0$ one finds $\lambda = 2$ and $s^2 = 1$ [22] which is formally identical to the linearized Holstein-Primakoff (LHP) approach [19,20].

In Fig. 3 results for $a_c(b)$ as obtained from (4) are included for the MFT and LHP approach. At the dimer to singlet-product phase-boundary the agreement with ED is very good, both for LHP and MFT. In principle, the singlet condensate restricts the bond-boson approaches to the dimer phase. In fact, the LHP spin-gap closes at $b = 3/8$ confining the LHP to $b < 3/8 < b_c$. The MFT can be continued from the dimer into the Haldane regime, even though the ground-state symmetries are different, yielding a transition line qualitatively still comparable to ED.

Within the dimerized $S = 1$-chain sector an additional second-order quantum phase transition exists between the dimer phase for $b < b_c$ and the Haldane phase for $b > b_c$. This transition has been studied extensively (see eg. [24] and refs. therein), resulting in $b_c \simeq 3/5$ from DMRG calculations [25] and finite-size scaling analysis [24]. We have performed an ED analysis of the finite-size scaling of the spin gap in the dimerized $S = 1$-chain sector as function of b. Our results are identical to those which have been obtained earlier by Kato and Tanaka [25].

Next we consider excitations in the dimer phase. The excited states may (i) remain in the spin-1 chain sector, or (ii) involve transitions into sectors with *localized* edge-singlets. As has been pointed out in [6], for a single tetrahedron, the energy of a pair of two type-(ii) excitations resides in the spin-gap of sector (i) for $1/2 \leq a \leq 1$. While analogous, *dispersionless* singlets are found in the spin gap of the dimer phase of the lattice model [26], we focus on the one- and two-triplet excitations of type (i) in the following. Figure 4 displays the dispersion of the first triplet, comparing ED, LHP, and MFT. The agreement is very good.

Two-particle excitations in the dimerized $S = 1$-chain sector deriving from the states of Fig. 4 are experimentally relevant, in particular for magnetic Raman scattering, i.e. if combined into total momentum zero and spin zero, e.g. as created by the $q \to 0$-limit of the operator [22]

$$R_{LHP}(q) = -\frac{2C}{3} \sum_k \cos(k + q/2)(t^\dagger_{k+q\,\alpha} + t_{-k-q\,\alpha})(t^\dagger_{-k\,\alpha} + t_{k\,\alpha}) . \quad (5)$$

The two-particle states are renormalized beyond the quadratic MFT and LHP approximations by two-particle interactions. The latter occur both, upon insertion of (2) into (1) and because of the hard-core (3) which has to be incorporated in the two-particle sector by introducing an additional contribution to the Hamiltonian [27]

$$H_U = U \sum_l t^\dagger_{l\,\alpha} t^\dagger_{l\,\beta} t_{l\,\alpha} t_{l\,\beta} , \quad (6)$$

with a summation convention on the Greek indices. Correlation functions have to be evaluated with H_U at finite U and the limit of $U \to \infty$ is taken at the end. For $b \ll 1$ the triplet density induced by quantum fluctuations $n_t = \langle t^\dagger_{l\,\alpha} t_{l\,\alpha} \rangle = 3 \sum_k h_k^2$ is a small parameter. Therefore the two-particle

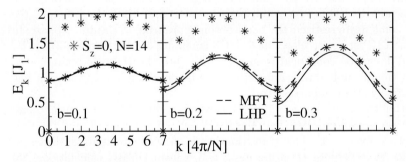

Fig. 4. Dashed (solid) line: E_k for MFT (LHP). Stars: first two total-$S_z = 0$ excitations of dimerized spin-1 chain from ED with PBC

Green function $\chi(q,\tau) = \langle T_\tau [\tilde{R}_{LHP}(q,\tau)\tilde{R}_{LHP}(q,0)]\rangle$ can be evaluated exactly by summing all ladder-diagrams, i.e. the T-matrix of Fig. 5. After several intermediate steps one finds

$$\chi^p(q,z) \stackrel{b \ll 1}{=} \frac{6(\text{sign}(\text{Re}(\nu))\sqrt{\nu^2-1}-\nu)}{b[\text{sign}(\text{Re}(\nu))\sqrt{\nu^2-1}-\nu-8\cos(q/2)/3]} \,. \tag{7}$$

where $\nu = 3(z-2)/(8b\cos(q/2))$. The corresponding spectrum is shown in Fig. 6 which, apart from a continuum of two-triplet scattering states displays a total-spin zero *bound state*. For $q < q_c$ the bound state turns into a resonance shortly above the lower edge of the continuum. This resonance feature should be contrasted against bound-states occurring in other dimerized and frustrated low-dimensional quantum spin systems where $S = 0$ collective modes have been detected rather as sharp excitations *within* the spin gap *at* $q = 0$ [28,29,30]. The actual location of the bound state with respect to the two-triplet continuum is significantly affected by the hard-core repulsion U. Eg., for $U = 0$ the short-range attraction of triplets is overestimated yielding a bound state *below* the lower edge of the continuum for all q.

An interesting issue for future studies is the effect of inter-chain coupling on the tetrahedral system, not only since this seems to be of relevance to the Tellurates but also since additional phases with magnetic LRO may result.

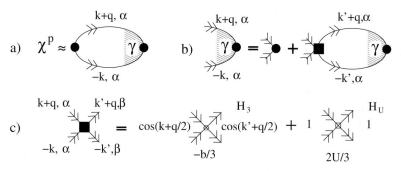

Fig. 5. T-matrix approximation for $\chi(q,\tau)$: thin, doubly-directed lines label normal bare one-triplet Green functions. The solid dot is the two-triplet part of (5). Summation on k' and α is implied in all bare triplet-bubbles. γ(solid square) refers to two-triplet (ir)reducible vertex

3 The Checkerboard Magnet

Recently numerical analysis [11] has given strong evidence in favor of a valence bond crystal (VBC) ground state on the checkerboard lattice of Fig. 2a). While this has been concluded from correlation of the $S = 0$ quadrumer

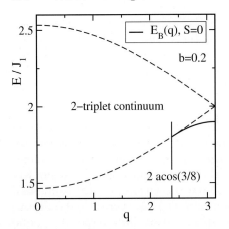

Fig. 6. Two-triplet continuum and dispersion of the $S=0$ bound state from (7). Note the y-axis offset

shown in this figure at $j=1$ it emphasizes the need for a perturbative investigation of the checkerboard magnet starting from the limit $j \to 0$ in Fig. 2a). In the following we describe the evolution of the spin spectrum of the checkerboard spin-$1/2$ magnet as a function of the inter quadrumer coupling j, contrasting exact diagonalization (ED) against high-order series expansion (SE).

Written in a form adapted to the VBC symmetry breaking and normalized to an overall unit of energy the Hamiltonian of the checkerboard magnet reads

$$H = \sum_\mathbf{l} \left[\frac{1}{2}(\mathbf{P}^2_{1234\mathbf{l}} - \mathbf{P}^2_{13\mathbf{l}} - \mathbf{P}^2_{24\mathbf{l}}) + j\,(\mathbf{P}_{34\mathbf{l}}\mathbf{P}_{12\mathbf{l}+\mathbf{x}} + \mathbf{P}_{23\mathbf{l}}\mathbf{P}_{14\mathbf{l}+\mathbf{y}}) \right] \quad (8)$$

$$= H_0 + j \sum_{n=-N}^{N} T_n \quad (9)$$

where $\mathbf{P}_{i...j\mathbf{l}} = \mathbf{S}_{i\mathbf{l}} + \ldots + \mathbf{S}_{j\mathbf{l}}$ and $\mathbf{S}_{i\mathbf{l}}$ refers to spin-$1/2$ operators residing on the vertices $i = 1 \ldots 4$ of the quadrumer at site $\mathbf{l}$, c.f. Fig. 2a). H_0 is the sum over local quadrumer Hamiltonians the spectrum of which, c.f. Fig. 2b), consists of four equidistant levels which can be labeled by spin S and the number of local energy quanta $q_\mathbf{l}$. H_0 displays an equidistant ladder spectrum labeled by $Q = \sum_\mathbf{l} q_\mathbf{l}$. The $Q = 0$ sector is the *unperturbed* ground state $|0\rangle$ of H_0, which is the VBC of quadrumer-singlets. The $Q = 1$-sector contains local $S = 1$ single-particle excitations of the VBC with $q_\mathbf{l} = 1$, where $\mathbf{l}$ runs over the lattice. At $Q = 2$ the spectrum of H_0 has total $S = 0, 1$, or 2 and is of multiparticle nature. For $S = 0$ at $Q = 2$ it comprises of one-particle singlets with $q_\mathbf{l} = 2$ and two-particle singlets constructed from triplets with $q_\mathbf{l} = q_\mathbf{m} = 1$ and $\mathbf{l} \neq \mathbf{m}$. Consequently the perturbing terms $\propto j$ in (8) can be written as a sum of operators T_n which *non-locally* create(destroy) $n \geq 0$ ($n < 0$) quanta within the ladder spectrum of H_0.

The structure of the problem (9) allows for high-order SE using a continuous unitary transformation generated by the flow equation method of Wegner [31,32,33,34]. The unitarily rotated effective Hamiltonian H_{eff} reads [32,34]

$$H_{\text{eff}} = H_0 + \sum_{n=1}^{\infty} j^n \sum_{\substack{|\mathbf{m}|=n \\ M(\mathbf{m})=0}} C(\mathbf{m}) \, T_{m_1} T_{m_2} \ldots T_{m_n} \tag{10}$$

where $\mathbf{m} = (m_1 \ldots m_n)$ is an $n = |\mathbf{m}|$-tuple of integers, each in a range of $m_i \in \{0, \pm 1, \ldots, \pm N\}$. In contrast to H of (8), H_{eff} conserves the total number of quanta Q. This is evident from the constraint $M(\mathbf{m}) = \sum_{i=1}^{n} m_i = 0$. The amplitudes $C(\mathbf{m})$ are rational numbers computed from the flow equation method [32,34]. Here we present results up to $O(7)$ for the ground state energy and the elementary triplet as well as up to $O(6)$ for the $Q = 2$ singlets. The $C(\mathbf{m})$-table is available on request [35]. Q-number conservation enforces the ground state and the elementary triplet energies, i.e., $E_g = \langle 0 | H_{\text{eff}} | 0 \rangle$

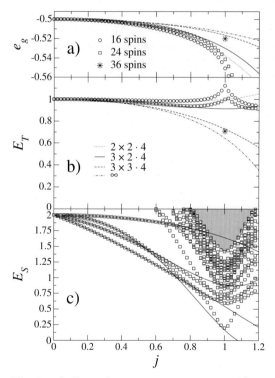

Fig. 7. a) Ground state energy per spin, b) spin gap and c) singlet excitations with $\mathbf{k} = \mathbf{0}$ as a function of j. Lines correspond to the series, symbols to exact diagonalization. The numerical data at $j = 1$ for 36 spins is taken from Ref. [11]. The grey shaded area in panel c) indicates that our numerical data is not complete in this region

and $E_\mu(\mathbf{k}) = \sum_{lm} t_{\mu,lm} e^{i(k_x l + k_y m)}$ to result from the $Q = 0$ and $Q = 1$ sector respectively. Here $t_{\mu,lm} = \langle \mu, lm | H_{\text{eff}} | \mu, 00 \rangle - \delta_{lm,00} E_g^{obc}$ are hopping matrix elements from site $(0,0)$ to site (l,m) for a quadrumer excitation μ inserted into the unperturbed ground state. *A priori* single-particle states from sectors with $Q > 1$ will not only disperse via H_{eff}, but can decay into multi-particle states. *A posteriori* however such decay may happen only at high order in j leaving the excitations almost true one-particle states.

Figure 7 summarizes our results for the ground state energy *per spin* $e_g(j)$, the spin gap $E_T(j)$, and the spectrum of the low lying singlets. The figure compares the SEs with ED data which we have obtained on $2(3) \times 2 \cdot 4$ systems, *i.e.* 16(24) spins, as a function of j, as well as available ED data [11] on $3 \times 3 \cdot 4$ sites, *i.e.* 36 spins, at $j = 1$. SEs are shown both for the *identical* system sizes and also for the thermodynamic limit to $O(j^7)$ for e_g and E_T as well as to $O(j^6)$ for the $Q = 2$-singlets. All SE data refers to the actual series, no Padé continuations have been applied.

First we focus on the ground state energy in Fig. 7a). Practically quantitative agreement is found for $j \lesssim 0.7$ between ED and SE for $2(3) \times 2 \cdot 4$ sites. At $j \sim 1$ very small deviations occur, which however decrease with system size (they are only 0.7% on the $3 \times 3 \cdot 4$ site system [11]). This is consistent with the VBC to incorporate the proper ground state correlations for $j \leq 1$ rather than the tetrahedral limit of [9]. Next we discuss the elementary triplet excitations. As one explicit example of the SE the $O(j^7)$-result for the thermodynamic limit of the triplet dispersion reads

$$E_T^\infty(\mathbf{k},j) = 1 - \frac{7 j^2}{36} - \frac{41 j^3}{864} - \frac{329887 j^4}{6531840} - \frac{580309487 j^5}{21946982400}$$
$$- \frac{16957803829 j^6}{790091366400} - \frac{7822020675129119 j^7}{557488468131840000} - \left(\frac{53 j^5}{1036800}\right.$$
$$\left. + \frac{59527 j^6}{1306368000} + \frac{74504581093 j^7}{948109639680000}\right)(\cos(k_x) + \cos(k_y))$$
$$+ \left(\frac{1679 j^6}{32659200} + \frac{2039741 j^7}{438939648000}\right) \cos(k_x) \cos(k_y). \quad (11)$$

Remarkably, the hopping amplitude is exceedingly small and sets in only at $O(j^5)$ hinting at a fairly extended polarization cloud necessary in order to allow for triplet motion [36]. From (11) the spin gap occurs at $\mathbf{k} = \mathbf{0}$. The spin gap is shown in Fig. 7b). Again, for $j \lesssim 0.7$ the agreement between ED and $O(j^7)$-SE is very good. A cusp is observable in $E_T(j)$ close to $j = 1$ on the $2(3) \times 2 \cdot 4$ systems. This is related to extra symmetries of the 16(24)-spin systems at $j \approx 1$ and is not captured by the SE [11]. While no j-scan of E_T is available on $3 \times 3 \cdot 4$ sites it is remarkable that at $j = 1$ the SE agrees better with ED for 36 spins than for 16(24) spins and that the SE behaves qualitatively different, both on the $3 \times 3 \cdot 4$ lattice *and* in the thermodynamic limit as compared to the $2(3) \times 2 \cdot 4$ lattices. *I.e.*, on the larger systems the gap decreases monotonously, while it increases beyond $j \gtrsim 1$ on the smaller two. This suggests an absence of discontinuities close to $j = 1$ on thermodynamically large systems. Finally Fig. 7c) shows the evolution of all eigenvalues of H from (8) in the singlet sector on the $3 \times 2 \cdot 4$ system at $\mathbf{k} = \mathbf{0}$, i.e. the location of the singlet gap, as well as the spectrum of H_{eff} at $O(j^6)$ in the $S = 0$, $Q = 2$ sector. Regarding the latter this involves solving a two-particle problem coupled to a one-particle problem. Details of this will be given elsewhere [37]. For $j \lesssim 0.7$ the agreement between ED and SE is satisfying. Starting at $j \gtrsim 0.7$ a large number of singlets from higher Q-sector enter the low energy spectrum. At $j = 1$ several of these singlets reside in the spin gap. While a finite size analysis of the singlet gap remains restricted to the SE-method due to presently available system sizes, we have found the gap in the $Q = 2$-sector to *increase* with system size.

In conclusion, our comparison of ED and SE shows the decoupled quadrumer limit to be an excellent starting point for a perturbative treatment of the spin-$\frac{1}{2}$ checkerboard magnet even at $j \approx 1$. This indicates that the points $j = 0$ and 1 belong to the same phase and is consistent with a VBC picture of the planar pyrochlore. Future studies of higher Q-sectors will be necessary to further corroborate this finding.

Acknowledgments

It is a pleasure to thank B. Büchner, C. Geibel, C. Gros, C. Jurecka, E. Kaul, P. Lemmens, F. Mila, and R. Valenti for stimulating discussions and comments. This research was supported in part by the Deutsche Forschungsgemeinschaft under Grant No. BR 1084/1-1 and BR 1084/1-2 and trough SFB 463.

References

1. P. Schiffer and A.P. Ramirez, Comments Cond. Mat. Phys. **18**, 21 (1996).
2. C. Lhuillier and G. Misguich, preprint cond-mat/0109146; C. Lhuillier, P. Sindzingre, and J.-B. Fouet, Can. J. Phys. **79**, 1525 (2001).
3. I. Bose and S. Gayen, Phys. Rev. B **48**, 10653 (1993).
4. A. Ghosh and I. Bose, Phys. Rev. B **55**, 3613 (1997).
5. A. Honecker, F. Mila, M. Troyer, Eur. Phys. J. B. **15**, 227 (2000).
6. M. Johnsson, K.W. Törnross, F. Mila, and P. Millet, Chem. Mater. 12, 2853 (2000).
7. P. Lemmens, et al., Phys. Rev. Lett. **87**, 227201 (2001).
8. S.E. Palmer and J.T. Chalker, Phys. Rev. B **64**, 094412 (2001).
9. M. Elhajal, B. Canals, and C. Lacroix, Can. J. Phys. **79**, 1353 (2001).
10. R. Moessner, O. Tchernyshyov, and S.L. Sondhi, preprint cond-mat/0106286.
11. J.-B. Fouet, M. Mambrini, P. Sindzingre, and C. Lhuillier, preprint cond-mat/0108070.
12. We have checked $e(b, N)$ to increase monotonously with N by evaluating the ground-state energy per site of a dimerized $S = 1$ chain with *open* boundary conditions as a function of system size.

13. T. Sakai and M. Takahashi Phys. Rev. B **43**, 13383 (1991).
14. S.R. White and D.A. Huse, Phys. Rev. B **48**, 3844 (1993).
15. O. Golinelli, T. Jolicoeur, and R. Lacaze, Phys. Rev. B **50**, 3037 (1994).
16. S. Sachdev and R. N. Bhatt, Phys. Rev. B **41**, 9323 (1990).
17. A. V. Chubukov and Th. Jolicoeur, Phys. Rev. B **44**, 12050 (1991).
18. S. Gopalan, T. M. Rice, M. Sigrist, Phys. Rev. B **49**, 8901 (1994).
19. O. A. Starykh, M. E. Zhitomirsky, D. I. Khomskii, R. R. P. Singh, and K. Ueda, Phys. Rev. Lett. **77**, 2558 (1996).
20. R. Eder, Phys. Rev. B **57**, 12832 (1998).
21. W. Brenig, Phys. Rev. B **56**, 14441 (1997).
22. W. Brenig and K.W. Becker, Phys. Rev. **B** 64, 214413 (2001).
23. $M_{\alpha\hat{\beta}\hat{\gamma}}$ and $N_{\alpha\hat{\beta}\hat{\gamma}}$ have been tabulated in [22].
24. K. Totsuka, Y. Nishiyama, N. Hatano, and M. Suzuki, J. Phys. Condensed Matter **7** 4895 (1995).
25. Y. Kato and A. Tanaka, J. Phys. Soc. Jp. **63**, 1277 (1993).
26. K. Totsuka, priv. commun., *unpublished*.
27. O.P. Sushkov and V.N. Kotov Phys. Rev. Lett. **81**, 1941 (1998).
28. P. Lemmens, M. Fischer, G. Güntheroth, C. Gros, P. van Dongen, M. Weiden, W. Richter, C. Geibel, and F. Steglich, Phys. Rev. B **55**, 15076 (1997).
29. P. Lemmens, M. Fischer, G. Els, G. Güntherodt, A.S. Mishchenko, M. Weiden, R. Hauptmann, C. Geibel, and F. Steglich, Phys. Rev. B **58**, 14159 (1998).
30. P. Lemmens, M. Grove, M. Fischer, G. Güntherodt, V.N. Kotov, H. Kageyama, K. Onizuka, and Y. Ueda, Phys. Rev. Lett. **85**, 2605 (2000).
31. F. Wegner, Ann. Physik **3**, 77 (1994).
32. J. Stein, J. Stat. Phys. **88**, 487 (1997).
33. A. Mielke, Eur. Phys. J. B **5**, 605 (1998).
34. C. Knetter and G.S. Uhrig, Eur. Phys. J. B **13**, 209 (2000).
35. These results are available on request from the authors.
36. This is reminiscent of similar findings for the 2D Shastry-Sutherland model: C. Knetter, A. Bühler, E. Müller-Hartmann, and G.S. Uhrig, Phys. Rev. Lett. **85**, 3958 (2000).
37. W. Brenig and A. Honecker, Phys. Rev. B **65**, 140407(R) (2002).

Part VIII

Applications

SiGe:C Heterojunction Bipolar Transistors: From Materials Research to Chip Fabrication

H. Rücker, B. Heinemann, D. Knoll, and K.-E. Ehwald

IHP, Im Technologiepark 25, 15236 Frankfurt (Oder), Germany

Abstract. Incorporation of substitutional carbon ($\approx 10^{20}\,\mathrm{cm}^{-3}$) into the SiGe region of a heterojunction bipolar transistor (HBT) strongly reduces boron diffusion during device processing. We describe the physical mechanism behind the suppression of B diffusion in C-rich Si and SiGe, and explain how the increased thermal stability of doping profiles in SiGe:C HBTs can be used to improve device performance. Manufacturability of SiGe:C HBTs with transit frequencies of 100 GHz and maximum oscillation frequencies of 130 GHz is demonstrated in a BiCMOS technology capable of fabricating integrated circuits for radio frequencies with high yield.

1 Introduction

SiGe heterojunction bipolar transistors (HBTs) have extended the field of application of Si microelectronics into the radio frequency (RF) range of several GHz. Integrated circuits combining the high-frequency performance of SiGe HBTs with the high level of integration of complementary metal-oxide-semiconductor (CMOS) technologies have opened an attractive route to cost-effective single-chip solutions for broadband and wireless communication systems [1].

A key problem in *npn* SiGe HBT technology is retaining the narrow as-grown boron profile within the SiGe layer during post-epitaxial processing. Thin, highly doped SiGe base layers are used in HBTs to realize low intrinsic base resistance and short base transit times. However, diffusion during device processing can broaden the boron profile into the adjoining Si regions, and thus significantly degrade device performance [2]. Incorporation of substitutional carbon is a way to suppress outdiffusion of B from the SiGe layer [3,4]. Compared to SiGe technologies, the addition of carbon provides greater flexibility in process design, and increases process margins. Suppressed B diffusion allows one to realize steeper doping profiles and smaller device dimensions in SiGe:C HBTs, resulting in significantly enhanced device performance [5,6].

In the following section, we discuss experimental results on suppressed B diffusion in C-doped Si and SiGe and explain the underlying physical mechanism. Major consequences for the design of SiGe:C HBTs are explored in section 4. The modular integration of SiGe:C HBTs in a CMOS platform

(BiCMOS) is discussed in section 5. Device results are presented for the SiGe:C BiCMOS technology developed at the IHP. We demonstrate the capability of this technology for fabrication of integrated RF circuits with excellent reproducibility and high yield.

2 Effect of Carbon on Boron Diffusion

Diffusion of boron atoms in Si and SiGe can be suppressed significantly by incorporation of substitutional carbon at concentrations of the order of 10^{20}cm^{-3}. Such C concentrations can be obtained in device-grade Si and SiGe layers by chemical vapor deposition or molecular beam epitaxy [7].

2.1 Diffusion Experiment

The effect of supersaturated C on B diffusion has been studied in B doping superlattices [8]. Diffusion profiles of B were measured by secondary ion mass spectroscopy (SIMS) for C background concentrations of 1×10^{20}, 1×10^{19}, and $<5\times10^{17}$cm^{-3}. Fig. 1 summarizes the measured diffusion coefficients of B as a function of carbon concentration and temperature [8]. Boron diffusion in the sample with the lowest C concentration is well described by the established diffusion coefficient of B in Si [9]. However, a strong suppression of the B diffusivity was found for increasing C concentrations. The measured reduction of the B diffusivity was almost independent of temperature (see Fig. 1).

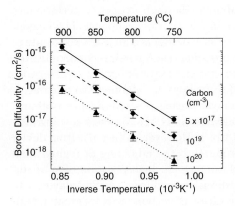

Fig. 1. Intrinsic boron diffusion coefficient as a function of temperature for three different carbon concentrations. The intrinsic diffusion coefficient $D_B = 0.76\exp(-3.46\text{eV}/kT)\text{cm}^2\text{s}^{-1}$ of B in Si is shown for comparison (solid line). Dashed and dotted lines represent $0.3D_B$ and $0.06D_B$, respectively [8]

2.2 Coupled Diffusion of C and Si Point Defects

The effect of C on B diffusion is due to non-equilibrium densities of self-interstitials (I) and vacancies (V) caused by coupled diffusion of C and Si point defects [10,11,8]. During anneals, C diffuses out of C-rich regions via a substitutional-interstitial exchange mechanism [12]. Immobile substitutional C atoms (C_s) are transformed into mobile interstitial carbon (C_i) through the kick-out reaction with Si self-interstitials

$$C_s + I \rightleftharpoons C_i. \tag{1}$$

In addition, interstitial C can be formed via the dissociative Frank-Turnbull reaction

$$C_s \rightleftharpoons C_i + V. \tag{2}$$

Eqs. (1) and (2) do not change the net number of interstitial defects (I+C_i-V). Consequently, the flux of interstitial C atoms out of the C-rich region has to be compensated by a flux of Si self-interstitials into this region and/or a flux of vacancies outwards. The individual fluxes are determined by the products of the diffusion coefficient D and the concentration. For C concentrations $C_C > 10^{18}$cm^{-3}, the C transport coefficient ($D_C C_C$) exceeds the corresponding transport coefficients of Si self-interstitials and vacancies

$$D_C C_C > D_I C_I^{eq} \quad \text{and} \quad D_C C_C > D_V C_V^{eq}, \tag{3}$$

where C_I^{eq} and C_V^{eq} are the equilibrium concentrations of Si self-interstitials and vacancies, respectively. Consequently, outdiffusion of C from C-rich regions becomes limited by the compensating fluxes of Si point defects, which leads to an undersaturation of self-interstitials and a supersaturation of vacancies in the C-rich region. Boron diffusion in Si occurs primarily via an interstitial mechanism. The effective diffusion coefficient of B is proportional to the normalized concentration of self-interstitials. Accordingly, interstitial undersaturation results in suppressed diffusion of B.

Calculated diffusion profiles B and C are plotted in Fig. 2. The coupled diffusion and reaction equations for B, C, and Si point defects were solved numerically, using established experimental data for the diffusion parameters $D_I C_I^{eq}$, $D_V C_V^{eq}$ [13], and D_C [14]. The only unknown parameters are the reaction constants of Eq. (1) and (2). While the kick-out reaction (1) is generally believed to be fast enough to establish local equilibrium [10], the dissociative reaction (2) is much slower due to its high activation energy. We follow here the approach suggested in Ref. [11] and fit the reaction rate to the measured diffusion profile of C (Fig. 2).

From this numerical analysis we find a reduction of the density of self-interstitials in the C-rich region by about a factor of 20 in agreement with the experimentally observed suppression of B diffusion. Further, we find an enhanced density of vacancies in the C-rich layer. This result is consistent with the measured enhancement of vacancy-mediated diffusion of Sb in C-rich Si [15].

The impact of C on B diffusion in SiGe can also be described by coupled diffusion of point defects and C, augmented with the reduction of B diffusion in SiGe [16], and the effect of the electric field due to hole confinement in SiGe [17].

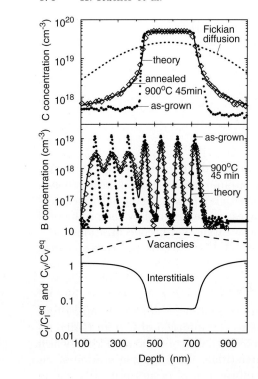

Fig. 2. Diffusion profiles of carbon (upper panel) and boron (middle panel) measured by SIMS before (full circles) and after annealing at 900°C for 45 minutes (open diamonds). Solid lines are calculated profiles. A calculated C profile for Fickian diffusion is shown for comparison (dotted line). Densities of Si self-interstitials (solid line) and vacancies (dashed line) normalized to there equilibrium values are plotted in the lower panel

2.3 Suppression of Transient Enhanced Diffusion

A substantial fraction of dopant redistribution during device processing is caused by transient enhanced diffusion (TED) during the first annealing step after ion implantation. This effect is also strongly suppressed in C-rich Si [18,19]. The impact of C on TED of B is shown in Fig. 3. Depth profiles of B were measured by SIMS for as-grown and annealed samples with and without implantation of BF_2 ions. We found strongly enhanced diffusion of B spikes in the C-poor region of the implanted sample. In contrast, TED was suppressed almost completely for B spikes located in the C-rich region.

The source of TED is an enhanced density of Si self-interstitials. Each implanted ion creates a trail of crystal defects through collisions with the lattice. These point defects can in turn interact with substitutional impurities and enhance their mobilities during subsequent high temperature steps. In C-doped Si, interaction with substitutional C reduces the densities of excess point defects. The suppression of TED as a function of C concentrations can be described quantitatively combining the semi-empirical "+1" model [20] for implantation damage with coupled diffusion of C and Si point defects [19].

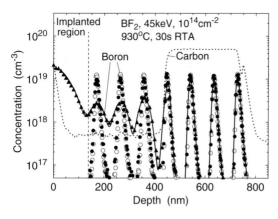

Fig. 3. SIMS profiles of annealed B doping superlattice with (triangles) and without BF_2 ion implantation (circles). Open circles are the as-grown B profile. The C concentration was enhanced to $5 \times 10^{19} \text{cm}^{-3}$ at about 450 nm below the surface (dotted line)

3 Heterojunction Bipolar Transistors

Design and fabrication of SiGe(:C) HBTs are based on two fundamental material properties. First, the electronic mode of operation of HBTs is built on the use of the smaller bandgap material SiGe for the base of the transistor. Second, the suppression of B diffusion due to C allows control of steep doping profiles during device processing. The impact of both aspects on devices characteristics is discussed next.

3.1 Operation of HBTs

The basic feature of a heterojunction bipolar transistor is the use of materials with different bandgaps for emitter and base. The emitter is formed of the material with the larger band gap. This is illustrated in Fig. 4 by the band diagram of an *npn* HBT with a SiGe base and a Si emitter.

The consequence of the smaller bandgap of the SiGe base layer is a lower barrier for injection of electrons from the emitter into the base, resulting in an increased electron injection current and an increased current gain. The

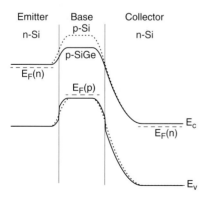

Fig. 4. Band diagram of a SiGe HBT (full lines) for base-emitter voltage $V_{BE} = 0.5$ V and collector-emitter voltage $V_{CE} = 1.0$ V. The band edges of a Si bipolar transistor with identical doping profiles are shown for comparison (dotted lines). Dashed lines indicate the Fermi levels for majority carriers

increased electron injection at a heterojunction allows the realization of a required current gain with higher base doping. Higher base doping leads to reduced base resistance and improved high-frequency characteristics.

3.2 Effect of B Outdiffusion from the SiGe Layer

Diffusion of boron out of the SiGe layer has dramatic consequences on electronic transistor parameters. As shown in the left panel of Fig. 5, potential barriers are formed at the heterojunctions. The conduction band barrier at the base-collector junction hinders the flux of electrons from the base to the collector. Consequently, electrons accumulate in the base region, as can be seen from the calculated electron density shown in the left panel of Fig. 5. As a result, transit frequencies are drastically reduced. The B profiles plotted in Fig. 5 were obtained from SIMS measurement at SiGe and SiGe:C HBTs which have undergone identical post-epitaxial processing [4]. The HBTs without C showed strong broadening of the B profile caused by transient enhanced diffusion due to selective collector implantation. For SiGe:C HBTs, B outdiffusion from the SiGe layer is strongly suppressed.

The effect of B outdiffusion on maximum cutoff frequencies f_T of SiGe and SiGe:C HBTs is shown in Fig. 6. Two transistors, with and without C-doping in a thin SiGe base layer are compared (see [17] for details). The effect of 1 minute annealing in inert atmosphere at temperatures between 850 and 1050°C was simulated using the diffusion model outlined in section 2.

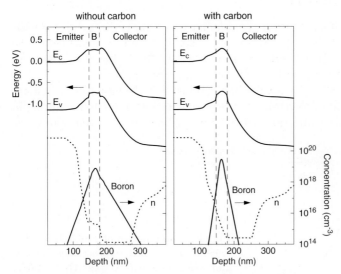

Fig. 5. Band diagrams of SiGe (left) and SiGe:C (right) HBTs for a base voltage $V_{BE} = 0.7$ V and a collector voltage $V_{CE} = 1.0$ V. The shown boron doping profiles were used for the calculation of the band edges and electron densities n (dotted lines). Dashed lines indicate the $Si_{0.8}Ge_{0.2}$ layer

Calculated doping profiles were used as input for device simulation within the drift-diffusion model. The calculated maximum transit frequency drops from above 80 GHz to below 10 GHz when B outdiffusion from the SiGe layer occurs. The simulation shows that the C incorporation increases the onset temperature for device degradation due to B outdiffusion from SiGe by about 100°C.

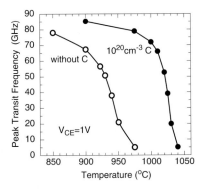

Fig. 6. Simulated transit frequencies of SiGe (open symbols) and SiGe:C (filled) HBTs as a function of the annealing temperature for 1 minute annealing in N_2 atmosphere. The SiGe:C HBTs have a carbon concentration of 10^{20}cm^{-3} within the SiGe layer and a carbon concentration below 10^{18}cm^{-3} outside the SiGe layer. The SiGe layer was 25 nm wide with a maximum Ge concentration of 20% [17]

4 SiGe:C BiCMOS Technology

Next-generation BiCMOS technologies are focused on the fabrication of high density CMOS-based integrated digital circuits, together with RF, analog, and mixed-signal sections on a single chip. This results in the following primary challenges for HBT process integration. 1) Integration of HBTs with excellent DC and RF characteristics into a CMOS platform in a modular way that allows full re-use of existing CMOS libraries. 2) Combining high HBT performance with low process complexity, and hence low additional cost, compared to the original CMOS process. Here, we focus on these challenges, and show that incorporating C in the base of SiGe HBTs is a key facilitator in meeting the requirements of next-generation HBT BiCMOS technologies.

4.1 Modular Integration of High-Speed HBTs

The major challenge for integrating state-of-the-art CMOS technologies with high-performance HBTs is the control of the impact of one process on the characteristics of the other device type. Shifting the HBT integration as far as possible to the front of a BiCMOS flow (right column in Fig. 7) eliminates the impact of the HBT thermal steps on the CMOS doping profiles. However, this increases the influence of the CMOS thermal budget on the HBTs profiles, which can lead to B outdiffusion from the SiGe layer, thus degrading the HBT characteristics. Best HBT performance would be achieved after all CMOS

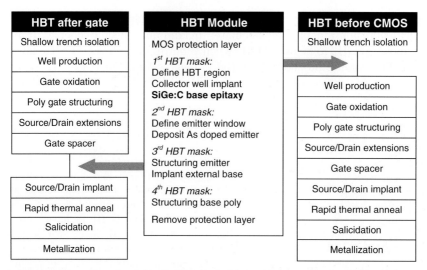

Fig. 7. Process flow for SiGe:C HBT BiCMOS. HBT integration is carried out after formation of CMOS gates or before CMOS preparation has started

thermal steps. But the HBT thermal steps make it difficult to preserve the MOS device parameters.

Taking advantage of the increased thermal stability of C-doped HBTs, we have demonstrated two approaches to modular HBT BiCMOS integration (Fig. 7). In the 0.25 μm gate length BiCMOS technology of the IHP [21,5], the HBT module was inserted in the CMOS process after gate formation (left column in Fig. 7). In this process, activation and diffusion of $n+$ and $p+$ source, drain, and gate regions were performed after HBT fabrication. Due to the high thermal budget of this step, C-doping is essential to prevent outdiffusion of B from the thin (25nm), highly doped ($R_{SBi} = 1.6$kΩ) SiGe layer.

For an easy migration of the HBT module from one CMOS generation to another it is desirable to integrate the HBT module even earlier in the CMOS flow. This is because CMOS technologies with gate length $< 0.25 \mu$m typically use shallow source/drain extensions which are sensitive to any additional thermal step. We have recently demonstrated a novel HBT-before-CMOS integration scheme (Fig. 7) to integrate SiGe:C HBTs with a 0.13 μm gate length CMOS process [22]. In this scheme, the impact of the HBT thermal budget on CMOS characteristics is entirely eliminated, making it possible to integrate the HBT module in a wide class of CMOS technologies without any changes in the CMOS process. The C-doped HBT withstands the full thermal budget of the 0.13μm CMOS flow without parameter degradation.

The impact of a further increased thermal budget of the CMOS process on HBT characteristics is demonstrated in Fig. 8. Transit frequencies markedly

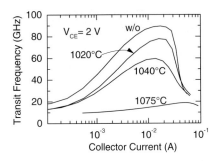

Fig. 8. Transit frequency vs. collector current for HBTs fabricated in the HBT-before-CMOS flow. After gate oxidation, some wafers were additional annealed for 60 sec at different temperatures to simulate the impact of CMOS processes with increased thermal budgets

degrade only for 1 minute annealing steps at temperatures above 1020°C in agreement with the simulations shown in Fig. 6 above.

4.2 HBT Device Characteristics

In the IHP BiCMOS process, various types of SiGe:C HBTs were simultaneously fabricated, covering a wide range of breakdown voltages BV_{CEO} and transit frequencies f_T [5]. Figs. 9 and 10 show data for high-speed SiGe:C HBTs with $BV_{CEO} = 2.6$ V, $f_T = 100$ GHz, and $f_{max} = 130$ GHz. The devices have a constant current gain of 130-140 over more than four current decades. The low leakage currents of large transistor arrays demonstrate the absence of electrically active defects in the C-doped epitaxial layer. For the realization of the outstanding f_T and f_{max} values shown in Fig. 10, the scalability of lateral device dimensions and the high B doping of the SiGe:C base were essential.

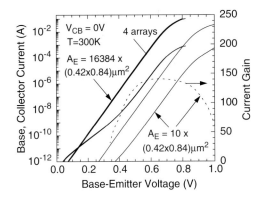

Fig. 9. Gummel plots of arrays of 16384 HBTs in parallel and of HBTs with 10 emitters with areas of $(0.42 \times 0.84 \mu m^2)$

4.3 Emitter Scaling

Next, we show that incorporation of C into the SiGe base can eliminate constrains for scaling of the emitter size of HBTs. Reducing the active emitter

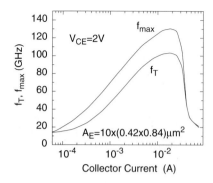

Fig. 10. Transit frequency f_T and maximum oscillation frequency f_{max} vs. collector current for 10 emitter SiGe:C HBTs

width and minimizing the overlap of the emitter poly-silicon layer result in lower power consumption and reduced parasitics, respectively. Lowering the base resistance due to reduced emitter overlap causes higher maximum oscillation frequencies f_{max} (Fig. 11). However, the TED effect of the high-dose implant, used to dope the external base regions, can degrade the device characteristics as the dimensions shrink.

Figure 11 shows a strong f_T drop of C-free HBTs as the overlap decreases below 0.4 μm. This drop is due to B outdiffusion from SiGe on the collector side caused by TED due to the external base implant, which increasingly influences the perimeter of the active HBT regions as the overlap decreases [23]. As a result of suppressed TED, SiGe:C HBTs show markedly improved behavior (Fig. 12). This allows one to reduce both the active emitter dimensions and the emitter overlap with no degradation in f_T.

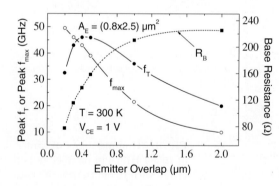

Fig. 11. Peak f_T, peak f_{max}, and base resistance of C-free SiGe HBTs as function of emitter overlap. The base resistance of these reference devices is higher then that of the SiGe:C HBTs due to a much lower B doping

4.4 Yield

Here, we present results from the qualified 0.25 μm BiCMOS process of the IHP with SiGe:C HBTs featuring 3.2V BV_{CEO}/60GHz f_T and 2.5V BV_{CEO}/80GHz f_T. Fig. 13 shows statistical device data, taken from 31

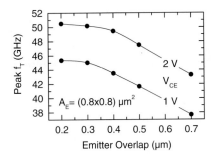

Fig. 12. Peak f_T vs. emitter overlap for SiGe:C HBTs

wafers. Fig. 13a demonstrates the high reproducibility of the epitaxial SiGe:C HBT base doping regime. Fig. 13b proves high wafer yields of typically more than 90% for arrays with 4096 transistors in parallel. Finally, high RF circuit performance with excellent reproducibility is demonstrated in Fig. 13c, showing current mode logic (CML) ring oscillator delay times with a mean value of 8.05 ps and standard deviation of only 0.16 ps.

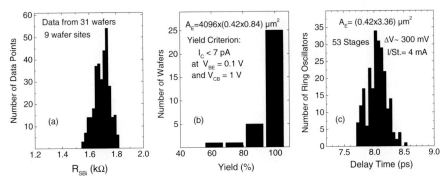

Fig. 13. Statistical device data, obtained on 31 wafers from 7 lots. (a) Histogram of HBT internal base resistance. (b)Wafer yield of arrays with 4096 HBTs in parallel. (c)Histogram of CML ring oscillator delay times

5 Conclusions

The strong suppression of thermal and transient enhanced diffusion of boron due to carbon doping can be used for significant improvement of device performance and manufacturability of SiGe HBTs. High-frequency HBTs with excellent static characteristics were integrated in a modular way in a state-of-the-art CMOS platform, allowing full re-use of CMOS libraries. The SiGe:C BiCMOS technology is capable of fabricating integrated RF circuits with excellent reproducibility and high yield. The RF performance of the HBTs and the described modular integration scheme take advantage of the improved thermal stability through carbon doping, which we believe is becoming an indispensable component of advanced SiGe technologies.

References

1. D. L. Harame et al.: IEEE Trans. Electron Devices **48**, 2575 (2001).
2. E.J. Prinz, P.M. Garone, P.V. Schwartz, X. Xiao, J.C. Sturm: IEEE Electron Device Lett. **12**, 42 (1991).
3. L. D. Lanzerotti et al.: in *Technical Digest, International Electron Device Meeting*, (IEEE, Piscataway, NJ, 1996) p. 249.
4. H. J. Osten et al.: in *Technical Digest, Int. Electron Device Meeting*, (IEEE, Piscataway NJ, 1997), p. 803.
5. B. Heinemann et al.: in *Technical Digest, Int. Electron Device Meeting*, (IEEE, Piscataway NJ, 2001), p. 348.
6. S. J. Jeng et al.: IEEE Electron Device Lett. **22**, 542 (2001).
7. H. J. Osten: *Carbon-Containing layers on Silicon, Growth, Properties, and Device Application* (Transtech Publications, Zürich 1999).
8. H. Rücker, B. Heinemann, W. Röpke, R. Kurps, D. Krüger, G. Lippert, H.J. Osten: Appl. Phys. Lett. **73**, 1682 (1998).
9. R. B. Fair: in *Impurity Doping Processes in Silicon*, F. F. Y. Wang (Ed.) (North-Holland, Amsterdam, The Netherlands, 1981), p. 315.
10. R. Scholz, U. Gösele, J.-Y. Huh, T. Y. Tan: Appl. Phys. Lett. **72**, 200 (1998).
11. R. F. Scholz, P. Werner, U. Gösele, T. Y. Tan: Appl. Phys. Lett. **74**, 392 (1999).
12. J. P. Kalejs, L. A. Ladd, U. Gösele: Appl. Phys. Lett. **45**, 268 (1984).
13. H. Bracht, N. A. Stolwijk, H. Mehrer: Phys. Rev. B **52**, 16542 (1995).
14. F. Rollert, N. A. Stolwijk, H. Mehrer: Materials Science Forum **38-41**, 753 (1989).
15. H. Rücker, B. Heinemann, R. Kurps: Phys. Rev. B **64**, 073202 (2000).
16. N. E. B. Cowern, P. C. Zalm, P. vander Sluis, D. J. Gravensteijn, W. B. de Boer: Phys. Rev. Lett. **72**, 2585 (1994).
17. H. Rücker, B. Heinemann: Solid-State Electronics **44**, 783 (2000).
18. P. A. Stolk, D. J. Eaglesham, H.-J. Gossmann, J. M. Poate: Appl. Phys. Lett. **66**, 1370 (1995).
19. H. Rücker, B. Heinemann, D. Bolze, R. Kurps, D. Krüger, G. Lippert, H. J. Osten: Appl. Phys. Lett. **74**, 3377 (1999).
20. M. D. Giles: J. Electrochem. Soc. **138**, 1160 (1991).
21. K. E. Ehwald et al.: in *Technical Digest, Int. Electron Device Meeting*, (IEEE, Piscataway NJ, 1999), p. 561.
22. D. Knoll et al.: in *Technical Digest, Int. Electron Device Meeting*, (IEEE, Piscataway NJ, 2001), p. 499.
23. B. Heinemann et al. in: *Proceedings 27th Europ. Solid-State Device Research Conf.*, H. Grünbacher (Ed.) (Editions Frontiers, Paris, 1997), p. 544.

Transition Edge Sensors for Imaging X-ray Spectrometers

H. F. C. Hoevers

SRON National Institute for Space Research
Sorbonnelaan 2, 3584 CA Utrecht, The Netherlands
h.hoevers@sron.nl

Abstract. The development of cryogenic radiation detectors based on superconducting to normal phase thermometers, generally called transition edge sensors (TESs), has shown a steady progress over the last years. Single pixel sensors of this type are amongst the most sensitive detectors today. Their application ranges from ground and space based astronomy to materials science from sub-mm to γ-ray wavelengths. This paper presents X-ray TES microcalorimeters to illustrate and quantify the performance of these detectors and to discuss the development of an imaging array. Critical aspects in the development of imaging X-ray spectrometers, such as filling factor, stopping power, thermal cross talk and electronical read-out will be discussed.

1 Introduction

In 1995 a new class of cryogenic microcalorimeters and bolometers was introduced [1] which have seen a rapid and succesful development since then. These detectors use a voltage-biased superconducting to normal phase transition thermometer, generally called a transition edge sensor (TES) as a temperature to current transducer. Dependent on the wavelength of the radiation that is to be detected, the TES is coupled to an appropriate absorber/thermalizer. For the detection of X-ray or γ-ray radiation, an absorber with a high cross section for X- or γ-rays is used. In the optical regime, the TES itself can serve as absorber. For sub-mm and infra-red radiation a spider web structure, quarter wavelength cavity or antenna structure is applied. The conductivity of the web structure is engineered such that it matches the space impedance for optimum sub-mm absorption while having a low cross section for cosmic particles. The performance of TES based detectors for X-ray detection ranges from an energy resolution ΔE of 2 eV at a photon energy E of 1.5 keV energy [2] to 3.9 eV at E equal to 5.9 keV [3], the resolving power $E/\Delta E$ is equal to 750, and 1,500, respectively. In the optical regime a resolving power of about 70 has been demonstrated at 1 eV photon energy [4]. For spider web bolometers a noise equivalent power of 1×10^{-17} W/$\sqrt{\text{Hz}}$ [5] has been achieved.

Imaging arrays with TES-based detectors are being developed for several applications [6]. The upgrade of the Superconducting Common User Bolometer Array (SCUBA-2) [7] for the detection of sub-mm radiation at the James

Clerk Maxwell Telescope, will have 12,800 pixels each equipped with a TES and operating at about 100 mK [8]. For X-ray astronomy, the development of arrays with at least 1024 pixels for Constellation-X (Con-X, NASA) [9] and the XEUS mission (X-rays of the Evolving Universe Spectrometer, ESA) [10] has started. TES-based microcalorimeters have already been used succesfully in materials science [2] and the availability of (small) arrays will lead to an increase of the count rate capabilities which is especially relevant in this field.

2 Principles of a Voltage Biased Detector with a TES

The principle of a voltage biased detector is shown in Fig. 1. The detector consists of an absorber and a TES thermometer (with a total heat capacity C), which are connected to a heat bath (with temperature T_{bath}) by a weak thermal link with a thermal conductance G. The intrinsic thermal time constant of the detector is $\tau = C/G$. The TES is a superconductor with a critical temperature T_c, operated at its transition from the superconducting to normal phase. It is characterized by its operation temperature T, the bias point resistance R, and temperature coefficient of resistance $\alpha \equiv \mathrm{d}(\log R)/\mathrm{d}(\log T)$. The current through the TES is usually measured with a low-impedance Superconducting Quantum Interference Device (SQUID) amplifier.

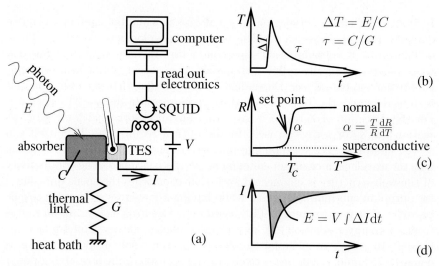

Fig. 1. Schematic of a TES-based radiation detector (a); the example depicted here applies for a microcalorimeter that absorbs a photon of energy E. As a result of the incident radiation the temperature of the detector increases (b). The steep superconducting transition of the TES converts the small temperature change into a change in electrical resistance (c). The increase of resistance results in a decrease of the current in the voltage biased circuit (d)

The applied bias voltage V determines the operating point on the superconducting transition of the TES, since the cooling by the heat link G equals the Joule power $P = V^2/R$ dissipated in the TES. Voltage bias, in combination with a positive α, introduces negative electro-thermal feedback (ETF). ETF stabilizes the device. If the detector temperature increases, the TES resistance increases and the current through the device decreases as well as the dissipated Joule power: the detector cools and returns to its set-point. ETF can be quantified by the loop gain $L_0 \equiv P\alpha/GT$. In the regime where $T \gg T_{\text{bath}}$, $L_0 = \alpha/n$ (strong ETF). As a result of strong ETF, the time constant of the device reduces to $\tau_e = \tau/(1 + \alpha/n)$; here, n is the power law exponent of the thermal transport in G. In the limit of strong ETF, and for frequencies below $f = 1/(2\pi\tau_e)$ the responsivity S_I [A/W] of a voltage biased detector approaches $S_I = -1/V$. The theoretical aspects of the performance of a voltage biased detector are described in Refs. [1,11]

The detector depicted in Fig. 1 has two principal sources of noise: the TFN noise in G and the Johnson noise in the resistive TES. Figure 2 shows the spectral distribution of the signal [A] as the result of the absorption of a photon and the TFN and Johnson noise [A/$\sqrt{\text{Hz}}$] in absence of ETF. As can be seen, the shape of the signal (as a result of an absorption of a photon) and TFN are identical with a corner frequency set by the intrinsic time constant: $f_c = 1/(2\pi\tau)$. The signal to noise ratio can now be optimized by having the ratio of $i_{\text{TFN}}/i_{\text{Johnson}} = \alpha\sqrt{P/GT} \approx \alpha/\sqrt{n}$ as high a possible. A detector with a large α and with a high value of P/GT, which basically means $T \gg T_{\text{bath}}$, will have a high forward gain which enhances the TFN level above the Johnson noise from the TES. In that case the detector performance can be limited by the TFN originating in the thermal link rather than Johnson noise. In order not to lose information in the read-out electronics these should allow for a bandwidth of f_{band}, i.e. the frequency at which the TFN becomes equal to the Johnson noise contribution; note from Fig. 2 that ETF does not affect the signal to noise of the sensor, neither the bandwidth required for the read-out electronics [12].

For a voltage biased microcalorimeter in the limit of strong ETF, the energy of an incident photon follows from $E = S_I^{-1} \int dI(t)dt$, with $dI(t)$ the change of the current during the pulse, see Fig. 1 (d). The energy resolution can be expressed as $\Delta E = 2.36(k_B T^2 C)^{1/2}\xi$, with $\xi = 2[0.5(n/\alpha^2) + (n/\alpha^2)^2]^{1/4}$. For optimum energy resolution a high value of α is preferred.

Moreover, low values of C and T are beneficial. The maximum photon energy E_{max} that is to be detected sets a lower limit to $C > \alpha E_{\text{max}}/T_c$. Regarding the operating temperature: for a metallic absorber $C \approx T$ and $\Delta E \approx T^{3/2}$. Practical cooler specifications, based on the use of a dilution refrigerator or an adiabatic demagnitzation cooler, limit T_{bath} to 30 – 50 mK. The typical operating temperatures range from 80 to 120 mK. It is noted that a TES-type microcalorimeter used so far has $\alpha \approx 20 - -50$ under typical biasing conditions, and its energy resolution can be better than that of a

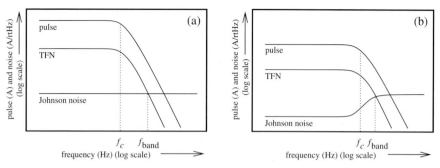

Fig. 2. Spectral distribution of the response of the detector of Fig. 1 on a pulse, the TFN from the link to the heat bath and the Johnson noise without ETF (a). The same spectra in the presence of ETF are shown in (b). The signal and noise levels in (b) are lower than in (a) and the corner frequency f_c is increased from $(2\pi\tau)^{-1}$ to $(2\pi\tau_e)^{-1}$. The frequency f_{band} where the levels of TFN and Johnson noise are equal is unaltered by ETF.

detector with a doped semiconductor thermometer, which typically has $\alpha \approx 5$ and due to electron-phonon decoupling cannot be pumped high above the heat bath temperature.

3 Design and Performance of an X-ray TES Microcalorimeter

To indicate the magnitude of the different physical parameters a simple, though representative, design of an X-ray detector is used. Morever, the engineering and production of a TES and absorber are discusssed followed by experimental performance data.

The design starts assuming $\alpha = 50$, $n = 3$ (for a thermal link of Si_xN_y) in combination with a requirement (count rate ability) of $\tau_e < 100$ μs. This yields $\tau = C/G < 2$ ms. The minimum allowable heat capacity in case $E_{\max} = 10$ keV leads to $C = 1$ pJ/K at a typical operating temperature of $T = 100$ mK and, consequently, $G = 0.5$ nW/K. The theoretical energy resolution of this detector is 0.7 eV.

The TES is, in general, not made from an elementary superconductor, since this would restrict the operating temperatures to a very limited number of T_cs. Instead, a superconductor-normal metal bilayer is used. Through the proximity effect and choice of materials, the T_c and normal state resistance can be tuned by adjusting the layer thicknesses of the superconducting layer and the normal metal layer. The groups at the Space Research Organization Netherlands (SRON) [3] and the University of Jyväskylä use TiAu bilayers while the groups at the National Institute of Standards and Technology (NIST) [13] and Goddard Space Flight Center (GSFC) [14] work with MoCu and MoAu bilayers, respectively; all four groups have been able to produce devices with T_cs in the required range from 80 to 120 mK and with normal state resistances between 10 mΩ and 1 Ω. Figure 3 shows the resistance-temperature curve of a TiAu bilayer. SRON, NIST and GSFC use photolithographical techniques for patterning of their devices, while Jyväskylä uses e-beam lithography.

In some cases specific engineering is required to obtain a steep and reproducible transition of the TES. The Mo-based TESs perform better with well-defined edges. For this purpose, NIST deposits normal metal banks on the edges of their TESs and GSFC deliberately underetches the Mo, causing the Au layer to overhang [14,15]. The effect of the normal metal edges is that they suppress spatial variations in the T_c that deteriorate the quality of the superconducting transition. The TiAu TESs of SRON have a transition width of a few mK (see Fig. 3), and the Ti is not intentionally underetched. An additional benefit of well-defined boundaries of the TES is the reduction of excess noise due to phase slipping as was observed by NIST in their first AlAg TESs which were produced with shadow masks and had uneven boundaries. An advantage of the Mo-based TESs over the Ti-based TESs is that they do not form intermetallic compounds, whereas the Ti might diffuse through the Au over time. Aging tests at SRON with TiAu TESs have shown, however, neither changes of the T_c nor changes in the width of the transition over a period of more than three years.

The performance of several single pixel X-ray TES microcalorimeters at 5.89 keV photon energy is summarized in Table 1. The groups at NIST and GSFC use their TES as absorber. Jyväskylä used a Bi absorber covering the entire TES. SRON tested with both a small Cu absorber and with a mushroom-shaped Cu/Bi absorber, yielding energy resolutions of 3.9 and 5.2 eV, respectively. An X-ray spectrum of a TES with a Cu/Bi mushroom shaped absorber is shown in Fig. 4.

A comparison of the experimental and theoretical values of ΔE in Table 1 indicates the present microcalorimeters have not yet reached their theoretical

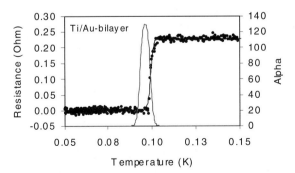

Fig. 3. Resistance versus temperature of a TiAu TES with a T_c of 105 mK (left axis). Thicknesses of the Ti and Au layer are 18 and 30 nm, respectively. The drawn line is a fit to the data which facilitates the derivation of the value of α (right axis)

Table 1. Comparison of microcalorimeter performance at 5.89 keV; the stopping power of the absorber at 5.89 keV is indicated between brackets in the last column. Fig. 4 shows the actual 5.2 eV spectrum

Group	ΔE_{meas} (eV)	ΔE_{theor} (eV)	Time constant (ms)	TES material (stopping power)
SRON [3]	3.9	1.4	0.15–0.20	TiAu+Cu absorber (35%)
NIST [13]	4.5	1.9	0.75–2.00	MoCu; TES acts as absorber (3%)
SRON	5.2	1.4	0.14	TiAu+Cu/Bi absorber (92%)
GSFC [14]	6.1	1.6	0.31	MoAu; TES acts as absorber (12%)
Jyväskylä	9.0	3.6	0.26	TiAu+Bi absorber (78%)
Con-X/XEUS	5	[-]	0.1	32×32 pixel array (>90% at 10 keV)

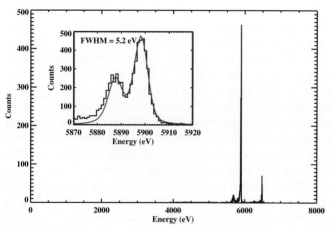

Fig. 4. Spectrum of an ^{55}Fe source measured with a microcalorimeter with a TiAu TES, operating at 100 mK and using a mushroom shaped Cu/Bi absorber. The Mn K$_\alpha$ and Mn K$_\beta$ lines at 5.89 and 6.40 keV respectively can be seen. The inset shows the Mn K$_\alpha$ line in detail and the drawn line is a weighted least squares fit to the data resulting in a ΔE for this detector of 5.2 eV. Below 5875 eV the spectrum does not return to zero, which is presumably due to photons absorbed in the substrate. The fitted line does not account for this effect and, as a result, the quality of the fit is rather poor at these energies

energy resolution as a result of the presence of excess noise. SRON identified thermal fluctuation noise in the TES (because of its finite thermal conductance) as a source of noise in their devices [16]. This type of noise shows up due to the particular design of the SRON microcalorimeter and cannot account for the noise observed by GSFC and it does not fully explain the performance of the SRON devices in terms of energy resolution. Both SRON and GSFC studied the excess noise of X-ray microcalorimeters in detail [3,17].

It is noted that the current performance in terms of energy resolution and thermal time constant is close to the requirement for future space missions.

4 Development of Imaging Arrays for X-ray Spectroscopy

The 32 × 32 pixel arrays for Con-X and XEUS, with an operating temperature of 100 mK, are developed using the expertise gained from working with single pixel microcalorimeters. Four major problems need to be tackled. Firstly, to avoid thermal cross-talk, pixels need to be positioned on individual, slotted membranes well-coupled to the heat bath. Secondly, optimization with respect to filling factor and stopping power for X-ray radiation is required. Thirdly, the array of pixels needs to be supported and be well-coupled to the heat bath. Fourthly, the electronical read-out of an array should be as simple as possible, not impair on the detector performance, and have a small dissipation given the limited cooling power that is available.

4.1 Single Pixel Optimization

The present, single pixel, devices are produced on a closed membrane. If a closed membrane, with all pixels positioned on it, should be used major thermal cross talk between the pixels would occur. This cross talk can be avoided by placing each pixel on its own slotted membrane which should be well-coupled to the heat bath. An example of a single pixel microcalorimeter on a slotted membrane is shown in Fig. 5 (a). The thermal conductance of the pixel to the heat bath can be tuned by the aspect ratio of the support beams and should be similar to that of the present closed membranes i.e. $G \approx 0.5$ nW/K. The filling factor of an imaging array for X-ray applications should be 90% or more. This can be achieved using mushroom shaped absorbers on top of a smaller TES: the hat of the mushroom extents outside the TES, wiring and the slotted membranes. An example of a mushroom-shaped Cu/Bi absorber (although still on a large-area TES) is shown in Fig. 5 (b). Bi is chosen as absorber material because of its low specific heat (only 5% of that of Cu) and its high stopping power. The performance of the detector in Fig. 5, in terms of energy resolution (see Fig. 4) and thermal reponse time, is almost identical to the performance of a detector without a Bi layer, indicating that the Bi adds no measurable heat capacity. The area of the 160 × 160 μm^2 absorber is already 45% of the area of a future XEUS pixel. The stopping power of the 5 μm thick Bi layer of the absorber in Fig. 5 is 92% at 5.89 keV. The group at GSFC follows a similar route with regard to the development of slotted membranes and (arrays of) mushroom-shaped Bi absorbers [18,19]. As an alternative to Bi, a superconducting absorber such as Ta can be considered.

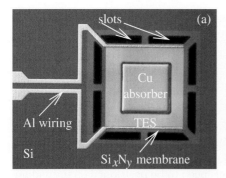 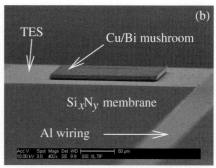

Fig. 5. (a) Optical photograph of a single pixel microcalorimeter on a slotted Si_xN_y membrane. Eight support beams suspend the Si_xN_y membrane with a TiAu TES of 160×160 µm^2 and central Cu absorber of 100×100 µm^2. (b) Scanning Electron Microscope image of a TiAu TES (area 310×310 µm^2) with a mushroom shaped Cu/Bi absorber on a closed Si_xN_y membrane. The stem of the mushroom is about 1.4 µm high and has an area of 100×100 µm^2. The hat of the absorber has sides of 160×160 µm^2. This detector was used in the measurement which is shown in Fig. 4. Images SRON-MESA$^+$ collaboration

4.2 Micromachining of the Pixel and Array Support Structure

Several options to support the pixels of an array exist. SRON is developing a structure where the individual membranes are suspended by parallel Si-ribs, fabricated by wet etching deep (111)-oriented trenches in a Si wafer, see Fig. 6. For proper coupling to the heat bath, radiative transport in the Si bars is essential: the thermal conductance is set by the cross section of the ribs rather than their length. In principle, the surface roughness of the sides of the Si-bars is smaller than 0.1 µm, which should allow for radiative transport of phonons (with a mean free path of about 1–10 µm). It is estimated that this geometry yields G of about 5 µW/K for a rib; in case of diffusive transport this would be an order of magnitude smaller. Assuming that each pixel dissipates 10 pW, the temperature of a radiatively cooled array of 1024 pixels will rise by about 10 mK (for the central pixels). In case of diffusive transport this would be 100 mK which is, of course, not acceptable. The low heat capacity of Si at 100 mK remains a potential problem since it might result in thermal cross talk between neighboring pixels.

As an alternative for the support structure with Si-bars a surface micromachined structure is envisaged. The individual pixels positioned on top of slotted membranes that cover so-called shallow boxes in the Si wafer, as schematically shown in Fig. 7. The production of this structure is more complicated and requires more steps in the micromachining process. A possible advantage is that most of the Si wafer remains in tact, adding mechanical strength to the imaging array and increasing the heat capacity, thus reducing the thermal cross talk.

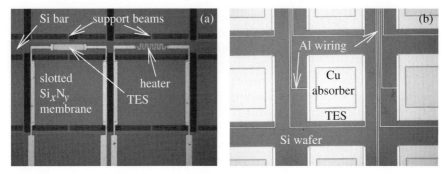

Fig. 6. (a) Photograph of an array support structure with slotted Si_xN_y membranes (210 × 250 µm^2). Each membrane is suspended by six support beams hanging from paralled Si-bars of 40 µm width. One of the (111) bars that suspends the membranes is visible (running from left to right). For thermal characterization of the Si bar a TES and a heater have been processed on top of it; the wiring to the TES and heater can be seen as well. (b) Array of TiAu TESs (180 × 130 µm^2) with Cu absorbers (130 × 130 µm^2) and electrical wiring on top of a solid Si wafer. Images SRON-MESA$^+$ collaboration

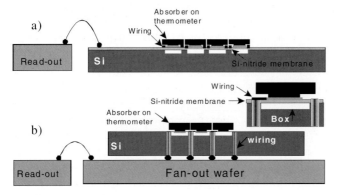

Fig. 7. Schematics of surface micromachined structures with pixels on slotted membranes above shallow boxes in a Si wafer. Option (a) shows wiring to the pixels on the top side of the detector wafer and option (b) shows wiring through the wafer a the connection (via bump bonding) of the detector wafer on top a fan-out wafer

4.3 Electrical Read-Out of an Imaging Array

The integration of the microcalorimeter array with the read-out electronics has two critical components. Firstly, the wiring to the pixels on the detector chip. Secondly, the electronical read-out concept. Figure 7 shows two options for the wiring. The wiring can be processed on the top side of the detector, i.e. in between the pixels. There is, however, only limited space available resulting in possible electrical cross talk and the risk of exceeding the critical current of the narrow superconducting wires, if these are deposited on the Si(111) support bars. Through-the-wafer metallization of the detector chip and bump bonding on a fan-out chip with read-out electronics might provide an alte rnative.

Three options for the electronical read-out of a detector array are considered. First of all, each pixel can be connected to a separate SQUID. This option has a high degree of redundancy but will require a large number of electrical connections to the detector chip which is non-trivial given the limited cooling power available. Also, a large number of (dissipating) SQUIDs at low temperatures are required along with their many connections to the room temperature electronics. To circumvent these problems two multiplexing schemes are being pursued: Time Division Multiplexing (TDM) and Frequency Division Multiplexing (FDM), see Fig. 8 [20]. In TDM rows of SQUIDs are switched on and off, and each column has its output SQUID. In FDM the rows of detectors are encoded by running them at different frequencies and each column is read out with its own SQUID. The minimum number of SQUIDs that is needed for the read-out of an $m \times m$ array is m^2 in case of direct read-out, $m(m+1)$ for TDM and m in case of FDM. As can be seen, the number of SQUIDs can be reduced by choosing an appropriate read-out technique. Another advantage of FDM is the smaller number of wires that runs to the detector (each wire imposes at heat load to the cooler): for reading out and biasing a 32×32 pixel detector FDM requires about 150 wires (minimum). For TDM and direct read out the number minimum of wires

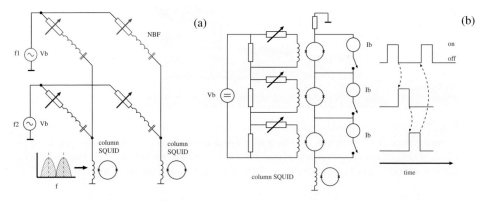

Fig. 8. (a) Schematic of frequency division multiplexing of 2×2 detectors. The rows are biased at frequency f_1 and f_2. The noise blocking filters (NBF) prevent that wide-band Johnson noise adds to the noise in neighbouring rows which would lead noise level which is too high. Each column has only one read out SQUID. (b) Schematic of a time division multiplexed 3×1 (row $\times$ column) detector array. The SQUIDs in the detector bias circuit are alternatingly switched on and off in time by switching their bias current. The time multiplexed signal of a column is read out with one SQUID

is in the order of 1,500 and 4,000. Both TDM and FDM require advanced electronics to retrieve the signals from the detector; these electronics needs, however, not to be operated at cryogenic temperatures.

The SCUBA-2 array will be read-out using time multiplexing, as is currently developed by NIST. Although the thermal response time of Con-X is shorter than that of SCUBA-2, it is expected that TDM is also applicable for Con-X. NIST has already shown that the SQUID response of a 8 channel-multiplexed read-out can be retrieved [21]. The groups at SRON and VTT Microsensing [22], Berkeley [23], ISAS [24], and LLNL [25] are working on FDM multiplexing. Recently it has been demonstrated that a microcalorimeter can be biased at 46 kHz and still deliver an energy resolution of 6.9 eV (at 5.89 keV). This is very close to the performance of the same detector under (conventional) dc-bias (5.5 eV). The biasing frequency in this experiment was set by practical considerations and it will increased. It is expected that the energy resolution will improve at higher baising frequencies [26].

Acknowledgements

The author gratefully acknowledges the information and discussions with colleagues from NIST, GSFC, the university of Jyväskylä, the MESA$^+$ Institute, VTT Microsensing and the people of the SRON microcalorimeter group. This work was supported by the Nederlandse Organisatie voor Wetenschappelijk Onderzoek (NWO).

References

1. K.D. Irwin, Appl. Phys. Lett. **66** 1998 (1995).
2. D.A. Wollman, S.W. Nam, D.E. Newbury, G.C. Hilton, K.D. Irwin, N.F. Bergren, S. Deiker, D.A. Rudman, and J.M. Martinis, Nucl. Instrum. and Meth. Phys. Res. A **444**, 145 (2000).
3. W.M. Bergmann Tiest, H.F.C. Hoevers, W.A. Mels, M.L. Ridder, M.P. Bruijn, P.A.J. de Korte, M.E. Huber, American Institute of Physics Conference Proceedings **605**, 199 (2002).
4. A.J. Miller, B. Cabrera, R.W. Romani, E. Figueroa-Feliciano, S.W Nam, and R.M. Clarke, Nucl. Instrum. Meth. A **444**, 445 (2000).
5. M.J. Griffin, Nucl. Instrum. Meth. A **444**, 397 (2000).
6. H.F.C. Hoevers, American Institute of Physics Conference Proceedings **605**, 193 (2002).
7. http://www.jach.hawaii.edu/JACpublic/JCMT/Continuum_observing/SCUBA-2/home.html.
8. W. Duncan, W. Holland, D. Audley, D. Kelly, T. Peacock, P. Hastings, M. MacIntosh, K. Irwin, S.W. Nam, G. Hilton, S. Deiker, A. Walton,A. Gundlach, W. Parkes, C. Dunare, P. Ade, and I. Robson, American Institute of Physics Conference Proceedings **605**, 581 (2002).
9. http://constellation.gsfc.nasa.gov/docs/main.html.
10. http://astro.esa.int/SA-general/Projects/XEUS/main/xeus_main.html.

11. K.D. Irwin, G.C. Hilton, and J.M. Martinis, J. Appl. Phys. **83**, 3978 (1998).
12. D. McCammon, R. Almy, E. Apocada, S. Deiker, M. Galeazzi, S.-I. Han, A. Lesser, W. Sanders, R.L. Kelley, S.H. Moseley, F.S. Porter, C.K. Stahle, A.E. Szymkowiak, Nucl. Instrum. and Meth. Phys. Res. A **436**, 205 (1999).
13. K.D. Irwin, G.C. Hilton, J.M. Martinis, S. Deiker, N.F. Bergren, S.W. Nam, D.A. Rudman, D.A. Wollman, Nucl. Instrum. Meth. A **444**, 184 (2000).
14. C.K. Stahle, P. Brekosky, E. Figueroa-Feliciano, F.M. Finkbeiner, J.D. Gyax, M.J. Li, M.A. Lindeman, F. Scott Porter, N. Tralshawala, SPIE **4140**, 367 (2000).
15. G.C. Hilton, J.M. Martinis, K.D. Irwin, N.F. Bergren, D.A. Wollman, M.E. Huber, S. Deiker, S.W. Nam, IEEE Trans. on Appl. Superconductivity **11**, 739 (2001).
16. H.F.C. Hoevers, A.C. Bento, M.P. Bruijn. L. Gottardi, M.A.N. Korevaar, W.A. Mels, and P.A.J. de Korte, Appl. Phys. Lett. **77**, 4422 (2000).
17. M.A. Lindeman, R.P. Brekosky, E. Figueroa-Feliciano, F.M. Finkbeiner, M. Li, C.K. Stahle, C.M. Stahle, American Institute of Physics Conference Proceedings **605** 219 (2002).
18. C.K. Stahle, M.A. Lindeman, E. Figuero-Feliciano, M.J. Li, N. Tralshawala, F.M. Finkbeiner, R.P. Brekosky, and J.A. Chervenak, M.A. Lindeman, R.P. Brekosky, American Institute of Physics Conference Proceedings **605** 223 (2002).
19. N. Tralshawala, S. Aslam, R.P. Brekosky, T.C. Chen, E. Figueroa-Feliciano, F.M. Finkbeiner, M.J. Li, D.B. Mott, C.K. Stahle, and C.M. Stahle, Nucl. Instrum. Meth. A **444**, 188 (2000).
20. M. Kiviranta, H. Seppä, J. van der Kuur, and P.A.J. de Korte, American Institute of Physics Conference Proceedings **605** 295 (2002).
21. J.A. Chervenak, K.D. Irwin, E.N. Grossman, J.M. Martinis, C.D. Reintsema, and E. Huber, Appl. Phys. Lett. **74** 4043 (1999).
22. SRON-VTT trade-off report "Read-out concepts for an array of microcalorimeters for XEUS", unpublished (2001).
23. J. Yoon, J. Clarke, J.M. Gildemeister, A.T Lee, M.J. Myers, P.L. Richards, and J.T. Skidmore, Appl. Phys. Lett. **78** 371 (2001).
24. T. Miyazaki, Ph.D. thesis, ISAS, unpublished (2001).
25. M.F. Cunningham, J.N. Ullom, T. Miyazaki, O. Drury, A. Loshak, M.L. van den Berg, and S.E. Labov, American Institute of Physics Conference Proceedings **605** 317 (2002).
26. J. van der Kuur, P.A.J. de Korte, H.F.C. Hoevers, M. Kiviranta, H. Seppä, Appl. Phys. Lett. submitted for publication.

Charge Injection in Polymer Light-Emitting Diodes

T. van Woudenbergh[1], P. W. M. Blom[1], and J. N. Huiberts[2]

[1] Materials Science Center and DPI, University of Groningen,
 Nijenborgh 4, 9747 AG Groningen, The Netherlands
[2] Philips Research Laboratories,
 Professor Holstlaan 4, 5656 AA Eindhoven, The Netherlands

Abstract. The injection-limited hole current from Ag into poly-dialkoxy-p-phenylene vinylene (PPV) exhibits a weak dependence on temperature, in spite of the presence of a large injection barrier of 1 eV. The measured field- and temperature dependence of the hole injection is explained by a hopping model in which energetic disorder is taken into account. In a PPV-based light-emitting diode it is demonstrated that the hole injection is enhanced by the presence of electrons. As a mechanism for this enhanced hole injection an increase of the electric field at the hole injecting contact due to trapped electrons is proposed.

1 Introduction

Directly after the discovery of polymer light emitting diodes (PLEDs) [1], charge injection has been recognized as an important process for the performance of a PLED [2]. An unbalanced charge injection leads to an excess of one of the two charge carrier types, leading to poor device efficiencies. However, the mechanisms of charge injection into conjugated polymers are poorly understood, compared with the knowledge of inorganic semiconductors. For classical semiconductors the current density J for injection from an electrode into a semiconductor or insulator has been described by thermionic or Richardson-Schottky emission [3], given by

$$J = A^* T^2 \exp\left(\frac{-q\phi_B}{kT}\right) \tag{1}$$

arising from the band offset between the semiconductor and the electrode. However, it was pointed out by Simmons [4] that for low mobility materials this expression is not valid: due to the low mobility a large amount of charge builds up at the contact and as a result back-flow from the semiconductor to the electrode will occur. In that case, the diffusion-limited regime, J is

$$J = qp(0)\mu E(0) = qN_C \exp\left(\frac{-q\phi_B}{kT}\right)\mu E(0) \tag{2}$$

with $p(0)$ the charge carrier density at the contact, μ charge carrier mobility, and N_C the effective density of states.

Contrary to inorganic semiconductors like Si, with charge carrier mobilities of typically 1000 cm^2/Vs, the charge transport in conjugated polymers is determined by tightly bound charge carriers on transporting sites that are subject to energetic and spatial disorder [5]. Typical mobilities for PPV-based polymers are in the range 10^{-6} to 10^{-7} cm^2/Vs. As a result of this low mobility, it is expected that the charge injection into PPV is diffusion-limited, indicating that the charge carrier mobility will play an important role in the injection process. As a result, in order to disentangle the contributions from the mobility μ and injection barrier ϕ_B to the injection process, the field and temperature dependence of both the mobility and the energy barriers have to be known.

2 Hole Mobility of PPV

The hole mobility μ_p of PPV is strongly dependent on electric field E and temperature T [6], given by

$$\mu_p(E,T) = \mu_0(T) \exp\left(\gamma \sqrt{E}\right) \tag{3}$$

with

$$\mu_0(T) = \mu_0 \exp\left(-\frac{\Delta}{kT}\right) \tag{4}$$

and

$$\gamma = B\left(\frac{1}{kT} - \frac{1}{kT_0}\right) \tag{5}$$

with $\Delta = 0.50$ eV, $B = 3.1 \cdot 10^{-5}$ eV(V/m)$^{-1/2}$, $T_0 = 420$ K, and $\mu_0 = 1.0 \cdot 10^{-2}$ m^2/Vs. This functional form of the field E and temperature T dependence of the charge carrier mobility (3) is an intriguing feature of disordered organic semiconductors. The stretched exponential form has first been observed for poly(N-vinyl carbazole) by Gill in 1972 [7]. Numerous experimental studies on molecularly doped polymers, pendant group polymers and amorphous molecular glasses have revealed a similar behavior [8,9,10].

Charge transport in disordered organic conductors is thought to proceed by means of hopping in a Gaussian site-energy distribution. This density of states (DOS) reflects the energetic spread in the charge transporting levels of chain segments due to fluctuation in conjugation lengths and structural disorder. Bässler and co-workers [5,11] have performed numerical simulations of charge transport in a regular array of hopping sites with a Gaussian distribution of site energies. In this Gaussian disorder model (GDM) the following

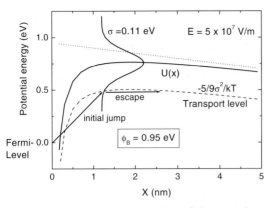

Fig. 1. Schematic representation of the initial carrier jumps at the metal-polymer interface. The dotted line shows the potential distribution due to the applied electric field, the solid line includes the potential lowering due to the image force. The dashed line represents the transport energy level

functional dependence of μ has been proposed [5],

$$\mu_{GDM} = \mu_\infty \exp\left[-\left(\frac{2\sigma}{3kT}\right)^2 + C\left(\left(\frac{\sigma}{kT}\right)^2 - 2.25\right)\sqrt{E}\right], \qquad (6)$$

with μ_∞ the mobility in the limit $T \to \infty$, σ the width of the Gaussian DOS and C a constant (depending on, e.g., the site spacing). The simulations revealed that in principle the transport in the Gaussian DOS is governed by two energy levels: an equilibrium level to which the charge carriers relax, located at $-\sigma^2/kT$ (zero energy is the maximum of the Gauss), and a transport level E_{tr} located at $-5/9\sigma^2/kT$, as shown in Fig. 1. As a result, the charge transport (6) is governed by an activation energy of $4/9\sigma^2/kT$. Using (6) the phenomenological parameters Δ and γ (3) may be related to the microscopic material parameter σ. From the zero-field mobility of our PPV a width of the Gaussian DOS of $\sigma = 0.11$ eV and a typical hopping distance of 1.2 nm have been obtained [12].

3 Mechanism of Charge Injection

In order to discriminate between the contributions of the charge carrier mobility and the energy barrier at the interface the temperature dependence of contact-limited currents in PPV has been investigated [13]. As an electrode silver (Ag) has been used, which for hole injection has an energy barrier ϕ_B of nearly 1 eV. As a result, from the diffusion-limited injection model (2) a thermally activated behavior is expected, according to $\sim \exp(-(\Delta + \phi_B)/kT)$, with $\Delta + \phi_B \sim 1.5$ eV. However, the experimental injection-limited J-V characteristics revealed a very weak temperature dependence, even weaker than the thermal activation Δ, which is in strong contrast to the diffusion-limited injection model. However, an injection model based on thermally assisted hopping from the electrode into the localized states of the polymer [14] consistently describes the experimental results [13].

The mechanism for charge injection into a disordered conductor is schematically depicted in Fig. 1. The potential distribution U in which the charge carrier is injected is the sum of the barrier height ϕ_B, the image potential and the external potential relative to the Fermi level of the metal

$$U(x) = \phi_B - \frac{e^2}{16\pi\epsilon_0\epsilon_r x} - eFx \qquad (7)$$

where the potential is given as a function of the distance x, measured from the metal/polymer interface, F is the applied electric field. The essential assumption of the analytical model is, that the first upward jump is rate limiting. The next jumps can in that case be treated as a diffusive escape from the interface. The minimum distance for a carrier to pass from the contact is limited by the spacing of transport sites and will therefore be close to the nearest neighbor distance a.

As the carriers will jump immediately to the transport energy E_{tr} for an upward jump, the criterion for the regime of upward jumps is given by

$$U(a) - E_{tr} = \phi_B - \frac{e^2}{16\pi\epsilon_0\epsilon_r a} - eFa - E_{tr} \geq 0 \qquad (8)$$

The current j_{inj} is given by the integral of the net hopping onto states at x multiplied with the total injection probability to all energy states at x, given by the Gaussian distribution $g(U(x) - E)$ of the target sites, E being the extra site energy measured from the center of the Gauss, and assuming a Boltzmann occupation of energies, Bol(E). This gives [14]

$$j_{inj} = ev_0 \int_a^\infty dx \exp(-2\gamma x) w_{esc}(x) \int_{-\infty}^\infty dE \mathrm{Bol}(E) g\left(U(x) - E\right). \qquad (9)$$

The exponential factor $2\gamma x$ determines the hopping probability to a distance x, where γ is the inverse localization radius, which is $\gamma \simeq 10/\mathrm{nm}$. This probability is multiplied by the escape probability w_{esc} from x. In order to compare this hopping based model with experiments it should be realized that the injection-limited current is determined by four parameters: the energetic width σ, the dielectric constant ϵ, the nearest neighbor distance a, and the energy barrier ϕ_B at the Ag/PPV interface. From the field- and temperature dependence of the hole mobility of our $OC_1C_{10} - PPV$, $\sigma = 0.11$ eV and $a = (1.2 \pm 0.1)$ nm have been extracted [12].

Furthermore, from impedance measurements $\epsilon = 2.1$ has been found [15]. The only remaining parameter in the random hopping model is the potential barrier ϕ_B. This barrier can be estimated from the difference between the HOMO level of the PPV (5.3 eV) and the Fermi level of Ag of (-4.3 eV) [16],

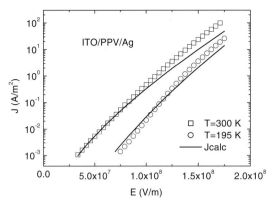

Fig. 2. Injection limited hole current density J versus electric field E for an ITO/Ag/PPV/Ag device at $T = 300$ K and $T = 195$ K. The calculated J-V characteristics for a barrier height $\phi_B = 0.95$ eV are plotted as solid lines. For our $OC_1C_{10} - PPV$ an energy width of $\sigma = 0.11$ eV, a nearest neighbor distance of $a = (1.2 \pm 0.1)$ nm, and a dielectric constant of $\epsilon = 2.1$ have been used

thus $\phi_B = 1.0$ eV. As a result all input parameters are fixed. In Fig. 2 it is demonstrated that the observed field- and temperature dependence of the hole injection from Ag into PVV is in good agreement with the predictions of the model.

4 PLED with an Injection Limited Hole Contact

Charge injection is an important process with regard to the performance of PLEDs. Especially for materials with a large energy gap, as applied for blue PLEDs, large energy barriers at the injecting interface are expected. So far, experimental results on PLEDs with Ohmic electron- and hole contacts have been modeled [17]. By incorporating the injection model based on thermally assisted hopping into the PLED device model also PLEDs with strongly hindered hole injection can be investigated. The injection-limited PLED devices that have been investigated consist of dialkoxy-PPV ($OC_1C_{10} - PPV$) sandwiched between two electrodes on top of a glass substrate. The $OC_1C_{10} - PPV$ polymer is spin coated on top of a silver (Ag) bottom electrode and is covered by a Ca contact. The Ca top electrode has a work function which is close to the conduction band energy of $OC_1C_{10} - PPV$ [16], resulting in an Ohmic contact for the electron injection. The Ag-contact at the other hand, makes an injection barrier of 1 eV with the valence band of the PPV [16]. As a result, the hole injection into PPV from the Ag contact is strongly hindered.

Furthermore, for comparison also bulk-limited PLED devices have been made, where the $OC_1C_{10} - PPV$ has been spin coated on top of an ITO contact. As the device current of a PLED based on PPV is hole dominated [18], it

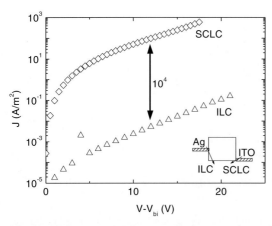

Fig. 3. Hole current density J versus voltage V at room temperature of an ITO/PPV/Ag hole-only device with thickness $L = 240$ nm. For hole injection from ITO the current is space-charge limited (SCLC), for hole injection from Ag the current is injection limited (ILC)

is expected that a reduction of the hole current by a high hole contact barrier will strongly reduce the device current. In Fig. 3 it is demonstrated that for hole injection from Ag the hole current is reduced by 4 orders of magnitude as compared to the bulk space-charge limited hole current. Therefore, the number of holes in the injection-limited PLED (IL-PLED) is also reduced by a factor 10^4.

The current-density voltage (J-V) characteristics for both the IL-PLED and the PLED are shown in Fig. 4a, measured at room temperature. It is observed that the current-density of the IL-PLED is, as expected, strongly reduced compared with the current density of the PLED. The electron current in the PLED is about two to three orders of magnitude lower than the SCLC hole current [18], and consequently larger than the injection-limited hole current. As a result it is expected that the current of the IL-PLED will behave as a space-charge limited electron-only device. From Fig. 4a it is observed that the IL-PLED indeed follows the electron-only current at low voltages. However, at an applied bias V of typically 7 V, the current starts to increase rapidly from the electron current.

The current of the IL-PLED has been calculated by numerically solving the current density equation together with the Poisson equation and applying the proper boundary conditions for the Ohmic electron contact and the injection-limited hole contact. The result of the calculation, also shown in Fig. 4a, confirms that the calculated current is nearly equal to the electron-only current. Due to the strong reduction of the number of holes the electron space-charge is not compensated by holes and as a result the current of the electron-only device is the maximum current a device can support. In Fig. 4b the experimental and calculated light-output of a PLED and an IL-PLED

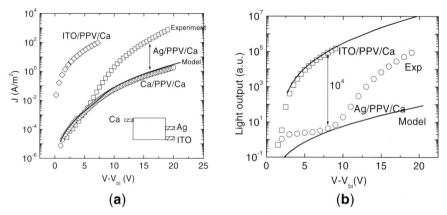

Fig. 4. (a) Current density J versus voltage V at room temperature for an ITO/PPV/Ca polymer light emitting diode (PLED), a Ag/PPV/Ca injection limited PLED (IL-PLED) and a Ca/PPV/Ca electron-only device, all with thickness $L = 240$ nm. Also shown is the numerically calculated J-V characteristic for the IL-PLED, using a barrier height $\phi_B = 0.95$ eV for the hole injection. (b) Light output of the ITO/PPV/Ca (PLED) device and the Ag/PPV/Ca (IL-PLED) device at room temperature, together with the calculated light output (*solid lines*)

are compared. Again, for $V > 7$ V the experimental light-output of the IL-PLED is strongly enhanced and strongly exceeds the predictions from the model. Clearly, the experimental J-V and light-output characteristics of the IL-PLED strongly disagree with the predictions of the device model.

As stated above, with a limited number of holes the space-charge limited electron current is the maximum possible current in the IL-PLED. As a result, the observed increase for $V > 7$ V must originate from an enhanced hole current. From the hopping injection model [14] it is found that the injection current grows rapidly with increasing electric field. A possible origin of an enhanced electric field at the hole-injecting contact might be tunneling through an interface barrier or the trapping of electrons at the interface, as schematically indicated in Fig. 5. Enhancement of charge injection by a tunnel barrier has recently been demonstrated by Murata et al. [19]. Such a tunnel barrier will prevent the electrons to flow into the hole injection contact. Consequently, a large electric field across the tunnel barrier builds up, which gives rise to an increased hole injection. However, the presence of such an electron-blocking tunnel barrier is not in agreement with the fact that we observe the bulk-limited electron current at low voltages in our IL-PLEDs.

An alternative explanation is the presence of electron traps at the Ag/PPV interface. The trapped electrons will increase the electric field at the Ag/PPV interface, leading to an enhanced hole injection. Furthermore, in a hole-only device, as is used in our study of hole injection from Ag into PPV, these electron traps remain unfilled and therefore do not play a role. In

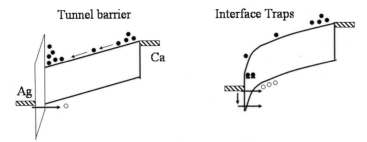

Fig. 5. Possible origin of the enhanced hole injection: (**a**) electron blocking barrier at the hole contact, (**b**) small interfacial region with electron traps near the hole contact. In case (a) the hole tunneling through the barrier is enhanced by strong band banding of the tunnel barrier, in case (b) the hole injection is increased by the extra lowering of the effective barrier

order to model the influence of electron interface traps we incorporate in our model a small interfacial region of a few nm which contains interface traps. In this region the relation between trapped electrons n_t and free electrons n is

$$n_t = \frac{n}{\theta}. \qquad (10)$$

Thus, we have added one additional parameter θ to our PLED device model. In Fig. 6a the calculated J-V characteristics are shown for $\theta = 5 \times 10^{-5}$.

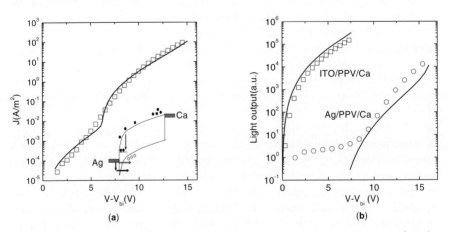

Fig. 6. (**a**) Current density J versus voltage V at room temperature for the Ag/PPV/Ca injection limited PLED (IL-PLED). The calculation (*solid line*) includes an electron trap in a small interfacial region near the hole contact, with a ratio θ between free and trapped electrons, $\theta = 5 \times 10^{-5}$. (**b**) Light output of the ITO/PPV/Ca (PLED) and the Ag/PPV/Ca (IL-PLED) device at room temperature, together with the calculated light output (*solid lines*) from the device model incorporating an electron trap near the hole contact

The calculated J-V characteristics consistently describe the experimental results of the IL-PLED. Furthermore, in Fig. 6b the light-output is shown, inclusion of an interface trap also gives good agreement between model and experiments. In order to find out more about the nature of the interface traps the temperature dependence of the IL-PLED will be investigated, which is a subject of further study.

5 Conclusions

In conclusion, it is found that the injection-limited hole current in a polymeric LED is significantly enhanced by the presence of electrons. The increase of the hole current is quantitatively explained by an electron trap near the PPV interface, which enlarges the electric field at the interface resulting in a strong enhancement of the hole injection.

References

1. J.H. Burroughes, D.D.C. Bradley, A.R. Brown, R.N. Marks, K.Mackey, R.H. Friend, P.L. Burn, and A.B. Holmes, Nature (London) **347**, 539 (1990).
2. R.N. Marks and D.D.C. Bradley, Synth. Metals **57**, 4128 (1993).
3. H.A. Bethe. MIT Radiat. Lab. Rep. **43**, 12 (1942).
4. J.G. Simmons, Phys. Rev. Lett. **15**, 967 (1965).
5. H. Bässler, Phys. Status Solidi B **175**, 15 (1993).
6. P.W.M. Blom, M.J.M. de Jong, and M.G. van Munster, Phys. Rev. B **55**, R656 (1997).
7. W.D. Gill, J. Appl. Phys. **43**, 5033 (1972).
8. L.B. Schein, A. Peled, and D. Glatz, J. Appl. Phys. **66**, 686 (1989).
9. P.M. Borsenberger, J. Appl. Phys. **68**, 6263 (1990).
10. M.A. Abkowitz, Phil. Mag. B **65**, 817 (1992).
11. L. Pautmeier, R. Richert, and H. Bässler, Synth. Met. **37**, 271 (1990).
12. H.C.F. Martens, P.W.M. Blom, and H.F.M. Schoo, Phys. Rev. B **61**, 7489 (2000).
13. T. van Woudenbergh, P.W.M. Blom, M.C.J.M. Vissenberg, and J.N. Huiberts, Appl. Phys. Lett. **79**, 1697 (2001).
14. V.I. Arkhipov, E.V. Emelianova, Y.H. Tak, and H. Bässler, J. Appl. Phys. **84**, 848 (1998).
15. H.C.F. Martens, H.B. Brom, and P.W.M. Blom, Phys. Rev. B **60**, R8489 (1999).
16. I.H. Campbell, T.W. Hagler, D.L. Smith, and J.P. Ferraris, Phys. Rev. Lett. **76**, 1900 (1996).
17. P.W.M. Blom and M.J.M. de Jong, IEEE J. Sel. Top. Quantum Electron. **4**, 105 (1998).
18. P.W.M. Blom, M.J.M. de Jong, and J.J.M. Vleggaar, Appl. Phys. Lett. **68**, 3308 (1996).
19. K.A. Murata, S. Cinà, and N.C. Greenham, Appl. Phys. Lett. **79**, 1193 (2001).

Sensors and the Influence of Process Parameters and Thin Films

Hans-Reiner Krauss

Robert Bosch GmbH, Geschäftsbereich AE, Sensortechnologiezentrum
Tübingerstraße 123, 72760 Reutlingen

Abstract. Design of semiconductor silicon sensors is based on physical principles and the variation of geometry and material parameters. Normally in MEMS design a conversion method from non electrical input to an electrical output signal is chosen to get a good linear sensor. During miniaturization and increase of the accuracy the device behaves no more linear. For different function (e.c. sensitivity, resolution) and quality parameters (e.c. drift, burst pressure) of a device temperature coefficients (TC) nonlinearities, hysteresis and so on take place. Additional during characterization you find offsets and other effects, not described in your physical model of the device. The reason for this is that in thin films values, like the k-factor of metallic resistance, are influenced of process parameters, inhomogeneities in the capacity are influenced by process steps or geometry parameters have influence to void generation. For sensor design this all has to be modeled, that means we have to investigate microscopic effects. The influence of thin layers to function of a sensor is shown for Yaw rate sensors, pressure sensors, mass flow sensors and chemical sensors.

1 Introduction

Semiconductor sensors have become very important in automobile application in the last decade [1,2,3,4]. This is due to the development of MEMS processes and cost advantage of silicon technology in mass production. [5]. To establish the advantage of high yield and performance quickly, sensors have to be designed based on simulation. For this, many microscopical effects caused by process distribution and especially thin film behavior have to be taken into account. In the following the influence of these effects on the behavior of typical MEMS sensors will be shown.

2 Design Method

Our method has three basic components

- Define and optimize the transducer principle for each input parameter (see Fig. 1)
- Analytic based simulation of the sensor.
- Only process and material parameters are allowed input parameters for the simulation models.

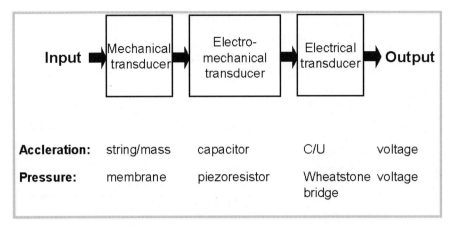

Fig. 1. Transducer principle

The simulation based on process parameters is important due to cost aspects. Normally a design is made for one set of parameters. But the devices are produced in mass production processes, that means there is not only one set of input parameter (geometry, material parameters), because each of these parameters are produced with a nominal value and with a distribution around this nominal value. How to include process distributions in simulation? For this we allow only process parameters as input parameters to the simulation. For instance the input parameter to the simulation could be the volume of the seismic mass in an accelerometer. To allow only process dependent parameters we have to use, instead of the volume, three one dimensional parameters (i.e. length of the seismic mass) including etch bias of the trench etch. With that we get a distributed output signal for the sensor due to process distributions (see. Fig. 2). That means we can predict a yield for each design in a given process.

Many material parameters of bulk materials are process independent parameters. For thin films we have a total different situation. Here many material parameters become process dependent, like i.e. the specific resistance of thin metal films.

For device modeling this process dependent behavior has to be understood. In the following some examples for the influence of process steps and thin film behavior on the function parameters of sensor devices are shown.

3 Examples

3.1 Pressure Sensors

3.1.1 Temperature Hysteresis

In a bulk micro mechanical process a membrane is formed by an KOH etch step. Before this step a piezo resistive bridge is built by diffusion and Alu-

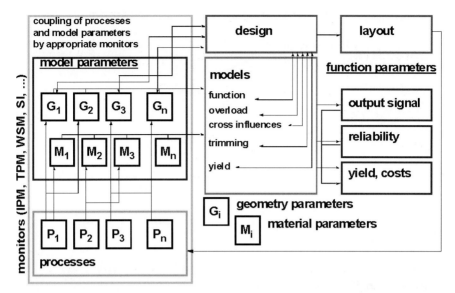

Fig. 2. General schematic chart of the design flow for micro sensors used at Robert Bosch: the models use only such parameters (geometry or material) which can be correlated to the process influences and the process tolerances by appropriate monitors (SPC)

minum metalization. Local and global stress [6] cause temperature hysteresis (Fig. 3) in the output signal of the sensor. Global stress causes hysteresis effects inside the integrated electrical circuit, local stress inside the membrane and its resistors (Fig. 4 [6]).

3.1.2 Gage Factor of Thin Metal Films

Our high pressure sensors are built on steel membranes with a Wheatstone bridge formed by NiCrSi metal film resistors (process licensed by Wika, NKS). The sensitivity of this sensor depends on geometry parameters, the Young's modulus of the membrane and on the gage factor $\Delta R/R = k\, \Delta l/l$. In the literature [7] the gage factor for metals is known to be about 2. During characterization of our sensors we found a difference between the models and the measurements. Assuming a gage factor of 2.2 for the NiCrSi films (sputtered with a thickness of 250 nm) we can reproduce the measurement results with our models. Investigations of the gage factor [8] showed a dependence of surface roughness and impurity concentration of the metal films. Direct measurements show also a k-factor of 2.2 (Fig. 5).

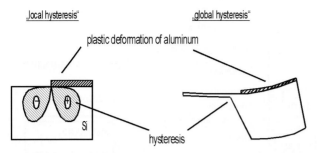

Fig. 3.

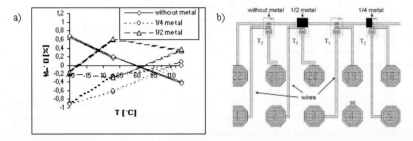

Fig. 4. Hysteresis caused by different additional metal-geometries on npn transistors and temperature cycles. a) measured deviation of the current gain regarding to a current gain of 100 and b) layout of the transistor structures

3.2 Air Quality Sensor

On a thin membrane a gas sensitive thick film is formed to measure different gas concentrations in air. This type of sensor needs for detection a temperature difference ΔT according to room temperature. To reach an homogeneous detection temperature inside the area covered with the gas sensitive film, a membrane is built to control the thermal flow. Inside the membrane different thin metal stripes has to be formed (i.e. for heater, contact electrodes,...). These metal films are covered with a thin passivation layer. Others, for instance contacts to the sensitive thick film, have to be passivation free (Fig. 6).

After process related temperature treatment voids are formed in a Pt-film, covered by an passivation layer, near the passivation window. Investigations showed, local stress in the membrane near the passivation edge is responsible for formation of hillocks and voids [9]. In Fig. 7 the behavior at the edge of the passivation layer is shown. In the region with lower stress there was no void formation inside the Pt film.

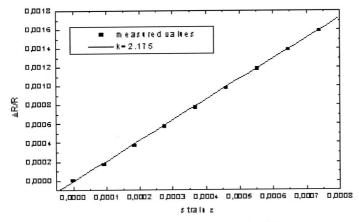

Fig. 5. Direct measured k-factor measurements [8]

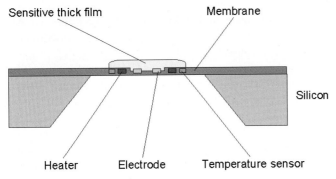

Fig. 6. Cross section of a air quality sensor element

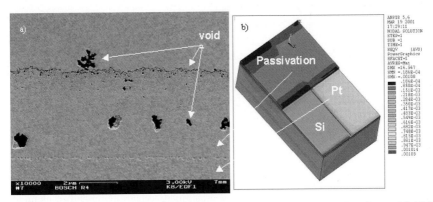

Fig. 7. a) SEM picture of a Pt film at the edge of the passivation layer. b) FEM simulation of the local stress in the membrane

3.3 Acceleration Sensor

A capacitive accelerometer (Fig. 8) is built in surface micro mechanical process. That means a wafer is processed in a normal CMOS process to get doping, metalization and passivation layers. Additional to these process steps the mechanical structure is formed by deposition and patterning of a sacrificial layer, deposition, patterning (trench) of the functional polysilicon layer and removal of the sacrificial layer.

For capacitive sensors one part of the model parameters for sensitivity are geometrical parameters after the trench process. To simulate the accelerometer it is important to know the shape of the beams (string, capacitive fingers). Current methodologies for extraction of process parameters are either based on test structures [10], or on destructive tests, like SEM investigations of sensor element cross sections. Commonly, these methods deliver only data for a specific wafer position. To design based on process parameters it is necessary to get statistical distributions for these parameters. Therefore it is required to measure in-process-parameters on wafer level. This become possible by combining analytical models and electrical stimulation. By combining capacity measurements in x-, y- and z-direction we can extract parameters directly from surface micro machined devices.

The test method is based on the measurement of CV-characteristics. The sensor itself can be considered as "test structure" to extract process dependent geometrical parameters of the polysilicon layer such as layer thickness, lateral structural over etch and sidewall difference angle. A bias voltage applied to the fixed comb drive electrodes results in a lateral displacement of the sensor mass due to the attractive electrostatic force. The capacitance change ΔC_x due to an applied bias voltage V can be solved from an implicit equation given by

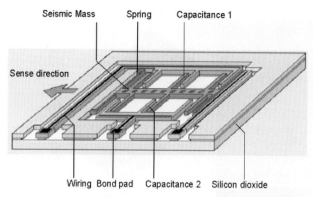

Fig. 8. Working principle of a surface micro machined accelerometer sensor element

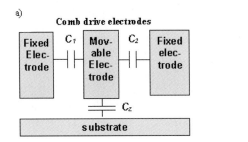
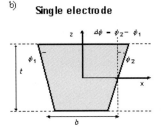

Fig. 9. a) Cross section and equivalent electrical circuit of the comb drive electrodes b) Cross section geometry of the beams

$$\Delta C_x = \left(\frac{2k_x g_x \alpha_x^2 \epsilon N t l}{\beta_x}\right)^{1/3} \left(\frac{\Delta C_x}{V^2}\right)^{1/3} - \alpha_x \frac{\epsilon N t l}{g_x}$$

where ϵ is the permittivity of the surrounding gas, g_x is the gap between the electrodes, x is the in plane displacement of the movable comb drive electrodes, z is the out of plane displacement, l is the length of the overlapping comb finger section, N represents the number of comb drive electrodes, t is the thickness of the polysilicon layer, V is the bias voltage across the comb electrodes, α_x and β_x are form factors that include fringe-field effects and k_x is the spring stiffness. The process related parameter t/g_x can be obtained from a linear fit to the measured CV-characteristic. The gap g_x is given by the sum of the layout dimension and the lateral sidewall over etch as a result of the surface micro machining process.

Because of the difference angle $\Delta \Phi$ (Fig. 9b) this movement in x-direction leads to additional movement in vertical direction.

$$x = \frac{\Delta \Phi}{2} \left(\frac{t}{b}\right)^2 z$$

To study the effect of the lateral sidewall over etch on the CV-characteristic, sensor elements with various gaps between fixed and movable comb drive electrodes were fabricated. The verification of the method is shown in Fig. 10.

From yield prediction of a sensor in a given process it is necessary to know the process distribution. With the method described here, we are able to measure at the device itself direct on a wafer. For that we get wafer mappings and a good statistical base. In Fig. 11 such a wafer mapping is shown for the sidewall etch loss in our trench process.

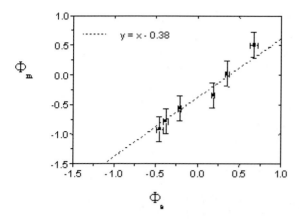

Fig. 10. Normalized measured sidewall angle (SEM) versus Normalized extracted sidewall angle (measured described method) [10]

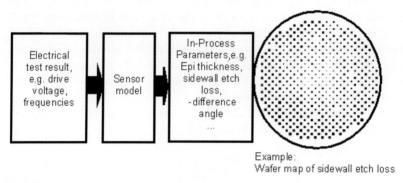

Fig. 11. Concept for MEMS testing

4 Conclusion

It was shown that effects of thin layers and process tolerances have to be taken into account in the simulation of MEMS devices. Miniaturization and increase in accuracy of the sensors require more and detailed knowledge of material parameters. In the near future it will be necessary to include effects down to atomic scale to our sensor models. New measurement methods have to be developed to characterize in atomic scale regions and to get process tolerances by appropriate monitors.

References

1. J. Marek, F. Bantien, H.-P. Trah: *Sensoren und Aktoren in Silizium-Mikromechanik*, in "Halbleiter in Forschung und Technik - Grundlagen, Anwendungen und Perspektiven", J. Werner, J. Weber, W. Rühle (Eds.), p. 107, (Expert-Verlag, Ehningen, 1991)
2. H. Braun: *Sensorik im Kraftfahrzeug*, Tribologie Fachtagung, Goettingen, 28.-30.9.1998.
3. J. Marek, M. Möllendorf: Mikrotechniken im Automobil, mikroelektronik me, **Vol. 3**, 10 (1995).
4. J. Marek: *Microsystems in Automotive Applications*, Micro System Technologies **98**, 43 (1998).
5. M. Offenberg: Industrial Foundry for Surface Micromachined Sensors, MST-News, **Vol. 15/96**, (1996).
6. S. Finkbeiner, J. Franz, S. Hein, A. Junger, J. Muchow, B. Opitz, W. Romes, O. Schatz, H.-P. Trah, Simulation of Nonideal Behaviour in Integrated Piezoresistive Silicon Pressure Sensor, *DTP 1999* (1999).
7. G. Gerlach, W. Dötzel: *Grundlagen der Mikromechanik* (Carl Hauser Verlag, Müchen 1997).
8. J. Muchow, W. Bernhard, et. al, unpublished (2002).
9. M. Wallrauch: *PhD thesis*
10. M. Maute, S. Kimmerle, J. Franz, D. Schubert, H.-R. Krauss, D. P. Kern: Parameter Extraction for Surface Micromachined Devices Using Electrical Charakterization of Sensors, *MSM Proceedings 2001* (2001).

Index

ab initio methods, 138
AlB$_2$, 293
AlCuFe, 151
Alexe, G., 13
alloy, 433, 439, 440, 443
– multi-phase, 207
AlN, 219, 220
Alsabbagh, N., 241
aluminum, 219
amorphous solid, 335
Andreev reflection, 107, 163, 169, 172
angle-resolved photoemission, 311
antiferromagnetism, 419, 433
application
– optoelectronic, 67
Arrigoni, E., 307
astronomical instrumentation, 483
atomic tunneling system, 335
atomic-size contact, 107
Au, 107

band
– gap
– – photonic, 41
– structure, 189, 434, 445
– – photonic, 41
Banys, J., 241
Bauch, H., 241
Be chalcogenide
– quaternary, 67
Becker, K. W., 457
Belzig, W., 107, 163
Bennemann, K. H., 319
Bertram, F., 81
Beryllium Chalcogenide, 67
Beschoten, B., 419
betaine
– phosphate, 241

– phosphite, 241
BeTe, 67
Bethe-Salpeter equation, 189
BiCMOS, 471
Bimberg, D., 27
bipolar transistor, 471
Bloch-Floquet theorem, 43
Blom, P. W. M., 495
Bohnen, K.-P., 293
Böttcher, R., 241
boundary motion, 207
Bragg reflector, 13
break-junction, 107
Brenig, W., 457
buckyball, 144
Busch, K., 41

C$_{60}$, 144
carbide, 219
carbon nanotube, 140
carburization, 219
Carpene, E., 219
CdSe, 13
CdSe/BeTe, 67
cementite, 219
channel noise, 359
charge injection, 495
charge-density wave, 308
chemical bond, 434
Christen, J., 81
cluster, 151
clustering, 371
Co, 419
coating, 219
coherence peak, 267
coherence resonance, 359
coherent echo, 335
colloid, 347

colloidal suspension, 347
complex conductivity, 267
conductance, 134
– differential, 175
– molecular, 133, 134
conductance quantum, 135
conjugated polymer, 495
contact, 133, 495
CoO, 419
correlation, 55, 459, 463
– function, 462
– ground state, 466
Coulomb
– blockade, 175
– interaction, 175
counterion releaes, 347
counting statistics, 163, 164, 166, 170, 172
critical current, 281
cryptography, 383
crystal growth, 207
Crystal Orbital Hamilton Population (COHP), 433
Cuniberti, G., 133
cuprate, 319
Curie temperature, 407
Curie-Weiss law, 407
current-voltage characteristics, 107

defect, 43, 52
– mode, 49
– photonic, 41
– structure, 46, 48
Deneke, Ch., 231
density functional
– algorithm, 133
– theory (DFT), 138
density functional approximation, 407
density functional theory (DFT), 189, 293, 434, 441
depletion interaction, 347
device
– optoelectronic, 67
DFT, 67, 189, 293, 407
diboride
– transition-metal, 293
dielectric property, 335
DIFFOUR model, 241
diffusion, 471

– anomalous, 371
dimethylammonium gallium sulfat, 241
DMAGaS, 241
DMFT
– linearized, 121
domain, 420–422
– antiferromagnetic, 419
– state, 420
– state model, 422
– structure, 445
– wall, 422
doping
– p-type, 27
Dronskowski, R., 433
dynamical mean-field theory, 121

Eder, W. , 307
Ehwald, K.-E., 471
elastic
– property, 335
– scattering, 134
electro-reflectance, 67
electroluminescence, 95
electron, 189
– paramagnetic resonance, 241
– turnstile, 172
electron-doped superconductor, 319
electron-phonon
– coupling, 293
– spectral function, 293
Eliashberg equation, 319
ENDOR, 241
energy gap, 267
Enss, C., 335
EXAFS, 221
exchange
– bias, 419
– interaction, 457
exciton, 189
– binding energy, 81

Fe, 95
Fe-C, 219
Fe-N, 219
FEL, 219
FEM, 47, 509
femtosecond
– pulses, 220
– regime, 219

Fermi-edge singularity, 3
ferroelectric, 241
ferromagnet, 441
ferromagnetic
– metal, 95
ferromagnetism, 255, 419, 433, 445
fluid
– classical, 347
Forster, F., 81
Frahm, H., 3
Frank, M., 41
Free Electron Laser, 219, 220
friction, 335
fullerene, 144

g-factor, 3
(Ga,Mn)As, 445
GaAs, 95
Garcia-Martin, A., 41
gas-fluid coexistence, 347
Geurts, J., 67
glass, 335, 393
Goldacker, W., 281
Goychuk, I., 359
granular gas, 371
Green function
– techniques, 136
– theory, 133
ground state, 407
Großmann, F., 133
Gruber, Th., 81
Güntherodt, G., 419
Gutiérrez, R., 133
Gutjahr, M., 241
Gutowski, J., 13
GWA, 189

Han, M., 219
Hänggi, P., 359
Hanke, W. , 307
Hao, H.-Y., 95
Hapke-Wurst, I., 3
harmonic approximation, 394
Haug, R. J., 3
Heid, R., 293
Heinemann, B., 471
Heinke, H., 13
Heitz, R., 27
Hermann, D., 41

heterojunction bipolar transistor, 471
heterostructure
– type-II, 67
high-TC materials, 307, 308
Hodgkin-Huxley mode, 359
Hoevers, H. F. C., 483
Hoffmann, A., 27
hole injection, 495
hole-doped superconducto, 319
Hommel, D., 13, 27
HOMO-LUMO, 146
– gap, 139, 145
Honecker, A., 457
hopping model, 495
Hoyer, W., 55
Hubbard Hamiltonian, 319
Hubbard model, 121
Huiberts, J. N., 495
HYSCORE, 241
hysteresis, 371

II-VI compound, 67
III-V semiconductor, 95
imaging detector array, 483
InAs, 3
incommensurate phase, 241
interaction, 335
interface, 67, 207
– dynamics, 207
– ferromagnet-semiconductor, 95
– ferromagnetic/antiferromagnetic, 422
– trap, 495
ion channel, 359
itinerant electron, 407

Jahn-Teller distortion, 433
Jansen, A. G. M., 3
Jin-Phillipp, N. Y., 231

Kahle, M., 219
Kanter, I., 383
Kawaharazuka, A., 95
Keller, J., 419
key exchange, 383
Kinzel, W., 383
Kira, M., 55
Kirchner, Ch., 81
Kivelson, S. A., 307
Klarer, D., 81

Klude, C., 13
Klude, M., 27
Knoll, D., 471
Koch, S., 55
Kondo
– quantum dot, 175
– temperature, 175
König, J., 445
Krüger, P., 189
Kröger, R., 13
Krauss, H.-R., 505
Kroha, J., 175
Kübler, J., 407

Landau model, 241
Landauer approach, 134
Landauro, C. V., 151
LaNdSrCuO, 307
laser
– diode, 13, 67
– reactive treatment, 219
laser plasma, 219
– Synthesis, 219
lattice dynamics, 293
LDA, 189, 407
lead self-energy, 136
lead-molecule contact, 133
LED, 67, 95, 495
Leonardi, K., 13
Li_3N, 27
light-emitting diode, 495
Lischka, K., 27
local density approximation, 189
Lohse, D., 371
low dimensional system, 457
low energy dynamics, 393

magnet, 457
– quantum, 457
magnetic
– field, 175, 335
– – high, 3
– impurity, 81, 255
– moment, 407
– semiconductor
– – diluted, 445
magnetism, 407, 433
magnon, 407
Manske, D., 319

many-body interaction, 55, 347
material
– optical, 41
Mazur, A., 189
MBE, 81
mesoscopic system, 134
metal-insulator transition, 121
$MgAlB_4$, 293
MgB_2, 281, 293
MgB_2, 267
Michel, D., 241
Michler, P., 13
microcalorimeter, 483
microcavity, 13, 55
micromagnetic, 445
microstructure, 207
miniaturization, 133
MnAs, 95
MnSi, 255
molecular
– devices, 133
– electronics, 133
– wires, 138
Monte-Carlo simulation, 445
Mössbauer spectroscopy, 219
Mott transition, 121
MOVPE, 81
Müller, C., 231

Nakamura, Y., 231
nano-circuit, 231
nanofabrication, 107
nanostructure, 231
nanotube
– Carbon, 231
– semiconductor, 231
Naveh, Y., 107
NbB_2, 293
Nestler, B., 207
neural network, 383
neutron scattering, 393
NiPdP, 393
nitride, 219
nitriding, 219
non-Fermi liquid, 255
nonlinear dynamics, 371
nonlinearity, 50
Nowak, U., 419
Nuclear Reaction Analysis, 220

one-dimensional, 307, 308, 312
optical
– gain, 13
– properties, 41, 49, 189, 267
– spectroscopy
– – time-resolved, 27
optoelectronics, 81
organic semiconductor, 495

p-contact, 67
Paaske, J., 175
Passow, T., 13
Peierls distortion, 433
Pfleiderer, C., 255
phase
– behavior, 347
– coherence, 134
– field model, 207
– transition, 207
– – antiferromagnetic, 319
phonon, 189
– dispersion, 293
photoemission, 307
photoluminescence, 55
– time-resolved, 95
photon
– antibunching, 13
photonic crystal, 41, 46, 52
Pierz, K., 3
Pimenov, A., 267
PLED, 495
Ploog, K.-H., 95
Pohl, U. W., 27
point defect, 471
Pollmann, J., 189
poly-dialkoxy-p-pheny-lene vinylene, 495
Pöppl, A., 241
Potthoff, M., 121
PPV, 495
proton glass, 241
proximity effect, 107, 171
pumped laser
– electrically, 27
– optically, 27

quantum
– chemistry, 433, 434
– critical point, 255

– dot, 13, 27, 67
– – Kondo, 175
– – semiconductor, 231
– optics, 55
– phase transition, 459–461
– well, 55
quantum dot, 3, 175
quasicrystal, 151

Raman scattering, 67
Ramsteiner, M., 95
reentrance, 347
Reiner, H., 281
Renker, B., 293
renormalization group, 175
reservoir, 134
resonant tunnelling, 3
Rohlfing, M., 189
Rosch, A., 175
Rücker, H., 471

Sarkar, D., 3
Sauer, R., 81
scattering
– incoherent, 394
Schaaf, P., 219
Scheer, E., 107
Schikora, D., 27
Schlachter, S. I., 281
Schmid, G., 359
Schmidt, O. G., 231
Schulz, O., 27
Sebald, A., 13
self-similarity, 371
semiconductor, 55
– magnetic, 81
– surface, 189
sensor, 505–512
– simulation, 505, 506, 509, 512
shot noise, 169
Si_3N_4, 219
SiC, 219
SiGe, 471
sodium wires, 138
Solbrig, H., 151
solidification, 207
spectroscopy
– electromodulation, 67
– Raman, 67

spin
- density wave, 308
- dynamics, 457
- electronics, 81
- fluctuation, 407
- Hamiltonian, 241
- injection, 95
- polarization, 434, 437
- relaxation, 95
- spiral, 407
- transport, 3, 95
- wave, 445
spintronics, 95
squeezing, 55
statistical physic, 371
stochastic resonance, 359
strain, 281
Strassburg, M., 27
stress, 507, 508
stripe phase, 307, 308
strongly-correlated system, 307, 457
structural defect, 419
structure, 293
Suck, J.-B., 393
superconductivity, 107, 255, 267, 281, 293, 307–310, 312, 319
- numerical results, 308, 311, 317
superconductor
- ferromagnetic, 255
- high-TC, 307
superlattice, 67
surface, 219
- exciton, 189
- phase, 121
- state, 121
- structure, 189
synchronization, 359, 383

thin film, 505, 506
- analysis, 219
- production, 219
Thonke, K., 81
TiB_2, 293
TiC, 219
tight-binding Hamiltonians, 137
Tillmanns, A., 419
TiN, 219
Tkeshelashvili, L., 41
transition edge sensor, 483

transition metal, 439
- diborides, 293
transition temperature, 319
transmission, 134, 135, 140, 145
- coefficient, 139
- spectrum, 139
transport, 163, 166–169
- Andreev, 172
- channel, 107
- Cooper pair, 163
- electron, 164
- electronic, 151
- mesoscopic, 163, 165
- nonequilibrium, 175
- quantum, 163
- statistics, 163
- time dependent, 172
trapped electron, 495
tunneling, 95
- model, 335

Ulrich, S., 13
Urbina, C., 107
Usadel, K. D., 419

van der Meer, D., 371
van der Weele, K., 371
van Hove singularities, 144
van Woudenbergh, T., 495
Völkel, G., 241
von Grünberg, H.-H., 347

Waag, A., 67, 81
Wagner, V., 67
Wannier function
- photonic, 41
wetting, 207
Wigner-Seitz cell, 47
wire
- MgB_2, 281
Wölfle, P., 41, 175

X-ray
- absorption spectroscopy, 219
- diffraction, 219
X-ray detector, 483

Zacher, M. G., 307
Zapf-Gottwick, R., 231

Zeeman splitting, 3
Zeitler, U., 3
Zhu, H., 95
Zimmer, S., 281

ZnCdSe, 27
ZnO, 81
ZnSe, 13, 67
$ZrZn_2$, 255

Printing (Computer to Film): Saladruck Berlin
Binding: Stürtz AG, Würzburg